Drug Design Strategies
Quantitative Approaches

RSC Drug Discovery Series

Titles in the Series:

1: Metabolism, Pharmacokinetics and Toxicity of Functional Groups: Impact of Chemical Building Blocks on ADMET
2: Emerging Drugs and Targets for Alzheimer's Disease; Volume 1: Beta-Amyloid, Tau Protein and Glucose Metabolism
3: Emerging Drugs and Targets for Alzheimer's Disease; Volume 2: Neuronal Plasticity, Neuronal Protection and Other Miscellaneous Strategies
4: Accounts in Drug Discovery: Case Studies in Medicinal Chemistry
5: New Frontiers in Chemical Biology: Enabling Drug Discovery
6: Animal Models for Neurodegenerative Disease
7: Neurodegeneration: Metallostasis and Proteostasis
8: G Protein-Coupled Receptors: From Structure to Function
9: Pharmaceutical Process Development: Current Chemical and Engineering Challenges
10: Extracellular and Intracellular Signaling
11: New Synthetic Technologies in Medicinal Chemistry
12: New Horizons in Predictive Toxicology: Current Status and Application
13: Drug Design Strategies: Quantitative Approaches

How to obtain future titles on publication:
A standing order plan is available for this series. A standing order will bring delivery of each new volume immediately on publication.

For further information please contact:
Book Sales Department, Royal Society of Chemistry, Thomas Graham House, Science Park, Milton Road, Cambridge, CB4 0WF, UK
Telephone: +44 (0)1223 420066, Fax: +44 (0)1223 420247, Email: books@rsc.org
Visit our website at http://www.rsc.org/Shop/Books/

Drug Design Strategies
Quantitative Approaches

Edited by

David J. Livingstone
ChemQuest, Sandown, Isle of Wight, UK

Andrew M. Davis
AstraZeneca, Loughborough, UK

RSC Publishing

RSC Drug Discovery Series No. 13

ISBN: 978-1-84973-166-9
ISSN: 2041-3203

A catalogue record for this book is available from the British Library

Published by The Royal Society of Chemistry,
Thomas Graham House, Science Park, Milton Road,
Cambridge CB4 0WF, UK

Registered Charity Number 207890

For further information see our web site at www.rsc.org

Printed in Great Britain by CPI Group (UK) Ltd, Croydon, CR0 4YY

Preface

We will never know what motivated our ancestors to grind/powder/boil parts of plants/insects/animals and then administer them to sick people. No doubt there were many failures, but over millennia, the successes have given us useful drugs such as morphine ($\sim$4000 BC), reserpine ($<$1000 BC), aspirin ($<$200 BC) and ephedrine ($\sim$1 AD). Changes in attitude to the value of human life, the development of sciences and medicine, commercial pressures and the effect of regulators have all led to the abandonment of this strategy for drug discovery. So what has taken its place? The development of synthetic organic chemistry and methods of structure determination coupled with the recognition that changes in chemical structure lead to changes in biological activity had a profound effect on the search for new medicinal compounds. The source of drugs has developed from complex mixtures derived from natural products, to pure, well-characterised molecules produced by synthetic methods. Testing procedures took longer to change. For example, in the early 1900s an antimalarial research programme at the Bayer research institute used, as their "guinea pigs", patients who had been rendered insane and paralysed by the final stages of syphilis and who were then deliberately infected with malaria. The anti-malarial pamaquine emerged from these studies and was marketed in 1926.

Most of the functions of a modern pharmaceutical company developed as the industry itself matured, and they often started out as empirical "rules of thumb" and folklore, known only to a few specialists. These various facets of drug discovery gradually became established as specialist disciplines in their own right. As with most sciences, hand in hand with this development came the application of quantitative approaches. By "quanitative approaches", we mean numerical analysis to enable robust predictions to be made. This book aims to show how these different parts of the drug discovery process have developed, with particular emphasis on quantitative aspects, and how they might progress

RSC Drug Discovery Series No. 13
Drug Design Strategies: Quantitative Approaches
Edited by David J. Livingstone and Andrew M. Davis

Published by the Royal Society of Chemistry, www.rsc.org

in the future. Our authors are influential, authoritative experts in their respective fields, with a combined total of over 30 000 citations to their published work. We have asked our authors to show how the topic used to be approached, how quantitative methods have been developed, what problems remain to be addressed and possible future directions.

After an overview of quantitative approaches to drug design, we describe the development of the concepts of "drug-like properties", quantitative structure–activity relationships and molecular modelling, and in particular, structure-based design approaches to guide lead optimisation. How we manage and describe chemical structures underpins all quantitative approaches to drug design, and this is described in Chapters 6–8. The value, and also the challenge, of a quantitative approach is to describe the confidence in any prediction, and methods to assess predictive model quality are described in Chapter 9. The later chapters describe the application of quantitative approaches to describing and optimising potency, selectivity, drug metabolism and pharmacokinetic properties and toxicology, as well as the design of chemical libraries to feed the screening approaches to lead generation that underpin modern drug discovery. Finally, we describe the impact of bioinformatics, the current status of predicting ligand affinity directly from the protein structure and the application of quantitative approaches to predicting environmental risk.

We hope this book will provide a useful summary of the current state-of-the-art in quantitative approaches to drug design, as well as future opportunities, and will serve as a good way-in to the original literature. But more importantly, we hope it will provide inspiration to drug design practitioners to apply careful design and to make best use of the quantitative methods that are available, while continuing to improve them. While we like to think we have gone a long way beyond grinding and boiling random insects, plants and animals to make "medicines" to administer to the sick in the hope of a miracle cure, drug discovery still relies heavily on random screening and empirical screening cascades to identify leads and drugs. The process has gone through many failures to deliver only a small handful of drugs. With the rapidly escalating costs of drug discovery and development, together with spiralling delivery, quantitative approaches hold the promise of shifting the balance of success, to enable drug discovery to maintain its economic viability.

Contents

RSC Drug Discovery Series No. 13
Drug Design Strategies: Quantitative Approaches
Edited by David J. Livingstone and Andrew M. Davis

Published by the Royal Society of Chemistry, www.rsc.org

CHAPTER 1

The Evolution of Quantitative Drug Design

KEITH JAMES

The Scripps Research Institute, 10550 N. Torrey Pines Road, BCC-483, La Jolla, CA 92014, USA

1.1 Introduction

Drug designers occupy centre stage in the bio-pharmaceutical research enterprise. Their role is to repeatedly pose, and then answer, one of the most important questions facing a drug discovery program: '*what compound(s) should we make next?*'. Although they have asked themselves this central question since the very beginnings of 'medicinal chemistry', the body of information brought to bear upon answering it has changed enormously over the last 30 years, both in scope and degree of precision. Furthermore, the data analysis and visualization tools used to explore this information have been transformed, as computing power and the sophistication of desktop applications have grown. This chapter will attempt to explore the nature of these changes, assess where the current leading edge of drug design stands, and try to project how these evolutionary changes might continue into the future.

Successful drug design has always required a blend of chemistry know-how (in its broadest sense), an appreciation of the principles of medicinal chemistry, access to quality data, analytical skills and creative flair. It is a team enterprise, with the range of disciplines involved continually expanding as new technologies emerge and become assimilated into the mainstream. This chapter will also

RSC Drug Discovery Series No. 13
Drug Design Strategies: Quantitative Approaches
Edited by David J. Livingstone and Andrew M. Davis

Published by the Royal Society of Chemistry, www.rsc.org

attempt to characterize these changes and consider what future skills might become critical to drug discovery success.

Finally, the very nature of the molecular targets being addressed, and our understanding of them, has changed steadily over time. The notion of a well prescribed 'druggable' genome, comprised of familiar gene families, is increasingly in conflict with the drive to break new ground in order to achieve 'breakthrough' therapies. Again, the implications of these changes for the application of quantitative methodologies in drug design strategies will be considered.

1.2 A Framework for Integrated Drug Design

1.2.1 A Simple Model

In order to understand the nature of the changes in drug design over time, it is helpful to have a conceptual framework against which these changes can be mapped. Figure 1.1 depicts such a high level framework, illustrating a simple interaction between a drug and its molecular target.

The drug might be a synthetic small molecule, a modified natural product, an oligonucleotide, a peptide *etc*. The molecular target is most often a protein, but could in principle be an oligonucleotide. The key point is that successful drug design requires the integration of information from all three domains depicted: the drug, the target and the drug–target complex.

One of the most striking changes in the world of drug discovery has been the steady growth in availability of data across all these categories, and especially in information concerning the details of the drug–target complex. Furthermore, multiple streams of data are increasingly available simultaneously for a given drug molecule and in parallel for a set of related molecules.

For simplicity, the term 'drug' will be used throughout this chapter in reference to molecules at any stage of optimization, and not restricted to agents

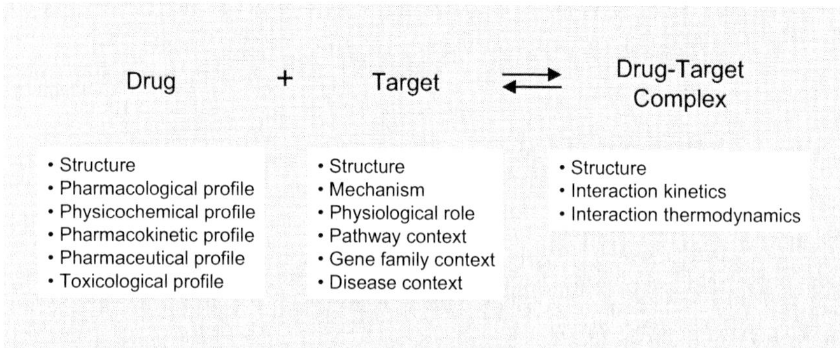

Figure 1.1 A framework for categorizing the information used in drug design programs.

that have been approved. Additional descriptors, such as 'ligand', 'hit', 'fragment', 'lead' and 'candidate' will also be used where they aid clarity.

1.2.2 Drug-Related Data

Even in the absence of detailed information about the molecular target being addressed, or the nature of the interaction with that target, the profile of the drug under study can be characterized extensively along the dimensions outlined in Figure 1.1. Below each of these dimensions exist yet further fine structure, enabling a deep, rich and quantitative description of the drug to be constructed. They encompass both the intrinsic properties of the molecule, which are governed by its structure (including, for example, its molecular weight, melting point and molar refractivity), and also its extrinsic properties, as determined by its interaction with other systems. Thus, its physicochemical profile is defined by its behavior upon interaction with physical systems, *e.g.* its ability to dissolve in a given medium (solubility), its ionization state according to the pH of that medium (pK_a) and how it partitions between media with different dielectric properties (lipophilicity). Its pharmacological profile is determined by how it interacts with biological systems designed to measure potency, efficacy and selectivity, *e.g.* its ability to inhibit an enzyme, maximally activate a receptor, lower the blood pressure of an animal or selectively block the function of one ion channel *versus* another. Its pharmacokinetic profile is determined by how its concentration changes with time when exposed to biological and physical systems related to absorption, distribution, metabolism and excretion, in either model systems, specific tissues or compartments (such as plasma), or entire organisms. Its pharmaceutical profile is determined by its behavior upon exposure to the processes involved in formulation and use as a product, *e.g.* its polymorphic forms, solubility, interaction with excipients and stability. Its toxicological profile is determined by its effect on biological systems such as cells or organs, upon repeated and elevated exposure, *e.g.* multi-week or multi-year studies, at concentrations typically higher than those which evoke a pharmacological effect.

Even this simplified description illustrates the range of parameters ideally taken into consideration when optimizing the profile of a potential drug candidate. All of them can be measured or calculated to varying degrees of precision, illustrating the central place that *quantitative* approaches have in the drug design process. They all follow from the structure of the molecule, which can usually be characterized rapidly and comprehensively using analytical tools that have been available to synthetic chemists for decades. The relationship between the structure of the molecule and a particular aspect of its profile (*e.g.* its activity in a biological screen) has given rise to the concept of structure-activity relationships (SARs). Furthermore, given the possibility of quantitatively characterizing the key structural aspects that govern this activity, a natural progression has led to the emergence of quantitative structure activity relationships (QSAR). As will become evident, when surveying the drug discovery scene of thirty years ago, the term *structure-activity relationship* was both useful and apt at that time, since the focus of drug designers was almost exclusively on the optimization of biological activity. However, as drug design has become more sophisticated, and a wider range of parameters have been taken into consideration, the term *structure–property relationship* has been coined, in order to better capture the broader set of factors that determine drug discovery success. However, the ability describe these relationships in quantitative terms remains critically important. Over the last thirty years, drug designers have also learned that these critical elements of a drug's profile cannot be optimized in a serial fashion, but must be optimized simultaneously, in order to avoid the disastrous situation where, late in a program, a highly potent and selective agent is found to be completely insoluble or cleared rapidly from the plasma or too unstable to achieve a 2 year shelf life at room temperature. Hence, the concept of *multi-parametric optimization* has emerged as a key feature of modern drug discovery. Finally, the ongoing pursuit of greater speed and efficiency in the drug discovery enterprise, the drive to calculate (rather than measure) many of these parameters and the steady increase in available computing power has brought *in silico* methods (and therefore the computer) to the forefront of modern drug design.

1.2.3 Target-Related Data

Unlike the drug molecule to which it binds, the structure of the target in Figure 1.1, at a molecular level of resolution, is usually not straightforward to determine, although from a drug designer's perspective, it is the item of information they would most like to possess. Even now, in many cases this information can remain obscure for the duration of the drug discovery program. However, even in the absence of this detailed structural information, there remain other critical aspects of the target which can be determined. Thus a detailed understanding of the mechanism of the target, *e.g.* whether it is an enzyme, receptor, ion channel, transcription factor or transporter, is of key consideration in designing agents that will modulate its function. Its physiological role enables assessment of the implications of modulating it, as well as the concentration/potency of drug required in order to effect that modulation and the likely challenges faced in accessing it (for example, if it is located in the central nervous system). Its pathway context can reveal its involvement in feedback or feed-forward loops, or the existence of endogenous modulators. Its gene family context can reveal useful information on the modulation of related targets, or highlight closely related family members which might pose selectivity challenges. Its disease context enables an assessment of its place in a pathological, as opposed to physiological, process and how this might impact expression levels and consequently abundance and location.

Although many of these factors appear inherently qualitative in nature, there are intense efforts underway to codify these types of information within an integrated framework and to apply quantitative approaches to the analysis and simulation of entire systems, in the growing field of systems biology. Although these methodologies are more likely to impact on target selection decisions before they impact on compound design decisions, they represent yet another evolution in drug design, and an ambitious example of applying quantitative approaches to improve drug design effectiveness.

1.2.4 Drug–Target Complex-Related Data

In terms of design impact, the ability to determine the high-resolution structure of the complex between a drug and its cognate molecular target represents the pinnacle of technological advances in the field of drug discovery to date. Although it still does not constitute a complete description of the molecular interaction, the ability to understand the spatial arrangement of interacting functional groups in both target (*e.g.* protein) and ligand (*e.g.* small molecule hit) enables drug designers to answer the key question: *'which molecule(s) should we make next?'*, with much greater insight.

However, two additional types of information are critical to a full understanding of the nature of this complex and how its affinity can be optimized. Thus the interaction kinetics (the rate at which the complex both forms and dissociates, in the presence or absence of substrates or ligands) yields important information regarding the competitiveness of the interaction and the duration of biochemical effect. Additionally, the interaction thermodynamics (the underlying contribution of enthalpic and entropic changes to the measured affinity of the interacting species) indicates, for example, whether an emphasis should be placed upon optimizing particular H-bond interactions, or increasing the 'hydrophobic-binding' between lipophilic aromatic residues on the protein and ligand. Drug designers have long recognized that the ability to accurately calculate (and therefore predict) the binding energy for an interaction between a target and prospective, or virtual, ligand would constitute a powerful method for prioritizing which compound(s) to make next. Accordingly, this has been, and continues to be, a focus of substantial research effort within the drug design community.

1.2.5 Drug Design Evolution

Using this simple model as a lens through which drug design strategies can be viewed, one can compare the prevailing 'state-of-the-art' at any point along the evolutionary continuum depicted in Figure 1.2. As a means of consistently sampling the vast body of drug discovery literature and identifying the key developments in drug design capabilities, the following sections will use snapshots of publications from *The Journal of Medicinal Chemistry*, from 1981, 1991, 2001 and the present day, as a basis for judging where the 'leading edge' of drug design stood at that time. This period covers some of the most profound changes in drug discovery approaches, which are illustrated graphically in Figure 1.2. It also happens to be within the personal recollection of the

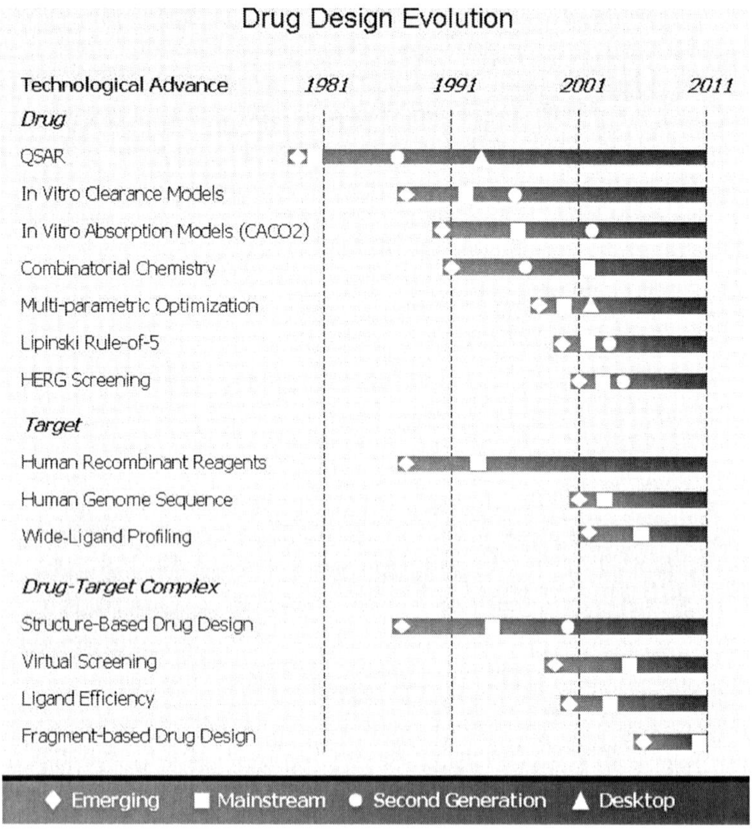

Figure 1.2 The evolution of drug design technologies.

author. Although there is an inevitable lag between conducting and publishing research, the assumption has been made throughout that the lag is similar for each snapshot and the elapsed time between snapshots is sufficiently large that it can be disregarded.

1.3 Drug Design 1981

1.3.1 QSAR-Driven Drug design

This was a period in drug discovery when cardiovascular, gastrointestinal and anti-infective research featured prominently; where assays were based typically on animal tissues;[1] where it was commonplace to find no clear molecular mechanism underlying the activity observed;[2] and where potency was the predominant focus of optimization programs.[3] In this era, biological effect was not routinely related to drug concentration, and pharmacophore and active-site models were typically depicted by hand-drawn curves, angles and charges on sheets of paper.[4] There might have been some communal 'computer terminals' in the medicinal chemistry department, linked to a mainframe system, but structures were entered as SMILES (simplified molecular input line entry specification) strings and computational chemistry was still an emerging discipline.

Drug designers were therefore working with a much more limited palette of quantitative information than they are in 2011. Nevertheless, some outstanding drugs were discovered around this time, and the first glimmerings of a more molecular understanding of drug-target interactions were beginning to appear, driven by pioneers who were determined to reach a deeper understanding of how their drugs worked. Two particularly significant examples of the advance in drug-design expertise during this era were recognized in the pages of *J. Med. Chem.* in 1981: the discovery and introduction of cimetidine as the first H_2-receptor antagonist treatment for gastric ulcers,[5] and the discovery of captopril as the first orally-active angiotensin converting enzyme inhibitor.[6] These are considered further in the following sections.

1.3.2 Drug-Related Data

The analytical tools used to determine or confirm the structure of drug molecules were essentially in place at this time (primarily NMR spectroscopy, mass spectrometry, IR/UV spectroscopy and X-ray crystallography), although at considerably lower levels of resolution, throughput and accessibility to bench scientists. The pharmacological profile of a drug was built using primarily animal tissue preparations (*e.g.* cell membrane preparations or segments of tissue suspended in organ baths) and whole animal experiments. Even thirty years ago however, it was recognized that in order to fully describe those attributes of a molecule which govern its biological activity it was necessary to take into account its physicochemical properties. Thus reviews of the current state of QSAR approaches,[3] outlined the objective of relating structure driven properties to (almost exclusively) biological potency. A detailed treatment of QSAR and the pioneering contributions of Hansch and his contemporaries can be found elsewhere in this book, but there was clearly very active use of these methodologies throughout the pages of *J. Med. Chem.*[7] Lipophilicity, as denoted by the term, LogP, was the pre-eminent physicochemical property, and higher throughput methods of measuring LogP were actively being sought at this time.[8]

The discovery of the H_2-receptor antagonist cimetidine (Tagamet) by scientists at SmithKline French is an outstanding example of how the team integrated structural data, pharmacological data and physicochemical data, together with a good dose of intuition and creative flair (in the absence of any knowledge of the H_2-receptor structure) to design a truly breakthrough agent. Figure 1.3 depicts a few of the milestones along the pathway from histamine to cimetidine, described in detail in an ACS Medicinal Chemistry Award lecture.[5] In particular, the nature and pK_a of the basic side chain of antagonists was carefully optimized in order to avoid partial agonism, increase affinity and minimize adverse effects.

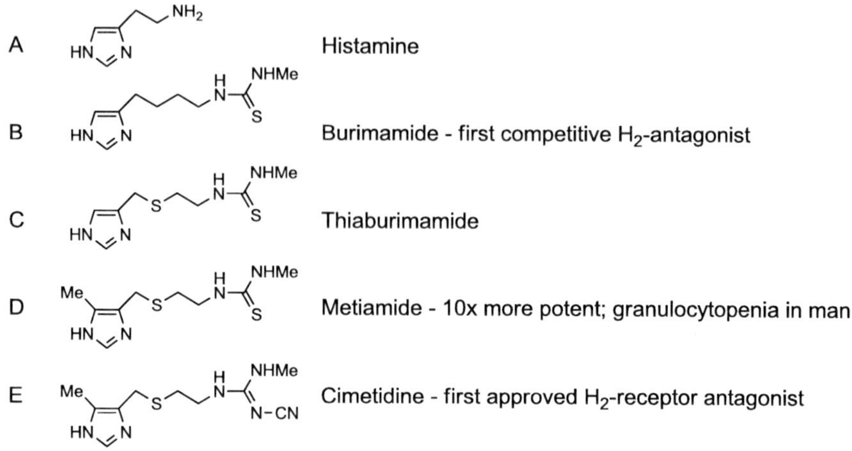

Figure 1.3 Design milestones on the path from histamine to cimetidine.

It is striking how few examples of pharmacokinetic studies were being published at this time, perhaps because the 'drug metabolism' groups in most companies were embedded in development rather than discovery departments. Duration of pharmacological effect is therefore used as a basis for determining how long the target is exposed to the drug.

1.3.3 Target-Related Data

Whilst many programs were based upon a discrete molecular target, whose mechanism was well understood, there were still many programs where the target of the drug remained obscure. However, even for those targets which could be named, the era of recombinant human proteins and cell lines was literally just beginning. Pioneer biotechnology companies such as Genentech and Genetics Institute had only recently been launched, and access to human reagents for screening was a significant challenge.

Whilst the notion of receptor sub-types was familiar, *e.g.* α- *versus* β-adrenergic receptors, and even α_1- *versus* α_2-adrenergic receptors, scientists were unaware that yet a further subdivision of receptor subtypes, *e.g.* α_{2A}- *versus* α_{2B}-adrenergic receptors, was to be established, nor had they any inkling of the number of dopaminergic, serotonergic or non-aminergic receptor subtypes yet to be identified.

1.3.4 Drug–Target Complex-Related Data

A quote from workers at Roche[9] beautifully captures both the hopes of drug designers of the time and the current state of affairs with regard to the structure of drug-target complexes: '*The methodology of drug design could be greatly improved if receptors and their mode of interaction with active substances were known in precise molecular detail. The molecular features of important pharmacologic receptors, however, are presently unknown beyond their pharmacological and biochemical classification.'* Although the final statement, unfortunately, still holds true today for many receptors, the chances of possessing detailed molecular information about a drug target complex in 1981 were extremely remote.

The first X-ray crystal structures of drugs complexed with their protein molecular targets had begun to appear several years before this point.[10] However, these types of data were not readily available. In fact, companies such as Agouron were soon to be founded based on the potential of protein X-ray crystallographic data to revolutionize drug discovery. Enterprising drug designers had therefore used the next best thing, the X-ray structure of a related protein, to guide their design efforts. A brilliant example of such an approach is the discovery of the ACE (angiotensin converting enzyme)-inhibitor, captopril, by Ondetti and Cushman, who used the structure of the related zinc-dependent carboxypeptidase A, in complex with carboxyl-based peptide-mimetics, to refine their design.[6] They reasoned that since ACE was a dipeptidyl-carboxypeptidase, cleaving substrates to release a dipeptide unit from the C-terminal of its substrate—as opposed to release of a single amino acid from the C-terminal of a substrate, in the case of carboxypeptidase A—that inhibitors based upon dipeptide fragments should be effective. Their design approach culminated in captopril, based upon an ala-pro-dipeptide incorporating a thiol moiety to ligate the active site zinc ion. Their graphical description of the presumed active site provides an excellent sense of the way drug designers thought about their molecular targets at that time (Figure 1.4). Interestingly, it was not until over 10 years later that the primary sequence of human ACE was published[11] and it was over 25 years before the protein X-ray crystal structure of an ACE-inhibitor complex was published![12]

In the absence of crystallographic data, drug designers channeled their energies into building models based upon deductions they could make from their SAR findings, incorporating stereochemical information as best they could.[4,7]

1.4 Drug Design 1991

1.4.1 The Emergence of Structure-Based Drug Design

This was a period of enormous progress in drug discovery, whether measured in terms of the eventual therapeutic impact of the drugs then being discovered, or the increased sophistication in drug design strategies being employed. This is no better exemplified than by the concerted impact the drug design community was poised to have on the treatment of HIV/AIDS, enabled by the emergence of powerful new structure-based drug design (SBDD) approaches. Thus the urgency of the therapeutic fight against AIDS, coupled with the applicability of X-ray crystallographic analysis to compelling anti-HIV molecular targets (such

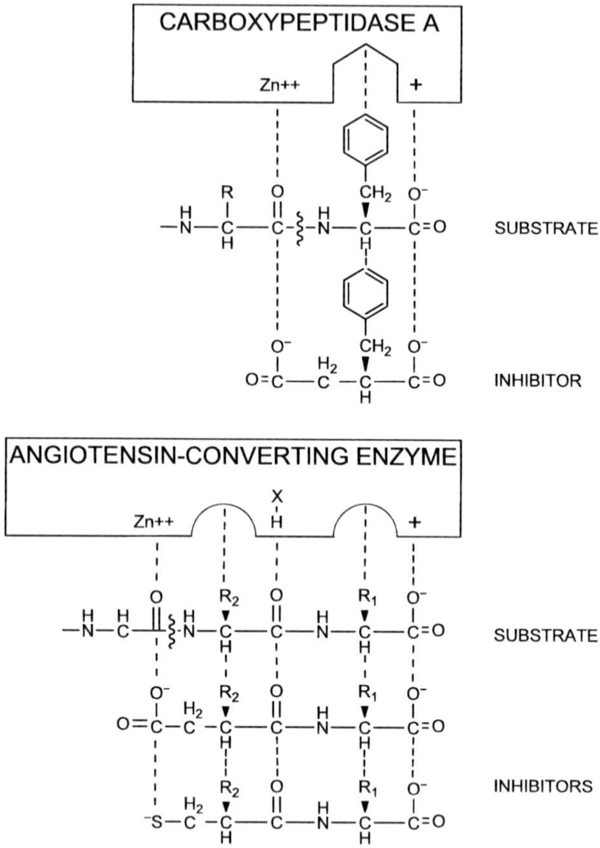

Figure 1.4 Diagrammatic model of the active sites of carboxypeptidase-A and angiotensin converting enzyme and putative interaction modes with substrates and inhibitors (adapted from ref. 6).

as the soluble enzymes, reverse transcriptase and HIV protease), at a time when molecular visualization tools were improving rapidly, helped catalyze these advances and establish a new, high water mark for the drug design field.

There was also tremendous progress across a range of other therapeutic areas, most notably cardiovascular, where blockbuster therapies such as angiotensin II (AII) antagonists,[13] calcium-channel blockers[14] and HMG-CoA reductase inhibitors,[15] were being discovered and developed. In the inflammation field, leukotriene antagonist programs were yielding future asthma treatments[16] and elsewhere in the infectious disease area, novel quinolones[17] and erythromycin analogs[18] were being designed.

Elsewhere in the world of drug design, landmark papers were appearing in the embryonic field of combinatorial chemistry,[19,20] but it would be some years before the applications of these concepts would feature within the pages of *J. Med. Chem.* (see Section 1.5.1).

1.4.2 Drug-Related Data

The use of quantitative methods for building structure activity relationships had continued to evolve, even though, in the words of Leo Hansch, '... *the glimmerings of the promised land of "rational drug design" loom near the horizon...*'.[21] There were now efforts to take into account three-dimensional structure in assessing drug properties, as exemplified the by the use of comparative molecular field analysis (CoMFA) to determine the pK_a for a series of aminoimidazoline derivatives.[22] Furthermore, computational chemists were beginning to explore the use of neural networks to establish QSAR models, with dihydrofolate reductase inhibitors continuing to provide a useful benchmark dataset with which to establish the usefulness of new methodologies.[23] QSAR analyses and pharmacophore models continued to be the principal means of understanding selectivity issues in many programs, particularly those related to CNS disorders.[24] Access to molecular information on such targets remained in the distant future, and were at this time based upon computed conformations and thermodynamics and depicted *via* computer graphics generated images.[25]

1991 was a banner year for publications on the discovery of a number of hugely important drugs. In each of these cases, the designers accomplished their spectacular feat through varying degrees of QSAR analysis, in the complete absence of data on the molecular details of the drug-target complex. The final stages of the SAR program leading to the discovery of the HMGCoA reductase inhibitor, atorvastatin (see Figure 1.5), were described by Roth and colleagues, with this compound featuring as example (+)-33 in their SAR table.[15] Sold as 'Lipitor', atorvastatin went on to become the world's most successful drug to date. Careful optimization of a penta-substituted pyrrole-based system, for affinity towards rat liver derived HMGCoA reductase, yielded a very potent series of inhibitors, with atorvastatin demonstrating a 5-fold increase in potency *versus* the prototypical natural product inhibitor, compactin. Typical of this period, there were no drug concentration data presented in this paper. Even a subsequent paper from Roth *et al.*, which demonstrated, *via* a QSAR

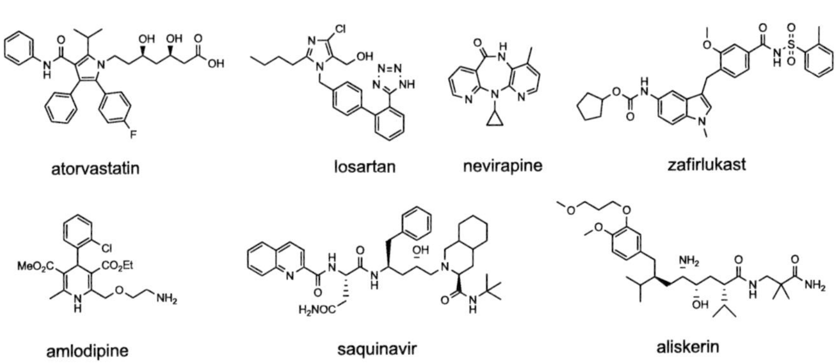

Figure 1.5 Examples of important drugs.

analysis, that there was a strong relationship between lipophilicity and tissue selectivity, used enzyme inhibition data rather than tissue concentration data from which to draw their conclusions.[26]

Later in this same volume of *J. Med. Chem.*, Carini and colleagues described the discovery of the angiotensin II antagonist, DUP-753/losartan (Cozaar, one of the world's most successful antihypertensive agents, see Figure 1.5), which features as compound **17** in their SAR table.[13] Their accomplishment is remarkable for a number of reasons: firstly, the ability to successfully antagonize a peptidergic GPCR (where the agonist is an octapeptide) with an orally active small molecule was a hugely encouraging step forward for the field, given the plethora of peptide hormones of therapeutic importance which could constitute potential therapeutic targets. Secondly, this receptor antagonist strategy for modulation of the renin–angiotensin system proved more successful than the alternative enzyme inhibition approach for targeting renin, which would probably not have been predicted ten years earlier, based upon availability of leads and information on the molecular target. Finally, the introduction of oral activity in this series required several structural modifications that are still somewhat difficult to rationalize, including introduction of a biphenyl *versus*, for example, a biphenyl ether, and replacement of a carboxylic acid with a tetrazole. Again, as in the case of atorvastatin, no data on *in vivo* drug concentrations were presented, therefore the optimization of oral bioavailability was based primarily upon pharmacological evaluation. Thus even in 1991 real time access to pharmacokinetic data was rare and drug designers would have to await future developments in analytical techniques, particularly mass spectrometry, in order to precisely quantify low concentrations of drug without the need for radiolabeled material.

Yet another major drug discovery detailed in this volume of *J. Med. Chem.* was the non-nucleotide reverse transcriptase inhibitor (NNRTI) program that led to nevirapine (Viramune, the first approved NNRTI (see Figure 1.5)), which features as example **96** in the SAR table in this paper.[27] Unlike the HIV protease inhibitor approaches (discussed later) this program did not benefit from details of the drug-target complex, although this was published several years later. It was a classical SAR program based on a screening lead originally derived from an M_1 muscarinic antagonist project. Although the paper does not discuss how many compounds were screened in order to find the original hit, this is a notable early example of a program based upon high throughput screening (HTS), a strategy that would become a key theme across the industry in the decade ahead.

The respiratory therapeutic area was also witnessing the emergence of an important new class of drugs, the leukotriene antagonists as outlined in a review by Shaw and Krell, published that year.[16] Drug designers had made impressive progress in the discovery modulators of the leukotriene pathway and, in particular, leukotriene LTD4 antagonists, which were destined to become a significant new treatment for asthma. One of the drugs from this class, zafirlukast (Accolate, ICI-204,219, see Figure 1.5), is identified as compound **20** in Shaw and Krell's review, whilst a predecessor of the most successful drug in this class, montelukast (Singulair) was identified as a promising clinical candidate (MK-571, compound **11** in the review). Their review provides an excellent sense of the classical drug design strategy of hybridizing SAR information from diverse lead series, inspired also by the structure of the peptidoleukotriene agonist ligands themselves, in order to identify agents with superior *in vitro* and oral *in vivo* activity.

In 1991, a notable advance on the exclusive focus on biological activity evident ten years earlier was the recognition of the critical importance of drug interactions with the cell membrane lipid bilayer in determining therapeutic performance. Thus, Herbette outlined the importance of drug interactions with lipid bilayers for those agents which interacted with membrane-bound receptors *via* a lipid-exposed, rather than an aqueous-exposed, site.[28] He exemplified his thinking by referring to drug classes such as the dihydropyridine calcium channel blockers, and to the the ability of this membrane interaction model to rationalize the unusual pharmacokinetic and pharmocodynamic behavior of amlodipine (Figure 1.5), which had been discovered in the preceding decade but was *en route* to becoming another of the world's most successful antihypertensive drugs.[29] This work highlighted the anisotropic nature of the cell membrane and the inevitable shortcomings of trying to represent it *via* bulk octanol–water partition coefficients. This thinking therefore presaged the future use of cell monolayer models to better represent membrane flux *in vitro*.

1.4.3 Target-Related Data

By this time, the biological complexity, therapeutic opportunities and selectivity challenges resulting from a plethora of receptor sub-types had been recognized, particularly in the CNS, where aminergic ligand-based programs were still a dominant theme. The serotonergic system provides a good example of the growing diversity of CNS programs based upon receptor sub-types, where programs pursuing $5HT_{1A}$, $5HT_2$ and $5HT_3$ receptors were all described.[30–32] The existence of the $5HT_4$ receptor was also known at this time, but the $5HT_5$, $5HT_6$ and $5HT_7$ receptors were not, nor was the full subdivision of the $5HT_1$ receptors. This steady growth in receptor heterogeneity would lead in due course to the more comprehensive selectivity profiling, which has become a feature of modern day programs and which adds considerably to the body of quantitative data to be visualized and analyzed by drug designers. The assays for these programs were still based upon animal, rather than human, receptors and often on tissue preparations (*e.g.* rat brain homogenates), rather than cellular reporter systems. There were also still CNS programs being described, where there was no underlying mechanism of action known for the agents under evaluation, and where assays were based upon whole animal pharmacology.[33] In addition to the aminergic-based CNS programs, there was growing recognition of the importance of glutamatergic and GABAergic transmission in CNS disorders, and there are examples of active research programs in these areas, most of which would not come to fruition for over a decade.[34] Like the

G-protein Coupled Receptor (GPCR)-based programs, the assays described were based upon non-human tissue and the molecular details of these targets were a complete mystery.

Non-CNS programs illustrated a greater diversity of molecular target types, including not just GPCRs but also a wide range of enzymes. Thus, several early kinase inhibitor programs were described[35,36] where, in contrast to the CNS programs, human cell-based or recombinant materials are used as assays. Incredibly, the investigators noted that no X-ray crystallographic information on protein kinases was available at that time and, whilst they recognized that there were selectivity issues to be addressed, the scale of the kinome was hugely under-appreciated. Between 1991 and 2011 there has been an explosion in quantitative information available to drug designers. The kinase inhibitor field is an extreme example of this.

1.4.4 Drug–Target Complex-Related Data

A landmark perspective from scientists at Agouron Pharmaceuticals introduced the notion of a systematic approach to drug optimization: using serial X-ray crystallographic analysis of ligand-protein complexes, exemplifying their strategy through application to a thymidylate synthase inhibitor program.[37] The era of structure-based drug design (SBDD) had arrived. Whilst there had been sporadic examples of X-ray crystallographic analysis applied in drug design before this point, the use of this technology for 'real-time' optimization of drug leads had now begun and was set to become a hugely important factor in the future of drug design. As discussed earlier, the design of HIV protease inhibitors was accelerated by the application of SBDD to reveal details of the drug target complex, as evidenced by the plethora of papers in this volume of *J. Med Chem.* A perspective article,[38] written within a year of publication of the first structures of HIV protease inhibitor complexes, highlighted advances in understanding the mechanism and the homo-dimeric structure of the protease. Progress in inhibitor design and evaluation moved rapidly as drug designers drew upon the lessons learned from the extensive investments in renin inhibitor design, coupled with these new structural insights.[39,40–42] This system also provided a powerful test bed for emerging molecular dynamics simulations of protein–ligand complexes, which helped shed light on, for example, the protonation state of the two catalytic aspartyl residues in the enzyme active site.[43]

It is ironic to note the publications describing the design of renin inhibitors[44–47] in the same *J. Med. Chem.* volume as publications describing the design of HIV protease inhibitors. Both enzymes are aspartyl proteases, but the earlier renin inhibitor campaigns did not benefit from structural information on ligand–target complexes in the way that the subsequent HIV protease inhibitor programs did and almost all were unsuccessful. In fact, the only marketed renin inhibitor, aliskiren (Figure 1.5), was first approved in 2007, over a decade after the first HIV protease inhibitor, saquinavir (Figure 5.1), which was approved in 1995.

Whilst the use of X-ray crystallographic data on target–ligand complexes was to become an important theme in drug design, NMR-based analysis of target–protein complexes was also becoming a viable technology, driven by novel, hetero-nuclear, multi-dimensional NMR techniques together with parallel advances in the molecular biology techniques required to produce large quantities of labeled proteins.[48] NMR-based analysis of ligand–protein complexes has become especially important in the growth of fragment-based drug design (see Section 1.6.4).

1.5 Drug Design 2001

1.5.1 Parallel Medicinal Chemistry

The notion of parallelism as a strategy for increasing the speed and/or success rate in drug discovery programs had firmly taken hold by 2001. Compounds were being synthesized in parallel, the resultant libraries of compounds were being screened in parallel and the multiple streams of data generated on a given compound were being analyzed in parallel. As far back as the mid 1980s, Geysen and Houghten independently disclosed the concept of synthesizing a library of peptides through a *parallel synthesis* strategy.[49,50] Although under-appreciated at the time, within these papers were the seeds of what would in time become a major thrust in drug design strategy: *parallel medicinal chemistry*. Subsequently, in 1991, Fodor and colleagues at Affymax described 'light-assisted, spatially-addressable parallel chemical synthesis', using a considerably more sophisticated solid-phase synthesis strategy, again using peptide synthesis to exemplify their approach.[20] They highlighted the combinatorial power of their approach to generate chemical diversity. That same year, Lam and colleagues further energized the emerging field of combinatorial chemistry with their publication of a 'low-tech' *mix and split* solid-phase synthesis approach, yielding a one-compound-one-bead peptide library.[19] Building upon these concepts, the field evolved over the next 10 years as issues such as library size, library scale, library quality *etc.* were explored. By 2001, the majority of companies had a library group in place and programs based upon library-generated leads were now featuring in *J. Med. Chem.* Even before compound libraries were built, back in the 1990s, HTS of corporate compound collections was being undertaken, and these approaches too became more parallelized and miniaturized over time. The need to evaluate libraries of compounds, rather than individual compounds, added further impetus to parallel screening strategies. The third area of parallelism—the integration of multiple (*e.g.* pharmacological, physicochemical and pharmacokinetic) data streams related to a given compound or ensemble of compounds—was also well underway at this time and was driving the adoption of more sophisticated data analysis tools, such as 'Spotfire'.

Some aspects of this parallel drug design environment were evident in the pages of *J. Med. Chem.* in 2001, for example, leads from HTS of designed

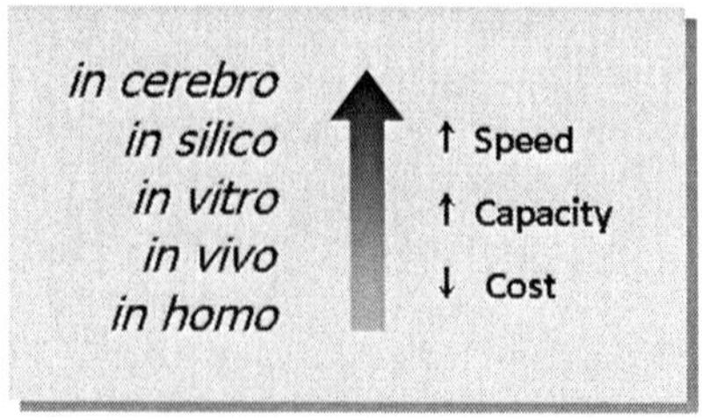

Figure 1.6 The hierarchy of drug design data efficiency.

compound libraries. Others, such as the analysis of multi-parametric datasets, were apparently not.

In addition to the adoption of parallel approaches, the growing pressure on the pharmaceutical/biotechnology industry to operate more efficiently, together with an internal drive to find better ways to discover drugs, had led drug designers to drive the generation of 'drug design data' steadily higher up the hierarchy depicted in Figure 1.6. For example, the likelihood of failure, or 'attrition', relating to lack of oral bioavailability in humans (*in homo*) can be mitigated by preclinical evaluation in animal (*in vivo*) bioavailability models. Better still, in terms of speed and cost, would be to assess the oral absorption potential *in vitro*, *via* a cell-flux assay. Moving yet further up this 'efficiency hierarchy' would be to compute the likelihood of oral absorption *in silico*, based upon calculated properties. The highest aspiration, and the ultimate in efficiency, would be to have some type of rule embedded in the mind of the drug designer (*in cerebro*), which would enable them to avoid even writing down on paper (or a computer screen), the structure of an agent that was highly unlikely to be orally absorbed.

In fact, this is exactly the status achieved by 2001 with regard to oral absorption potential, as discussed in the next section.

1.5.2 Drug-Related Data

As prefaced earlier in this chapter, the era of *in vitro* permeability screens to assess oral absorption potential—as opposed to relying exclusively on *in vivo* experiments—had been ushered in with the introduction of the CACO-2 screen, a colon carcinoma cell line grown as a monolayer on a filter support, allowing measurement of drug concentrations on either side of the monolayer, thus determining cellular flux rates.[51–53] The use of this system grew extensively during the 1990s and it could be regarded as one of the most significant technological advances in terms of predicting *in vivo* and even *in homo* behavior from an *in vitro* assay. Other cell lines, and even non-biological permeability assays, were also developed,[54] as the field strived for more straightforward, but still predictive, *in vitro* absorption models.[55] Even more significantly, these *in vitro* developments were transcended by the 'Rule of 5' (Ro5) framework introduced by Lipinski in 1997, which offered drug designers a powerful *in cerebro* tool for the design of agents with increased prospects of oral absorption.[56] The Ro5 framework has proven to be one of the simplest yet most powerful 'quantitative' drug design methods devised. It has penetrated deeply, albeit slowly, into the consciousness of drug designers, providing a quantitative framework to explain what they intuitively suspected to be the case. Although there is only modest reference to it in *J. Med. Chem.* in 2001 (van de Waterbeemd[57]), those drug designers who already appreciated the importance of physicochemical property co-optimization embraced it enthusiastically. In 2011, it is now the stuff of undergraduate courses.

An excellent commentary in this volume of *J. Med. Chem.* highlights how far the drug design field had progressed in the preceding ten years with regard to the incorporation of multiple parameters, rather than exclusively *in vitro* biological activity, into the design of superior molecules.[57] Thus, the steady progression of physicochemical, pharmaceutical and pharmacokinetic measurements into high throughput *in vitro* formats, as discussed earlier, was beginning to enable a comprehensive and timely survey of a molecule's drug properties and therefore simultaneous *multi-parametric optimization*. Although the authors focus primarily upon the optimization of pharmacokinetic properties, they emphasize the very clear link between physicochemical (*e.g.* lipophilicity and pK_a) and pharmaceutical (*e.g.* solubility and dissolution rate) properties in governing the interaction of drugs with the key physicochemical (*e.g.* lipid bilayers) and biological (*e.g.* P450 enzymes and drug transporters) systems that govern their pharmacokinetic profiles.

An emerging data stream encompassed in the model depicted in Figure 1.1, and an increasingly important one in terms of improving drug survival, is toxicological profile. Historically, apart from *in vitro* mutagenesis studies conducted at the time of candidate nomination, toxicology assays were conducted on compounds of particular interest using animals. However, 2001 was notable for the first appearance in *J. Med. Chem.* of the term, 'hERG' in reference to a toxicological liability as opposed to a therapeutic objective, and also with regard to the *in vitro* evaluation of hERG activity as part of a drug optimization program.[58,59] This followed the market withdrawal of a series of medicines, including the antihistamine, terfenadine and $5HT_4$ partial agonist, cisapride because of arrhythmias resulting from inadvertent blockade of the inwardly rectifying cardiac I_{KR} channel, encoded by the hERG gene. As a result, rapid identification of agents with a potential hERG liability became an imperative for the drug discovery field, and the speed of industry-wide response to this new regulatory concern/attrition risk was truly impressive. Within a few years of the issue being recognized, screening strategies were in place (following the hierarchy outlined in Figure 1.6), encompassing sophisticated electrophysiology studies in human volunteers, *in vivo* studies in dogs and *in vitro* assays in Purkinje fibres and oocytes, as well as *in vitro* binding assays in human cell lines expressing the I_{KR} channel.[60] This rapid response reflected not only both the prevailing state of ion channel molecular biology and the availability of cloned human ion channel constructs, but also the impressive ability of the industry to learn and adapt rapidly.

An area of intense drug design efforts at this time, which faced the challenge of navigating through this new threat of hERG inhibition alongside the more prosaic challenges of pharmacological potency and oral bioavailability, was antagonism of the CCR_5 co-receptor involved in HIV infection of immune cells. During the previous decade, a number of companies had launched programs in response to the finding that individuals who lacked a functional CCR_5 receptor were resistant to HIV infection, even upon repeated exposure. The strength of the genetic data supporting CCR_5 blockade as a compelling anti-HIV approach, together with the higher than average prospect of finding leads against this GPCR target, fuelled considerable interest across the industry. The first, and thus far only, approved CCR_5 antagonist, maraviroc (see Figure 1.7), was launched in 2006. The antagonists disclosed in this volume of *J. Med. Chem.* represented early forays into the clinic and were yet to encounter the hERG related QT-prolongation issues which beset the field.[61,62]

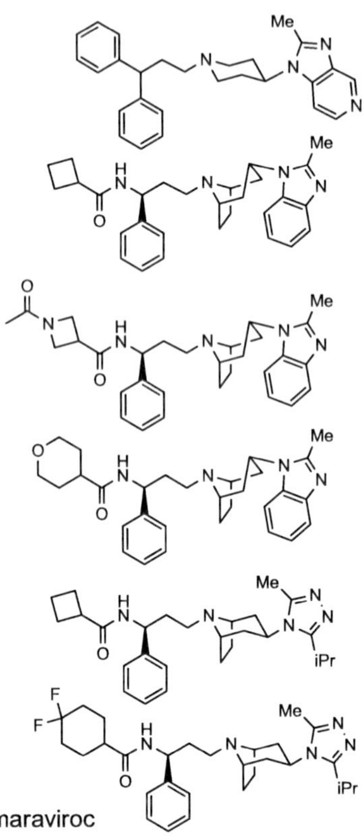

	Antiviral IC_{90} nM	hERG Inhibition	CaCo-2 Flux Papp x10^{-6} cm/s	HLM $t_{1/2}$ (min)
A	Inactive	(High Throughput Screening Hit)		
B	3	99% @ 1 μM		
C	1	0% @ 300 nM	<1	
D	0.6	0% @ 100 nM	23	low
E	8	14% @ 100 nM 30% @ 300 nM	4.5	55
F maraviroc	1	0% @ 300 nM		51

HLM = Human Liver Microsome

Figure 1.7 Elimination of hERG activity in the design of the CCR_5 antagonist, maraviroc.

In contrast to typical receptor antagonist programs set 10 years earlier, these workers secured their leads *via* HTS, assayed their compounds using human recombinant reagents and routinely characterized the pharmacokinetic profiles of their agents. They also dealt with the inevitable selectivity challenges associated with deriving their leads from a series designed to address a completely different pharmacology (*e.g.* muscarinic acetylcholine receptor antagonism). The design of maraviroc highlights the successful negotiation of a multi-parametric optimization challenge, encompassing cell-based anti-viral activity, *in vitro* absorption, *in vitro* clearance and, in particular, hERG inhibition. The HTS hit **A** is transformed through a sequence of structural changes (**B**–**E**) to achieve the optimized profile, culminating in compound **F** (maraviroc). Although no structural information was available regarding the drug–target complex, just as in the discovery of cimetidine decades earlier, these drug designers applied their wit and ingenuity to a carefully assembled body of data in order to design a superb agent.[63]

1.5.3 Target-Related Data

The first draft of the human genome was announced in 2000 and the complete genome sequence was published in 2003. Drug design projects published in 2001 therefore benefitted from a much greater appreciation of the scope of the genome than projects conducted a decade earlier and in particular were much more likely to be aware of the full scope of the gene family of interest. A window into what would in due course become a genome-wide scan of selectivity issues is provided by a paper from researchers at Cerep.[64] They highlighted the opportunity to build unique fingerprints, describing the pharmacological profile of a given chemical lead across a panel of selectivity assays. This *wide ligand profiling* strategy is now an essential step in establishing the credentials of a putative lead or HTS hit, enabling drug designers to select molecular starting points which offer the best prospect of delivering a highly selective candidate most efficiently.

The full range of receptor sub-types was now appreciated across, for example, the serotonergic receptor family, where a wider range of programs were described compared to 10 years prior: $5HT_{1A}$,[65] $5HT_{2A}$,[66,67] $5HT_{2A/C}$,[68] $5HT_{1B/1D}$,[69] $5HT_{1F}$,[70] and $5HT_6$.[71] Many of these programs were CNS related and explored the ability of selective receptor agonists/antagonists to achieve a given pharmacological profile. However, it was clear even then that many CNS drugs exhibit their therapeutic effects through interaction with multiple receptor sub-types.[39] Therefore, ironically and despite the advances in human genome sequencing, screening for CNS active agents in a reductionist manner, seeking exquisitely receptor sub-type selective agents, does not necessarily offer advantages over the 'old fashioned' approach of using animal models, which had the benefit of possessing a fully integrated CNS circuitry. Interest in a more exhaustive exploration of the effects of a drug on a biological system would grow over the coming years, as the field of systems biology matured.

In addition to the steady growth of GPCR-based targets, interest in ion channels as therapeutic targets had grown considerably, as knowledge of their sub-types, molecular structure and physiological role had grown. Thus, there were comprehensive reviews of both potassium channel and sodium channel modulator drug design available.[72,73] As with GPCR-based programs, the membrane-bound, multimeric nature of these ion channels precluded ready access to structural information on the drug–target complex and drug design programs still relied heavily on QSAR analysis. Additional challenges posed by ion channel targets included the potential for 'state dependent' interaction with modulators and the more sophisticated (and therefore typically lower throughput) screening strategies, such as patch clamp electrophysiology, which meant that HTS against ion channel targets lagged other target classes. These reviews provide a good sense of the current state-of-the-art regarding screening technologies, *e.g.* radio-ligand binding, radioactive flux, electrophysiology, fluorescence-based assays (such as FRET), voltage sensitive dyes and ion-selective dyes. Many of these assay systems also benefitted from advances in fluorescence imaging technology and the advent of devices such as FLIPR, from Molecular Devices Corp.

The efficiency hierarchy referred to in Figure 1.6 is equally applicable to pharmacological data, therefore a natural progression from high throughput *in vitro* screening—which had now become an industry standard tool—was *virtual screening*, where the test agent was not exposed to the molecular target in a wet experiment, but was exposed to the target *in silico*. This offered the prospect of priority ranking a collection of compounds for *in vitro* screening, both speeding up the evaluation process and conserving expensive screening capacity. The technology appeared in the late 90s and Stahl *et al.* review the performance of a series of scoring functions in by ranking a library of over 7000 compounds for affinity *versus* 7 diverse molecular targets.[74] They highlighted the differences in effectiveness of the scoring functions depending upon their ability to handle target sites rich in H-bonding or hydrophobic interactions.

1.5.4 Drug–Target Complex-Related Data

By 2001, X-ray crystallographic data on drug–target complexes had become an integral aspect of programs based upon inhibition of soluble protein targets, such as enzymes. Two contrasting types of enzyme under active investigation in this volume of *J. Med. Chem.* were secreted proteases, such as cathepsin K for osteoporosis, and a variety of intracellular kinases for the treatment of cancer. Cathepsin K, a cysteine protease of the papain family, is expressed almost exclusively in osteoclasts and is essential for the process of bone resorption. It therefore represented a compelling approach to the treatment of osteoporosis, where an imbalance of bone production by osteoblasts *versus* bone resorption by osteoclasts results in progressive bone loss. At the time of writing, there are several cathepsin K inhibitors approaching regulatory submission, but in 2001 drug designers were exploring a variety of strategies for cysteine protease inhibition, each of which typically involved trapping the active site cysteine thiol nucleophile as either a reversible or irreversible covalent complex with the inhibitor.[75,76] Impressively, X-ray crystallography studies were integrated into the program alongside biological evaluation (using assays which were now conducted with human recombinant enzymes), mechanistic studies (involving detailed characterization of reaction kinetics), absorption studies (employing CACO-2 cell lines) and animal pharmacokinetic studies (describing clearance, volume and half-life data). Judging by the agents in advanced clinical development, use of a nitrile as an electrophile to trap the catalytic thiol was deemed to offer the optimum design approach.

Interest in kinase inhibition as a strategy for addressing oncogenically driven cancers was growing rapidly at this time and drug designers were using knowledge of the inhibitor–protein complexes in order to address the challenging selectivity issues in this enzyme class. A series of papers on the checkpoint dependent kinase (CDK) family provides a beautiful example of how X-ray structural data on drug–target complexes was being used in conjunction with protein sequence data on other tyrosine and threonine/serine kinases.[77,78] This allowed identification of those residues involved in interactions with the inhibitor, which were least homologous to other family members, and which therefore offered the best prospect for selective inhibition as drug-protein interactions, were optimized. This integration of crystallographic data for a ligand–protein complex with informatics data from the rapidly growing protein sequence databank would become a key drug design strategy in the kinase inhibitor field. This emerging *structural informatics* approach adds yet another dimension to the 'Cerep' parallel screening philosophy discussed in Section 1.5.3. Furthermore, the generation of compound libraries specifically designed to probe these key interactions aligned the empirical power of parallel synthesis with the precision of structural biology data, providing a more efficient way of optimizing inhibitor selectivity.

1.6 Drug Design 2011

1.6.1 The Omics Era

As we enter the second decade of this century, the application of quantitative approaches in drug design, as judged by the latest publications in *J. Med. Chem.*, appears in good health and continues to advance steadily. There is a wealth of large and rapidly growing datasets from which drug designers can draw lessons, new insights and valuable information. These include the vast drug design literature, not just from *J. Med. Chem.* but also the plethora of journals which have appeared over the last several decades in the fields of drug design, computational chemistry, bio-organic chemistry and, more recently, chemical biology. In addition, there are now large public domain datasets available, encompassing the genomes of a myriad species and transformed cell types, protein sequences, protein structures, protein–ligand complexes, small molecule structures and biological data on large ensembles of molecules. In fact, the integration, interrogation and interpretation of these data are a key aspect of modern drug design teams, further driving their diversification beyond the traditional chemistry and biology disciplines.

It is also clear that drug design tools have developed sufficiently to move beyond the domain of large pharmaceutical and biotech companies and become accessible to the academic world. Thus, drug design is increasingly being undertaken within universities and academic medical centers, many of which are exploring opportunities for therapeutic intervention in areas of new biology which have yet to attract the attention of large companies. This diversification of research base, together with the availability of genomic information on parasitic organisms, has enabled a much greater investment in potential therapies for diseases of the developing world, as evidenced by the striking number of publications in this area in the 2010 volume of *J. Med. Chem.*

As with each of the decades surveyed, the level of drug design sophistication encountered today varies widely. Incredibly, there are still publications based upon programs where the molecular target is completely unknown! However, at the leading edge, the current 'state-of-the-art' in drug design is now very high. Thus, with a few important exceptions, the ability to generate leads against therapeutically important targets *via* HTS of sophisticated compound libraries, virtual screening *in silico* or, increasingly, by use of fragment-based approaches, is impressive. Furthermore, the real-time integration within the drug design environment of all the data streams outlined in Figure 1.1 is evident in numerous examples. The depth of understanding now possible at the level of the drug lead, the molecular target and the drug–target complex would have seemed miraculous to the drug designer of 1981.

1.6.2 Drug-Related Data

Although there is exciting progress in the structural biology of GPCRs, it is likely to be some time before most integral membrane proteins are amenable to structure-based drug design. Consequently, pharmacophore-based modeling remains a very important drug-design strategy in programs based upon modulation of targets such as GPCRs and ion-channels. The state-of-the-art in pharmacophore modeling for rationalizing SAR, conducting virtual screening and enabling compound design has recently been reviewed.[79] The authors highlight both the progress that has been made in this area over the last 30 plus years, as well as the opportunities for future improvements. A complementary, multi-lateral survey of computational approaches to describing and comparing the shapes of molecules also highlights the progress in this area, asserting that '*knowing the structure of an active ligand is better than knowing the structure of its molecular receptor*'.[80] This is a telling comment, and both emphasizes the continued importance (and success) of drug design based entirely on the structure/properties of the ligand, as well as alluding to the challenges in *de novo* ligand design based only upon the structure of the complementary active site.

Although many modeling tools require the expertise of computational chemistry specialists, there has been a welcome transfer of such tools into the hands of medicinal chemists, enabling a more rapid cycle of virtual design and analysis.[81]

The relationship between drug structure and properties continues to be fertile ground for analysis, as drug-designers search for new overarching frameworks that match the power of Lipinski's Ro5, and the growing body of drug related data generates new substrates for interrogation. For example, a recent analysis of marketed drugs identifies the physicochemical drug space most likely to yield good oral bioavailability, based upon intestinal absorption and first pass clearance considerations.[82] The importance of managing rotatable bond count is strongly emphasized in this analysis. The need to determine oral absorption potential successfully *in vitro* continues to fuel retrospective analyses of cell-based permeation screen performance,[83] and these analyses continue to support the use of systems based upon CACO2 and MDCK cell lines.

Although now well understood, the importance of managing physicochemical properties, especially lipophilicity, in transforming an early lead into a clinical candidate continues to be reinforced.[84] A powerful strategy for influencing physicochemical properties is the manipulation of hydrogen-bond donors and acceptors within a molecule, which strongly influence solubility, lipophilicity and membrane permeability. Impressively, there is yet more understanding to be wrung from consideration of H-bonding behavior, as evidenced by a superb analysis of intramolecular H-bonds and their impact on properties, based upon a review of large crystallographic databases.[85] This is another great example of a computational analysis of a large dataset driven by a desire to codify information in a way that is useful to drug designers.

Oligonucleotide-based therapeutics represent a specialized area of drug design, however, the relationship of structure to biological function and properties remains just as relevant. Despite the extensive literature available on nucleotide modifications, there are still innovations possible, as designers grapple with the simultaneous optimization of target affinity, RNAse-H susceptibility of the drug–RNA complex (in the case of antisense oligonucleotides), nuclease stability, pharmacokinetics and cellular uptake.[86]

An excellent example of modern day multi-parametric lead optimization, in the absence of target protein structure, is described by Pinard *et al.*, who balanced pharmacological (*in vitro* hGlyT1 uptake inhibition), physicochemical (calculated LogP (cLogP) and polar surface area (PSA)), pharmacokinetic (mouse brain *versus* plasma concentration) and safety (hERG) profiles in order to deliver an impressive GlyT1 inhibitor clinical candidate.[87]

1.6.3 Target-Related Data

Since the complete sequencing of the human genome was accomplished a decade ago, the genomes of many organisms have also been sequenced, opening up a new era in comparative genomics and anti-infective therapy. Thus, the genomes of many human pathogenic bacteria, fungi and viruses have been completely sequenced, as have those of parasitic organisms—such as malaria

(*Plasmodium falciparum* and *Plasmodium vivax*), Chagas disease (*Trypanasoma cruzi*), sleeping sickness (*Trypanosoma brucei*), leishmaniasis (*Leishmania major*) and trichomonas vaginalis (*Trichomonas*)—responsible for diseases of the developing world. The ability to study the consequences of genetic knock-outs in these organisms and apply well established drug design strategies for common gene families, together with the growth in academic-based drug design efforts, has led to a large increase in publications on the design of drugs for diseases of the developing world. For example, there are over twenty publications relating to projects targeting malaria in the 2010 volume of *J. Med. Chem.* Interestingly, cysteine proteases, including falcipain (from *Plasmodium falciparum*)[88] and cruzain (from *Trypanosoma cruzi*)[89,90] have emerged as high profile targets for the treatment of these infections, even though there are no marketed therapies targeting the mammalian members of this enzyme class yet. The cruzain program has benefitted from the NIH Molecular Libraries Probe Production Network (MLPCN), through which a new molecular target can be screened against the NIH compound collection in order to identify small molecule hits. Although the molecular targets of these programs are non-mammalian, they nevertheless benefit from all the drug design tools that have been developed *versus* mammalian systems, including recombinant cell-lines and assays, wide selectivity profiling and X-ray crystallographic analysis of the protein target.

A further exciting new area of target space now being explored is that of mechanisms involved in directing the fate of stem cells.[91] This is both a highly promising but also challenging field, which offers the prospect of regenerating cells and tissues through small molecule intervention. However, the pathways involved, such as *wnt*, *hedghog* and *notch* are poorly understood and involve many protein-protein interactions (PPIs) in their signaling networks. This is therefore an opportunity to use chemical genomics to decipher these signaling pathways and to devise novel strategies for modulating 'tough targets', such as PPIs and transcription factors.

A further development in the area of drug targets has been the recognition that 'polypharmacology' is a good thing! Now referred to as *network pharmacology*, it is proposed that in order to achieve good efficacy in a complex biochemical system, interaction with multiple molecular targets in a specific manner might be necessary.[92] This has already been alluded to in the area of CNS pharmacology (discussed in Section 1.5.3), but is also increasingly important in other areas, particularly oncology, where simultaneous inhibition of multiple kinases is important in driving the efficacy of a number of marketed kinase inhibitors (see Section 1.6.4). Although the successful co-optimization of multiple biological activities remains one of the biggest challenges facing drug designers, generating and analyzing the necessary biological data on multiple targets in parallel is now a straightforward proposition. Interestingly, there is still room for innovation in methods for determining inhibitor selectivity across a panel of kinase assays, for example, a thermodynamic partition index, which measures the extent to which a test agent binds to a given kinase when exposed to an equimolar mixture of kinases present in excess.[93] This approach has the attraction of generating an easily assimilated SAR table with an emphasis on selectivity rather than raw potency.

1.6.4 Drug–Target Complex-Related Data

As referred to in the previous section, the large body of information contained in the protein data bank, together with proprietary protein structure databases, is now enabling extremely comprehensive analyses of molecular interactions. In the 2011 volume of *J. Med. Chem.* Stahl *et al.* provide a superb overview of molecular interactions of importance to drug designers, which has been assembled through analysis of such databases.[94] The quality of such scholarship can surely only enhance the design skills of those who study this compilation.

Fragment-based drug design (FBDD) has emerged as an important strategy for lead generation in programs where the molecular structure of the target is available,[95] and FBDD approaches feature prominently in this volume of *J. Med. Chem.*[96,97] This approach relies heavily upon the integration of biophysical and spectroscopic measurements in order to detect the low affinity interactions between small molecular fragments and the target protein. However, computational drug design approaches are also very important in both design of fragment libraries and in design of growth strategies to optimize active fragments. A key principle in FBDD is ligand efficiency (LE), a comparative measurement which has been central in HTS hit selection and lead optimization programs for some time, but which has been further identified as a critical determinant of success in FBDD programs.[98]

The field has matured to the point where leads against high profile targets are being identified[99] and scholarship on the field is emerging, for example, addressing issues such as fragment specificity, which fragments work best for which targets and whether deconstruction of active ligands yields fragments that bind the molecular target.[97,100,101] Figure 1.8 provides an impressive example of a lead optimization program from Astex, a pioneering company in

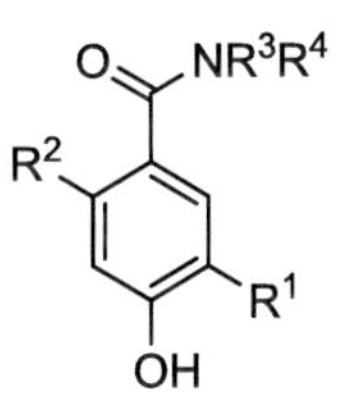

	R^1	R^2	NR^3R^4	Hsp90 Affinity (μM)
A	OMe	H	NEt_2	760
B	^iPr	H	NEt_2	7
C	^iPr	OH	NEt_2	0.068
D	^iPr	OH	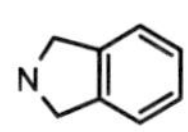	0.00054
E	^iPr	OH	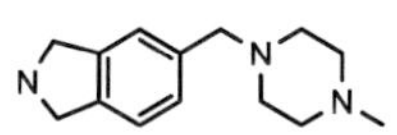	[0.0007 (Hsp90 K_d)]

Figure 1.8 Fragment-based discovery and optimization of an Hsp90 inhibitor lead.

this field, based upon a 760 μM fragment (**A**). Affinity was increased 10^6-fold by three successive 100-fold increments, each adding only modest molecular weight, to yield lead **D**, a sub-nanomolar lead (MW ~300), which was optimized for solubility, cell-penetration, minimal hERG inhibition and pharmacokinetic properties to yield the clinical candidate, **E**.[96]

A very recent and most significant advance in the area of drug–target complex data has been the structure determination of membrane-bound GPCRs, which has now opened the way to SBDD programs on this hugely important gene family. Thus, structure-based design programs on a GPCR, the A_{2A}-receptor, feature for the first time in this volume of *J. Med. Chem.*[102,103] The potential impact of these new data is illustrated by the virtual screening of a 4 million compound database and the discovery of novel ligands with 60 nM affinity and high ligand efficiency.[102]

Kinase inhibition continues to be an intense focus of structure-based drug design efforts. However, the emphasis has shifted over time to the design of 'multi-target' kinase inhibitors (MKIs), a much more ambitious objective, which involves inhibiting a specific subset of kinases whilst minimizing interaction with the remainder of the kinome.[104,105] Thus, the challenges of multiparametric lead optimization now include a large, parallel set of biochemical assays, as well as the standard physicochemical and pharmacokinetic assays. This strategy is rendered viable by the availability of structural information on a large number of kinase inhibitor complexes, as well as the efficiency of current structure determination techniques. There is also continued interest in kinase inhibitor strategies that involve alternative binding modes, which exploit interactions beyond the ATP-binding pocket, for the inhibitor,[106] Again, this objective is enabled by the efficiency with which the structures of drug–target complexes can now be determined.

1.7 Drug Design in the Future

Having witnessed the impressive evolution of drug design over the last 30 years and the growing role played by computational methods, it is possible to speculate on the trajectory these changes might follow as we look to the future. It seems certain that drug designers will continue to play their centre stage role, and that a quantitative understanding of potential drug molecules, their molecular targets and the complex between them will remain a basis for thinking about drug design.

We should expect new ways of characterizing and visualizing molecular properties, ways which steadily improve our ability to represent how biological and physicochemical systems actually 'see' drug molecules. We should also expect new data analysis tools that extract trends, insights and even 'rules' from very large public domain or proprietary datasets, which will continue to grow rapidly. There are also likely to be more examples of intelligent compound libraries, either targeting specific gene families or representing completely novel areas of chemical space, such as macrocycles, and where drug property considerations are embedded within the library members.

In the target domain, as the network of interactions between every protein in a given cell is mapped out, we should expect to see more sophisticated systems biology approaches, providing detailed, spatio-temporal simulations of signaling systems, under both physiological and pathological conditions, which could enable selection of therapeutic intervention points with much greater precision. As next generation sequencing brings the cost of genome sequencing down, we will see a growth in the number of entire human genome sequences available, as well as the sequences of many tumor genomes. This will drive yet more novel targets, and test the ability of drug designers to tackle new target classes. Thus, although the range of druggable gene families has expanded to some extent over the decades, and it is reasonable to expect this trend to continue, in order to translate this compelling new biology into breakthrough therapeutics it will be essential to address more of the 'tough' molecular targets, which have been avoided by drug designers to date. Protein protein interactions, for example in intracellular signaling cascades, as well as protein-DNA interactions, represent a new and challenging frontier waiting to be conquered. To date, many of these targets have been viewed as thermodynamic 'no go' areas for small molecule intervention. However, 'easy' targets will eventually become an endangered species, and the pressure to make progress in these areas will continue to grow, eventually overcoming the reluctance of the drug design community to invest in developing new strategies. As powerful and valuable as the Ro5 construct has been, it will be necessary to move beyond the rule of five and establish a new drug design framework, perhaps learning lessons from the natural product domain, where there are intriguing examples of orally bioavailable, high molecular weight agents which penetrate cells and modulate protein–protein interactions. The *diversity-oriented synthesis* movement, which aims to create libraries of much more densely functionalized and stereochemically rich molecules, has been motivated by this thinking.[107] These tough targets are also likely to drive the fusion of small molecule design with biological macromolecule design. There are already examples of this, where helical peptide sequences can be stabilized through synthetic bridging strategies and enabled to penetrate cells,[108] and where biologically active peptides can be protected from plasma degradation by conjugation to an antibody *via* a synthetic linker.[109] The opportunity for drug design to encompass macromolecular systems and fuse the small molecule world with the big-molecule world appears compelling.

The principle of using drug–macromolecule conjugation strategies to achieve tissue selective targeting[110] and cellular uptake[111] of therapeutics has been appreciated for some time. However, this does seem to be a heavily underutilized strategy, which could well provide a way forward in the modulation of tough intracellular targets. In these cases, one component of a fusion construct interacts selectively with cell surface recognition elements and the other component delivers a therapeutic intervention.

Drug targeting is also of interest in the field of oligodeoxynucleotide (ODN)-based therapeutics, including antisense, siRNA and micro-RNA approaches, where the issues of tissue targeting and cellular uptake (and subsequent release) currently limit the full therapeutic potential of this area. Although ODNs have not featured prominently in this review, they represent another opportunity for drug design advances and are a potentially important component of the concerted effort needed to address these tough intracellular targets.

One could argue that the single most important technological development in drug discovery over the last 30 years has been protein X-ray crystallography. At some point in the future we will also know, at the molecular level, the structure of every protein encoded by the genome and be able to mount structure-based drug discovery projects against any target of interest. Although it is difficult to project when this might be achieved, the recent progress with integral membrane proteins and transporters suggests that this is only a matter of time.

Given the potential for structural information of this type to advance the entire global drug design enterprise, it is encouraging to see more examples of open source research, such as the Structural Genomics Consortium, which is committed to placing novel structural biology data into the public domain. Hopefully there will be many more examples of open source, pre-competitive consortia across the drug design continuum, enabling access to drug design information at much lower cost.

Finally, the outsourcing of components of the drug discovery process to lower cost vendors, particularly in India and China, is likely to result in a growth in innovation-driven drug discovery companies in these countries in due course. So, just as drug design know-how and tools have been disseminated within the academic world over the last decade, it is likely that the geographic boundaries of drug design will expand in the years ahead, bringing fresh approaches and new concepts to further enrich the field.

References

1. J. M. Caroon, R. D. Clark, A. F. Kluge, J. T. Nelson, A. M. Strosberg, S. H. Unger, A. D. Michel and R. L. Whiting, *J. Med. Chem.*, 1981, **24**, 1320.
2. J. A. Bristol, E. H. Gold, R. G. Lovey and J. F. Long, *J. Med. Chem.*, 1981, **24**, 927.
3. Y. Martin, *J. Med. Chem.*, 1981, **24**, 229.
4. G. L. Olson, H. C. Cheung, K. D. Morgan, J. F. Blount, L. Todaro, L. Berger, A. B. Davidson and E. Boff, *J. Med. Chem.*, 1981, **24**, 1026.
5. R. Ganellin, *J. Med. Chem.*, 1981, **24**, 913.
6. M. A. Ondetti and D. W. Cushman, *J. Med. Chem.*, 1981, **24**, 355.
7. G. Crippen, *J. Med. Chem.*, 1981, **24**, 198.
8. S. H. Unger and G. H. Chiang, *J. Med. Chem.*, 1981, **24**, 262.
9. G. L. Olson, H.-C. Cheung, K. D. Morgan, J. F. Blount, L. Todaro, L. Berger, A. B. Davidson and E. Boff, *J. Med. Chem.*, 1981, **24**, 1026.
10. D. A. Matthews, R. A. Alden, J. T. Bolins, S. T. Freer, R. Hamlin, N. Xuong, J. Kraut, M. Poe, M. Williams and K. Hoogsteen, *Science*, 1977, **197**, 452.

11. L. Wei, E. Clauser, F. Alhenc-Gelas and P. Corvol, *J. Biol. Chem.*, 1992, **267**, 13398.
12. R. Natesh, S. L. U. Schwager, E. D. Sturrock and K. R. Acharya, *Nature*, 2003, **421**, 551.
13. D. J. Carini, J. V. Duncia, P. E. Aldrich, A. T. Chiu, A. L. Johnson, M. E. Pierce, W. A. Price, J. B. Santella III, G. J. Wells, R. R. Wexler, P. C. Wong, S.-E. Yoo and P. B. M. W. M. Timmermans, *J. Med. Chem.*, 1991, **34**, 2525.
14. D. Alker, S. F. Campbell and P. E. Cross, *J. Med. Chem.*, 1991, **34**, 19.
15. B. D. Roth, C. J. Blankley, A. W. Chucholowski, E. Ferguson, M. L. Hoefle, D. F. Ortwine, R. S. Newton, C. S. Sekerke, D. R. Sliskovic, C. D. Stratton and M. W. Wilson, *J. Med. Chem.*, 1991, **34**, 357.
16. A. Shaw and R. D. Krell, *J. Med. Chem.*, 1991, **34**, 1235.
17. D. T. W. Chu, C. W. Nordeen, D. J. Hardy, R. N. Swanson, W. J. Giardina, A. G. Pernet and J. J. Plattner, *J. Med. Chem.*, 1991, **34**, 168.
18. P. A. Lartey, S. L. DeNinno, R. Faghih, D. J. Hardy, J. J. Clement, J. J. Plattner and R. L. Stephens, *J. Med. Chem.*, 1991, **34**, 3390.
19. K. S. Lam, S. E. Salmom, E. M. Hersh, V. J. Ruby, W. M. Kazmierski and R. J. Knapp, *Nature*, 1991, **354**, 82.
20. S. P. A. Fodor, J. L. Read, M. C. Pirrung, L. Stryer, A. T. Lu and D. Solas, *Science*, 1991, **251**, 767.
21. C. D. Selassie, R.-L. Li, M. Poe and C. Hansch, *J. Med. Chem.*, 1991, **34**, 46.
22. K. H. Kim and Y. C. Martin, *J. Med. Chem.*, 1991, **34**, 2056.
23. T. A. Andrea and H. Kalayeh, *J. Med. Chem.*, 1991, **34**, 2824.
24. M. Froimowitz and S. Riimsby, *J. Med. Chem.*, 1991, **34**, 1707.
25. J. J. Kaminski, B. Wallmark, C. Briving and B.-M. Andersson, *J. Med. Chem.*, 1991, **34**, 533.
26. B. D. Roth, T. M. A. Bocan, C. J. Blankley, A. W. Chucholowski, P. L. Creger, M. W. Creswell, E. Ferguson, R. S. Newton, P. O'Brien, J. A. Picard, W. H. Roark, C. S. Sekerke, D. R. Sliskovic and M. W. Wilson, *J. Med. Chem.*, 1991, **34**, 466.
27. K. D. Hargrave, J. R. Proudfoot, K. G. Grozinger, E. Cullen, S. R. Kapadia, U. R. Patel, V. U. Fuchs, S. C. Mauldin, J. Vitous, M. L. Behnke, J. M. Klunder, K. Pal, J. W. Skiles, D. W. McNeil, J. M. Rose, G. C. Chow, M. T. Skoog, J. C. Wu, G. Schmidt, W. W. Engel, W. G. Eberlein, T. D. Saboe, S. J. Campbell, A. S. Rosenthal and J. Adams, *J. Med. Chem.*, 1991, **34**, 2231.
28. R. P. Mason, D. G. Rhodes and L. G. Herbette, *J. Med. Chem.*, 1991, **34**, 869.
29. J. E. Arrowsmith, S. F. Campbell, P. E. Cross, J. K. Stubbs, R. A. Burges, D. G. Gardiner and K. J. Blackburn, *J. Med. Chem.*, 1986, **29**, 1697.
30. R. K. Raghupathi, L. Rydelek-Fitzgerald, M. Teitler and R. A. Glennon, *J. Med. Chem.*, 1991, **34**, 2633.
31. J.-L. Malleron, M.-T. Comte, C. Gueremy, J. F. Peyrone, A. Truchon, J. C. Blanchard, A. Doble, O. Piot, J.-L. Zundel, C. Huon, B. Martin,

P. Mouton, A. Viroulaud, D. Allam and J. Betschart, *J. Med. Chem.*, 1991, **34**, 2477.
32. C. J. Swain, R. Baker, C. Kneen, J. Moseley, J. Saunders, E. M. Seward, G. Stevenson, M. Beer, J. Stanton and K. Watling, *J. Med. Chem.*, 1991, **34**, 140.
33. H. Kohn, K. N. Sawhney, P. LeGall, D. W. Robertson and J. D. Leander, *J. Med. Chem.*, 1991, **34**, 2444.
34. P. L. Ornstein, D. D. Schoepp, M. B. Arnold, J. D. Leander, D. Lodge, J. W. Paschal and T. Elzey, *J. Med. Chem.*, 1991, **34**, 90.
35. A. Ricouart, J. C. Gesquiere, A. Tartar and C. Sergheraert, *J. Med. Chem.*, 1991, **34**, 73.
36. P. M. Traxler, O. Wacker, H. L. Bach, J. F. Geissler, W. Kump, T. Meyer, U. Regenass, J. L. Roesel and N. Lydon, *J. Med. Chem.*, 1991, **34**, 2328.
37. K. Appelt, R. J. Bacquet, C. A. Bartlett, C. L. J. Booth, S. T. Freer, M. A. M. Fuhry, M. R. Gehring, S. M. Herrmann, E. F. Howland, C. A. Janson, T. R. Jones, C.-C. Kan, V. Kathardekar, K. K. Lewis, G. P. Marzoni, D. A. Matthews, C. Mohr, E. W. Moomaw, C. A. Morse, S. J. Oatley, R. C. Ogden, M. R. Reddy, S. H. Reich, W. S. Schoettlin, W. W. Smith, M. D. Varney, J. E. Villafranca, R. W. Ward, S. Webber, S. E. Webber, K. M. Welsh and J. White, *J. Med. Chem.*, 1991, **34**, 1925.
38. J. R. Huff, *J. Med. Chem.*, 1991, **34**, 2305.
39. T. A. Lyle, C. M. Wiscount, J. P. Guare, W. J. Thompson, P. S. Anderson, P. L. Darke, J. A. Zugay, E. A. Emini, W. A. Schleif, J. C. Quintero, R. A. F. Dixon, I. S. Sigal and J. R. Huff, *J. Med. Chem.*, 1991, **34**, 1230.
40. D. H. Rich, C.-Q. Sun, J. V. N. Vara Prasad? A. Pathiasseril, M. V. Toth, G. R. Marshall, M. Clarel, R. A. Mueller and K. Houseman, *J. Med. Chem.*, 1991, **34**, 1222.
41. J. P. Vacca, J. P. Guare, W. M. Sanders, E. A. Giuliani, S. D. Young, P. L. Darke, J. Zugay, I. S. Sigal, J. W. A. Schleif, J. C. Quintero, E. A. Emini, P. S. Anderson and J. R. Huff, *J. Med. Chem.*, 1991, **34**, 1225.
42. S. Thaisrivongs, A. G. Tomasselli, J. B. Moon, J. Hui, T. J. McQuade, S. R. Turner, J. W. Strohbach, W. J. Howe, W. G. Tarpley and R. L. Heinrikson, *J. Med. Chem.*, 1991, **34**, 2344.
43. D. M. Ferguson, R. J. Radmer and P. A. Kollman, *J. Med. Chem.*, 1991, **34**, 2654.
44. A. M. Doherty, J. S. Kaltenbronn, J. P. Hudspeth, J. T. Repine, W. H. Roark, I. Sircar, F. J. Tinney, C. J. Connolly, J. C. Hodges, M. D. Taylor, B. L. Batley, M. J. Ryan, A. D. Essenburg, S. T. Rapundalo, R. E. Weishaar, C. Humblet and E. A. Lunney, *J. Med. Chem.*, 1991, **34**, 1258.
45. S. H. Rosenberg, H. D. Kleinert, H. H. Stein, D. L. Martin, M. A. Chekal, J. Cohen, D. A. Egan, K. A. Tricarico and W. R. Baker, *J. Med. Chem.*, 1991, **34**, 469.
46. R. H. Bradbury and J. E. Rivett, *J. Med. Chem.*, 1991, **34**, 151.
47. S. Thaisrivongs, D. T. Pals, D. W. DuCharme, S. R. Turner, G. L. DeGraaf, J. A. Lawson, S. J. Couch and M. V. Williams, *J. Med. Chem.*, 1991, **34**, 633.

48. S. W. Fesik, *J. Med. Chem.*, 1991, **34**, 2937.
49. H. M. Geysen, R. H. Meloen and S. J. Barteling, *Proc. Natl. Acad. Sci. U.S.A.*, 1984, **81**, 3998.
50. R. A. Houghten, *Proc. Natl. Acad. Sci. U.S.A.*, 1985, **82**, 5131.
51. I. J. Hidalgo, T. J. Raub and R.T. Borchardt, *Gastroenterology*, 1989, **96**, 736.
52. G. Wilson, I. F. Hassan, C. J. Dix, I. Williamson, R. Shah and M. Mackay, *J. Controlled Release*, 1990, **11**, 25.
53. C. S. Schasteen, M. G. Donovan and J. N. Cogburn, *J. Controlled Release*, 1992, **21**, 49.
54. F. Wohnsland and B. Faller, *J. Med. Chem.*, 2001, **44**, 923.
55. P. Stenberg, U. Norinder, K. Luthman and P. Artursson, *J. Med. Chem.*, 2001, **44**, 1927.
56. C. A. Lipinski, F. I. Lombardino, B. W. Dominy, P. J. Feeney, *Adv. Drug Deliv. Rev.*, 1997, **23**, 3.
57. H. van de Waterbeemd, D. A. Smith, K. Beaumont and D. K. Walker, *J. Med. Chem.*, 2001, **44**, 1313.
58. I. M. Bell, S. N. Gallicchio, M. Abrams, D. C. Beshore, C. A. Buser, J. C. Culberson, J. Davide, M. Ellis-Hutchings, C. Fernandes, J. B. Gibbs, S. L. Graham, G. D. Hartman, D. C. Heimbrook, C. F. Homnick, J. R. Huff, K. Kassahun, K. S. Koblan, N. E. Kohl, R. B. Lobell, J. J. Lynch Jr, P. A. Miller, C. A. Omer, A. D. Rodrigues, E. S. Walsh and T. M. Williams, *J. Med. Chem.*, 2001, **44**, 2933.
59. M. Rowley, L. J. Bristow and P. H. Hutson, *J. Med. Chem.*, 2001, **44**, 477.
60. R. Netzer, A. Ebneth, U. Bischoff and O. Pongs, *Drug Discovery Today*, 2001, **6**, 78.
61. A. Palani, S. Shapiro, J. W. Clader, W. J. Greenlee, K. Cox, J. Strizki, M. Endres and B. M. Baroudy, *J. Med. Chem.*, 2001, **44**, 3339.
62. J. R. Tagat, R. W. Steensma, S. W. McCombie, D. V. Nazareno, S.-I. Lin, B. R. Neustadt, K. Cox, S. Xu, L. Wojcik, M. G. Murray, N. Vantuno, B. M. Baroudy and J. M. Strizki, *J. Med. Chem.*, 2001, **44**, 3346.
63. D. A. Price, D. Armour, M. de Groot, D. Leishman, C. Napier, M. Perros, B. L. Stammen and A. Wood, *Current Topics in Medicinal Chemistry*, 2008, **8**, 1140.
64. R. Poulain, D. Horvath, B. Bonnet, C. Eckhoff, B. Chapelain, M.-C. Bodinier, and B. Déprez, *J. Med. Chem.*, 2001, **44**, 3391.
65. R. Perrone, F. Berardi, N. A. Colabufo, M. Leopoldo, E. Lacivita, V. Tortorella, A. Leonardi, E. Poggesi and R. Testa, *J. Med. Chem.*, 2001, **44**, 4431.
66. M. Rowley, D. J. Hallett, S. Goodacre, C. Moyes, J. Crawforth, T. J. Sparey, S. Patel, R. Marwood, S. Patel, S. Thomas, L. Hitzel, D. O'Connor, N. Szeto, J. L. Castro, P. H. Hutson and A. M. MacLeod, *J. Med. Chem.*, 2001, **44**, 1603.
67. J. B. Rangisetty, M. Dukat, C. S. Dowd, K. Herrick-Davis, A. DuPre, S. Gadepalli, M. Teitler, C. R. Kelley, N. A. Sharif and R. A. Glennon, *J. Med. Chem.*, 2001, **44**, 3283.

68. J. J. Chambers, D. M. Kurrasch-Orbaugh, M. A. Parker and D. E. Nichols, *J. Med. Chem.*, 2001, **44**, 1003.
69. K. S. Jandu, V. Barrett, M. Brockwell, D. Cambridge, D. R. Farrant, C. Foster, H. Giles, R. C. Glen, A. P. Hill, H. Hobbs, A. Honey, G. R. Martin, J. Salmon, D. Smith, P. Woollard and D. L. Selwood, *J. Med. Chem.*, 2001, **44**, 681.
70. Y.-C. Xu, K. W. Johnson, L. A. Phebus, M. Cohen, D. L. Nelson, K. Schenck, C. D. Walker, J. E. Fritz, S. W. Kaldor, M. E. LeTourneau, R. E. Murff, J. M. Zgombick, D. O. Calligaro, J. E. Audia and J. M. Schaus, *J. Med. Chem.*, 2001, **44**, 4031.
71. M. G. N. Russell, R. J. Baker, L. Barden, M. S. Beer, L. Bristow, H. B. Broughton, M. Knowles, G. McAllister, S. Patel and J. L. Castro, *J. Med. Chem.*, 2001, **44**, 3881.
72. M. J. Coghlan, W. A. Carroll and M. Gopalakrishnan, *J. Med. Chem.*, 2001, **44**, 1627.
73. T. Anger, D. J. Madge, M. Mulla and D. Riddall, *J. Med. Chem.*, 2001, **44**, 115.
74. M. Stahl and M. Rarey, *J. Med. Chem.*, 2001, **44**, 1035.
75. J.-P. Falgueyret, R. M. Oballa, O. Okamoto, G. Wesolowski, Y. Aubin, R. M. Rydzewski, P. Prasit, D. Riendeau, S. B. Rodan and M. D. Percival, *J. Med. Chem.*, 2001, **44**, 94.
76. R. W. Marquis, Y. Ru, S. M. LoCastro, J. Zeng, D. S. Yamashita, H.-J. Oh, K. F. Erhard, L. D. Davis, T. A. Tomaszek, D. Tew, K. Salyers, J. Proksch, K. Ward, B. Smith, M. Levy, M. D. Cummings, R. C. Haltiwanger, G. Trescher, B. Wang, M. E. Hemling, C. J. Quinn, H.-Y. Cheng, F. Lin, W. W. Smith, C. A. Janson, B. Zhao, M. S. McQueney, K. D'Alessio, C.-P. Lee, A. Marzulli, R. A. Dodds, S. Blake, S.-M. Hwang, I. E. James, C. J. Gress, B. R. Bradley, M. W. Lark, M. Gowen and D. F. Veber, *J. Med. Chem.*, 2001, **44**, 1380.
77. H. N. Bramson, J. Corona, S. T. Davis, S. H. Dickerson, M. Edelstein, S. V. Frye, R. T. Gampe, Jr., P. A. Harris, A. Hassell, W. D. Holmes, R. N. Hunter, K. E. Lackey, B. Lovejoy, M. J. Luzzio, V. Montana, W. J. Rocque, D. Rusnak, L. Shewchuk, J. M. Veal, D. H. Walker and L. F. Kuyper, *J. Med. Chem.*, 2001, **44**, 4339.
78. T. Honma, T. Yoshizumi, N. Hashimoto, K. Hayashi, N. Kawanishi, K. Fukasawa, T. Takaki, C. Ikeura, M. Ikuta, I. Suzuki-Takahashi, T. Hayama, S. Nishimura and H. Morishima, *J. Med. Chem.*, 2001, **44**, 4628.
79. A. R. Leach, V. J. Gillet, R. A. Lewis and R. Taylor, *J. Med. Chem.*, 2010, **53**, 539.
80. A. Nicholls, G. B. McGaughey, R. P. Sheridan, A. C. Good, G. Warren, M. Mathieu, S. W. Muchmore, S. P. Brown, J. A. Grant, J. A. Haigh, N. Nevins, A. N. Jain and B. Kelley, *J. Med. Chem.*, 2010, **53**, 3862.
81. S. W. Muchmore, J. J. Edmunds, K. D. Stewart and P. J. Hajduk, *J. Med. Chem.*, 2010, **53**, 4830.
82. M. V. S. Varma, R. S. Obach, C. Rotter, H. R. Miller, G. Chang, S. J. Steyn, A. El-Kattan and M. D. Troutman, *J. Med. Chem.*, 2010, **53**, 1098.

83. A. Avdeef and K. Y. Tam, *J. Med. Chem.*, 2010, **53**, 3566.
84. E. Perola, *J. Med. Chem.*, 2010, **53**, 2986.
85. B. Kuhn, P. Mohr and M. Stahl, *J. Med. Chem.*, 2010, **53**, 2601.
86. T. P. Prakash, A. Siwkowski, C. R. Allerson, M. T. Migawa, S. Lee, H. J. Gaus, C. Black, P. P. Seth, E. E. Swayze and B. Bhat, *J. Med. Chem.*, 2010, **53**, 1636.
87. E. Pinard, A. Alanine, D. Alberati, M. Bender, E. Borroni, P. Bourdeaux, V. Brom, S. Burner, H. Fischer, D. Hainzl, R. Halm, N. Hauser, S. Jolidon, J. Lengyel, H.-P. Marty, T. Meyer, J.-L. Moreau, R. Mory, R. Narquizian, M. Nettekoven, R. D. Norcross, B. Puellmann, P. Schmid, S. Schmitt, H. Stalder, R. Wermuth, J. G. Wettstein and D. Zimmerli, *J. Med. Chem.*, 2010, **53**, 4603.
88. J. M. Cotero, D. Catterick, J. Castro, M. J. Chaparro, B. Díaz, E. Fernández, S. Ferrer, F. J. Gamo, M. Gordo, J. Gut, L. de las Heras, J. Legac, M. Marco, J. Miguel, V. Muñoz, E. Porras, J. C. de la Rosa, J. R. Ruiz, E. Sandoval, P. Ventosa, P. J. Rosenthal and J. M. Fiandor, *J. Med. Chem.*, 2010, **53**, 6129.
89. K. Brak, I. D. Kerr, K. T. Barrett, N. Fuchi, M. Debnath, K. Ang, J. C. Engel, J. H. McKerrow, P. S. Doyle, L. S. Brinen and J. A. Ellman, *J. Med. Chem.*, 2010, **53**, 1763.
90. B. T. Mott, R. S. Ferreira, A. Simeonov, A. Jadhav, K. Kean-Hooi Ang, W. Leister, M. Shen, J. T. Silveira, P. S. Doyle, M. R. Arkin, J. H. McKerrow, J. Inglese, C. P. Austin, C. J. Thomas, B. K. Shoichet and D. J. Maloney, *J. Med. Chem.*, 2010, **53**, 53.
91. A. I. Lukaszewicz, M. K. McMillan and M. Kahn, *J. Med. Chem.*, 2010, **53**, 3439.
92. A. L. Hopkins, *Nat. Chem. Biol.*, 2008, **4**, 682–690.
93. A. C. Cheng, J. Eksterowicz, S. Geuns-Meyer and Y. Sun, *J. Med. Chem.*, 2010, **53**, 3439.
94. C. Bissantz, B. Kuhn and M. Stahl, *J. Med. Chem.*, 2010, **53**, 5061.
95. M. Congreve, G. Chessari, D. Tisi and A. J. Woodhead, *J. Med. Chem.*, 2008, **51**, 3661.
96. a) C. W. Murray, M. G. Carr, O. Callaghan, G. Chessari, M. Congreve, S. Cowan, J. E. Coyle, R. Downham, E. Figueroa, M. Frederickson, B. Graham, R. McMenamin, M. A. O'Brien, S. Patel, T. R. Phillips, G. Williams, A. J. Woodhead and A. J.-A. Woolford, *J. Med. Chem.*, 2010, **53**, 5492; b) A. J. Woodhead, H. Angove, M. G. Carr, G. Chessari, M. Congreve, J. E. Coyle, J. Cosme, B. Graham, P. J. Day, R. Downham, L. Fazal, R. Feltell, E. Figueroa, M. Frederickson, J. Lewis, R. McMenamin, C. W. Murray, M. A. O'Brien, L. Parra, S. Patel, T. Phillips, D. C. Rees, S. Rich, D.-M. Smith, G. Trewartha, M. Vinkovic, B. Williams and A. J.-A. Woolford, *J. Med. Chem.*, 2010, **53**, 5956.
97. S. Barelier, J. Pons, K. Gehring, J.-M. Lancelin and I. Krimm, *J. Med. Chem.*, 2010, **53**, 5256.
98. A. L. Hopkins, C. R. Groom and A. Alex, *Drug Discov. Today*, 2004, **9**, 430.

99. Y.-S. Wang, C. Strickland, J. H. Voigt, M. E. Kennedy, B. M. Beyer, M. M. Senior, E. M. Smith, T. L. Nechuta, V. S. Madison, M. Czarniecki, B. A. McKittrick, A. W. Stamford, E. M. Parker, J. C. Hunter, W. J. Greenlee and D. F. Wyss, *J. Med. Chem.*, 2010, **53**, 942.
100. A. W. E. Chan, R. A. Laskowski and D. L. Selwood, *J. Med. Chem.*, 2010, **53**, 3086.
101. S. Barelier, J. Pons, O. Marcillat, J.-M. Lancelin and I. Krimm, *J. Med. Chem.*, 2010, **53**, 2577.
102. V. Katritch, V.-P. Jaakola, J. R. Lane, J. Lin, A. P. IJzerman, M. Yeager, I. Kufareva, R. C. Stevens and R. Abagyan, *J. Med. Chem.*, 2010, **53**, 1799.
103. J. Carlsson, L. Yoo, Z.-G. Gao, J. J. Irwin, B. K. Shoichet and K. A. Jacobson, *J. Med. Chem.*, 2010, **53**, 3748.
104. R. Morphy, *J. Med. Chem.*, 2010, **53**, 1413.
105. T. P. Cho, S. Y. Dong, F. Jun, F. J. Hong, Y. J. Liang, X. Lu, P. J. Hua, L. Y. Li, Z. Lei, H. Bing, Z. Ying, L. F. Qiong, F. B. Bei, L. L. Guang, G. A. Shen, S. G. Hong, S. W. Hong and M. X. Tai, *J. Med. Chem.*, 2010, **53**, TBD.
106. F. Zuccotto, E. Ardini, E. Casale and M. Angiolini, *J. Med. Chem.*, 2010, **53**, 2681.
107. S. Shang and D. S. Tan, *Curr. Opin. Chem. Biol.*, 2005, **9**, 248.
108. L. D. Walensky, A. L. Kung, I. Escher, T. J. Malia, S. Barbuto, R. D. Wright, G. Wagner, G. L. Verdine and S. J. Korsmeyer, *Science*, 2004, **305**, 1466.
109. K. E. Bower, S. N. Lam, B. D. Oates, J. R. del Rosario, E. Corner, T. F. Osothprarop, A. G. Kinhikar, J. A. Hoye, R. R. Preston, R. E. Murphy, L. A. Campbell, H. Huang, J. Jimenez, X. Cao, G. Chen, Z. W. Ainekulu, A. B. Datt, N. J. Levin, V. R. Doppalapudi, S. R. Pirie-Shepherd, C. Bradshaw, G. Woodnutt and R. W. Lappe, *J. Med. Chem.*, 2011, **54**, 1256.
110. I. Niculescu-Duvaz, *Curr. Opin. Mol. Ther.*, 2000, **2**, 691.
111. C. P. Leamon and P. S. Low, *Proc. Nat. Acad. Sci. U.S.A.*, 1991, **88**, 5572.

CHAPTER 2

Drug-Like Physicochemical Properties

PAUL D. LEESON[a] AND TUDOR I. OPREA[b]

[a] AstraZeneca R&D Charnwood, Bakewell Road, Loughborough, Leics., LE11 5RH, UK; [b] Division of Biocomputing, Department of Biochemistry and Molecular Biology and University of New Mexico (UNM) Center for Molecular Discovery, UNM School of Medicine, MSC11 6145, Albuquerque, New Mexico 87131, USA

2.1 Introduction

Physical properties play a central role in drug action. It has long been recognized that optimizing the interaction with a macromolecular target requires modification of specific properties of potential small molecule drugs: size, charge, hydrogen bonding and lipophilic characteristics. For a drug to be successful, it must also be sufficiently well absorbed, metabolically stable, safe and have appropriate duration of action in the disease setting. Over the past two decades, it has become evident that these pharmacokinetic, metabolic and toxicity properties of drug molecules are also driven by the same bulk physical properties. Absorption, distribution, metabolism, excretion and toxicity (ADMET) profiling has joined primary and secondary biological testing in drug discovery project screening cascades. The ADMET properties of molecules are generic in that they represent 'the action of the body on the drug,' in contrast to the drug's biological action through specific targets, which can be described as 'the action of the drug on the body.' Thus the properties driving ADMET are broadly similar for most small molecule classes, whereas

RSC Drug Discovery Series No. 13
Drug Design Strategies: Quantitative Approaches
Edited by David J. Livingstone and Andrew M. Davis

Published by the Royal Society of Chemistry, www.rsc.org

biological target potency and selectivity requires specific, fine-tuned, molecular interactions. The challenge for drug discoverers is to combine these attributes, preferably in the form of a once daily oral drug, for use in human disease. The medicinal chemist now employs multiple parallel optimization of biological and ADMET properties, from the initial lead generation phase through to optimizing *in vivo* profiles, prior to candidate drug selection.

Profiling drug databases, especially those of oral drugs, has shown that they occupy distinct regions of physicochemical space, and that risks to the ADMET profile and subsequent attrition in clinical drug development appear to be linked. The control of physical properties in drug candidates is an achievable goal in reducing the dramatically high levels of attrition seen currently in drug discovery pipelines.[1] In this review, we discuss the evolution and current status of 'drug-like' and 'lead-like' concepts, with a focus on physicochemical drug properties. We believe that emphasizing physicochemical profiles in ways that medicinal chemists can respond to in the design of new molecules and can easily reduce to synthetic practice is essential if the drug discovery community is to improve overall compound quality. The application of user-friendly, predictive chemoinformatic tools, *prior to synthesis of new molecules*, is a critically important development in the development of these ideas.[2,3]

2.2 Drug-likeness

How does one differentiate a 'drug' from a 'non-drug'? From a practical standpoint, drugs are actually medicines that provide benefits to patients. Drugs are approved around the world by regulatory agencies, which may operate with different priorities with respect to the risk–benefit balance provided by each new drug application. Drugs result from diverse research efforts, with the protein targets often originating from academia, and are usually invented by medicinal chemists in different pharmaceutical and biotechnology companies. Clinical efficacy, safety and differentiation from other drugs are paramount drug-like properties; to achieve these goals there is no simple formula for success, despite the considerable work on drug-likeness that is discussed in the following pages. The approaches taken by companies, even when pursuing the same target, can be quite different, and at the level of compound physical properties it has been shown that major pharmaceutical companies operate different practices.[4] Chemical motifs used in drugs have evolved over time. However, many recent drugs and much current drug design practice have their structural origins in molecules found in the natural world.[5] In addition, while medicinal chemists themselves have a tendency to recycle similar but synthetically accessible structures,[6] they also have different views on the merits and demerits of compounds, based on their knowledge, intuition and prejudice.[7,8] Some of these trends drive towards uniformity of drug structures, others towards diversity. All of these issues confound the construction of an underlying, absolute 'drug-like' molecular space. Despite this, the use of physicochemical properties of drugs, supported by empirical ADMET observations, has gained increasing support. This thinking was stimulated by the seminal 'rule of five' publication from Lipinski and co-workers in 1997.[9]

2.2.1 The Impact of the Rule of Five

The rule of five (Ro5) paper[9] was based on an examination of the early results of high throughput screening (HTS) and combinatorial chemistry, as practiced at Pfizer in the mid-1990s. Many hits had high molecular weight (mol. wt) and lipophilicity (cLog$P$, the calculated 1-octanol – water partition coefficient), which was making their progression difficult. By analyzing drugs in phase II or later development, Lipinski and co-workers proposed that drug absorption and permeability are more likely when the mol. wt is <500, Log$P$ is <5, hydrogen bond acceptors (HBA, a count of oxygen and nitrogen atoms) is <10 and hydrogen bond donors (HBD, a count of OH plus NH groups) is <5.[9] The Ro5 is so called because each of the cut-off criteria is divisible by five. The cut-off values represent the 90 percentile values of the drugs examined; for each property, the 10% failing will be a different set of drugs. Failing a single Ro5 is generally considered acceptable, but failing two is not. In our oral drug database[4] 79% of drugs pass all four rules but only 2.9% fail both the mol. wt and LogP criteria. The distribution of Ro5 violations by drug formulation, summarized in Table 2.1, shows that only 9% of all drugs ($n=113$) violate two or more Ro5 criteria; of these, nearly half ($n=49$) are formulated orally. Only 56.1% of all drugs ($n=704$) are formulated orally and have zero violations.

The Ro5 is useful where drugs are passively transported across membranes, since exceptions include antibiotics, antifungals, vitamins and cardiac glycosides, molecules that are known to make use of transporters for their permeation. However, there has been increased recent interest in the role of transporters, and it has been proposed that, given the large number of transporter systems, most, if not all, compounds may be transporter substrates.[10]

Table 2.1 Distribution of rule of 5 (Ro5) violations according to drug formulation for a set of 1225 marketed drugs (source: WOMBAT-PK).[105] Percentage values are given for each formulation and against total drug count (in brackets). i.m. = intramuscular; i.v. = intravenous; s.c. = sub-cutaneous.

		Number of Ro5 Violations			
Drug Formulation	*Count*	*0*	*1*	*2*	*>2*
Oral	881	79.9% (56.1%)	14.5% (10.2%)	3.5% (2.5%)	2.0% (1.4%)
Injectable (i.m., i.v., s.c.)	221	57.5% (10.1%)	20.4% (3.6%)	9.9% (1.8%)	12.2% (2.2%)
Topical	153	71.2% (8.7%)	18.9% (2.3%)	6.6% (0.8%)	3.3% (0.4%)

The structure–activity relationships for passive and active transport, dependent on bulk and specific molecular properties, may overlap.

The simplicity and ease of calculation of the Ro5 has made it an attractive tool for medicinal chemists to use in design, and evidence of its essentially unanimous acceptance by the scientific community comes from >2400 literature citations in Scifinder® as of August 2010. This work has clearly enhanced awareness regarding the existence of a chemical space that may restrict useful drug properties. As a result, Ro5 filters are widely used to enhance screening compound collections and in the design of libraries for parallel and combinatorial synthesis.[11] Ro5 compliance has been viewed by many drug discovery practitioners as *the* primary 'drug-like' rule. But is this actually the case? While most drugs are clearly accommodated by the Ro5 chemical space, so too are many 'non-drugs.' Thus a set of 190 000 chemicals from the Available Chemical Directory (ACD) were almost as likely to pass the Ro5 as a set of 400 oral drugs from the Physician's Desk Reference (PDR).[12] This is a result of the massive size of chemical space; it has been estimated that there are 10^{20}–10^{24} synthetically accessible molecules within drug-like space.[13] Restricting the molecular size to 13 heavy atoms still provides 10^9 accessible drug-like molecules.[14] Application of the Ro5 is therefore best considered in the context of active drug discovery projects, as an aid to decision making in HTS, lead selection and optimization.

The Ro5 has stimulated a large number of other studies on the broad topic of drug physical properties and compound quality. A wide range of machine learning techniques—from neural networks to decision trees and support vector machines, based on a variety of molecular descriptors and fingerprints, together with filters using molecular properties and functional groups—have been applied to distinguish between a number of 'drug' and 'nondrug' data sets. This area has been reviewed in detail elsewhere.[15] Other recent work has focused on the statistical appearance of drug-like molecular fragments amongst drug and non-drug compounds.[16] Here, our focus will be on developments related to the application of those physical properties of molecules, readily calculable by the medicinal chemist[2] in advance of synthesis and not requiring expert computational input. Table 2.2 summarizes some of the key work in this area, applied to the question of drug-likeness, as well as drug or compound quality.

2.2.2 Oral and Non-Oral Drug Properties

The profiles of three large oral drug databases, derived from the PDR,[17] the US Food and Drug Administration (FDA) approved drugs[18] and worldwide drug approvals[19] are in good agreement with each other and with the Ro5. A study by Wenlock and co-workers[17] tested the Ro5 further by additionally examining compounds in different phases of development, either progressing or discontinued. The results were clear: compounds had mol. wt ranking in the order phase I > phase II > phase III > marketed drugs, and drugs progressing in phases I, II and III had lower cLogP values than those that were discontinued

Table 2.2 A Selection of key physicochemical studies on drug-likeness and compound quality.

Measure	*Properties and Guidelines*	*Reference(s)*
Rule of 5	90 percentile values for drugs: Mol. wt <500; *c*Log*P* <5; H-bond donors (OH + NH) <5; H-bond acceptors (O + N) <10	Lipinski (1997),[9] Wenlock (2003),[17] Vieth (2004),[18] Proudfoot (2005)[19]
Lead-likeness	Mol. wt <350; *c*Log*P* <3	Teague (1999)[100]
Fragment-likeness (Rule of 3)	Mol wt. <300; *c*Log*P* <3; rot.bonds <3; H-bond acceptors and donors <3	Congreve (2003)[123]
Oral bioavailability	(a) Rot. bonds <10; PSA <140 $Å^2$	Veber (2002)[45]
	(b) Ro5, PSA and ion class	Martin (2005)[46]
	(c) Ro5, PSA, rot. bonds for absorption & clearance	Varma (2010)[44]
Permeability and metabolic stability	'Golden triangle' with mol. wt <500, apex & Log*D* 0–3 base	Johnson (2009)[34]
Membrane permeability	(a) PSA <120 $Å^2$ for oral absorption and <70 $Å^2$ for blood brain barrier penetration	Kelder (1999)[33]
	(b) Mol. wt <500 & variable Log*D* (range 0.5–3.5)	Waring (2009)[36]
Receptor promiscuity	*c*Log*P* <3; bases >acids and neutrals; ligand lipophilicity efficiency (LLE) = p(Activity) – *c*Log*P* >5	Leeson & Springthorpe (2007)[4]
In vivo toxicity	*c*Log*P* <3; PSA >75 $Å^2$	Hughes (2008)[88]
Molecular shape	(a) Fsp3 = 0.47 in drugs *versus* 0.36 in research compounds; increasing Fsp3 increases solubility	Lovering (2009)[25]
	(b) [Ar atom–sp3 atom] count <11 for 90% oral drugs	Leeson (2011)[23]
Aromaticity & Developability	(a) Mol. wt and aromatic fraction (Ar atoms/heavy atoms) influence solubility	Lamanna (2008)[24]
	(a) #Aromatic rings ≤3 increases solubility, reduces plasma protein binding and Cyp & hERG inhibition	Ritchie & MacDonald (2009)[26]
	(b) *c*Log*D* + #Aromatic rings <5 increases solubility	Hill & Young (2010)[27]

at those phases. It was suggested that the drug development pressures may therefore limit the progression of compounds with poorer physicochemical properties, providing more support for the Ro5. The link with drug pipeline attrition[17] suggests that better control of physical properties in the design process may improve overall drug development productivity. Vieth and co-workers examined physical properties by various routes of administration,[18] showing that oral drugs have substantially lower mol. wt, higher *c*Log*P* and fewer HBDs and HBAs than injectable drugs. Based on data summarized in Table 2.1, 22% of injectable drugs have two or more Ro5 violations, *versus* only 5.5% of oral drugs. However, with the observed variance, physical properties alone could not distinguish between orals and injectables.[18] Inhaled drug discovery is in its relative infancy, with many inhaled drugs[20] being initially designed for other routes of administration. The design of lung-retentive compounds that have low systemic exposure, high systemic clearance and low bioavailability points to the deliberate design of 'anti-Ro5' molecules,[21] with higher mol. wt and polar surface area (PSA).

Proudfoot examined the progression of Ro5 properties over time, using the publication dates of oral drugs.[19] Increases in mol. wt, HBA and HBD, but not *c*Log*P*, over time were noted. A narrow range of acceptable HBD values was evident, with only 1.1% of drugs having HBD >5. Combining this with mol. wt showed only 2% of drugs to have mol. wt >500 and HBD >3. Further studies showing the progression of oral drug properties by launch date[22] and by both publication and launch date[4] are in good agreement, showing that mol. wt is increasing over time and that *c*Log*P* is not. There was also no difference in HBD and % PSA in drugs launched pre- and post-1983. It was proposed that those physical properties showing least change over time are of greater underlying drug-like importance.[22] A study of launched oral drugs compared physical properties (by drug publication date) with patent compounds published in 2000-2009 from major companies,[23] focusing additionally on ion class trends. While all ion classes (acidic, basic or neutral compounds) showed increases in mol. wt and HBA over time, there were marked differences in lipophilicity trends; bases have shown declining *c*Log*P* values since the 1950s, while neutral compounds and acids showed increasing *c*Log*P* values from pre-1930 to the mid-1960s. However, since the mid 1960s, *c*Log*P* values in all ion classes have converged in the range 2–2.5, with the exception of acids published post-1990, where mean *c*Log*P* is 4.0. Acids had the highest PSA values and, in agreement with other studies, HBD was not changing[19,22] notably over time. These studies confirm that drug-likeness is not an 'absolute' property, but is determined by the drug discovery practices and targets at the time of invention, ion class and the pressures imposed by development pipeline activities.[23]

Recent additions to analyses of drug-like physical properties are simple descriptors of shape and aromaticity. The fraction of aromatic atoms and mol. wt were the two most important parameters in accounting for the solubility of a set of 3563 molecules.[24] The parameter Fsp^3, equal to the ratio of tetrahedral (sp^3 hybridised) carbon atoms to total carbon atoms,[25] was found to gradually increase in moving from research compounds ($Fsp^3 = 0.36$) through phases I–III to marketed drugs ($Fsp^3 = 0.47$). A similar trend was seen with chirality, where a higher proportion of drugs passing the Ro5 were chiral (61%) in comparison with the percentage of research compounds passing Ro5 (46%). These data suggest that three-dimensionality is an important drug-like attribute. A study of the GlaxoSmithKline pipeline[26] (the first such company report we are aware of) showed that the number of aromatic rings decreased in moving from pre-clinical candidate (mean 3.3 rings) to proof of concept (mean 2.2 rings), compared to a mean of 1.6 rings in marketed drugs. Fewer than three aromatic rings were recommended. The aromatic ring count correlated positively with *c*Log*P* and with some developability properties that are known to be *c*Log*P* dependent, namely Log*D*7.4, % serum albumin binding, CYP 3A4 inhibition and hERG (human ether-a-go-go related gene) inhibition. Aromatic ring count correlated negatively with solubility, an effect that was independent of *c*Log*P*. An additional analysis of GlaxoSmithKline compound solubility profiles[27] questioned the use of experimental Log*D* data with poorly soluble compounds and proposed a solubility forecast index (SFI), incorporating aromatic ring count, where SFI = *c*Log*D* - #Ar rings. Although % serum albumin binding was included as a developability property,[26] this is not the case, since measures of drug fraction bound to plasma protein are of little significance *in vivo*.[28] The more relevant *in vivo* measure is the free drug concentration that is available for binding to the drug target. The free drug concentration is independent of protein binding because higher free drug concentrations lead to increased clearance.[28]

Further studies of aromatic and sp^3 ring counts with oral drugs showed that these are inversely related, *i.e.* increasing the fractional molecular aromatic content (ratio of aromatic atoms to heavy atoms) is coincident with reducing aliphatic content (Fsp^3).[23] A new shape parameter, aromatic atoms–sp^3 atoms (Ar–sp^3), takes both properties into account; furthermore, Ar-sp3 is markedly lower in drugs *versus* recently patented compounds, independently of mol. wt or *c*Log*P*.[23] Taken together, the studies on molecular shape suggest that current chemistry is markedly 'flatter' and/or 'more aromatic' than marketed drugs, probably due to the facile synthetic chemistry procedures for aryl cross coupling reactions.[25] Furthermore, drugs that break the Ro5 *c*Log*P* and mol. wt guidelines tend to be sp^3 rich (having negative Ar–sp^3 values), suggesting aromaticity in contra-indicated in 'Ro5 exception' compounds.[23] Many drugs that possess a high sp^3 content are derived from natural products (*e.g.* steroids, vitamins and macrocycles) and it has been noted that in moving from phase I to the market, compounds are increasingly similar to biological metabolite structures.[29] A related concept of 'metabolite' likeness has been proposed[30] on the basis that in order to enter cells, drugs can employ transport proteins that have naturally occurring intermediary metabolites as their substrates.

The increases in mol. wt in newer oral drugs are reflected in the patent portfolios of large pharmaceutical companies, where significant inflation of bulk physicochemical properties is evident.[4,23] While oral drugs show little increase in *c*Log*P* over time (mean value 2.5), pharmaceutical industry patented compounds are significantly more lipophilic (mean value 4.0), a worrying trend that may be contributing to pipeline attrition rates.[4] There are key questions here that have to be considered but are difficult to answer: future drugs will emerge from current chemistry but will selection pressure during development[17] favor smaller and less lipophilic compounds? Will pursuit of large and lipophilic molecules result in a further decrease in the output of new medicines?[4] Or will future drugs simply reflect the properties of compounds that are entering the development pipeline?[31] There are no easy answers to these questions. However, the empirical evidence that inflated molecular properties carry higher risk of ADMET failure is now very strong and working at the periphery, rather than at the centre, of Ro5 chemical space requires the discovery of *exception* compounds. Drug discovery is a 'balancing act', where physical property relationships controlling potency and many ADMET properties may be opposed. In general, potency will increase as molecules become larger and more lipophilic,[32] however, this will be accompanied by risk of poor solubility, high metabolic clearance and toxicity.

2.2.3 ADME Profiles and Physical Properties

Soon after the Ro5 publication, PSA was implicated in both oral absorption (PSA < 120 Å^2) and in blood–brain barrier penetration (PSA < 70 Å^2).[33] Recent studies of the ADMET profiles of large datasets of compounds from GlaxoSmithKline,[34] Pfizer[35] and AstraZeneca[36] have provided strong empirical reinforcement for the Ro5. An extensive study of GlaxoSmithKline proprietary compounds covering solubility, permeability, bioavailability, volume of distribution, clearance, hERG inhibition, P-gp (P-glycoprotein) efflux and P450 inhibition led to the conclusion that for overall lower risk, it is optimal that compounds, in all ion classes, have both mol. wt <400 and Log*P* <4.[34] In CACO-2 screens at AstraZeneca, maintaining a 50% chance of high permeability as mol. wt increased required co-increased Log*D* 7.4 (distribution coefficient between 1-octanol and pH7.4 buffer), in the range Log*D* 0.5–3.5, and mol. wt <300–500.[36] Further increasing mol. wt to >500 substantially reduced the odds of finding acceptable permeability at all. In the Pfizer data set, several tests for permeability and *in vitro* metabolic stability were employed.[35] Compounds with acceptable combined permeability and metabolic stability fell into a 'golden triangle' with a mol. wt apex of ~450 and a Log*D* base of 0–3. Of the compounds at the centre of this space (Log*D* 1.5, mol. wt 350), 29% passed permeability and stability criteria, compared to 4% or less with mol. wt 450 and Log*D* 0 or 3. In agreement with the AstraZeneca data,[36] when mol. wt was >450, there were very few permeable and stable compounds. Certain outliers could be rationalized, for example compounds with metabolically vulnerable sites on aromatic rings blocked by halogen substitution. These important studies[34–36] are in good agreement with each other, despite covering a very wide range of chemistries in projects from three large pharmaceutical companies. Taken together with the vast literature on individual compounds and structure–activity data sets,[37] they reinforce the essential Ro5 properties of lipophilicity and size in determining acceptable ADMET properties, and provide a useful framework for understanding the risks associated with more extreme physical properties.

Further analyses of large proprietary databases have focused on matched pair analysis of phenyl substituent effects on metabolic stability[38,39] and on cytochrome P450, hERG inhibition, solubility and artificial membrane permeability.[40] These studies also show that substituent lipophilicity is the dominant driver of ADMET properties. However, because of the opposing effects of physical properties (for example on solubility and permeability) there are no 'perfect' substituents, where all ADMET properties are improved at once.[40] In addition, the performance of a particular substituent will also depend on the overall molecular framework and specific structure–activity relationships. However, these studies[38–40] usefully provide the medicinal chemist with further insight, substantially extending the Topliss decision tree[41] in the construction of aryl substituent design sets.

It is clear from the studies discussed here[34–40] that the lipophilicity of a compound or substituent is a dominant property for metabolic stability and inhibition assays. This is not surprising given that lipophilic compounds are substrates for two proteins, P-glycoprotein (P-gp) and CYP3A4, which engage in a dynamic interplay during absorption in the intestinal lumen.[42] P-gp mediates drug efflux back towards the intestinal lumen, which results in prolonged exposure of the drug to the enterocyte-located CYP3A4, increasing the chances of intestinal drug metabolism. Thus P-gp blockade would limit this interplay, reducing access of the drug to the metabolizing enzyme. This becomes specifically relevant for compounds that are dual substrates of both CYP3A4 and P-gp, as is the case for drugs with low solubility and high lipophilicity. These drugs have a higher oral bioavailability when co-administered with a high-fat content meal.[43]

Since it pertains to permeability and absorption, the Ro5 is often assumed to be linked to bioavailability. However, while Ro5 physical properties are clearly important for overall bioavailability, there are no useful all-embracing predictive models as yet. Satisfactory models have not been easy to generate because bioavailability is a multi-factorial process, dependent on dissolution, membrane permeability, potential active transport and *in vivo* clearance by hepatic pathways. Perhaps unsurprisingly, differing results have been obtained; the models generated seem to depend on the data set and animal species employed. Only recently have attempts been made to better understand overall human bioavailability by separately examining absorption and clearance physicochemical trends.[44] A model suggesting rotatable bond count of <10 and PSA of <140 Å^2 as limits for rat bioavailability for a proprietary set of GlaxoSmithKline compounds,[45] though widely cited, worked less effectively on other proprietary data sets.[46,47] In a set of Abbott compounds, the probability that rat bioavailability was $>10\%$ depended primarily on charge type, with acids displaying declining bioavailability with increasing PSA.[46] In neutral, basic and zwitterionic compounds the Ro5 was more predictive, but not impressively so, with 55% of compounds that passed having a bioavailability of $>10\%$ and 17% of those that failed having a bioavailability of $>10\%$. Attempts to generate models for humans have been problematic; it appears that multiple physical properties are involved.[48] Studies taking into account elimination routes show differing effects of physical properties on overall human bioavailability.[44] For example, higher mol. wt was associated with reduced absorption, whereas elimination decreased with increasing lipophilicity, and increases in rotatable bond count appeared to reduce both absorption and elimination. Overall, the apparently differing results for physicochemical influence on bioavailability may stem from the use of different proprietary data sets, coupled with the close correlations between bulk physical properties, such as mol. wt and rotational bond count.[18] This may lead to differing emphasis between closely related bulk physical properties in the resulting models.

The therapy class of a drug can play a role in determining its physical properties, especially where there are specific permeability requirements.[22] Thus drugs that act on the central nervous system (CNS) have to cross the blood–brain barrier and have lower PSA[33,49] and mol. wt than their non-CNS counterparts.[22,50] Preferred ranges of physical properties for CNS drugs have been proposed as follows: PSA <70 Å^2, HBA 0–1, *c*Log*P* 2–4, *c*Log*D* 2–4, and mol. wt <450,[51] and a multi-parameter 'desirability' scoring function for CNS drugs has been devised.[52,53] Anti-infective drugs act on non-human systems and may require bacterial cell wall penetration. In general, anti-infectives display lower *c*Log*P* and higher mol. wt than other therapy classes.[22] A detailed study[54] analyzed the physical properties of the different classes of historical anti-infective drugs, including both synthetic compounds and a significant number of natural product-derived agents. The largest classes of anti-infective drugs are cephems ($n=28$, mol. wt 452, cLogP 0.1) and quinolones ($n=24$, mol. wt 371, *c*Log*P* 1.3). The study showed that gram-positive drugs had higher mol. wt and Log*P* but lower PSA than gram-negative drugs.[54]

Screening has had less success in the anti-infective area than others and the natural world may remain a good source of future anti-infective agents.

2.2.4 Toxicity Risks and Physical Properties

The presence of overt electrophilic or nucleophilic chemical reactivity in potential drug candidates has long been recognized as a toxicity liability. Structural alerts for unwanted functional groups associated with reactive metabolites[55] and mutagenicity[56] are widely employed. Prediction of chemical reactivity induced by metabolism, leading to the formation of reactive metabolites, is often a challenge.[55,57] Reactive metabolites are implicated in idiosyncratic toxicity, which can appear in a minority of patients (1 in 10 000 to 1 in 100 000) after a drug has been marketed. However, there are many examples where reactive metabolites are formed but do not cause toxicity.[55] There is also little evidence that idiosyncratic toxicity is linked to a drug's bulk physical properties. However, despite these observations, the formation of reactive metabolites is generally considered a liability. The most appropriate strategy is avoidance of the issue, and helpful compilations of reactive organic functional groups are available.[56,58] Trapping screens can be employed in a screening cascade where a structural risk is likely,[59] but *in vivo* evaluation of covalent protein binding normally requires a specific radiolabel. Genetic toxicity testing[60] must take into account metabolism and degradation;[61] for example, embedded aniline structures that can be potentially be released by deacylation or dealkylation should be evaluated in genetic toxicity assays (*e.g.* the Ames test).

In addition to underpinning chemical reactivity, physical properties play a role in several aspects of toxicity,[61] for example: binding to unwanted targets, interaction with the hERG ion channel (leading to cardiac toxicity), induction of phospholidiposis and organ toxicity. Literature mining for adverse event reporting is increasingly being investigated, for example in the generation of structure–activity relationships in hepatotoxicity,[63,64] although the utility of these approaches for predicting toxicity of new structures has yet to be fully tested. The understanding that drugs often bind to multiple targets has been widely recognized; this understanding is opening up new therapeutic uses for existing medications and has spawned an area of research called 'network pharmacology' [65–67] or 'systems chemical biology'.[68] A need for non-selective, multiple-acting drugs has long been accepted in the therapeutic management of psychiatric disorders[69] and such drugs are now being used for kinase inhibition[70,71] in oncology.[72] The role of P-gp, in addition to the histamine H1 receptor, as a drug target for second generation antihistamines in order to avoid central H1 blockade (leading to drowsiness) has been suggested.[73] However, deliberate incorporation of multiple activities into single molecules provides major risks for drug-like properties, unless a common pharmacophore can be identified.[74,75] Moreover, unwanted polypharmacology may bring about undesired biological responses leading to toxicity; this type of polypharmacology is usually termed receptor 'promiscuity.' It does not include false hitters in screens, where compounds are active at multiple sites because of their physical aggregation properties. A useful compilation of structures associated with pan assay interference compounds (PAINS) is available.[76]

Several studies have suggested a key role for bulk physical properties in receptor promiscuity. A theoretical study concluded that charged groups, polarity and flexibility increase specificity, but hydrophobicity will decrease it.[77] A major problem in studying compound promiscuity is the availability of an appropriate data set, where all compounds are examined in the same assays. The best source of promiscuity data of this type, so far, is the Cerep Bioprint® database.[4,78] A Pfizer study[78] showed that the number of hits in Bioprint® increased with increasing compound lipophilicity, and in Pfizer HTS output, increasing mol. wt decreased hit frequency. The positive effect of lipophilicity on promiscuity was confirmed in an analysis of >2000 compounds in Bioprint®, where promiscuity showed a strong dependence on ion class. Bases were more promiscuous than acids and neutral compounds, although all increased promiscuity with lipophilicity; *c*Log*P* values >3 increased risk substantially.[4] This study[4] failed to confirm the Pfizer data[78] on mol. wt and promiscuity; instead, a non-linear influence of mol. wt on promiscuity, which was ion class dependent, was observed.[4] However similar mol. wt, ion class and lipophilicity trends were observed in a set of Roche compounds and it was suggested[79] that the results were sensitive to the assays employed. Thus the presence of aminergic G-protein coupled receptors[80] and transporters in the screening set might bias the results towards bases over other ion classes; it was observed for the Roche data set that, increasing basic pK_a to >8 increased promiscuity.[79] Researchers at Novartis also observed a trend towards higher lipophilicity and promiscuity; in this case, higher mol. wt led to higher promiscuity.[81] In agreement with the other studies, carboxylic acids had high selectivity and bulky lipophilic bases low selectivity. Novartis developed a predictive model, which showed a decline in promiscuity and an increase in selectivity in moving from lead optimization, through phases I-III, to launched drugs.[81] This trend is comparable with those seen using physical properties.[4,17] Overall, the risk of receptor promiscuity associated with lipophilic bases is clear. The differing results with mol. wt may depend on the data used, and have prompted studies on promiscuity–molecular topology relationships.[82] The parameter, fMF (equal to the number of heavy atoms in a sub-molecular framework divided by the total number of heavy atoms in the molecule) correlates with promiscuity in the BioPrint database. This trend is independent of lipophilicity and it was found that the number of terminal ring systems in a compound correlates with promiscuity.[82] Notably, structure–promiscuity studies to date have not looked at 'off-target' margins or potencies. Compounds may indeed be promiscuous if counting total numbers of hits, but margins of $>10^2$-fold greater than the preferred target affinity will make such interactions less important in compounds' biological activities at preferred doses.

Several drugs have been withdrawn because of cardiac toxicity linked to inhibition of the hERG ion channel. A large number of structure–activity studies have shown that binding to hERG is driven mainly by compound basicity, lipophilicity and the presence of aromatic rings.[83] Models of the hERG channel[84] have also been developed to aid structure–binding relationships, and with these insights, the risk of high *in vitro* hERG channel affinity should now be reduced. However, avoidance of *in vivo* prolongation of the QTc interval *via* hERG channel blockade calls for additional control of the maximum free concentration in plasma ($C_{\text{max free}}$). A study of 39 drugs concluded that cardiac risks were significantly reduced when the hERG IC_{50} to $C_{\text{max free}}$ ratio is >45.[85] Phospholipidosis, an excessive intracellular accumulation of phospholipids and drug, alone may not cause organ toxicity. However, because some compounds cause concurrent phospholipidosis and organ toxicity, avoidance of the issue is warranted. There are links between phospholipidosis and lipophilicity, basicity and amphiphilicity,[86] which allow for its prediction and control.

Lipophilicity is a key drug-like attribute and is important for desolvation and binding to drug protein targets. It is not surprising therefore, that high lipophilicity may be responsible for off-target receptor promiscuity and subsequent toxicity.[4,62,78,79,81,87] There is direct evidence linking lipophilicity to the *in vivo* toxicity outcomes in rats and dogs of 245 Pfizer compounds.[88] The methodology used selected a plasma exposure level of 1 μM, which maximized the toxic and non-toxic number of observations. This was necessary and is a key learning point, since most or all compounds became increasingly toxic with increasing exposures. Hence moderating exposure and dose should reduce toxicity outcomes *in vivo*; indeed it has been stated that doses of <10 mg in humans result in fewer idiosyncratic toxicities.[89] Compounds with $c\text{Log}P<3$ and PSA >75 (the 3/75 guideline) were found to be 6-fold less likely to exhibit *in vivo* toxicity *versus* compounds with $c\text{Log}P>3$ and PSA <75, and a selected set of promiscuous compounds from the Bioprint® database showed the same trend.[88] The results are intriguing in that both $c\text{Log}P$ and PSA parameters, both reflecting polarity, are needed and are actually capable of distinguishing whole animal toxic events to some extent. It appears that neither property on its own sufficiently reflects the hydrophobic properties of the molecules in their broad biological action. However, the performance of oral drugs *versus* the 3/75 guideline suggests that it may actually be too restrictive as drug-like filter.[23] Thus only 32% of oral drugs have *c*Log*P* <3 and PSA >75, and the fraction passing is not improving over time.[23] There are sufficient exceptions in the original data[88] to suggest that it is possible to find safe drugs violating 3/75. This work has been extended by combining *in vitro* cytotoxicity data with the 3/75 metric, which provided additional qualitative differentiation of the severity of *in vivo* toxicity observations.[90] Thus combination of data from panels of *in vitro* toxicity screens with physiochemical properties and prediction of activity[66] at other targets provides a way forward in assessing potential toxicity risks prior to conducting safety studies in animals.

2.2.5 Drugability

The terms 'drugability, or 'druggability', and 'drug-like' are often used in a similar sense. 'Drugability' is essentially an attribute of the macromolecular target; for example, if that target is responsive to Ro5 compliant compounds, then it is said to be 'drugable', in other words, it is feasible to find drug-like small molecule modulators of the target.[91] There is an apparent scale of drugability amongst the individual target classes, which is broadly related to the physical properties of the endogenous ligand. Thus GPCR receptors responding to peptides require larger and more lipophilic small molecule mediators than those responding to aminergic hormones, such as serotonin or dopamine.[92,93] The physical property profiles of marketed drugs are different by target class,[4,94,95] although it has been noted that the differences in lipophilicity are not statistically different (*i.e.* the lipophilicity variance within a target class is as great as that between target classes).[36] A survey of the literature in the early 2000s reveals similar differences by target class in current practice, as assessed by Ro5 properties.[95] In general, the highest mol. wt targets are peptidic GPCRs, integrins, transferases, proteases and phodphodiesrases; the most lipophilic are peptidic GPCRs, nuclear receptors, transferases, oxidases and esterases.[95] The mol. wt of ligands has been well correlated with the PSA of their binding sites using X-ray crystallography, leading to an accessible prediction of drugability within Ro5 space when the binding site on the protein structure can be identified.[96] Other predictive methods for protein drugability have followed.[97,98] A caveat to the use of protein structures to predict binding sites is that proteins often change conformation, undergoing induced fit on ligand binding, a process that is not easily predictable.[99]

2.3 Lead-likeness

The concept of 'lead-likeness' was developed at AstraZeneca in 1999.[100] This was based on experience with HTS, and the subsequent progression of screening hits. Hits usually had micromolar potency and those that had low mol. wt and *c*Log*P* tended to progress more rapidly towards more potent compounds, since they could be made larger and more lipophilic, yet remain in Ro5 space. The tendency of medicinal chemists to increase physical properties, especially mol. wt, in progressing hit and lead optimization programs was noted. This has been confirmed in a number of studies examining chemical starting point – optimized molecule pairs, as summarized in Table 2.3.

This concept was initially exemplified by a hand-picked group of 18 lead–drug pairs from the literature, and it was proposed that 'lead-like' combinatorial libraries for screening should have lower cut-offs than the Ro5, namely mol. wt <350 and *c*Log*P* <3.[100] This questioned the technologies prevalent at the time, where multi-component reactions and split and mix synthetic chemistry protocols, amongst others, were tending to produce libraries with much higher mol. wt ranges. Further analysis of a historical set of 470 lead–drug pairs,[101] extracted from Walter Sneader's book on drug prototypes,[102] and a set of 96 lead–drug pairs,[103] extending the earlier work,[100] were in good agreement and provided further confirmation that in optimization (*i.e.* in going from lead to drug), mol. wt, *c*Log*P* and a number of other physical properties all increased. From drugs launched in the year 2000, analysis of 25 lead–drug pairs found very small differences between leads and drugs,[104] since most of the drugs in that year were minor refinements of earlier drugs (*e.g.* chirally pure forms of earlier racemates, additions or isomerization of alkyl groups or single point alterations). A larger data set, based on 385 leads, 1651 launched drugs, plus many more drugs in phases I-III, and bioactive compounds from the literature, supports the essential observation that leads are significantly smaller and less lipophilic than other compound groups.[105] In line with other studies,[17,105] mol. wt, molecular complexity and aqueous solubility are all reduced in moving from bioactive compounds through the development phases to marketed drugs.[106]

A study of a total of 1680 pairs of starting compound–optimized compound, from the early 2000s literature,[95] is more indicative of current practice than the earlier historical studies.[101,103] Although properties were higher than in historical leads and drugs, changes in mean mol. wt (+42) and *c*Log*P* (+0.4) during optimization were entirely consistent with the earlier data. A comprehensive analysis of lead-finding approaches in the 2000s[107] also served to support the lead-like concept. This literature survey of 335 HTS hit–lead pairs showed that leads were larger and more lipophilic than hits; the high *c*Log*P* values of hits (mean 4.0) with other data led the authors to conclude that organizational changes in hit selection approaches were needed.[107] In an analysis of 60 lead–drug pairs from recent drug approvals, it was shown that mol. wt and potency increased but *c*Log*P* did not,[108] supporting the notion[4] that in successful lead optimization, increasing potency while maintaining or reducing lipophilicity is a key strategy.

The studies on lead molecules and their optimisation are consistent with each other and together provide strong support for the lead-like concept as a central contributing component in achieving ultimate optimization success.[32,109] The average changes in mol. wt and *c*Log*P* are consistent and in these studies are

Table 2.3 What happens in compound optimisation? Median mol. wt and *c*Log*P* values for chemical starting point and optimised compound pairs from five studies.

Compound pair	n	*Mol. wt of starting compound*	*Mol. wt of optimised compound*	*ΔMol. wt*	c*LogP of starting compound*	c*LogP of optimised compound*	*Δ*c*LogP*	*Reference*
Historical lead to drug	62	315	384	**69**	2.1	2.5	**0.4**	Oprea (2001)[103]
Historical lead to drug (*mean values*)	470	272	314	**42**	1.9	2.4	**0.5**	Hann (2001)[101]
Post-1990 drug launches, lead to drug	60	328	436	**108**	2.9	3.0	**0.1**	Perola (2010)[108]
2000s Literature, optimisation	1680	382	422	**40**	3.7	4.0	**0.3**	Morphy (2006)[95]
2000s Literature, HTS hit to lead	335	359	406	**47**	3.8	4.1	**0.3**	Keseru & Makara (2009)[107]

largely independent of the starting molecule properties. Choosing the chemical starting point is a key step in a drug discovery program, and hit–lead chemical quality is very likely to determine successful outcomes. The recognition of the fundamental importance of high lead compound quality[110] was contemporaneous with the lead-like concept and has led to the implementation, by many organizations, of strategic developments, focusing on lead generation and hit-to-lead optimization, to weed out the poorer molecules in advance of lead-to-candidate optimization. Although the principles are general, much of the thinking has been driven by the widespread application of HTS and subsequent hit triage; the strategies adopted, including the setting of specific lead-like criteria, have been extensively reviewed.[107,110–118] The composition of compound screening collections, in order to identify better hit- and lead-like chemical starting points and to remove 'ugly'[115] molecules, have consequently been enhanced and revamped in many organizations.[119–121] Perhaps the most compelling practical application of lead-like principles has come from the introduction and adoption of fragment-based drug discovery (FBDD), which is becoming a mainstream activity in drug discovery.[122] FBDD relies on biophysical methods, including nuclear magnetic resonance spectroscopy and X-ray crystallography, to detect low mol. wt fragment binding sites on protein targets. Often the biological activities of fragment hits will be very low (in the mM range) but the efficiency of binding of fragments, since they have low mol. wt, can be as high as that of optimized drugs with high mol. wt. Optimization of small fragments with defined property requirements[100,123] (Table 2.2), accommodating protein-specific binding pockets, is then guided directly by X-ray crystallography.[124] Application of this methodology to kinase targets is a major current area for FBDD, where there is notable room for improvement in drug-like profiles.[125]

2.4 Ligand Efficiency Metrics

The concept of ligand efficiency[126] has its origin in earlier studies examining contributions of functional groups to binding energy[127] and the maximum affinities achievable by ligands.[128] Essentially, ligand efficiency metrics are biological potency or affinity measures, corrected for physical properties. The higher the efficiency, the better use the molecule makes of its physical properties in protein binding interactions. Efficiency metrics can be based on molecular size, lipophilicity, polar surface area or indeed any physical property[129] or combinations of physical properties.[107] It has also been extended to the efficiency contributions of different groups[130] and to multiple targets.[131] The most frequently used metrics are based on heavy atom count (HA)[126] and lipophilicity:[4,132]

$$\mathrm{LE} = [\mathrm{pIC}_{50}\ (\mathrm{or\ p}K_\mathrm{i}) \div \mathrm{HA}] \times 1.37 \tag{2.1}$$

$$\mathrm{LLE} = \mathrm{pIC}_{50}\ (\mathrm{or\ p}K_\mathrm{i}) - c\mathrm{Log}P\ (\mathrm{or\ Log}D) \tag{2.2}$$

Equation (2.1) was derived to provide LE units of kcal/heavy atom. Most often, the correction factor (1.37) in equation (2.1) is not used, and instead of heavy atom count, mol. wt can be used.[129] In equation (2.2), ligand lipophilicity efficiency can be derived using calculated or experimental partition coefficients. The biological activity term in equation (2.1), according to the original definition,[126] should ideally be a *Ki* value, but a range of other measures–including pIC_{50} values for receptor affinities from radioligand binding studies, functional potencies or effective potencies in plasma (which takes into account drug free levels)–can in principle be used for LE and LLE calculation. Although recommended ideal LE and LLE values of drug candidates are $>\sim 0.3$ and $>\sim 5$ respectively,[4,129] there is insufficient data on marketed drug ligand efficiencies published to date[108] to draw clear conclusions. The relationship between LE and heavy atom count is not linear, since smaller ligands have higher LE values than larger ligands.[133] It has been proposed that this is a consequence of poorer quality of ligand fit to the binding site, combined with a reduction in ligand-accessible surface area, with increasing molecular size.[133] As a result, size-independent corrections to LE have been introduced, namely fit quality (FQ)[133] and size-independent ligand efficiency (SILE):[134]

$$\mathrm{FQ} = [\mathrm{pIC_{50}}\ (\text{or } \mathrm{p}K_\mathrm{i}) \div \mathrm{HA}] \div [0.0715 + (7.5328 \div \mathrm{HA}) + (25.7079 \div \mathrm{HA}^2) - (361.4722 \div \mathrm{HA}^3)] \tag{2.3}$$

$$\mathrm{SILE} = \mathrm{pIC_{50}}\ (\text{or } \mathrm{p}K_\mathrm{i}) \div \mathrm{HA}^{0.3} \tag{2.4}$$

In equations (2.3) and (2.4), HA is the heavy atom count. However, LE = affinity ÷ HA, being very simple to calculate, is the most widely used measure.

Perhaps the most useful applications of ligand efficiency metrics are in FBDD[124,135–137] and in hit selection,[107,138] where biologically weakly active fragments or hits can be equally efficient with more elaborated compounds. For example, a 10 nM ligand with 38 heavy atoms (mol. wt $\sim$500) and a 100 μM fragment with 13 heavy atoms (mol. wt $\sim$170) both have LE values of $\sim$0.3. Taking into account the size-dependency of LE,[133,134] low mol. wt fragments with LE values >0.3 are preferred chemical starting points. Such fragments are close to being optimal in their binding interactions, and even though affinities may be in the mM range, they are valid starting points for further optimization. Maintaining or increasing LE through optimization is a cornerstone of the strategy in FBDD.[122,135–137] Similarly, in hit selection from HTS, selection of hits with LE values of >0.3, rather than just the most potent hits, should allow for more facile optimization.[138] The tendency in the past to select only the most potent hits from screens has probably contributed to the increases in physical properties seen in recent medicinal chemistry projects.[4,139] A change in mind-set, focusing on ligand efficiency together with potency, should help reverse this trend.

Lipophilic ligand efficiency (LLE) is a more recent concept than LE and has been implicated in toxicity resulting from promiscuous receptor interactions, which are driven mainly by lipophilicity. It was proposed[4] that LLE values of

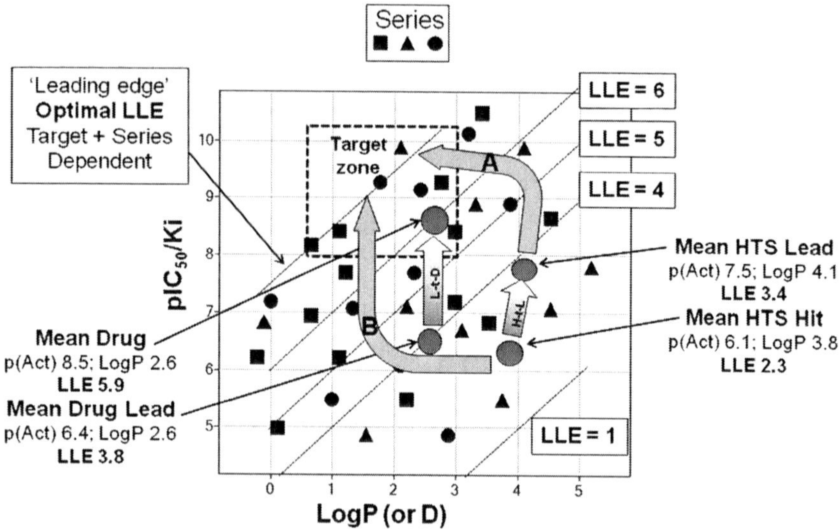

Figure 2.1 The application of LLE in hit optimization. The mean drug and drug lead data ($n = 60$; L-t-D = lead to drug) are from Perola (reference 108), and the mean HTS hit and lead data ($n = 335$; H-t-L = hit to lead) are from Keseru and Makara, (reference 107). The specific target criteria are illustrative and will depend on the nature of the project; in general, aiming for $pIC_{50} > 8$ and *c*Log*P* 1–3 combines drug-like potency with optimal *c*Log*P* for oral absorption and avoidance of unwanted ADMET issues. The leading edge boundary at the top left of the graph represents the maximum LLE achieved, which is dependent on the specific chemical series examined. Starting with an average HTS hit, with LLE ~2, the goal will often be to reduce *c*Log*P* and increase potency. Increasing potency, as in strategy **A**, means that highly potent lipophilic compounds will at some point have to be modified to lower lipophilicity, which may be a challenge if the hydrophobic binding is optimized. In contrast, lowering lipophilicity to meet the target goal, while searching for high affinity though optimizing H-bonding or polar interactions, as in strategy **B**, is a potentially superior approach. Following strategies **A** or **B** is likely to result in different structure–activity relationships and different optimized molecules.

>5 (*i.e.* potency <10 nM and *c*Log*P* <3) is a target value for reducing toxicity attrition and that useful hits should have LLE values of >2.[132] The increase in potency, but not *c*Log*P*, in the lead optimization of recent drug candidates supports LLE as a useful drug-like measure.[108] The application of LLE in drug discovery projects is very important because extracting maximal binding interactions for minimal lipophilicity will reduce the risk of unwanted lipophilicity-driven ADMET issues. The use of LLE measures in the design of chemokine antagonists lacking unwanted hERG activity illustrates the general approach.[140] The use of LLE thinking during optimization is illustrated in Figure. 2.1 and there are several recent examples of its application.[141–145] In essence, these examples show increasing binding to the target by finding new polar ligand–protein interactions, which also reduce lipophilicity.

2.5 Conclusions

Studies of the physical chemistry profiles of successfully marketed oral drugs, backed up by ever-increasing empirical ADMET data, point strongly to an area of physicochemical size/lipophilicity/shape space, where drugs are most likely to be found. Whilst the Ro5 metric is the simplest and most widely accepted measure of preferred drug-like space, current medicinal chemistry, as assessed from the patent literature, is largely operating close to Ro5 property extremities, where the risks are greater than at the centre of historical drug space. This fact may in part explain the reduced productivity[4] and increased pipeline attrition[146] now placing the pharmaceutical industry under considerable pressure.[147,148] Whilst other attritional factors operate,[147] the molecular properties of molecules entering drug development are controllable, being entirely at the discretion of the drug designer. Using lead-like thinking, where lead compounds with low physical properties can be 'grown' during optimization, whilst remaining in optimal drug-like space, is one approach to help solve the problem[4] of inflated molecular properties. The introduction of fragment-based drug discovery is a promising application of the lead-like concept. The routine application of ligand efficiency measures in lead generation and optimization programs, especially lipophilicity ligand efficiency, is a powerful and practical means of keeping physical properties under control.

Acknowledgements

TIO is supported by NIH grant 5U54MH084690-02.

References

1. P. D. Leeson and J. R. Empfield, *Annu. Rep. Med. Chem.*, 2010, **45**, 393.
2. S. W. Muchmore, J. J. Edmunds, K. D. Stewart and P. J. Hajduk, *J. Med. Chem.*, 2010, **53**, 4830.
3. G. F. Smith, *Prog. Med. Chem.*, 2009, **48**, 1.
4. P. D. Leeson and B. Springthorpe, *Nat. Rev. Drug Discovery*, 2007, **6**, 881.
5. W. Sneader, *A History of Drug Discovery*, John Wiley & Sons, Chichester, England, 2005.
6. A. H. Lipkus, Q. Yuan, K. A. Lucas, S. A. Funk, W. F. Bartelt III, R. J. Schenck and A. J. Trippe, *J. Org. Chem.*, 2008, **73**, 4443.
7. M. S. Lajiness, G. M. Maggiora and V. Shanmugasundaram, *J. Med. Chem.*, 2004, **47**, 4891.
8. P. D. Leeson, A. M. Davis and J. Steele, *Drug Discovery Today: Technol.*, 2004, **1**, 189.
9. C. A. Lipinski, F. Lombardo, B. W. Dominy and P. J. Feeney, *Adv. Drug Delivery Rev.*, 1997, **23**, 3.
10. P. D. Dobson and D. B. Kell, *Nat. Rev. Drug Discovery*, 2008, **7**, 205.
11. M. M. Olah, C. G. Bologa and T. I. Oprea, *Curr. Drug Discov. Technol.*, 2004, **1**, 211.

12. T. I. Oprea, *J. Comput.-Aided Mol. Des.*, 2000, **14**, 251.
13. P. Ertl, *J. Chem. Inf. Comput. Sci.*, 2003, **43**, 374.
14. L. C. Blum and J.-L. Reymond, *J. Am. Chem. Soc.*, 2009, **131**, 8732.
15. O. Ursu, A. Ryan, and A. Goldblum and T.I. Oprea, *WIREs Comp. Mol. Sci.*, 2010, submitted.
16. O. Ursu and T. I. Oprea, *J. Chem. Inf. Model.*, 2010, **50**, 1387.
17. M. C. Wenlock, R. P. Austin, P. Barton, A. M. Davis and P. D. Leeson, *J. Med. Chem.*, 2003, **46**, 1250.
18. M. Vieth, M. G. Siegel, R. E. Higgs, I. A. Watson, D. H. Robertson, K. A. Savin, G. L. Durst and P. A. Hipskind, *J. Med Chem.*, 2004, **47**, 224.
19. J. R. Proudfoot, *Bioorg. Med. Chem. Lett.*, 2005, **15**, 1087.
20. A. Tronde, B. Norden, H. Marchner, A.-K. Wendel, H. Lennernaes and U. H. Bengtsson, *J. Pharm. Sci.*, 2003, **92**, 1216.
21. T. J. Ritchie, C. N. Luscombe and J. F. Macdonald, *J. Chem. Inf. Model.*, 2009, **49**, 1025.
22. P. D. Leeson and A. M. Davis, *J. Med. Chem.*, 2004, **47**, 6338.
23. P. D. Leeson, S. A. St-Gallay and M. C. Wenlock, *Med. Chem. Commun.*, 2011, **2**, 91.
24. C. Lamanna, M. Bellini, A. Padova, G. Westerberg and L. Maccari, *J. Med. Chem.*, 2008, **51**, 2891.
25. F. Lovering, J. Bikker and C. Humblet, *J. Med. Chem.*, 2009, **52**, 6752.
26. T. J. Ritchie and S. J. F. MacDonald, *Drug Discovery Today*, 2009, **14**, 1011.
27. A. P. Hill and R. J. Young, *Drug Discovery Today*, 2010, **15**, 648.
28. D. A. Smith, L. Di and E. H. Kerns, *Nat. Rev. Drug Discovery*, 2010, **9**, 929.
29. D. Kong, W. Ren, W. Lu and H. Zhang, *J. Chem. Inf. Model.*, 2009, **49**, 2376.
30. P. D. Dobson, Y. Patel and D. B. Kell, *Drug Discovery Today*, 2009, **14**, 31.
31. H. Kubinyi, *Nat. Rev. Drug Discovery*, 2003, **2**, 665.
32. T. I. Oprea, *J. Comp.-Aided Mol. Des.*, 2002, **16**, 325.
33. J. Kelder, P. D. Grootenhuis, D. M. Bayada, L. P. Delbressine and J. P. Ploemen, *Pharm. Res.*, 1999, **16**, 1514.
34. M. P. Gleeson, *J. Med. Chem.*, 2008, **51**, 817.
35. T. W. Johnson, K. R. Dress and M. Edwards, *Bioorg. Med. Chem. Lett.*, 2009, **19**, 5560.
36. M. J. Waring, *Bioorg. Med. Chem. Lett.*, 2009, **19**, 2844.
37. H. van de Waterbeemd, *Chem. Biodivers.*, 2009, **6**, 1760.
38. M. L. Lewis and L. Cucurull-Sanchez, *J. Comp.-Aided Mol. Des.*, 2009, **23**, 97.
39. A. G. Dossetter, *Bioorg. Med. Chem.*, 2010, **18**, 4405.
40. P. Gleeson, G. Bravi, S. Modi and D. Lowe, *Bioorg. Med. Chem.*, 2009, **17**, 5906.
41. J. G. Topliss, *J. Med. Chem.*, 1972, **15**, 1006.

42. L. Z. Benet, C. L. Cummins and C. Y. Wu, *Int. J. Pharm.*, 2004, **277**, 3.
43. J. M. Custodio, C. Y. Wu and L. Z. Benet, *Adv. Drug Delivery Rev.*, 2008, **60**, 717.
44. M. V. S. Varma, R. S. Obach, C. Rotter, H. R. Miller, G. Chang, S. J. Steyn, A. El-Kattan and M. D. Troutman, *J. Med. Chem.*, 2010, **53**, 1098.
45. D. F. Veber, S. R. Johnson, H. Cheng, B. R. Smith, K. W. Ward and K. D. Kopple, *J. Med. Chem.*, 2002, **45**, 2615.
46. Y. C. Martin, *J. Med. Chem.*, 2005, **48**, 3164.
47. J. J. Lu, K. Crimin, J. T. Goodwin, P. Crivori, C. Orrenius, L. Xing, P. J. Tandler, T. J. Vidmar, B. M. Amore, A. G. E. Wilson, P. F. W. Stouten and P. S. Burton, *J. Med. Chem.*, 2004, **47**, 6104.
48. T. Hou, J. Wang, W. Zhang and X. Xu, *J. Chem. Inf. Model.*, 2007, **47**, 208.
49. D. E. Clark, *J. Pharm. Sci.*, 1999, **88**, 815.
50. H. Pajouhesh and G. R. Lenz, *J. Am. Soc. Exper. Neuro. Ther.*, 2005, **2**, 541.
51. S. A. Hitchcock and L. D. Pennington, *J. Med. Chem.*, 2006, **49**, 1.
52. T. T. Wager, R. Y. Chandrasekaran, X. Hou, M. D. Troutman, P. R. Verhoest, A. Villalobos and Y. Will, *ACS Chem. Neurosci.*, 2010, **1**, 420.
53. T. T. Wager, X. Hou, P. R. Verhoest and A. Villalobos, *ACS Chem. Neurosci.*, 2010, **1**, 435.
54. R. O'Shea and E. H. Moser, *J. Med. Chem.*, 2008, **51**, 2871.
55. A. S. Kalgutkar and M. T. Didiuk, *Chem. Biodivers.*, 2009, **6**, 2115.
56. S. J. Enoch and M. T. D. Cronin, *Crit. Rev. Toxicol.*, 2010, **40**, 728.
57. A. S. Kalgutkar, G. Fate, M. T. Didiuk and J. Bauman, *Expert Rev. Clin. Pharmacol.*, 2008, **1**, 515.
58. A. S. Kalgutkar, I. Gardner, R. S. Obach, C. L. Shaffer, E. Callegari, K. R. Henne, A. E. Mutlib, D. K. Dalvie, J. S. Lee, Y. Nakai, J. P. O'Donnell, J. Boer and S. P. Harriman, *Curr. Drug Metabol.*, 2005, **6**, 161.
59. S. Kumar and T. A. Baillie, *Drugs Pharm. Sci.*, 2009, **186**, 597.
60. R. K. Elespuru, R. Agarwal, A. H. Atrakchi, C. A. H. Bigger, R. H. Heflich, J. R. Jagannath, D. D. Levy, M. M. Moore, Y. Ouyang, T. W. Robison, R. E. Sotomayor, M. C. Cimino and K. L Dearfield, *Toxicol. Sci.*, 2009, **109**, 172.
61. S. P. Raillard, J. Bercu, S. W. Baertschi and C. M. Riley, *Org. Process Res. Dev.*, 2010, **14**, 1015.
62. D. A. Price, J. Blagg, L. Jones, N. Greene and T. Wager, *Expert Opin. Drug Metab. Toxicol.*, 2009, **5**, 921.
63. N. Greene, L. Fisk, R. T. Naven, R. R. Note, M. L. Patel and D. J. Pelletier, *Chem. Res. Toxicol.*, 2010, **23**, 1215.
64. (a) E. J. Matthews, C. J. Ursem, N. L. Kruhlak, R. D. Benz, D. Aragonés Sabaté, C. Yang, G. Klopman and J. F. Contrera, *Regul. Toxicol.*

Pharm., 2009, **54**, 23; (b) E. J. Matthews, N. L. Kruhlak, R. D. Benz, R. D. A. Sabate, C. A. Marchant and J. F. Contrera, *Regul. Toxicol. Pharm.*, 2009, **54**, 43.
65. A. L. Hopkins, *Nat. Chem. Biol.*, 2008, **4**, 682.
66. M. J. Keiser, V. Setola, J. J. Irwin, C. Laggner, A. I. Abbas, S. J. Hufeisen, N. H. Jensen, M. B. Kuijer, R. C. Matos, T. B. Tran, R. Whaley, R. A. Glennon, J. Hert, K. L. H. Thomas, D. D. Edwards, B. K. Shoichet and B. L. Roth, *Nature*, 2009, **462**, 175.
67. E. E. Schadt, S. H. Friend and D. A. Shaywitz, *Nat. Rev. Drug Discovery*, 2009, **8**, 286.
68. T. I. Oprea, A. Tropsha, J. L. Faulon and M. D. Rintoul, *Nat. Chem. Biol.*, 2007, **3**, 447.
69. L. Roth, D. J. Sheffler and W. K. Kroeze, *Nat. Rev. Drug Discovery*, 2004, **3**, 353.
70. R. Morphy, *J. Med. Chem.*, 2010, **53**, 1413.
71. F. Milletti and A. Vulpetti, *J. Chem. Inf. Model.*, 2010, **50**, 1418.
72. Z. A. Knight, H. Lin and K. M. Shokat, *Nat. Rev. Cancer*, 2010, **10**, 130.
73. F. Broccatelli, E. Carosati, G. Cruciani and T. I. Opera, *Mol. Inf.*, 2010, **29**, 16.
74. R. Morphy and Z. Rankovic, *J. Med. Chem.*, 2006, **49**, 4961.
75. C. Melchiorre, M. L. Bolognesi, A. Minarini, M. Rosini and V. Tumiatti, *J. Med. Chem.*, 2010, **53**, 5906.
76. J. B. Baell and G. A. Holloway, *J. Med. Chem.*, 2010, **53**, 2719.
77. M. L. Radhakrishnan and B. Tidor, *J. Phys. Chem. B*, 2007, **111**, 13419.
78. A. L. Hopkins, J. S. Mason and J. P. Overington, *Curr. Opin. Struct. Biol.*, 2006, **16**, 127.
79. J. Peters, P. Schnider, P. Mattei and M. Kansy, *ChemMedChem*, 2009, **4**, 680.
80. E. Gregori-Puigjane and J. Mestres, *Comb. Chem. High Throughput Screening*, 2008, **11**, 669.
81. K. Azzaoui, J. Hamon, B. Faller, S. Whitebread, E. Jacoby, A. Bender, J. L. Jenkins and L. Urban, *ChemMedChem*, 2007, **2**, 874.
82. Y. Yang, H. Chen, I. Nilsson, S. Muresan and O. Engkvist, *J. Med. Chem.*, 2010, **53**, 7709.
83. D. J. Diller, *Curr. Comput.-Aided Drug Des.*, 2009, **5**, 106.
84. A. Stary, S. J. Wacker, L. Boukharta, U. Zachariae, Y. Karimi-Nejad, J. Aqvist, G. Vriend and B. L. de Groot, *ChemMedChem*, 2010, **5**, 455.
85. G. Gintant, *Pharmacology & Therapeutics*, 2011, **129**, 109.
86. A. J. Ratcliffe, *Current Medicinal Chemistry*, 2009, **16**, 2816.
87. M. T. D. Cronin, *Curr. Comput.-Aided Drug Des.*, 2006, **2**, 405.
88. J. D. Hughes, J. Blagg, D. A. Price, S. Bailey, G. A. DeCrescenzo, R. V. Devraj, E. Ellsworth, Y. M. Fobian, M. E. Gibbs, R. W. Gilles, N. Greene, E. Huang, T. Krieger-Burke, J. Loesel, T. Wager, L. Whiteley and Y. Zhang, *Bioorg. Med. Chem. Lett.*, 2008, **18**, 4872.
89. J. Uetrecht, *Curr. Opin. Drug. Discovery Dev.*, 2001, **4**, 55.

90. N. Greene, M. D. Aleo, S. Louise-May, D. A. Price and Y. Will, *Bioorg. Med. Chem. Lett.*, 2010, **20**, 5308.
91. A. L. Hopkins and C. R. Groom, *Nat. Rev. Drug Discovery*, 2002, **1**, 727.
92. K. Beaumont, E. Schmid and D. A. Smith, *Bioorg. Med. Chem. Lett.*, 2005, **15**, 3658.
93. D. E. Grigoriadis, S. R. J. Hoare, S. M. Lechner, D. H. Slee and J. A. Williams, *Neuropsychopharmacology*, 2009, **34**, 106.
94. M. Vieth and J. J. Sutherland, *J. Med. Chem.*, 2006, **49**, 3451.
95. R. Morphy, *J. Med. Chem.*, 2006, **49**, 2969.
96. A. C. Cheng, R. G. Coleman, K. T. Smyth, Q. Cao, P. Soulard, D. R. Caffrey, A. C. Salzberg and E. S. Huang, *Nat. Biotechnol.*, 2007, **25**, 71.
97. P. Schmidtke and X. Barril, *J. Med. Chem.*, 2010, **53**, 5858.
98. J. Seco, F. J. Luque and X. Barril, *J. Med. Chem.*, 2009, **52**, 2363.
99. S. J. Teague, *Nat. Rev. Drug Discovery*, 2003, **2**, 527.
100. S. J. Teague, A. M. Davis, P. D. Leeson and T. Oprea, *Angew. Chem. Int. Ed.*, 1999, **38**, 3743.
101. M. M. Hann, A. R. Leach and G. Harper, *J. Chem. Inf. Comput. Sci.*, 2001, **41**, 856.
102. W. Sneader, *Drug Prototypes and their Exploitation*, Wiley, Chichester, 1996.
103. T. I. Oprea, A. M. Davis, S. J. Teague and P. D. Leeson, *J. Chem. Inf. Comput. Sci.*, 2001, **41**, 1308.
104. J. R. Proudfoot, *Bioorg. Med. Chem. Lett.*, 2002, **12**, 1647.
105. T. I. Oprea, T. K. Allu, D. C. Fara, R. F. Rad, L. Ostopovici and C. G. Bologa, *J. Comp.-Aided. Mol. Des.*, 2007, **21**, 113.
106. C. Tyrchan, N. Blomberg, O. Engkvist, T. Kogej and S. Muresan, *Bioorg. Med. Chem. Lett.*, 2009, **19**, 6943.
107. G. M. Keseru and G. M. Makara, *Nat. Rev. Drug Discovery*, 2009, **8**, 203.
108. E. Perola, *J. Med. Chem.*, 2010, **53**, 2986.
109. G. M. Rishton, *Curr. Opinion Chem. Biol.*, 2008, **12**, 340.
110. W. F. Michne, *Pharmaceut. News*, 1996, **3**, 19.
111. C. J. Manly, J. Chandrasekhar, J. W. Ochterski, J. D. Hammer and B. B. Warfield, *Drug Discovery Today*, 2008, **13**, 99.
112. R. A. Goodnow Jr and P. Gillespie, *Prog, Med. Chem.*, 2007, **45**, 1.
113. T. Wunberg, M. Hendrix, A. Hillisch, M. Lobell, H. Meier, C. Schmeck, H. Wild and B. Hinzen, *Drug Disc. Today*, 2006, **11**, 175.
114. A. Steinmeyer, *ChemMedChem*, 2006, **1**, 31.
115. A. M. Davis, D. J. Keeling, J. Steele, N. P. Tomkinson and A. C Tinker, *Curr. Top. Med. Chem.*, 2005, **5**, 421.
116. K. H. Bleicher, M. Nettekoven, J.-U. Peters, R. Wyler, *Chimia*, 2004, **58**, 588.
117. K. H. Bleicher, H.-J. Boehm, K. Mueller and A. I. Alanine, *Nat. Rev. Drug Disc*overy, 2003, **2**, 369.
118. A. Alanine, M. Nettekoven, E. Roberts and A. W. Thomas, *Comb. Chem. High Throughput Screening*, 2003, **6**, 51.

119. E. Jacoby, A. Schuffenhauer, M. Popov, K. Azzaoui, B. Havill, U. Schopfer, C. Engeloch, J. Stanek, P. Acklin, P. Rigollier, F. Stoll, G. Koch, P. Meier, D. Orain, R. Giger, J. Hinrichs, K. Malagu, J. Zimmermann and H.-J. Roth, *Curr. Top. Med. Chem.*, 2005, **5**, 397.
120. D. H. Drewry and R. Macarron, *Curr. Opin. Chem. Biol.*, 2010, **14**, 289.
121. M. A. Snowden and V. S. Green, *Curr. Opin. Drug Discovery Dev.*, 2008, **11**, 553.
122. C. W. Murray, D. C. Rees, *Nature Chem.*, 2009, **1**, 187.
123. M. Congreve, R. Carr, C. Murray and H. Jhoti, *Drug Discovery Today*, 2003, **8**, 876.
124. M. Congreve, G. Chessari, D. Tisi and A. J. Woodhead, *J. Med. Chem.*, 2008, **51**, 3661.
125. A. L. Gill, M. Verdonk, R. G. Boyle and R. Taylor, *Curr. Top. Med. Chem.*, 2007, **7**, 1408.
126. A. L. Hopkins, C. R. Groom and A. Alex, *Drug Discovery Today*, 2004, **9**, 430.
127. P. R. Andrews, D. J. Craik and J. L. Martin, *J. Med. Chem.*, 1984, **27**, 1648.
128. I. D. Kuntz, K. Chen, K. A. Sharp and P. A. Kollman, *Proc. Nat. Acad. Sci., U.S.A.*, 1999, **96**, 9997.
129. C. Abad-Zapatero, *Expert Opin. Drug Discovery*, 2007, **2**, 469.
130. M. L Verdonk and D. C. Rees, *ChemMedChem.*, 2008, **3**, 1179.
131. J. Mestres and E. Gregori-Puigjane, *Trends Pharmacol. Sci.*, 2009, **30**, 470.
132. A. R. Leach, M. M. Hann, J. N. Burrows and E. J. Griffen, *Mol. BioSyst.*, 2006, **2**, 429.
133. C. H. Reynolds, B. A. Tounge and S. D. Bembenek, *J. Med. Chem.*, 2008, **51**, 2432.
134. J. W. M. Nissink, *J. Chem. Inf. Model.*, 2009, **49**, 1617.
135. S. D. Bembenek, B. A. Tounge and C. H. Reynolds, *Drug Discovery Today*, 2009, **14**, 278.
136. D. C. Rees, M. Congreve, C. W. Murray and R. Carr, *Nat. Rev. Drug Discovery*, 2004, **3**, 660.
137. J. S. Albert, N. Blomberg, A. L. Breeze, A. J. H. Brown, J. N. Burrows, P. D. Edwards, R. H. Folmer, S. Geschwindner, E. J. Griffen, P. W. Kenny, T. Nowak, L.-L. Olsson, H. Sanganee and A. B. Shapiro, *Curr. Top. Med. Chem.*, 2007, **7**, 1600.
138. A. B. Reitz, G. R. Smith, B. A. Tounge and C. H. Reynolds, *Curr. Top. Med. Chem.*, 2009, **9**, 1718.
139. C. A. Lipinski, *J. Pharmacol. Toxicol. Methods*, 2001, **44**, 235.
140. I. Shamovsky, S. Connolly, L. David, S. Ivanova, B. Norden, B. Springthorpe and K. Urbahns, *J. Med. Chem.*, 2008, **51**, 1162.
141. T. Ryckmans, M. P. Edwards, V. A. Horne, A. M. Correia, D. R. Owen, L. R. Thompson, I. Tran, M. F. Tutt and T. Young, *Bioorg. Med. Chem. Lett.*, 2009, **19**, 4406.

142. K. N. Dack, S. Skerratt, P. S. Johnson, P. A. Bradley and I. R, Marsh, *Bioorg. Med. Chem. Lett.*, 2010, **20**, 3384.
143. C. E. Mowbray, R. Corbau, M. Hawes, L. H. Jones, J. E. Mills, M. Perros, M. D. Selby, P. A. Stupple, R. Webster and A. Wood, *Bioorg. Med. Chem. Lett.*, 2009, **19**, 5603.
144. I. Shamovsky, C. de Graaf, L. Alderin, M. Bengtsson, H. Bladh, L. Borjesson, S. Connolly, H. J. Dyke, M. van den Heuvel, H. Johansson, B.-G. Josefsson, A. Kristoffersson, T. Linnanen, A. Lisius, R. Mannikko, B. Norden, S. Price, L. Ripa, D. Rognan, A. Rosendahl, M. Skrinjar and K. Urbahns, *J. Med. Chem.*, 2009, **52**, 7706.
145. M. P. Edwards and D. A. Price, *Annu. Rep. Med. Chem.*, 2010, **45**, 381.
146. I. Kola and J. Landis, *Nat. Rev. Drug Discovery*, 2004, **3**, 711.
147. B. Munos, *Nat. Rev. Drug Discovery*, 2009, **8**, 959.
148. S. M. Paul, D. S. Mytelka, C. T. Dunwiddie, C. C. Persinger, B. H. Munos, S. R. Lindborg and A. L. Schacht, *Nat. Rev. Drug Discovery*, 2010, **9**, 203.

CHAPTER 3
Development of QSAR

YVONNE CONNOLLY MARTIN

Martin Consulting, 2230 Chestnut St., Waukegan IL 60087, USA

3.1 Introduction

3.1.1 Overview of QSAR

A quantitative structure–activity relationship (QSAR) analysis searches for a mathematical relationship between the relative biological properties of a set of molecules and one or more of their calculated or physical properties. The first QSAR publications described the use of regression analysis to investigate the relationship between physical and biological properties within series of closely related molecules. From this beginning in the early 1960s, the field has blossomed to expand the repertoire of molecular properties considered, some of which are calculated from the 3D structures of the molecules. In addition, new ways to investigate and validate the relationships between molecular and biological properties have also received widespread attention. All methods have in common the goal of establishing a mathematical relationship that can be used to forecast the biological properties of molecules that have not yet been tested.

The various properties of molecules that can be used in a QSAR are often designed to quantitate the tendency of the molecules to participate in one of the fundamental types of intermolecular interactions: electrostatic, hydrogen bonding, dispersion forces, and hydrophobic interactions. In addition, the possibility of steric interference with an interaction is considered. Other methods capitalize on the fact that the 2D structure of a molecule indirectly encodes its properties, instead generate descriptors without an explicit relationship to some physical property.

RSC Drug Discovery Series No. 13
Drug Design Strategies: Quantitative Approaches
Edited by David J. Livingstone and Andrew M. Davis

Published by the Royal Society of Chemistry, www.rsc.org

Some of the methods concentrate on the effects of variable substituents on the properties of a constant parent molecule in order to generate a local model. Other methods consider the properties of the whole molecules, often structurally diverse, to generate a global model. Forecasts of biological activity of new molecules from a local model require that the new molecules contain the features common to those used to derive the model. In contrast, forecasts from global models are considered to apply to any new molecule. Usually, predictions from global models are accompanied by some measure of similarity of the new molecules to those used for the model.

3.1.2 The Invention of QSAR

The publication of the first QSAR paper in April 1962 was a landmark in medicinal chemistry.[1] It suggested to the community that it is possible to forecast the biological potency of a compound before it is synthesized. This publication reported equation (3.1):

$$\log 1/C = 4.08\pi - 2.14\pi^2 - 2.78\sigma + 3.36 \tag{3.1}$$

In this equation, C is the concentration (in mol L^{-1}) of an auxin that induces 10 percent growth of plant cells, π is the change in octanol–water partition coefficient of the substituted derivative relative to the unsubstituted, and σ is the Hammett sigma constant for the electronic effect of the substituent to the *ortho* position of the ring. The biological potency of the compounds was encoded on a log scale to signify a free energy relationship, and the reciprocal of the concentration was used so that higher numbers signify higher potency.

The publication was novel in at least six respects: (1) It provided a quantitative relationship between the biological potency and physical properties of a compound; (2) It recognized the quantitative importance of hydrophobicity to affinity; (3) It put the hydrophobic effect of a substituent, π, on a scale similar to that of the Hammett σ constant for electronic effects; (4) It provided a mechanism to identify the optimum value of some property as it relates to potency, for example, the parabolic fit of π with π^2; (5) It considered the additive effect of different physical properties on potency; (6) The fit to the equation used a computer.

The work was formalized in a March 1964 publication, which correlated seven additional datasets to equation (3.2).[2]

$$\log(1/C) = k'\pi - k\pi^2 - \rho\sigma - k'' \tag{3.2}$$

They thereby demonstrated the generality of the method. They also emphasized that π appears to be an additive property, the value of which is relatively independent of the parent aromatic structure or the position of the substituent. Several of the publications by the Pomona group[2–4] are recognized as citation classics in view of the number of citations they have received.

Since the time of this publication, literally thousands of datasets have been correlated; the database of the equations generated by the Pomona group (BioLoom) contains more than six thousand equations.[5] In addition, such studies led to the commercialization of products.[6] Most recently, the European Community in its Registration, Evaluation, Authorisation and Restriction of Chemicals (REACH) program states that for each chemical circulating in the European territory, a complete dossier on physico-chemical, biological, and toxicological properties must be compiled. For this purpose, validated QSAR methods are accepted.[7]

The following sections discuss the components of a QSAR: the calculation of molecular properties, the evaluation of the mathematical relationship between these properties and bioactivity, and the validation of a QSAR model. The final section compares QSAR and traditional SAR.

3.2 Properties Calculated from 2D Structure of Molecules

These descriptors remain popular because one does not need to propose a 3D structure in order to generate a value. Additionally, the electronic, hydrophobic, and steric properties are based both on values determined experimentally and those available in printed tables;[8] no sophisticated computer programs are needed, although they are often used.

3.2.1 Effects of Substituents on the Electrostatic Properties of the Parent Molecule

3.2.1.1 Electronic Effects of Aromatic Substituents

In the 1930s Hammett formulated the equation named in his honor.[9] It is based on the observation of the constancy of the relative effects of substituents on aromatic rings, regardless of the reaction rate or equilibrium of interest. This relationship is expressed according to equation (3.3).

$$\text{for rate constants: } \log k = \log k^0 + \rho\sigma \tag{3.3}$$

$$\text{for equilibrium constants: } \log K = \log K^0 + \rho\sigma$$

In these equations, k^0 or K^0 refer to the rate or equilibrium constant for the unsubstituted molecule, ρ characterizes the sensitivity of the reaction to electronic substituent effects, and σ is the relative electronic effect of the substituent. A substituent at the *meta* position has a different σ value than that of the same substituent at the *para* position. By definition, $\rho = 1.0$ for the pK_a of substituted benzoic acids and σ is the substituent effect on this reaction. As an example, Figure 3.1 shows the relationship between Hammett σ and the log of the acid dissociation constant ($\log K_a$) of *meta*-substituted phenols. The correlation coefficient of this relationship is 0.97. The data is presented in Table 3.1.

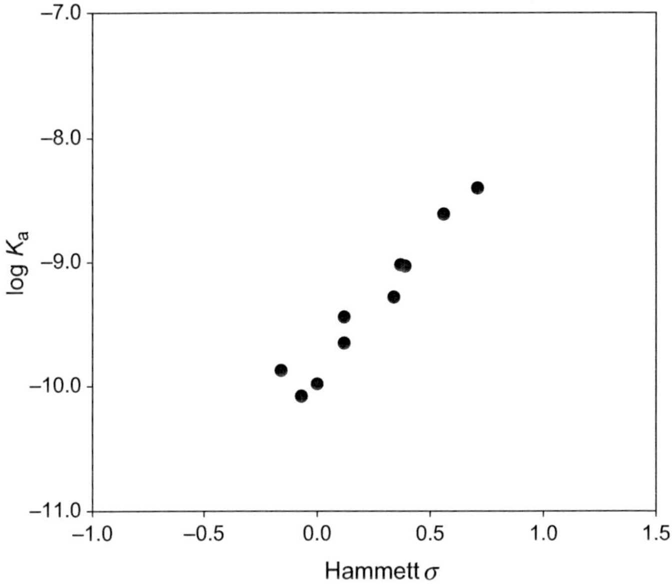

Figure 3.1 The relationship between Hammett σ and the log K_a of *meta*-substituted phenols.

Table 3.1 The logK_a values of substituted phenols, and values of various sigma constants.

Substituent	σ	σ-	*log* K_a	F	R	σ*
H	0.00	0.00	–9.98	0.00	0.00	0.49
m-amino	–0.16	–0.16	–9.87	0.08	–0.74	0.62
m-bromo	0.39	0.39	–9.03	0.45	–0.22	2.80
m-chloro	0.37	0.37	–9.02	0.42	–0.19	2.94
m-cyano	0.56	0.56	–8.61	0.51	0.15	3.64
m-fluoro	0.34	0.34	–9.28	0.45	–0.39	3.19
m-hydroxy	0.12	0.12	–9.44	0.33	–0.70	1.37
m-methoxy	0.12	0.12	–9.65	0.29	–0.56	1.77
m-methyl	–0.07	–0.07	–10.08	0.01	–0.18	0.00
m-nitro	0.71	0.71	–8.40	0.65	0.13	4.66
p-amino	–0.66	–0.15	–10.3			
p-bromo	0.23	0.23	–9.36			
p-chloro	0.23	0.23	–9.38			
p-cyano	0.66	1.00	–7.95			
p-fluoro	0.06	0.06	–9.95			
p-hydroxy	–0.37	–0.37	–9.96			
p-methoxy	–0.27	–0.26	–10.21			
p-methyl	–0.17	–0.17	–10.14			
p-nitro	0.78	1.27	–7.15			

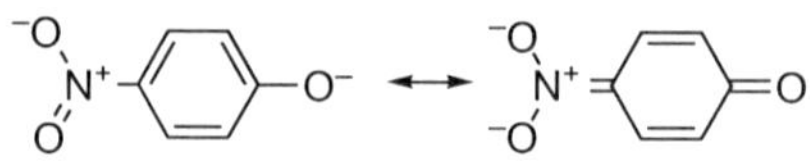

Figure 3.2 The resonance forms of *para*-nitro phenolate.

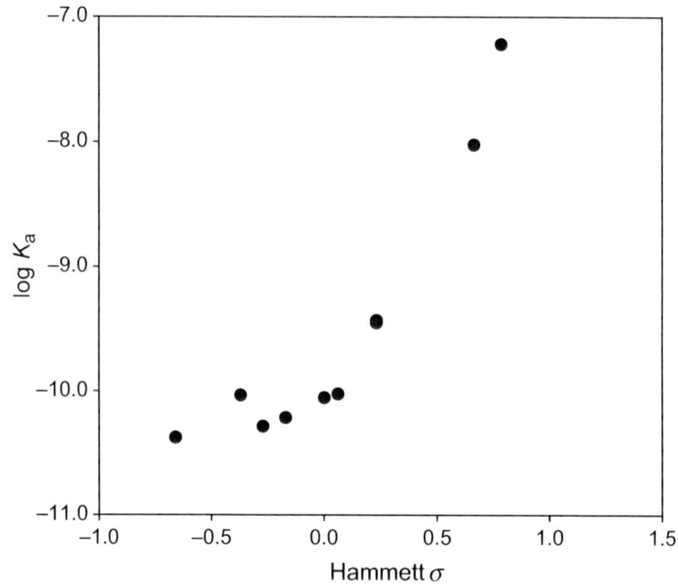

Figure 3.3 The relationship between Hammett σ and the log K_a of *para*-substituted phenols.

In the 1950s, Hammett equations and their extensions were active subjects of research by physical organic chemists.[10] Research revealed that for certain substituents the relationship did not hold for cases in which there can be direct resonance between the substituent and the reaction center. The classic example of this is the effect of a *para* nitro group on the log K_a of phenol. Figures 3.2 and 3.3 show the relationship between Hammett σ and the log K_a of *para*-substituted phenols. Notice the pronounced curve at the lower σ values.

To account for this effect, a second σ value, σ—(sigma minus), was assigned to each *para* substituent. Table 3.1 shows these values for the substituted phenols and Figure 3.4 shows that the resulting relationship is more linear.

The fact that two σ-constants are needed to explain the substituent effects led to the proposal that each σ value is a different linear combination of two separate factors: field and resonance effects, for example, the Swain-Lupton F and R (see Table 3.1), as shown in equation (3.4).

$$\sigma_i = a_i F + b_i R \tag{3.4}$$

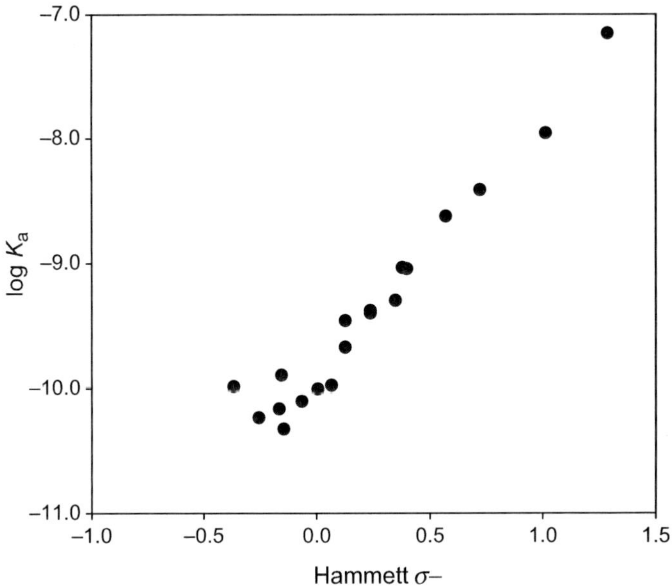

Figure 3.4 The relationship between $\sigma-$ and the log K_a of *meta*- and *para*-substituted phenols.

The σ value of a *para* substituent is estimated as the sum of the F and R values for the substituent, where $a = b = 1.0$, and that for a *meta* substituents as the sum of F plus 0.27 R, *i.e.* $a = 1.0$, $b = 0.27$. Hence the contribution of resonance to σ_m is only 27% that of σ_p. The importance of such a separation of field and resonance effects is that one can use these separate values directly in a regression equation without deciding which σ value to use. The problem with such an approach is that rather than one σ value (or sum of σ values) for the substituent(s) on a molecule, there are now twice as many descriptors as there are positions of substitution.

Ortho substituents pose a problem because as well as having an electronic effect on a reaction or equilibrium, they can sterically interfere with it. One strategy is to assume that $\sigma_o = \sigma_p$ and include a steric substituent constant (such as described in Section 3.2.2) for the *ortho* substituent.[11] Others have derived σ-constants for *ortho* substituents.[12–14] A fair summary is that the problem is complicated.

3.2.1.2 Electronic Effects of Aliphatic Substituents

Of course, not all substitutions of interest occur on aromatic systems. Robert Taft's studies in the 1950s on the kinetics of the formation and hydrolysis of carboxylic esters laid the groundwork for the separation of the polar (electronic) effect of a substituent and its steric effect.[12] He formulated equation

(3.5) to define σ^* (sigma star) for the aliphatic substituent R in an ester, RCH_2CO_2R'.

$$\sigma* = [\log(k/k^0)_B - \log(k/k^0)_A]/2.48 \quad (3.5)$$

In this equation, k/k^0 is the relative rate of reaction of the substituted *versus* the unsubstituted ($R = CH_3$) ester, and the subscripts B and A refer to basic and acidic conditions respectively. Equation (3.5) assumes that the reactivity in acid is independent of electronic effects, whereas in base the reactivity depends on both steric and electronic effects. The divisor of 2.48 is arbitrarily chosen to place σ^* values on a scale similar to that of normal σ values.[15] Table 3.1 lists σ^* values of typical substituents.

This discussion has barely touched on the variants of the Hammett equation. The interested reader is referred to several more detailed explorations of the topic.[10,16,17] However, a weary reader may wonder if there might not be a simpler method to calculate the relative differences in electronic properties of molecules.

3.2.2 Steric Effects of Substituents

3.2.2.1 *Taft* E_s *Values*

Taft pursued the separation of steric and electronic effects of substituents by introducing E_s, a substituent constant that describes the effect of a nearby substituent on the rate of reaction of a compound, specifically the effect of the R substituent on the rate of hydrolysis of an ester, RCH_2CO_2R'.[12] By assuming no electronic effect of substituents on the acid-catalyzed hydrolysis of esters, E_s values are calculated directly from this rate [equation (3.6)].

$$E_s = \log(k/k^0)_A \quad (3.6)$$

Table 3.2 lists some examples of E_s values.[8] Note that the E_s values of hydroxy and methoxy are identical, thus it can be seen that E_s values describe the minimum steric effect of a substituent.

Table 3.2 Values of various steric substituent constants and aromatic π values.

Substituent	E_s	L	B1	B5	0.1 MR	π
H	0.00	2.06	1.00	1.00	0.10	0
Cyano	−0.51	4.23	1.60	1.60	0.63	−0.24
Fluoro	−0.55	2.65	1.35	1.35	0.10	0.14
Hydroxy	−0.55	2.74	1.35	1.93	0.28	−0.67
Methoxy	−0.55	3.98	1.35	3.07	0.79	−0.35
Amino	−0.61	2.78	1.35	1.97	0.54	−0.42
Chloro	−0.97	3.52	1.80	1.80	0.60	0.68
Bromo	−1.16	3.82	1.95	1.95	0.89	0.92
Methyl	−1.24	2.87	1.52	2.04	0.57	0.58
Nitro	−2.52	3.44	1.70	2.44	0.74	−0.17

E_s^c values correct E_s values for the possibility of hyperconjugation.[18] In this equation [equation (3.7)], n represents the number of α-hydrogen atoms.

$$E_s^c = E_s + 0.306(n - 3) \tag{3.7}$$

Although E_s and E_s^c values describe steric hindrance of a substituent to the attack of a reagent on a nearby group, they do not necessarily describe the steric hindrance of a substituent to the binding of a substance to a biomolecule.

Taft recognized that the substituent effects in some reactions required both electronic and steric factors [equation (3.8)].

$$\log(k/k_0) = \rho^* \sigma^* + dE_s \tag{3.8}$$

It is this separation of electronic and steric effects that prompted the Pomona group to propose that a substituent constant for hydrophobicity should be included for biological QSAR.[19]

3.2.2.2 *Steric Constants Calculated from the 3D Structures of the Substituents*

There are two problems with E_s values: (1) Some substituents cannot be studied because the resulting esters are too unstable for the relevant measurement, and (2) They may not represent the steric effect for a biological system.

To solve the first problem, Charton recognized that E_s values are related to the size of the substituent and that for symmetric substituents, E_s can be calculated from the radius.[13] Kutter and Hansch modified this idea to use the average of the minimum and maximum radii of the group.[20]

As a way to solve both of these problems in using E_s to correlate biological properties, Verloop and co-workers expanded the approach of using the 3D structure of the substituent. This is implemented in the STERIMOL computer program.[21] They originally calculated five characteristic steric properties of a substituent from the van der Waals surface of the molecules. For this they used the atomic and bond properties that are used to construct the CPK physical models that were popular at the time. Later they decided that three descriptors were sufficient:

L: the length of the substituent to the van der Waals surface along the axis of the bond between the first atom of the substituent and the attachment point to the common core.

B1: the minimum width to the van der Waals surface of the substituent, measured at 90° to the axis of the bond to the common core.

B5: the maximum width to the van der Waals surface of the substituent, measured at 90° to the axis of the bond to the common core.

The examples of STERIMOLvalues[8] shown in Table 3.2 suggest that although E_s is roughly correlated with B1, these parameters are in fact independent.

3.2.2.3 *Molar Refractivity*

The inadequacy of steric descriptors derived from substituent effects on organic reactions led the Pomona group to revisit the classical studies of Pauling and Pressman[22] and those of Agin.[23] These researchers had shown that molar refractivity (MR) is a surprisingly good descriptor of the bulk of a substituent. Further analysis showed that MR is a linear function of the STERIMOL properties [equation (3.9):[24]

$$0.1\,\mathrm{MR} = 1.08(\pm 0.22)\,\mathrm{B1} + 0.19(\pm 0.11)\,\mathrm{B5} + 0.38(\pm 0.08)\,\mathrm{L} - 2.79(\pm 0.47) \qquad (3.9)$$

$$R^2 = 0.834,\ n = 101,\ s = 0.388$$

In this equation, the values in parentheses are the 95% confidence intervals of the fitted constant. MR has become a popular descriptor of bulk or steric effects. Typical values are noted in Table 3.2.[8]

In terms of intermolecular interactions, MR is a measure of the ability of a substituent to participate in dispersion interactions. Hence a positive coefficient of MR in a regression equation suggests dispersion bonding with the substituent, and a negative coefficient suggests a steric interference with binding.

3.2.3 Hydrophobic Properties of Molecules

3.2.3.1 *What is Hydrophobicity?*

Hydrophobic molecules are, in general, poorly soluble in water but more soluble in many non-polar solvents. Hydrophobicity refers to the property of molecules whereby they tend to associate with each other rather than with water, hence the term originates from "water hating". This association cannot be explained by ion-pairing or intermolecular hydrogen bonds between the molecules, but is instead explained by their tendency to avoid water. Thermodynamic measurements have led to a model in which the driving force of a hydrophobic interaction is the tendency of a hydrophobic substance dissolved in water to disrupt the strong hydrogen bonding network of the water. The result is that water repels a hydrophobic substance.[25,26] For alkanes, the strength of the hydrophobic interaction is proportional to the number of water molecules that would surround the alkane molecule in water.[25]

The observation of the hydrophobic effect on biological properties dates back at least a century, to the studies of the relationships between hydrophobicity and anesthetic potency conducted by Overton[27] and Meyer.[28] Although ignored by organic chemists, the subject remained an active interest of physical and biological chemists until the time of the original QSAR publication. Biochemists postulated that because membranes are hydrophobic, in order to transverse a membrane a molecule must itself be hydrophobic.[29] Protein chemists also realized that the stability of the 3D structure of proteins cannot be explained by hydrogen bonding, beautiful though α-helixes and β-sheets may be, because hydrogen bonds between residues within a protein are probably not much more stable than hydrogen bonds between the same residues and the water in which the protein is dissolved. Hence hydrophobic interactions must account for at least part of the stability of the 3D structures of proteins.[30]

In spite of this interest of physical and biochemists, prior to the work of Hansch and Fujita there appears to have been no effort to devise a scale of hydrophobicity analogous to the Hammett scale for electronic effects, or the Taft scale for steric effects. Part of this reluctance was due to the selection of the relevant solvent system for such a scale. In addition, biochemists who were interested in hydrophobic stabilization of protein structures needed only to consider the few hydrophobic amino acid side chains.

*3.2.3.2 Log*P *versus Log*D

A partition coefficient measurement involves the equilibration of a solute between an aqueous and an organic phase.[31] A compound that is ionizable will be present a combination of ionic and neutral species, determined by its pK_a values(s) and the pH of the aqueous phase. Hence, the equilibrium of a solute between the aqueous and organic phases will depend on pH in a manner that is dependent on the structure (and hence properties) of the solute. The measurement of such an equilibrium results in a distribution coefficient, which is defined as the ratio of the total concentrations of all species of the compound in the non-polar to the polar phase. The log of this is the log$D$ value. Because it is usually assumed that charged species do not partition to the non-aqueous phase, log$D$ values are lower than log$P$ values for the same compound. This is depicted in Figure 3.6.

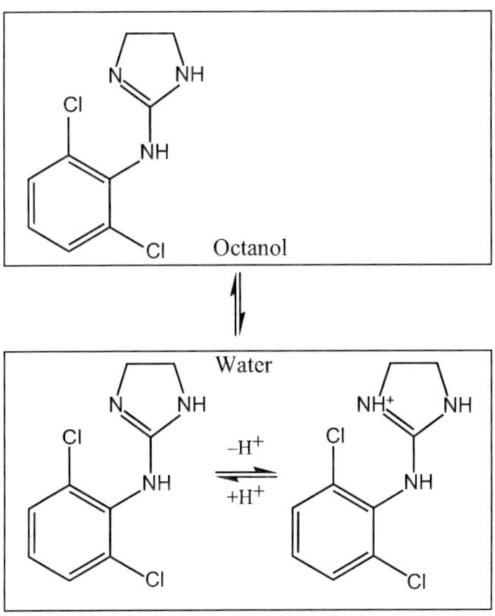

Figure 3.5 The model for the relationship between log$D$ and log$P$ for an amine, clonidine.

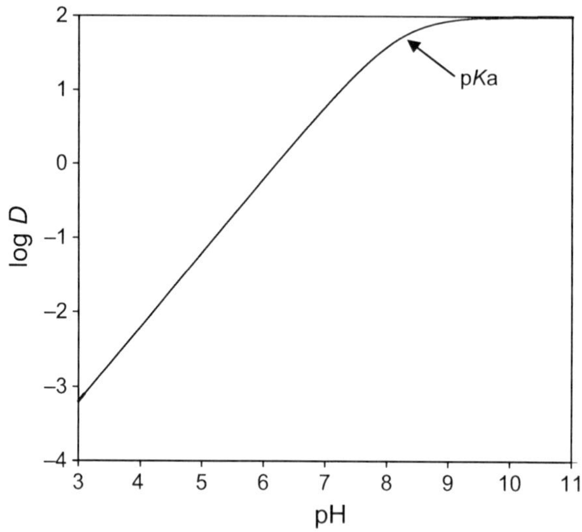

Figure 3.6 The relationship between $\log D$ and pH for an amine with a pK_a of 8.2 and a $\log P$ of 2.0.

The relationship between $\log D$ and $\log P$ is given by equation (3.10).

$$\log D_{\mathrm{pH}} = \log P + \log(1 - \alpha)_{\mathrm{pH}} \tag{3.10}$$

In equation (10), the term $(1 - \alpha)$ is the fraction of the molecules that are not protonated at the pH of interest. For a mono-basic compound it is calculated according to equation (3.11):

$$(1 - \alpha)_{\mathrm{pH}} = \frac{1}{1 + \frac{[H^+]}{K_a}} \tag{3.11}$$

To use a partitioning experiment to measure only the hydrophobicity of a solute, one adjusts the pH of the aqueous phase so that only the neutral form of the solute is present. This is typically pH 1 for acidic compounds and pH 12 or 13 for basic compounds. The result is a $\log P$ value. Clearly there is no such pH at which zwitterionic compounds exist in the uncharged form; in such cases the measurements might be made on neutral analogs.

3.2.3.3 Octanol as a Reference Solvent

Although Meyer and Overton used olive oil for their measurements of hydrophobicity, Hansch and Fujita rejected it because the composition varies from lot to lot. Instead they chose to do the measurements in 1-octanol, which is available commercially in pure form. Octanol has the advantage over pure hydrocarbons in that drug-like compounds are soluble enough in it that one can actually measure a partition coefficient.[32,33] In addition, if substances with polar functionality do partition into a hydrocarbon, they frequently carry along a water molecule of solvation. This complicates the interpretation of the results in terms of pure physical chemistry. However, because it contains a hydroxyl group, octanol can form hydrogen bonds with solutes; the result is that an octanol–water log$P$ is a composite of hydrophobicity and hydrogen bonding. Nonetheless, octanol–water log P measurements have become the benchmark for studies of hydrophobicity.

3.2.3.4 Relationships Between Partition Coefficients in Different Solvents

Recalling that hydrophobicity depends on the properties of water and not that of the non-polar phase, it is not surprising that there is a correlation between the relative log$P$ in octanol and that in other solvents. For various alcohols there is a 0.99 correlation, with only the intercept changing.[34] However, if the non-polar phase does not contain a hydroxyl group, then the log$P$ values of hydrogen bond donor and hydrogen bond acceptor solutes fall on different lines.[35] In fact, one experimental measure of hydrogen bond strength can be derived from the difference between the octanol–water log$P$ and the cyclohexane–water log$P$.[36] In addition, the relationships between log$P$ values of different solvents support the use of various HPLC methods to measure relative log$P$ values.

3.2.3.5 Hydrophobic Substituent Constants

The introduction emphasized that the QSAR breakthrough included considerations of the relative octanol–water log$P$ values of the compounds. Early measurements by the Pomona group revealed, as expected, that a particular substituent changes the octanol–water log$P$ by a constant magnitude, independent of the structure of the parent molecule. By analogy with the Hammett equation, the hydrophobic substituent constant π was defined according to equation (3.12).

$$\pi_X = \log P_{(R\text{-}X)} - \log P_{(R\text{-}H)} \tag{3.12}$$

Hundreds of π values were determined for both aromatic and aliphatic systems.[8] Table 3.2 shows some examples.[8]

3.2.3.6 Approaches to Calculating LogP

Although π values are useful for datasets that contain analogs of a parent structure, it is also of interest to develop QSARs with sets of more diverse structures. In the early 1970s, Nys and Rekker showed that to calculate an octanol–water log*P* from the structure of the molecule it wasn't necessary to start with the observed log*P* of a parent structure.[37–39] Their method starts with the hydrophobicity (*f*) of approximately 50 individual atom or group types and then applies correction factors for bond types, proximity effects, *etc*. The *f* values and correction terms are fitted by regression analysis to observed log*P* values. The results are embodied in the computer program, ProLogP.[40]

Subsequently, Leo and Hansch developed their own method to calculate log*P*, which has been programmed into CLOGP.[41] It uses many more atom and functional group types. The Leo *f* values and correction terms are assigned manually.

There are now many computer programs that calculate octanol–water log*P* from molecular structure.[42] They are usually based on a regression analysis fitting of observed log*P* values to a set of molecular descriptors, for example: counts of atoms of different atomic numbers,[43] counts of atoms of a specific atom type (distinguishing sp3, sp2, and sp1 carbon, *etc.*)[44,45] or substructures.[46,47]

We have correlated the antibacterial potency of alkyl esters of erythromycin (Figure 3.7), with the octanol–water log *P*, calculated with the aid of measurements on a few compounds.[48] Equation 3.13 describes the results.

$$\log(1/C) = 6.21 - 1.25 \log P \qquad (3.13)$$

$$R^2 = 0.79,\ s = 0.24,\ n = 28.$$

This indicates that less hydrophobic analogs might have increased potency.

Erythromycin A, R = OH
Erythromycin B, R = H

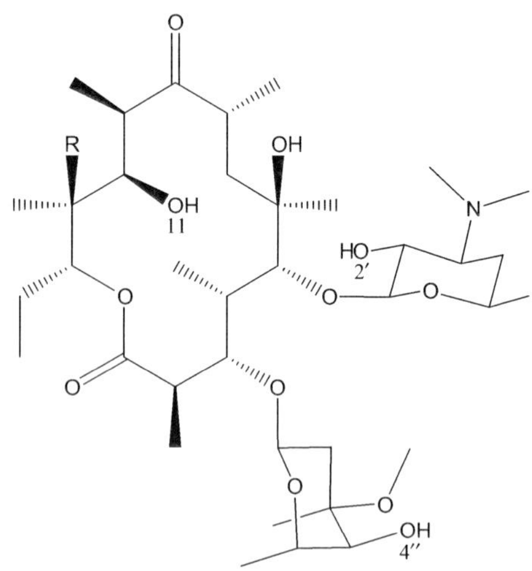

Figure 3.7 Structure of erythromycin.

3.2.4 2D Descriptors for Hydrogen Bonding

3.2.4.1 Substituent Effects on the Hydrogen Bonding Capability of a Parent Molecule

As discussed above, Hammett constants describe the effect of substituents on the transfer of a proton from the molecule to water (acidity) or to the molecule from water (basicity). However, ionization is just the extreme value in the continuum of the partial transfer of a proton in a hydrogen bond donating or accepting complex. Accordingly, Hammett constants also describe the influence of a substituent on the strength of a hydrogen bond to the parent molecule.[49]

3.2.4.2 Hydrogen Bonding Capability of Whole Molecules

Extensive experimental investigations of the factors that govern solvation have produced values for the hydrogen bonding donating and accepting ability of substituents.[50] The overall hydrogen bond donating or accepting ability of the molecule is the sum of the respective substituent constants. The complete solvatochromatic equation also includes terms for molecular volume and dipolarity-polarizability.[51]

The polar surface area of a molecule is considered a surrogate for the hydrogen bonding capability of the molecule. It is defined as the surface area of a molecule adjacent to a nitrogen or oxygen atom, or to a hydrogen atom attached to a nitrogen or oxygen atom.[52] In this definition, the relevant surface areas are calculated from a 3D structure of the molecule. However, the 3D polar surface area can be explained by an equation based on sub-structures present in the molecules.[53] The resulting value is the topological polar surface area (TPSA). The advantage of TPSA is that it is rapidly calculated and does not depend on the conformation of the molecule. Note that this calculation assumes that the difference in hydrogen bonding ability of different substituents depends only on the amount of surface area exposed by the group and not on the differences in the partial atomic charges of the substituents.

A recent parameterization for the hydrogen bonding properties of a molecule is to tabulate the number of hydrogen bond donors as the sum of the number of nitrogen or oxygen atoms that are bonded to a hydrogen atom, and to tabulate the number of nitrogen or oxygen atoms as a descriptor of the hydrogen bond accepting ability.[54] As with TPSA, this number is quickly calculated.

3.2.5 Other Molecular Descriptors

The electronic, steric, and hydrophobic descriptors described above are rooted in the physical chemistry of intermolecular interactions. It is also possible to calculate molecular descriptors that are not based on physical chemistry, but rather are just a consideration of the 2D structure of the molecule. Because the physical properties may be calculated from the 2D structure of the molecule, it follows that the structure contains this information and that consideration of structure alone should give a reliable QSAR. The problem is that it may be impossible to interpret such a model in terms of physical chemistry.

3.2.5.1 Substructure Descriptors

At approximately the same time as the original QSAR publications appeared, Free and Wilson published a regression analysis that quantified traditional SAR thinking—the constancy of the contribution to potency of a particular substituent at a particular position on a parent molecule.[55] The contribution is assumed to be independent of the effects of varying other substituents at other positions. For example, if substitution of a hydrogen by a chlorine atom increases potency 3-fold in one analog, it is assumed that such a substitution at this position will always increase potency 3-fold. However, it is also assumed that the contribution to potency of a particular substituent does depend on the position of substitution. These assumptions result in an equation that has, for each position of substitution, (n – 1) terms, where n is the number of substituents at that position. Hence the number of terms in the equation for p positions of substitution is given by equation (3.14).

$$X = \sum_{i=1}^{i=p} (1 - n_i) + 1 \tag{3.14}$$

The data is fitted by multiple regression analysis using a computer. The Free-Wilson equation for the erythromycin esters is shown in equation (3.15).[56]

$$\begin{aligned} \log(1/C) = 2.78 + 0.21A - 0.03FO2 - 0.17AC2 - 0.27FO4 - 0.69AC4 \\ - 0.76PR4 - 0.09FO11 - 0.56AC11 - 0.66PR11 \end{aligned} \tag{3.15}$$

$$R^2 = 0.986,\ s = 0.072,\ n = 28$$

In equation (3.15), A refers to a hydroxyl group at position 12, FO to a formyl ester, AC to an acetyl ester, PR to a propionyl ester, and the number to the position of substitution. The low value of s suggests that this relationship might over-fit the data because it is much lower than the standard deviation of replicate measurements.

The Free-Wilson method has the advantage that one does not need to estimate physical property descriptors for the compounds. This property also makes it difficult, if not impossible, to forecast the bioactivity of untested substituents. Moreover, the assumption of a constant effect on the potency of a particular substituent at a particular position breaks down in the case of a non-linear relationship of potency with log P.

If the substituents at a particular position of a molecule differ from each other in a qualitative way, for example if a particular group is esterified, then one may choose to include this in the matrix of properties, with an indicator variable set to 1.0 (if the property is present), or 0.0 (if it is not). For example, we obtained a better fit of the antibacterial activity of the erythromycin esters when we also included indicator variables A for the presence of a hydroxyl group at position 12, D4 for esterification of the 4″ hydroxyl group, and D11 for the esterification of the 11 hydroxyl group, as shown in the figure.[48]

$$\log(1/C) = 6.89 - 1.36\log P - 0.36A - 0.29D4 - 0.17D11 \quad (3.16)$$

$$R^2 = 0.95,\ s = 0.127,\ n = 28.$$

The coefficients of D4 and D11 suggest that there is more of a penalty for esterifying the 4″ hydroxyl than the 11 hydroxyl.

The assumption of a constant contribution to activity of a particular substituent is embodied in modern computational off-shoots of QSAR. For example, one report described a decision tree that separates active from inactive molecules by using the presence or count of pre-defined substructures as descriptors of the molecules.[57]

3.2.5.2 Topological Descriptors

The characteristics of the structure diagram of a molecule support the calculation of a number of descriptors. The most common of these are the various topological descriptors of Kier and Hall,[58] calculated with the computer program, MOLCONN-Z.[59] For example, the Molecular Connectivity Chi indices (${}^{m}X_i$) and the alternative Kappa indices (${}^{1}\kappa$, ${}^{2}\kappa$, and ${}^{3}\kappa$; and ${}^{1}\kappa_\alpha$, ${}^{2}\kappa_\alpha$, and ${}^{3}\kappa_\alpha$) encode information about the size and shape of the molecules, whereas the electrotoplogical state, *E*-State, indices (S_i) encode hydrogen bonding and other electrostatic and shape properties. It can be difficult to design molecules that are forecast to be as, or more, potent than the molecules used to develop a model based on topological indexes. This is called "the inverse QSAR" problem.[60]

3.3 Properties Calculated from the 3D Conformation of the Molecule

Because the interaction between a small molecule and a macromolecule occurs in 3D, there has been much effort to treat structure–activity relationships in 3D.[61–63] Information on the 3D structure of the binding site may be available from X-ray crystallography,[64] NMR,[65] or homology modeling,[66] or it might be inferred from the structure–activity relationships of the ligands. The properties of the binding site and the ligand are assumed to be complementary, the result being that ligand properties are often calculated in the space around the molecules.

One advantage of 3D QSAR methods is that once the conformation of the molecules is selected, it is straightforward to calculate the properties.

One doesn't need to decide which Hammett σ value to use, or try to estimate a π value for a substituent that isn't in the tables. Another advantage is that they provide a more appropriate description of steric effects of molecules.

3.3.1 Selection of the Conformation

Molecules exist as flexible objects in 3D space. Therefore, although many properties can be calculated and understood from the structure diagrams of molecules, it is possible that using 3D properties directly would provide more insight into structure–activity relationships. A complication here is that to calculate 3D properties one must decide which conformation of the molecules to use.[67] If there are structurally diverse molecules in the dataset, then one must usually also decide how to align the various molecules.

If one has the 3D structure of the binding site in the target biomolecule, then molecular graphics or docking methods[68] might suggest the bound conformations and how to compare the different molecules in 3D. However, because the hydrogen atoms are often not seen in a protein structure, the resolution of a crystal structure may not be sufficient to unambiguously assign a conformation of a molecule, for example, the rotation of terminal hydroxyl or amide groups. If the crystal structure of several analogs has been determined, for 3D QSAR it is common practice to refine the structures of the ligands so that the atoms in common between them are superimposed.

In the absence of the structure of the macromolecular target, for methods such as CoMFA,[69] Catalyst,[70] Compass,[71] and Phase,[72] the user or the computer programme uses potential energy fields or pharmacophore[73] notions to select the conformations of the molecules. For this purpose, the goal is to select conformations such that there is a correspondence of the distances between features common to all or most of the analogs. Typically, one would consider features such as permanently or potentially charged groups, hydrogen bond donors, hydrogen bond acceptors, and centers of aromatic rings. When the conformations are superimposed over these features, a further selection may try to maximize the overlap volume.[67] In the Topomer CoMFA method, the algorithm determines the conformation and overlap of each set of substituents.[74]

3.3.2 3D Property Calculation

For CoMFA and related methods, the descriptors are the calculated interaction energies of the molecule with various probe atoms, at hundreds of points arranged in a lattice surrounding the molecule. For steric energies, a methyl group is used as the probe; for electrostatic energies, the probe is a proton or a CH_3^+ group at those same points.[69] To calculate the electrostatic energies one must assign partial atomic charges to the atoms. Although the Gasteiger-Huckel method is frequently used for this purpose,[69] charges from quantum chemical calculations give better results if a strong electronic effect is involved.[75] It is also possible to use a hydrogen bonding or hydrophobic probe to more completely describe the hydrophobic properties of ligands.[76,77] Because steric energies grow exponentially positive as two atoms approach, it is necessary to truncate the steric energy at some "reasonable value", for example 10 kcal mol^{-1}. Similarly, very high electrostatic energies result for lattice points closer than the van der Waals radius to the atom, *i.e.* inside the molecule. Thus such energies are not considered in a CoMFA analysis.

The points need not be on a lattice, such as is used in CoMFA. The CoMSIA method places the points inside the union surface of the superimposed molecules.[78] The properties are not interaction energies, but rather the similarity of the molecular field of the ligand at a particular point with that of a probe atom. For this purpose steric, electrostatic, hydrophobic, and hydrogen bond acceptor and donor properties are considered. In the Compass method, probe atoms are placed along 162 equally spaced rays 2Å from the union surface of the molecules in their initial superposition.[71] The properties for analysis are the distance from the surface of the molecule to each of the probe atoms.

For Catalyst and Phase,[72] the descriptors are the degree to which each of the common 3D features are occupied, with a penalty for a less-than-optimal distance or angle to the feature.

We generated a CoMFA fit of the erythromycin esters, adding the substituents to the crystal structure of the unsubstituted compound. Only steric fields were significant. Table 3.5 (below) summarizes the statistics.

3.4 The Relationship between Calculated 2D and 3D Molecular Properties

Because both 2D and 3D descriptors are calculated from the molecular structure, they are not independent.[79] For example, the pK_a values of benzoic acids form the basis of the 2D Hammett σ values, but they are equally well described by CoMFA electrostatic fields[75] and by partial atomic charges, calculated by quantum chemical methods.[80] The relative hydrophobicity of molecules can be described by calculated octanol–water $\log P$ values, or by changes in the hydrogen bonding and steric potentials surrounding the molecules.[79,81] The most interesting results involve E_s values, proposed to contain only steric properties of substituents. Although a CoMFA analysis shows that steric fields are the predominant determinant of E_s, there is a statistically significant contribution of electrostatic fields also.[79] Thus, we can reject the original assumption of the absence of electronic effects on the acid-catalyzed hydrolysis of esters. As a result of the relationship between 2D and 3D descriptors of molecules, the choice of which to use for a QSAR will depend on which is easier to use or interpret for a particular application. It is also possible to use combinations of descriptors calculated from both the 2D and 3D representations of the molecules.

3.5 Computational Methods to Relate Molecular Properties to Bioactivity

It is necessary to use some mathematical method to relate the biological properties of molecules to their molecular properties. Such a method might be rooted in classical statistics or in the newer methods of artificial intelligence or data mining, an area of active research.

3.5.1 Statistical Approaches

These methods find the "best" estimate of the coefficients of an equation by minimizing the sum of the squares of the deviations (errors) between the observed and the fitted log $(1/C)$, or log of relative biological potency in the case of QSAR.[82] They generally assume that the errors are normally distributed. The validity of a statistical fit is usually evaluated by calculating the probability that the overall fit is significantly better than simply using the mean value and, if more than one descriptor is used, each descriptor contributes significantly to the quality of the fit.

3.5.1.1 Multiple Regression

As discussed previously, in multiple regression an equation in several unknowns is fitted. The fit is summarized by the following calculated values:[82]

R^2, which is the fraction of the variance in the dependent variable, log (1/C), that is explained by the overall equation,

s, which is the standard deviation of the difference between the fitted and observed log $(1/C)$, and the probability that each of the fitted coefficients is significantly different from zero.

In addition, the R^2 and s values can be evaluated to determine whether they are significantly different from zero.

Usually one does not know if each potential descriptor will be important for the fit, so the usual procedure is to calculate all possible equations in order to find the ones in which all terms are significant. If there are many possible equations then stepwise strategies may be used. At the end of the process the user will select one or more "best" equations, based on the overall quality of the fit and the contribution of each term to the final fit. The latter criterion is necessary because it sometimes happens that one term is statistically significant but increases R^2 and decreases s so little that how necessary it is to include it in later predictions is questionable.

3.5.1.2 Classification Methods

QSAR can also be used to investigate the properties that distinguish classes of molecules, active from inactive molecules or agonists from antagonists, for example.[83]

Table 3.3 The confusion matrix of the results of a discriminant analysis of the monoamine oxidase inhibitor dataset.[84]

	Fitted inactive	*Fitted active*	*Total*	*% Recovered*
Actual inactive	1444	91	1535	94
Actual active	16	72	88	82
Total	1460	163	1623	
Correct	1444	72	1516	
% Correct	99 (specificity)	44 (sensitivity)	93	

Table 3.4 The confusion matrix of the results of a recursive partitioning analysis with boosting of the monoamine oxidase inhibitor dataset.

	Fitted inactive	*Fitted active*	*Total*	*% Recovered*
Actual inactive	1425	110	1535	93
Actual active	5	83	88	94
Total	1430	193	1623	
Correct	1425	83	1508	
% Correct	99	43	93	

With discriminant analysis QSAR the same principles are used in the calculation, except that instead of fitting a line through the multidimensional space of the predictors and activities, one is fitting a plane that separates the two classes.[84–86] For the two-class case one can use a multiple regression program and simply substitute one value (for example, 1.0) for the active class and another (for example, 0.0 or −1.0) for the inactive class. If the dataset contains more than two classes, or if they are of greatly different sizes, then more elaborate calculations are necessary. The result of a discriminant analysis is a confusion table, also called a contingency or truth table. In this table, the first two rows refer to the observed class of the compounds and the first two columns the predicted class. The fewer the off-diagonal counts, the better the function. An example is shown in Table 3.3.[83]

Recursive partitioning also evaluates properties that distinguish classes.[57,83,87] It proceeds in a stepwise tree manner. The first step identifies the single predictor property that best separates the classes and divides the dataset on this basis. The second and succeeding steps investigate the resulting individual subsets in order to identify the single predictor property that best separates the classes within it. The process is stopped when the subsets are purely one class or another, or when the subsets are some predetermined minimum size. A confusion table can be used to summarize the quality results and a set of rules or a decision tree can be used to show the contribution of each property to the classification. More advanced methods generate multiple decision trees to identify the variables important for the classification, or iteratively calculate subsequent trees that improve the classification of outliers from the original main tree (known as boosting).[88] Table 3.4 shows it for the same dataset analyzed with recursive partitioning.[83]

3.5.1.3 Principal Component Analysis

The matrix of molecular descriptors frequently contains columns of values that are correlated with each other. This might arise from a poorly designed set in which there is a correlation between the MR and π of a substituent, or from descriptors that are intrinsically correlated, such as TPSA and the counts of nitrogen and oxygen atoms and the protons bonded to them. Principal component analysis considers the matrix of properties and extracts orthogonal components that contain a contribution (loading) from one or more of the properties.[82,89,90] The first principal component explains most of the variance in the descriptor matrix and each succeeding one, the largest fraction of that remaining. A useful result of a principal component analysis is the listing of the fraction of variance that is represented by each component. By examining these values one can estimate the number of significant properties represented by the data. Plots of the loadings on the principal components, or the scores of the molecules on the components, can reveal whether the data is clustered or more evenly distributed in property space.

3.5.1.4 Partial Least Squares

The molecular descriptors for a CoMFA analysis number in the hundreds or thousands, even for datasets of twenty or so compounds. A multiple regression equation cannot be fitted for such a dataset. In such cases, Partial Least Squares (PLS) is the appropriate method. PLS unravels the relationship between log $(1/C)$ and molecular properties by extracting from the data matrix linear combinations (latent variables) of molecular properties that best explain log $(1/C)$.[82,91] Because the individual properties are correlated (for example, steric properties at adjacent lattice points), more than one contributes to each latent variable. The first latent variable extracted explains most of the variance in log $(1/C)$; the second the next greatest degree of variance, *etc.* At each step R^2 and s are calculated to help one decide when enough variables have been extracted—the maximum number of extracted variables is found when extracting another does not decrease s substantially. Cross-validation, discussed in Section 3.5.3, is commonly used to decide how many latent variables are significant. For example, Table 3.5 summarizes the CoMFA PLS analysis

Table 3.5 Statistics of the CoMFA analysis of the potency of erythromycin esters, $n = 28$.

	Fitted		*Cross-validated*	
Latent Variables	*Standard error*	R^2	*Standard error*	q^2
1	0.160	0.903	0.192	0.860
2	0.129	0.939	0.153	0.914
3	0.121	0.948	0.150	0.921
4	0.109	0.960	0.140	0.934
5	0.098	0.969	0.134	0.942
6	0.069	0.985	0.108	0.964

of the erythromycin esters. Note the gradual increase in R^2 and q^2 and decrease in standard error as more latent variables are included in the model.

The result of a PLS analysis includes the contribution of each input property to the extracted latent variables. Because in CoMFA the input properties are located in 3D space, one can draw contour plots that enclose the regions in space that enclose these contributions.

3.5.2 Machine Learning and Other Non-Statistical Approaches

A later chapter will discuss these methods in more detail. For example, support vector machines and traditional neural networks are analogs of multiple regression or discriminant analysis that provide more flexibility in the form of the relationship between molecular properties and bioactivity.[83,92] Kohonen neural nets are a more flexible analog to principal component analysis.[83,93,94] Various Bayesian approaches are alternatives to the statistical methods described earlier.[92] A freely available program offers many of these capabilities.[95,96]

3.5.3 Methods to Assess the Predictivity of a QSAR

Although statistical tests can be used to rule out potential structure–activity relationships, they do not guarantee that models that pass such a test will have predictive value.[82]

One main strategy to test the predictivity of a relationship involves leaving some of the observations out of the modeling process and using the model to predict their potencies.[82] In the leave-one-out (LOO) cross-validation process, each compound is deleted once, a model is developed from the remaining compounds, and the potency of the left-out compound is predicted. The deviation of the observed from this predicted is then used to calculate q^2, analogous to R^2. Other cross-validation strategies leave out more of the data on each round of calculation.

With larger datasets one frequently uses a fraction of the molecules as a training set while setting aside a fraction of the data to use as a test set for any models derived from the training set. The limitation of this type of strategy is that if the training and test sets contain molecules of very similar properties and biological activity ("twins") then predicting the potency of one twin when the other is in the dataset will result in an accurate prediction, but it might not be generalizable.

One problem with developing QSARs can be that too many possible equations have been examined, with the result that a chance correlation has been accepted as a true correlation.[97] Y-scrambling is a method designed to uncover such a problem.[98] To perform a Y-scrambling[82] one starts with the original data matrix, which includes all of the molecular properties considered. The log $(1/C)$ values are then scrambled using a random number generator to assign their order. These scrambled log $(1/C)$ values are then used in the modeling

Table 3.6 Standard deviations of prediction errors of erythromycin QSAR models.

Analogues included in the model	*Standard error of fit*	*Number of predictions*	*Standard deviation of predictions*
QSAR, Equation (15), Alkyl esters, $n = 28$	0.13	44	0.911
QSAR Alkyl esters plus six more polar compounds	0.13	38	0.429
Alkyl esters, 1 component CoMFA model	0.16	44	0.512
Alkyl esters, 2 component CoMFA model	0.13	44	0.534
Alkyl esters, 6 component CoMFA model	0.07	44	0.564

process. This process is usually repeated at least a hundred times. If the relationships with the scrambled log $(1/C)$ values fit as well as when the true values are used, then the fits with the true values are suspicious.

Table 3.6 summarizes the ability of the various models to predict the potency of erythromycin analogs that were not used in the derivation of the model.[56] First, note that the linear model in log P has very poor predictivity, but adding a few more polar compounds to the model yielded the one with the best predictivity. Of the CoMFA models, the simplest one had the best predictivity.

3.6 The Contrast Between QSAR and Traditional SAR

With this background in mind, one can question how QSAR compares with traditional medicinal chemistry thinking. Medicinal chemists are responsible for deciding which new molecules will be synthesized. Most often their background is in synthetic organic chemistry and their interpretation of structure–activity relationships reflect this experience. As a result, they are more likely to interpret the effect of structure changes on bioactivity in a qualitative sense, with only a passing thought of the effect of such a change on the physical properties of the molecule. In addition, synthesis to probe an SAR often involves exploring which changes in structure destroy activity, but the resulting inactive molecules are not useful in QSAR except in classification models. Once the main structural requirements for activity are established, extensive synthesis of analogs aims to find the "best" molecule. This "best" molecule may not be the most potent one, but rather one that optimizes an extensive set of properties: selectivity for the preferred target; *in vitro* and *in vivo* potency; patentability; ease of synthesis or low cost of the final product; favorable absorption, distribution, and metabolism; and lack of side or toxic effects. If this design process includes thoughts about physical properties, it is more likely to be the rules of thumb, such as the Rule of Five[54] or the Golden Triangle.[99]

Medicinal chemists who use the 3D structure of the macromolecule–ligand complex to design molecules narrow their focus to those parts of the ligand that are located in regions amenable to substitution. Again, the design is based on structure, not properties. Any changes to the parts of the ligand structure that are in contact with the macromolecular structure attempt to preserve the interactions, often without precise calculations of the changes in energetics. If an unoccupied or water-filled non-polar region is observed in the binding site, then hydrophobic substituents would be considered. In a similar manner, it is common to attempt to displace bound water molecules by adding hydrogen bonding groups to the ligand.

Thus it can be seen that the viewpoints of QSAR and traditional medicinal chemistry are not identical. In a successful collaboration between the two, both disciplines learn to understand the viewpoint of the other.

3.7 The Future of QSAR

This chapter illustrates some of the many advances in QSAR since the original publications in the early 1960s. Currently, there is a much greater awareness of the role of hydrophobicity in all aspects of the biological properties of molecules. The growing power of computers, both the hardware and the algorithms, suggests that new approaches to QSAR and larger-scale studies will prompt even more use of the method. The press for better metrics of the quality of the model should prompt new thinking in this area as well. The future of QSAR is bright!

References

1. C. Hansch, P. P. Maloney, T. Fujita, and R. M. Muir, *Nature*, 1962, **194**, 178.
2. C. Hansch and T. Fujita, *J. Am. Chem. Soc.*, 1964, **86**, 1616.
3. T. Fujita, J. Iwasa, and C. Hansch, *J. Am. Chem. Soc.*, 1964, **86**, 5175.
4. A. Leo, C. Hansch, and D. Elkins, *Chem. Rev. (Washington, DC, U. S.)*, 1971, **71**, 525.
5. A. Leo, C. Hansch, and D. Hoekman, 'Bio-Loom', 2008, url:http://biobyte.com/bb/prod/bioloom.html, date accessed: January 11, 2011.
6. T. Fujita, *Quant. Struct.-Act. Relat.*, 1997, **16**, 107.
7. Anon., 'Regulatory Use of (Q)SARs under REACH', 2009, url:http://echa.europa.eu/doc/press/webinars/regulatory_use_of_qsars_under_reach_doris_hirmann_echa.pdf, date accessed: January 11, 2011.
8. C. Hansch, A. Leo, and D. Hoekman, 'Exploring QSAR: Hydrophobic, Electronic, and Steric Constants', American Chemical Society, Washington DC, 1995.
9. L. Hammett, 'Physical Organic Chemistry', McGraw-Hill, 1970.
10. H. H. Jaffé, *Chem. Rev. (Washington, DC, U. S.)*, 1953, **53**, 191.

11. C. Hansch and A. Leo, in 'Exploring QSAR: Fundamentals and Applications in Chemistry and Biology', American Chemical Society, Washington, DC, 1995, p. 89.
12. R. W. Taft, in 'Steric Effects in Organic Chemistry', ed. M. S. Newman, Wiley, New York, 1956, p. 556.
13. M. Charton, *J. Am. Chem. Soc.*, 1969, **91**, 615.
14. M. Charton, *Prog. Phys. Org. Chem.*, 1971, **8**, 235.
15. J. Shorter, in 'Correlation Analysis in Organic Chemistry: an Introduction to Linear Free-Energy Relationships', Clarendon Press, Oxford, 1973, p. 32.
16. N. B. Chapman and J. Shorter, 'Advances in Linear Free Energy Relationships', Plenum, New York, 1972.
17. C. Hansch and A. Leo, in 'Exploring QSAR: Fundamentals and Applications in Chemistry and Biology', American Chemical Society, Washington DC, 1995, p. 1.
18. C. K. Hancock, E. A. Meyers, and B. J. Yager, *J. Am. Chem. Soc.*, 1961, **83**, 4211.
19. T. Fujita, *Current Contents*, 1986, 15.
20. E. Kutter and C. Hansch, *J. Med. Chem.*, 1969, **12**, 647.
21. A. Verloop, W. Hoogenstraaten, and J. Tipker, in 'Drug Design', vol 7, ed. E. J. Ariens, Academic Press, New York, 1976, p. 164.
22. L. Pauling and D. Pressman, *J. Am. Chem. Soc.*, 1945, **67**, 1003.
23. D. Agin, L. Hersh, and D. Holtzman, *Proc. Natl. Acad. Sci. U. S. A.*, 1965, **53**, 952.
24. C. Hansch and A. Leo, in 'Exploring QSAR: Fundamentals and Applications in Chemistry and Biology', American Chemical Society, Washington, DC, 1995, p. 80.
25. C. Tanford, 'The Hydrophobic Effect: Formation of Micelles and Biological Membranes', Wiley-Interscience, New York, 1973.
26. C. Hansch and A. Leo, in 'Exploring QSAR: Fundamentals and Applications in Chemistry and Biology', American Chemical Society, Washington DC, 1995, p. 97.
27. E. Overton, *Z. Phys. Chem.*, 1897, **22**, 189.
28. H. Meyer, *Archiv fur Experimentell Pathologie und Pharmakologie*, 1899, **42**, 109.
29. C. A. M. Hogben, D. J. Tocco, B. B. Brodie, and L. S. Schanker, *J. Pharmacol. Exp. Ther.*, 1959, **125**, 275.
30. W. Kauzmann, *Adv. Protein Chem.*, 1959, **14**, 1.
31. C. Hansch and A. Leo, in 'Exploring QSAR: Fundamentals and Applications in Chemistry and Biology', American Chemical Society, Washington DC, 1995, p. 118.
32. R. N. Smith, C. Hansch, and M. Ames, *J. Pharm. Sci.*, 1975, **64**, 599.
33. Y. C. Martin, in 'Quantitative Drug Design. A Critical Introduction', CRC Press, Boca Raton, FL, 2010, p. 66.
34. R. Collander, *Acta Chem. Scand.*, 1951, **5**, 774.
35. A. Leo and C. Hansch, *J. Org. Chem.*, 1971, **36**, 1539.

36. P. Seiler, *Eur. J. Med. Chem.*, 1974, **9**, 473.
37. G. G. Nys and R. F. Rekker, *Chimica Therapeutica*, 1973, **9**, 521.
38. G. G. Nys and R. F. Rekker, *Chimica Therapeutica*, 1974, 361.
39. R. Mannhold, R. F. Rekker, K. Dross, G. Bijloo, and G. de Vries, *Quant. Struct.-Act. Relat.*, 1998, **17**, 517.
40. R. Mannhold and K. Dross, *Quant. Struct.-Act. Relat.*, 1996, **15**, 403.
41. A. Leo, CLOGP, 2007, Biobyte Corporation.
42. Y. C. Martin, in 'Quantitative Drug Design. A Critical Introduction', CRC Press, Boca Raton, FL, 2010, p. 58.
43. I. Moriguchi, S. Hirono, Q. Liu, I. Nakagome, and Y. Matsushita, *Chem. Pharm. Bull.*, 1992, **40**, 127.
44. A. Ghose and G. Crippen, *J. Comput. Chem.*, 1986, **7**, 565.
45. R. Wang, Y. Gao, and L. Lai, *Perspect. Drug Discovery Des.*, 2000, **19**, 47.
46. G. Klopman, S. Wang, P. S. Anderson, and J. R. Huff, *J. Comput. Chem.*, 1991, **12**, 1025.
47. W. M. Meylan and P. H. Howard, *J. Pharm. Sci.*, 1995, **84**, 83.
48. Y. C. Martin, P. H. Jones, T. Perun, W. Grundy, S. Bell, R. Bower, and N. Shipkowitz, *J. Med. Chem.*, 1972, **15**, 635.
49. R. S. Drago and T. D. Epley, *J. Am. Chem. Soc.*, 1969, **91**, 2883.
50. M. H. Abraham and J. A. Platts, *J. Org. Chem.*, 2001, **66**, 3484.
51. M. Kamlet, J.-L. Abboud, M. Abraham, and R. Taft, *J. Org. Chem.*, 1983, **48**, 2877.
52. K. Palm, K. Luthman, A. L. Ungell, G. Strandlund, and P. Artursson, *J. Pharm. Sci.*, 1996, **85**, 32.
53. P. Ertl, B. Rohde, and P. Selzer, *J. Med. Chem.*, 2000, **43**, 3714.
54. C. A. Lipinski, F. Lombardo, B. W. Dominy, and P. J. Feeney, *Adv. Drug Delivery Rev.*, 1997, **23**, 3.
55. S. M. Free and J. Wilson, *J. Med. Chem.*, 1964, **7**, 395.
56. Y. C. Martin, in 'Quantitative Drug Design. A Critical Introduction', CRC Press, Boca Raton, Florida, 2010, p. 187.
57. D. M. Hawkins, S. S. Young, and A. Rusinko, *Quant. Struct.-Act. Relat.*, 1997, **16**, 296.
58. L. B. Kier and L. H. Hall, 'Molecular Structure Description: The Electrotopological State', Academic Press San Diego, CA, 1999.
59. L. Hall and L. Kier, Molconn-Z, 2007, Edusoft.
60. L. B. Kier, L. H. Hall, and J. W. Frazer, *Jour. Chem. Inf. Comput. Sci.*, 1993, **33**, 143.
61. H. Kubinyi, eds. '3D QSAR in Drug Design. Theory Methods and Applications', ESCOM, Leiden, 1993.
62. H. Kubinyi, G. Folkers, and Y. C. Martin, eds. '3D QSAR in Drug Design. Vol 2. Ligand-Protein Interactions and Molecular Similarity', ESCOM, Leiden, 1998.
63. H. Kubinyi, G. Folkers, and Y. C. Martin, eds. '3D QSAR in Drug Design. Vol. 3. Recent Advances', ESCOM, Leiden, 1998.
64. J. L. Martin, *Curr. Med. Chem.*, 1996, **3**, 419.

65. K. Wuthrich, 'NMR Studies of Structure and Function of Biological Macromolecules', 2002, url:http://nobelprize.org/nobel_prizes/chemistry/laureates/2002/wutrich-lecture.pdf, date accessed: January 5, 2011.
66. T. Schwede, 'Homology Modeling', 2007, url:http://www.ch.embnet.org/CoursEMBnet/Pages3D07/documents/3DHomology.pdf, date accessed: January 5, 2011.
67. Y. C. Martin, in 'Quantitative Drug Design. A Critical Introduction', CRC Press, Boca Raton, FL, 2010, p. 31.
68. G. Warren, L., C. W. Andrews, A. M. Capelli, B. Clarke, J. La Londe, M. H. Lambert, M. Lindvall, N. Nevins, S. F. Semus, S. Senger, G. Tedesco, I. D. Wall, J. M. Woolven, C. E. Peishoff, and M. S. Head, *J. Med. Chem.*, 2006, **49**, 5912.
69. R. D. Cramer III, D. E. Patterson, and J. D. Bunce, *J. Am. Chem. Soc.*, 1988, **110**, 5959.
70. P. W. Sprague, *Perspect. Drug Discovery Des.*, 1995, **3**, 1.
71. A. N. Jain, T. G. Dietterich, R. H. Lathrop, D. Chapman, R. E. Critchlow Jr., B. E. Bauer, T. A. Webster, and T. Lozano-Perez, *J. Comput.-Aided Mol. Des.*, 1994, **8**, 635.
72. S. L. Dixon, A. M. Smondyrev, E. H. Knoll, S. N. Rao, D. E. Shaw, and R. A. Friesner, *J. Comp-Aid. Mol. Des.*, 2006, **20**, 647.
73. Y. C. Martin, in 'Comprehensive Medicinal Chemistry II', vol 4, ed. J. S. Mason, Elsevier, Oxford, 2007, p. 515.
74. R. D. Cramer III, R. D. Clark, D. E. Patterson, and A. M. Ferguson, *J. Med. Chem.*, 1996, **39**, 3060.
75. K. H. Kim and Y. C. Martin, *J. Org. Chem.*, 1991, **56**, 2723.
76. K. Kim, *J. Comput.-Aided Mol. Des.*, 1995, **9**, 308.
77. G. E. Kellogg, S. F. Semus, and D. J. Abraham, *J. Comput.-Aided Mol. Des.*, **1991**, **5**, 545.
78. G. Klebe, in '3D QSAR In Drug Design. Vol 3. Recent Advances', Kluwer Escom, Leiden, 1998, p. 87.
79. K. H. Kim, in '3D QSAR in Drug Design. Theory Methods and Applications', ed. H. Kubinyi, ESCOM, Leiden, 1993, p. 619.
80. P. Ertl, *Quant. Struct.-Act. Relat.*, 1997, **16**, 377.
81. K. H. Kim, *Curr. Top. Med. Chem.*, 1993, **1**, 453.
82. Y. C. Martin, in 'Quantitative Drug Design. A Critical Introduction', CRC Press, Boca Raton, FL, 2010, p. 131.
83. Y. C. Martin, in 'Quantitative Drug Design. A Critical Introduction', CRC Press, Boca Raton, FL, 2010, p. 254.
84. Y. C. Martin, J. B. Holland, C. H. Jarboe, and N. Plotnikoff, *J. Med. Chem.*, 1974, **17**, 409.
85. I. E. Frank, *Chemom. Intell. Lab. Syst.*, 1989, **5**, 247.
86. W. Albers, 'Discriminant Analysis', 2008, url:http://wwwhome.math.utwente.nl/~albersw/Discriminant%20Analysis.ppt, date accessed: May 6, 2008.
87. L. Breiman, 'Classification and Regression Trees', Chapman & Hall/CRC, Boca Raton, 1984.

88. C. Strobl, J. Malley, and G. Tutz, *Psychological Methods*, 2009, **14**, 323.
89. I. T. Jolliffe, 'Principal Components Analysis', Springer-Verlag, New York, 1986.
90. Anon., 'Principal Components and Factor Analysis', 2008, url:http://www.statsoft.com/textbook/stfacan.html, date accessed: January 11, 2011.
91. S. Wold, E. Johansson, and M. Cocchi, in '3D QSAR in Drug Design. Theory Methods and Applications', ed. H. Kubinyi, ESCOM, Leiden, 1993, p. 523.
92. I. H. Witten and E. Frank, 'Data Mining: Practical Machine Learning Tools and Techniques with Java Implementations', Morgan Kaufmann, San Francisco, 2005.
93. S. Anzali, G. Barnickel, M. Krug, J. Sadowski, M. Wagener, J. Gasteiger, and J. Polanski, *J. Comp-Aid. Mol. Des.*, 1996, **10**, 521.
94. J. Polanski, J. Gasteiger, M. Wagener, and J. Sadowski, *Quant. Struct.-Act. Relat.*, 1998, **17**, 27.
95. M. Hall, E. Frank, G. Holmes, B. Pfahringer, P. Reutemann, and I. H. Witten, *SIGKDD Explorations*, 2009, **11**.
96. 'Weka Wiki', 2011, url:http://weka.wikispaces.com/, date accessed: Jan 11, 2011.
97. J. G. Topliss and R. P. Edwards, *J. Med. Chem.*, 1979, **22**, 1238.
98. R. D. Clark and P. C. Fox, *J. Comput.-Aided Mol. Des.*, 2004, **18**, 563.
99. T. W. Johnson, K. R. Dress, and M. Edwards, *Bioorg. Med. Chem. Lett.*, 2009, **19**, 5560.

CHAPTER 4

The Development of Molecular Modelling Programs: The Use and Limitations of Physical Models

RICHARD A. LEWIS

Computer-aided Drug Discovery, Novartis Institutes for Biomedical Research, CH-4002, Basel, Switzerland

4.1 The Use and Limitations of Physical Models

Physical models are very widely used in molecular modelling and underpin our concepts of molecular recognition. These methods are also approximate. This concept of accuracy is at the core of this chapter; unfortunately computers work at a much higher level of precision, which can give the naive observer a false impression of how reliable the models are. The true representation of the Hamiltonian of an entire macroscopic system will always be beyond our grasp, so we settle for some simpler representations. The functional forms of equations that are commonly used to estimate energies are quite simple quadratic terms, initially to allow exact numerical treatment on slow computing resources (equation 4.1).

$$E = E_{\text{bond}} + E_{\text{angle}} + E_{\text{torsion}} + E_{\text{vdw}} + E_{\text{electrostatics}} \qquad (4.1)$$

With these simple models, we can estimate energies with limited accuracy. The first limitation arises from the experimental data that these models seek

RSC Drug Discovery Series No. 13
Drug Design Strategies: Quantitative Approaches
Edited by David J. Livingstone and Andrew M. Davis

Published by the Royal Society of Chemistry, www.rsc.org

to explain. Leaving aside the issue of data quality (a considerable source of error), most experiments return a macroscopic observation (for example a dielectric constant), whereas most models handle only a single microscopic instance (the energetics of one ligand interacting with one target). This gives a strong enthalpic slant to our thinking about molecular interactions. It has long been realised that, in the case of hydrogen-bonding, the number of bonds formed between the ligated partners in solution and after in complex is the same[2], but the quality of the bonds may vary, as may the entropic contribution. This balance is well illustrated by the work of Freire[3–5] on the balance between entropy and enthalpy observed in different inhibitors of HIV-1 protease. However, most scoring functions value such interactions too highly and without considering solvent (equation 4.2).

$$\text{Target}_{\text{aq}} + \text{ligand}_{\text{aq}} \leftrightarrow \text{target-ligand} + \text{aq-aq} \tag{4.2}$$

4.1.1 Development of Molecular Mechanics and Quantum Mechanics

Appreciation of the importance of the conformation of a molecule grew during the 1980s, driven by the increased availability of small molecule X-ray experimental data (curated in the Cambridge Structural Database[6]). Marshall and colleagues pioneered the concept of the bioactive conformation through their development of the "active analogue" approach.[7] The history of the development of molecular mechanics (MM) reflects the two schools of thought surrounding how the field ought to develop, physics-based and empirical. Physics-based approaches, with their promise of rigour and accuracy, have become the bedrock of many force fields. Parameters for the terms in the force field were derived either from interpretation of experimental data or from high-level quantum mechanical calculation of idealised systems. The different flavours of force field were trained to reproduce experimental geometries (MM2[8,9]), heats of hydration (OPLS[10]) and so on. In many cases, a simple harmonic oscillator model is used. The force fields were then used to make predictions on more complex structures, and extra terms were added to compensate for failures. Empirical force field models try instead to reproduce the conformations observed in crystal structures;[11,12] in this case, the environment is averaged and the results can be more robust. This use of experimental crystal data is still a very valuable way to check on the quality of interaction predicted by other methods.[13]

4.1.1.1 *The Parameter Paradox*

The true barriers to rotation, bond stiffness *etc.* are defined by the total electronic context of the molecule, including surrounding solvent molecules. In some cases, these barriers are not very sensitive to context and may be taken as constant, for example most Csp^3–Csp^3 bonds have the same barrier to torsional rotation. In other cases, for example N atoms with greater or lesser degrees of conjugation, the local environment up to several bonds away may influence the character of the atom. As a first approximation, many force fields operate with a limited number of atom types, and hence a limited ability to compute true conformational strain energies. Similarly, charge models affect the relative energies of conformation. At first, calculations of conformational strain were carried out *in vacuo*, leading to collapsed structures as the global energy minimum for polar structures, as the electrostatic term dominates. With better solvent models,[14,15] this is improving. However, one of the attractions of molecular mechanics approaches is their relative simplicity and interpretability. Introduction of many bespoke parameters reduces this advantage.

4.1.2 Quantum Mechanics

One defect in most force fields is that they do not handle well interactions driven by polarisability, for example π-π interactions. There are polarisable force fields based on distributed multipoles,[16] but quantum mechanics (QM) should handle these issues more robustly. QM, especially semi-empirical methods like MOPAC,[17] have been widely used as a more rigorous method for generating atomic charges and minimizing structures. They provide admirable accuracy in many situations, but they too contain many assumptions and parameters, for example the effect of solvent and element types. Hehre has nicely reviewed the many different types of QM calculations that are currently available, and their strengths and weaknesses.[18] QM calculations are particularly useful when one has doubts about the quality of parameters in a force field, for example when small changes in structure or conformation lead to large charges in energetics,[19] but even there QM can fall down. QM is needed for systems in which halogen–CH hydrogen bonds play a key role, or for groups whose energetic preferences are strongly dependent on the local environment, for example S atoms. π-Cation interactions are common binding motifs (for example in antibodies or nicotinic receptors) and these benefit from the extended representation.[20,21] A common ploy is to use MM for the protein (as that is generally well parameterized) and QM for the ligand/binding site.[22] Then the key choice is to decide which atoms belong to the MM region and which to the QM region.

4.1.3 Molecular Representations

The first driver for the adoption of modelling software was the ability to provide helpful visualisations; even now, a graphics workstation with a stereo display is a must-see for any open day. The appreciation of the numerical power of modelling calculations came later, as force fields improved. There is still the issue of the most effective way to represent a molecule; a receptor sees a dynamic cloud of electrons, but often we need to represent the same information in a coarser fashion. 2D descriptors, for example group counts or topological torsions, do provide a valuable way of comparing structure with activity (see chapter 3 for examples). The next level of sophistication is a pharmacophore, followed by field-based methods. The more realistic the representation, often the less precise the descriptor is.

4.1.4 Pharmacophores

The concept of a pharmacophore is one of the oldest in molecular modelling.[23] Before the structure of protein targets became readily available, one only had ligand structures. Starting from this point, one would hypothesise which parts of the structure were needed for activity, then make analogues. There was also a tendency in medicinal chemistry to make rigid analogues that not only fixed the pharmacophore features, but also their relative disposition. These pharmacophore models could then be used to search corporate databases for other potential hits. This was a very powerful tool when the size of corporate collections exceeded screening rates. A vast number of programs for pharmacophore elucidation are available.[24–27] This demonstrates that there is no definitive solution. The key issues have been reviewed elsewhere[23] but they can be seen as falling into two categories, representation and sampling. The representation must include as much as possible of the relevant information (hydrogen bonds, hydrophobes, charges) in an efficient yet meaningful way. Sampling needs to search the conformational space of flexible molecules to identify a set of low-energy conformers that hopefully include the bioactive conformation. This may lead to multiple solutions of equal feasibility, and identifying the correct solution can be extremely difficult. In contrast to a 2D descriptor, for example the number of acid groups, the pharmacophore description is much less precise, but nevertheless captures a necessary condition for ligand recognition.

4.1.4.1 Field-Based Methods

Field-based approaches[28] aim to quantify what the receptor "sees" in terms of charge distribution and shape, rather than focusing on the underlying structural skeleton. The molecular electrostatic potential (MEP) around a molecule depends on the distribution of atomic charge and is often modelled on a 3D grid. The MEP at each grid point surrounding a molecule is calculated as the interaction between a probe atom of unit positive charge (a proton) and partial charges which are centred on the atoms. The resulting grids of data points can be contoured to pick out key regions of interest, or modelled using atom-centred Gaussian functions.[29] The use of atom-centred charges can lead to an inadequate representation of the MEP. For example, a carbonyl oxygen will give rise to a single field maximum, extended outwards from the C=O double bond. The extended electron distribution (XED) force field attempts to model the distribution of charge[30] through multipoles. For example, the carbonyl oxygen is modelled by two maxima, which correspond to the positions of the lone pairs. Empirically-based programs (Isostar,[31,32] Relibase[33]) finesse this issue by giving probability plots, based on the assumption that geometries or interaction pairs that are frequently observed are favourable and should therefore be incorporated. They are based on true experimental observations and are therefore not fooled by false assumptions about electron pair positions, charges or the like. QM provides even better models of MEPs but the relative difficulty in handling them have made this more of a niche approach for comparing homologous series.

To look at the protein from a ligand-centric view, we can use the same principles, for example in the program GRID.[34] The algorithm can be described simply. The protein of interest is prepared (protonation, fixing broken side chains *etc.*), then a regular 3D lattice is created around the region of interest. Several probe groups can be selected, to represent water, bases, hydrophobes and so on. For each probe, at each grid point, an energy of interaction between the probe and the protein can be calculated. The energy is derived from a molecular mechanics calculation. The energies can then be contoured to visualise regions of strong interactions, water binding sites and greasy pockets. The energies are enthalpic in nature, so there is no guarantee that filling a pocket with a favourable probe will lead to an improvement in binding, nor do the authors of GRID ever claim this. However, through personal observation of how medicinal chemists interact with the results of such calculations and their successors (Ludi,[35] Sitemap,[36] MCSS[37]), there is often a direct expectation that the number given by GRID will translate directly to changes in affinity, at the full thermodynamic rate of 10-fold for every 1.5 kcal mol^{-1}. More quantitative values can be obtained if the probe energies are used as descriptors, as in the CoMFA method.[38]

4.1.5 Conformational Analysis and Global Energy Minima Prediction

Molecules exist in a Boltzmann-weighted equilibrium of states, relative to the global minimum. According to pre-organisation theory, any energetic cost paid in moving to the bioactive conformation will come out of the observed binding energy. In modelling, we generally assume that the bioactive conformers are a subset of the global energy conformers. The same is true when we build up multiconformational databases. We would like to be able to capture most of the populated conformational states in our database so that any structures retrieved by virtual screening are more likely to be active.

The first step in conformational analysis is to sample the conformational states of a molecule. To a first approximation, it was assumed that all bonds are equivalent and that sp^3–sp^3 bonds could be sampled at 120 degree intervals. This leads to 3n possible states for n rotatable bonds. If one adds in sp^2–sp^3 bonds at 6 intervals, sulfonamides at 4, the size of the problem quickly outstrips the computational power available to generate and score the states. Ring conformers add another layer of complexity. Possible solutions are to use random sampling, tabu searches and ring conformer databases, amongst other methods.[39] Many programs impose a hard upper limit of how many conformers can be stored for a molecule. With the advent of parallel computing, the sampling issue is much reduced but still occurs for methods such as pharmacophore determination.

As well as being able to sample conformational states, we also need to be able to rank them accurately. It is therefore important to know how reliably the global-energy minimum is identified, and how accurately the energy difference between this state and other minima is calculated. Too many conformations of relatively high energy will add to the noise, too few will lead to poorer recall. Initial studies looked to see how well structure generators recreated bioactive conformations, but this could be misleading if the bioactive conformation is not a clear global minimum. However, the general consensus is that tools to produce starting 3D structures[40] generally do a good job. Another advance is that with improved storage and partitioning, more conformers per structure can be explored and stored, although the calculation of relative energy is still problematic.

4.1.6 Molecular Dynamics Calculations

Molecular mechanics still gives an inherently enthalpic view of the world. Molecular dynamics (MD) calculations should give more reliable and rigorous predictions of thermodynamic properties, as they deal with entropy by averaging over time and temperature. Ideally, a full atomistic model, including solvent, should be used. This can be prohibitively expensive, so solvent–continuum models such as Poisson-Boltzmann[41] or GBSA[42,43] are used. Greater simulation times can be achieved but the results should be compared back to experiment. Even in atomistic models the influence of the water model used can be profound. The use of many replicate systems on linux farms can be used to enhance the sampling and hence give better thermodynamic ensembles. MD simulations are routinely used to probe changes in protein conformation, or to refine dockings, but are not fast enough to compute free energies for large numbers of states or systems.

4.1.6.1 Free-Energy Perturbation

Free energy perturbation (FEP) theory[44] is based on statistical mechanics, and computes the change going from state A to state B. Although this method has the "correct" physical basis, the calculations do not converge unless the difference between the two states is small. Calculations are therefore divided into a number of small steps that do not have to be physically realistic, for example mutating a Cl atom into a F atom (equation 4.3).

$$\Delta \mathrm{G}(\mathrm{A} \rightarrow \mathrm{B}) = \mathrm{G_B} - \mathrm{G_A} = -k_B T \ln \left\langle \exp\left(-\frac{E_B - E_A}{k_B T} \right) \right\rangle \qquad (4.3)$$

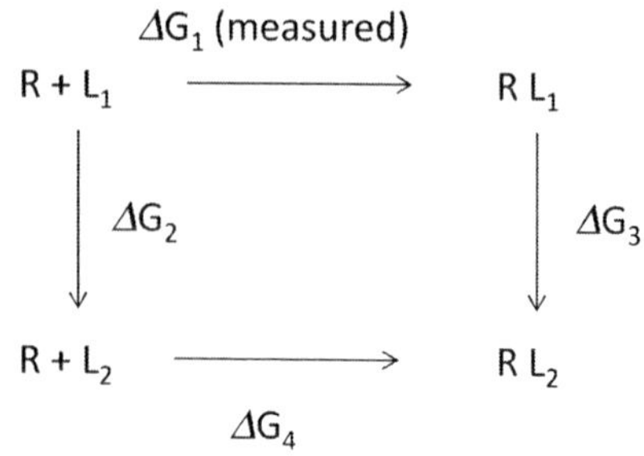

Figure 4.1 The binding affinity of L_2 (ΔG_4) is computed as $\Delta G_1 + \Delta G_3 - \Delta G_2$.

Each simulation requires a large amount of processor time but, as each step is independent, FEP can be run on a loosely coupled linux farm. In the example illustrated in Figure 4.1, the binding affinity of ligand L_2 to receptor R can be calculated by mutating L_1 to L_2 in solvent and in the binding pocket. The result is a free energy value relative to the binding of L_1; the accuracy of the final result depends in part of the quality of the sampling, but mainly on the force field. This protocol has been applied successfully in the design of HIV reverse transcriptase inhibitors[45] and in modelling a large series of analogues to a receptor site in a single simulation (Figure 4.1).[46]

4.2 Quantitative uses of Modelling

4.2.1 Virtual Screening and Rational Design

The process of virtual screening may be described as the scoring of a database of chemical structures according to some model, so that the most desired compounds (the most active, perhaps) are given a high score and the least relevant a poor score. The model might be a QSAR model (for example, rule-of-5), a substructural screen, a pharmacophore model or a sophisticated docking model. The score is usually a number; how one interprets the number in the context of the other scores is critical. Very rarely is there more than a passing correlation with the experimental values being modelled. It is more usual to look at the scores in terms of a receiver–operator characteristic (ROC) curve and to set thresholds for the score at which action will be taken. If one is in late-phase lead optimisation, looking for hERG or AMES signals, one is keen not to miss any compound with a potential liability so the threshold may be set quite low. On the other hand for high throughout screening follow-up, one is more interested in the strong actives, so the threshold would be raised. Better scoring metrics do give better enrichment of true (actives, negatives) over false ones, and this can be true regardless of the accuracy of the scoring function with regards to model reality. Enrichments of 10-fold to 1000-fold can be obtained; an interesting internal observation is that the hit lists from virtual screening and high throughput screening often do not overlap by more than 20%, making this workflow a valuable addition to the experimental approach.

4.2.2 Docking and Scoring

Docking programs are very widely used in support of hit finding and lead optimisation. This is true despite the fact that the correlation between the docking score and the observed affinity is weak, as entropy is poorly accounted for. What docking programs do well is predict the correct pose of a ligand in a binding site; that was what they were built to do. This can be used to support hypothesis-driven design; assuming a pose is correct, what changes can be made to improve activity or validate the hypothesis? This can be done even in the absence of iterative X-ray support. There is a wide range of docking programs,[47] and it is not in the scope of this chapter to discuss them (see Chapter 16, by Warren, in this book). Even the simple, empirically based scoring functions will provide a significant enrichment over random, and careful manual analysis of the binding poses will improve this further. It is ironic that this subjective method is needed, but often it captures a lot of prior local knowledge of the protein target that cannot be coded into a general purpose scoring function. The best successes have come from tailored scoring functions that use crystal structures and known structure–activity relationships (SARs) to adjust the scores.[48,49]

4.2.3 Modelling of Water

It has long been known that the role of water is vital to the energetics of binding,[50] but its role has always been ambiguous. Water can form hydrogen bonds with the protein and the ligand, or it can be displaced from the binding site by the ligand on formation of the complex.[51] The question is then: which waters can be displaced with a release of energy, and which should not be touched? In the case of HIV1 protease, a key discovery was that displacement of the flap water gave rise to an increase in affinity.[52] On the other hand, there are structural waters, or waters that modulate pK_a, that must be included to explain the activity of certain ligands.[53] GRID is able to predict sites of crystallographic water, but not which waters should be included or excluded in modelling the active site. Often the decision about waters in the binding site is a critical part of the manual preparation of a binding site for further studies. The problem can be approached stochastically, as in the approach of Verdonk *et al.* in GOLD,[54] in which water molecules can be switched on and off. The results can be mixed, as the increased number of states leads to increased noise. A more rigorous approach is to perform dynamics on the system, to obtain individual energies of desolvation, as performed in the program, Watermap.[55] However, rather than giving direct thermodynamic information, it seems that this approach gives a more kinetic picture of the binding event, allowing one to probe the rate constants of binding and dissociation.[56]

4.3 Impact of the Advent of Affordable Computing

It can be argued that the most progress has come from an increase in the size and speed of computational resources rather than advances in theory. Not only has the chip speed increased dramatically, the cost per CPU has dropped to the point where one can utilise 1000s of processors in linux farms or cloud clusters to run a calculation that used to be run on a single workstation. At the same time, the drug design cycle time remains at about 2–3 weeks. Therefore the scope of what can be done is much greater. As an example, MD simulations can be run for longer, and perhaps more realistic, times; there is no need to assume that either the ligand or protein is rigid during docking; and QM calculations can be run at higher levels of theory. However, the impact of modelling has not scaled in line with the degree of number-crunching power. The rate limiting step is often the time taken to set up a calculation and then to analyse the results. Running the calculations for longer does not lead to a more accurate result, but it has modified the nature of the questions that can be asked. What has not happened is the sudden advancement of modelling theory, leading to a rejection of the ingrained assumptions necessitated by the previous limitations in processor power: we have still only addressed two parts of the "fast, cheap, good" bottleneck. The availability of more nodes also poses a challenge to software vendors, whose licensing models are often predicated in the older "per CPU" paradigm. This has driven many modellers towards adopting open source code, and this trend is expected to grow.

4.4 Case Studies Showing the Impact of Modelling

One of the first early achievements of modelling was to assist in the solving of the X-ray structure of DNA.[57] The interpretation of tautomeric forms, the modelling of hydrogen bonds and conformational analysis all had a part to play in suggesting a structure consistent with the true experimental result, the diffraction pattern. As others have pointed out,[58] all atomic structures of proteins are in fact models of experimental data and require modelling tools (force fields) to build them.

4.4.1 Control Experiments

Senior managers like to be able to measure the influence of a particular technique on the drug discovery pipeline, to assess the return on investment. There is no doubt that in the early days, the return on molecular modelling was mixed due to the high cost of hardware and software and the limited accuracy of the models. It is also hard to tease apart the contribution of a technique, unless an equivalent team on the same project has tried to advance without the benefit of modelling. And, as the winners get to write the history (and the chemists the patents), the true story—such as the relative opportunity costs that were saved, or incurred, by a particular strategy—is often obscured. The most effective way in which modelling can contribute to a project is for the modeller to be a full member of the discovery team, explaining data, making hypotheses and designing compounds to test them with.[59] To have a true contribution to a drug, this partnership should continue all the way through lead optimisation so that the impact is not just the discovery of novel chemical matter by whatever means, but the maturation of this into a clinical candidate. Even then the project may fail for reasons unrelated to the modelling contribution. Have there been projects hindered by modelling? Probably as many as have been hindered by temperamental assays, dogmatic interpretations of structure–activity relationships or overly rigid adherence to rules for bioavailability. A recent survey of drugs accepted by the FDA since 1998 has been published.[60] A brief and unscientific study of the 215 low molecular weight compounds would point to at least 40 of these being significantly influenced by modelling, and that the same number again had benefited to some extent from modelling. What was more noticeable was that modelling (and structural biology) has had a much greater impact on first-in-class compounds.

4.4.2 Oral Bioavailability

The modelling of oral bioavailability is both a success and failure for the modelling community. In the 1990s, it was recognised that the numbers game being played by combinatorial chemistry and high throughput screening was leading to many large, greasy molecules being added to corporate screening decks. Movement across the gut wall is, to a first approximation, controlled by passive diffusion across a lipid membrane, a process that should be describable by simple surrogates for lipophilicity and hydrophobicity. Studies were conducted on orally bioavailable compounds to deduce simple rules of thumb.[61] These rules became absolutes, the numerical values of molecular weight, for example, being set in stone. In fact, the rules are more probabilistic and may weed out compounds with acceptable bioavailability; this again highlights the confusion between accuracy and precision. This relationship is made much more explicit in an alternative model.[62] So the models have been successful in that they have pushed thinking towards better properties, but unsuccessful in driving rational decision making.

4.4.3 The Hinge Binding Hypothesis for Kinase Inhibitors

Today, the design and discovery of ATP-competitive inhibitors for kinases seems almost like a solved problem, with many compounds in clinical trials and some drugs on the market (for example Gleevec®, Tasigna®, Iressa®). Undoubtedly part of the drive for discovery has come from the successes of structural biology in solving co-crystals, development of fragment-based screens and the like. However, the real breakthrough came from the development of the hinge-binding hypothesis by Furet *et al.*[63] Most kinase inhibitors of that time were based around the staurosporin/maleimide scaffold

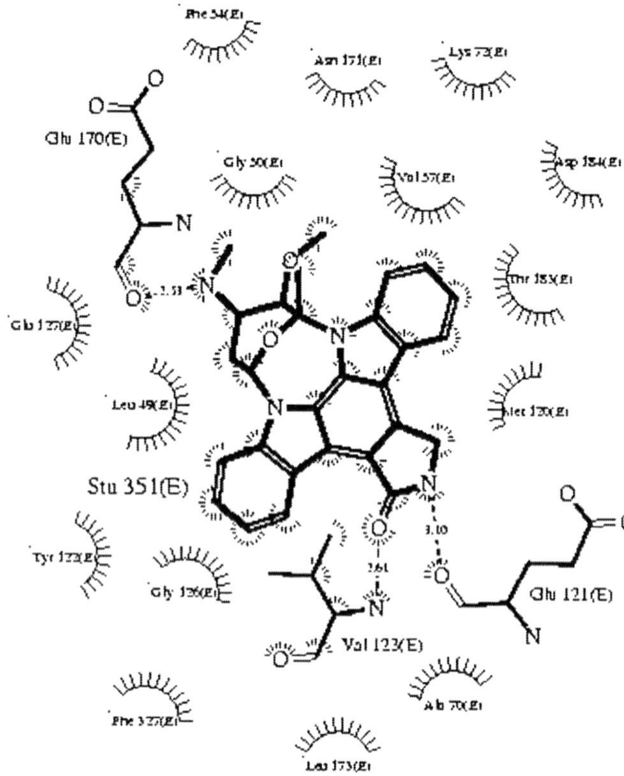

Figure 4.2 Ligplot representation of staurosporin binding to protein kinase A (pdb code 1stk). The maleimide binds to the hinge region backbone of Q121 and V123.

(see Figure 4.2). The only crystal structure was of protein kinase A, in its *apo* form.

Docking of staurosporin to this structure gave only incorrect poses. With the benefit of hindsight, we now know the reasons for this: the *apo* site was closed down, so that the correct pose had steric bumps, and the basic part of the binding site was incorrectly protonated. In addition, the assays were relatively inaccurate, giving a false picture of the SAR. Using manual docking, and a personal interpretation of the available SAR, Furet proposed the now familiar hinge-binding motif, which could explain the experimental results and allow design of alternative chemotypes. The docking pose could also be tested by predicting where solubilising groups could be added. The later solution of inhibitor–kinase complexes confirmed the hypothesis. While this modelling work did not lead directly to a drug itself, it triggered an immense amount of medicinal chemistry, some of which did lead, *via* a long and hard path, to clinical candidates and marketed drugs. A concept we now take for granted was based heavily on modelling alone.

4.4.4 Pharmacophores

A quick examination of any issue of the Journal of Medicinal Chemistry would reveal one or more papers describing the development and use of pharmacophore models. Pharmacophore models rarely lead directly to clinical candidates because of the relative coarseness of the representation.[23] They are very useful in finding alternative chemotypes that may have more attractive properties for optimisation. Pharmacophore modelling has always been particularly useful for membrane targets, for which direct structural information has been scarce.[64] One example is the discovery of novel ligands for the endothelin A receptor.[65,66] Two leads were known, a cyclic tetrapeptide BQ-123 and a natural product discovered by scientists at Shinogi. Conformational analysis combined with NMR experiments gave an initial two point pharmacophore model of an acid (indicated by a circle in Figure 4.3) separated from a ring (indicated by a rectangle) by 12 Å. A third, weaker hydrophobe (dashed parallepiped) was also postulated but was not used in the original virtual screen. Two leads emerged; these had much poorer affinity but much more attractive properties in terms of physicochemical properties and the ability to quickly explore the structure–activity relationships. It is interesting to note that a single pharmacophore cannot be generated from these four compounds; only partial

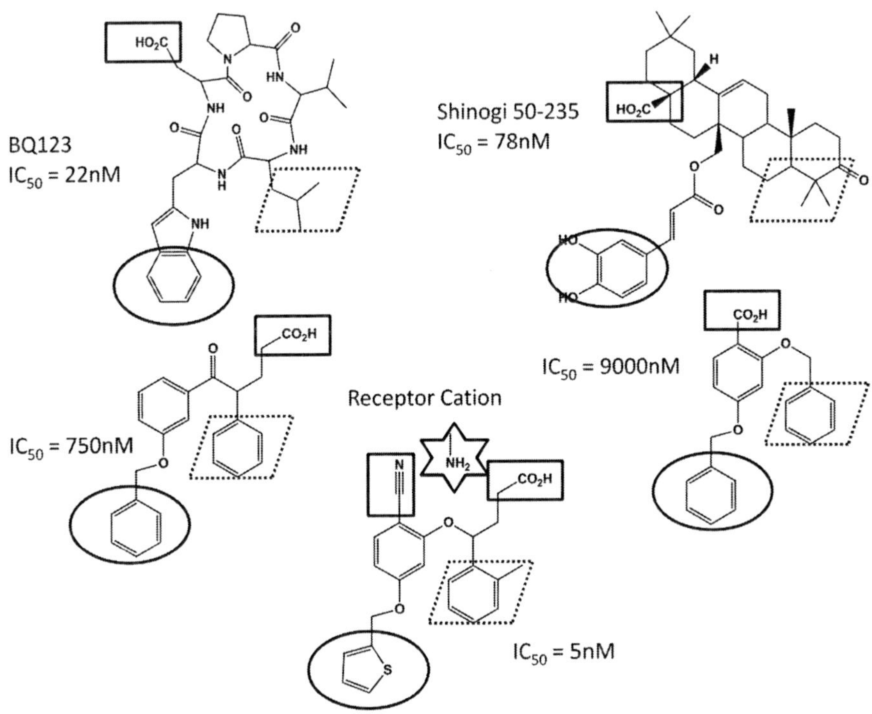

Figure 4.3 Development of a potent inhibitor of endothelin A through pharmacophore modelling.

overlays are possible. This implied that there was a dual binding mode; to explain the data, the team postulated a basic sidechain in the receptor that could lie between the different positions of the acid groups generated in the partial pharmacophore. It was also realised that the second aryl group should be tipped out of plane to fit the hydrophobe geometry. Conformational analysis indicated that an *ortho*-substituent on the aryl ring would favour this conformation. Combining all the requirements into a single molecule gave rise to a very potent inhibitor.

4.4.5 Drugs Discovered using Modelling

Structure-based drug design is the poster child for the impact of molecular modelling. Despite the issues with water, despite the relative inaccuracy of scoring functions, modelling has been able to utilise the three-way partnership with structural biology and medicinal chemistry to engineer some remarkable success stories. A key reason for this is the ability to iteratively solve X-ray structures so that there is a constant input of experimental data to confirm the design hypotheses. One can 'see' in the X-ray structures unfilled pockets that might affect binding, as well as parts of the ligand that do not contribute and therefore can be used to modulate other properties, such as solubility, bioavailability *etc.*, that are essential in optimising a lead into a drug. Drugs that have benefited strongly from such a paradigm are the carbonic anhydrase inhibitors, HIV1 protease inhibitors, COX2 inhibitors, neuraminidase inhibitors and kinase inhibitors.

4.4.5.1 Neuraminidase Inhibitors – the Story of Relenza (Zanamivir)

Neuraminidase was first suggested as a target for the treatment of influenza in the 1960s. The endogenous ligand, sialic acid was known and the biochemical reaction was elucidated by Meindl and Tuppy in 1969.[67] However, random screening and mechanism-based design turned up nothing of note. It could be argued that this is expected, given the relative paucity of sugar-based compounds in most commercial archives, but even random analoguing[68] of the natural ligands did not result in any good leads. Thus the traditional approaches of that time had been exhausted. In 1983, Colman *et al.*[69] solved the first structure of neuraminidase. The X-ray of the complex with a weakly binding sialic acid derivative, Nue5Ac2en was solved, and the binding site was analysed using GRID. Using many different probes, von Itzstein *et al.*[70] mapped the binding pocket to look for new and/or mismatched interactions. To quote from their work:

"Predictions of energetically favourable substitutions to the already known unsaturated sialic acid analogue Neu5Ac2en were made [...]. The most apparent of these was the replacement of the hydroxyl group at the four position on Neu5Ac2en by an amino group. GRID calculations with a ***protonated primary***

***amine** as the probe identified a hotspot around this accessible position at an energy contour level of -16 kcal/mol."*

To reiterate, GRID does not predict changes in free energy, rather enthalpy, but the predictions led to a strong hypothesis—that changing the hydroxyl to an amine would improve activity—which was confirmed by experiment as a 20-fold improvement in binding. Using an even stronger, amidine base gave a 5000-fold improvement and the final clinical compound. It should be noted that it is not just the increase in activity that turned a lead into a drug; this is a lesson that should also be taken from this example, to attend to the PK/PD profile of a compound as well as the siren song of potency. A competitor compound, Tamiflu®, was developed with far less input from modelling and much more traditional medicinal chemistry to build on these findings, but also to optimise the ADME profile. As a result, Relenza is given by inhalation, Tamiflu as a pill, and the latter has a larger market share.

4.4.5.2 The Discovery of Inhibitors of the Angiotensin 1 Receptor

The hormone angiotensin causes a rise in blood pressure through its action at the angiotensin receptors. The first compounds that were discovered to control hypertension by blocking this pathway were peptide analogues of angiotensin, for example the octapeptide, saralasin. As might be expected, these had poor

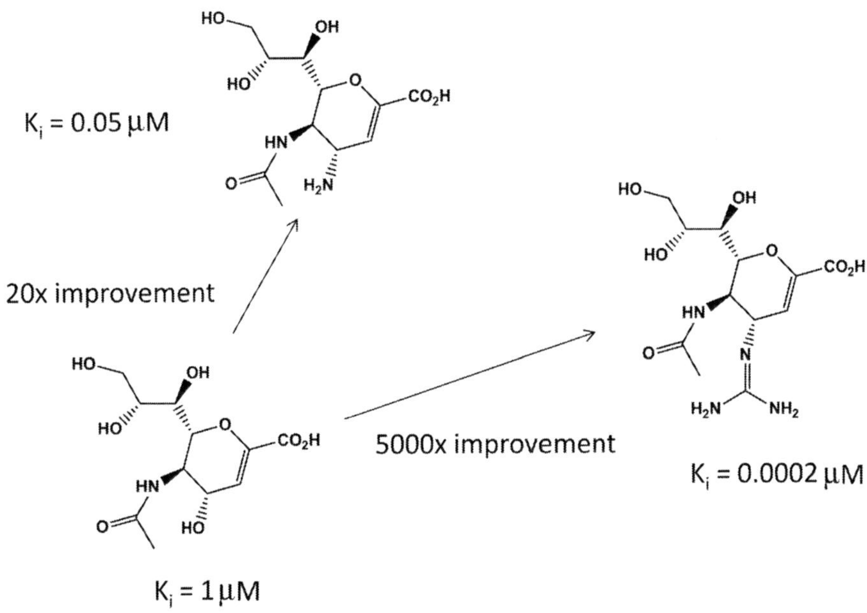

Figure 4.4 The evolution of Nue5Ac2en into zanamivir®.

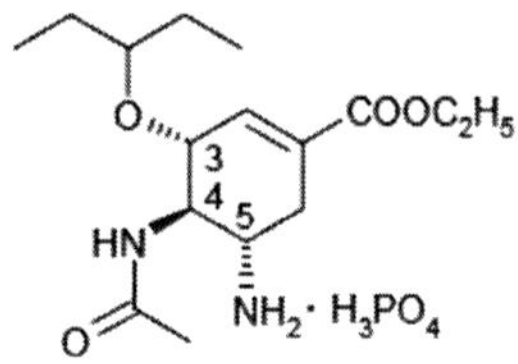

Figure 4.5 The orally active neuraminidase inhibitor, Tamiflu®.

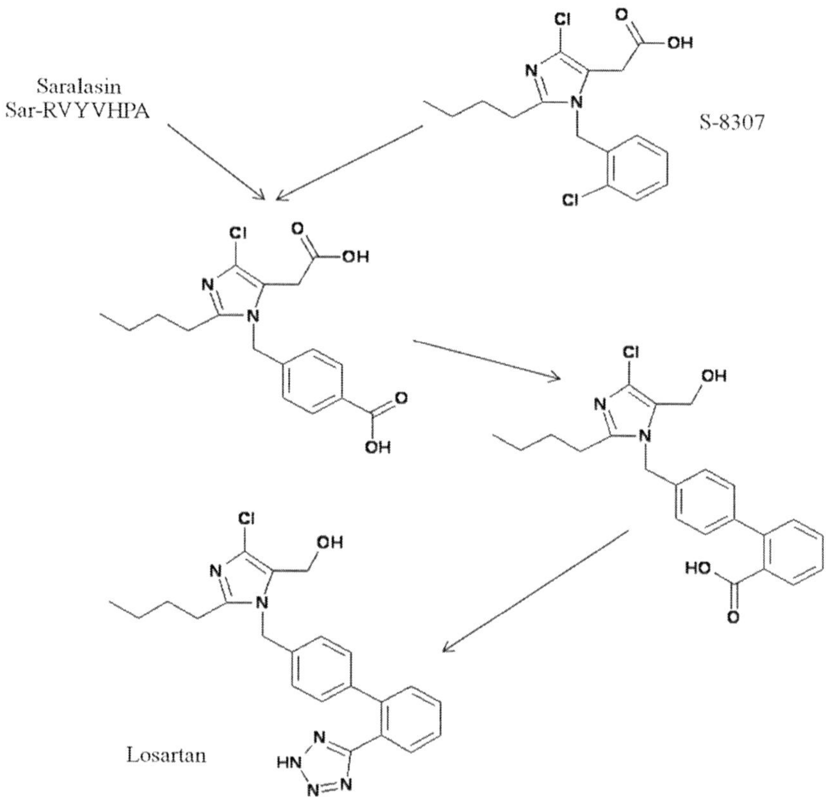

Figure 4.6 The discovery of losartan.

bioavailability. The path that went from here to the current classes of inhibitors, for example losartan, was strongly influenced by modelling, even though no structure of the GCPR target was available. The first non-peptide leads, for example S8307, were generated by screening. The next intellectual jump was made by scientists at Merck, who argued that angiotensin, saralasin and the non-peptide hits were competitive and bound in the same way to the receptor, that is, they had a common pharmacophore.[71] The bioactive conformation of angiotensin was elucidated using NMR, and the other structures were overlaid on this conformation. This immediately suggested the addition of some extra acidic residues; incorporation of this idea gave an immediate jump in potency. Carboxylic acids can lead to poorer bioavailability, so the more lipophilic tetrazole bioisostere was used instead.[72] Even though modelling tools available were rudimentary at that time, the modelling concepts for the bioactive conformation and pharmacophore of the angiotensin 1 receptor were used to drive discovery down a constrained pathway, leading to a more efficient discovery process.

4.5 Future Developments

There are still many unsolved problems in computational chemistry and, as discussed, many of them are not new. We still need a method for the quick and accurate calculation of free energies, and we need to understand the limits on the accuracy of any predictive methods. More broadly, what confidence limits can we put on any given prediction from any given level of theory? For virtual screening, what are the most important terms left out of the standard docking functions, and indeed higher simulation methods? Can we imagine a set of widely accepted well-validated model systems and benchmarks, both for experiment and theory?[73]

4.5.1 Sociology of Data Presentation and Interpretation

Most models have only limited accuracy. This may stem from a lack of data or errors in the data. For example, measurements of values of log P (log of the octanol–water partition coefficient) above 4.5 are accurate to 0.3–0.5 log units, depending on the instrumentation and so on. Therefore one should not expect to be able to derive a model of greater accuracy than this. A similar observation has been made for solubility.[74] On the other hand, computers are very precise and, without careful attention being paid to formatting, a model for estimating log P will return a real value with many decimal places. It is the author's experience that if the model predicts a value with precision of 4 significant figures, people will tend to believe the model also has an accuracy of that magnitude. Web pages that compute descriptors often fall into this trap.[75] This issue has been largely ignored by the vendors of modelling software intended for direct use by the bench chemist. As a result, docking calculations tend to be less believed even though the binding pose is probably right, because the score (or relative score) is not. Similar statements can be made about the relative energies of conformations, the rankings or classifications derived from QSAR models or even the results of FEP calculations. It is the duty of the modeller to interpret such numbers in the context of prior knowledge and experience, and to give an honest assessment of the robustness of the calculation. This should be done in the context of the design process, by suggesting molecules that can test the hypothesis.

4.5.2 Domain of Applicability

Another common area of confusion that can negatively impact the influence of modelling on drug discovery is not considering the scope and the ruggedness of the SAR being modelled.[76,77] Several groups have attempted to alleviate this issue for the medicinal chemist by at least showing the predicted and observed values for similar structures. Modellers also do not give an impression to the medicinal chemist about the rate of change of a prediction, for example that swapping a cyano to a chloro will cause a large change in a score. Cliff analysis is starting to bring this to the attention of the users.[78] Models are all trained in some way on experimental data. This means the models 'know' about the information contained in the training set but might be very poor at predicting new data. This information is often not passed back to the user and no distinction is made between predictions with a high degree of confidence and plain random noise.

In contrast to these scientific issues, we can look back to a prediction made by Snyder in 1991, at the start of the age of modelling.[79] He said that the greatest impact of molecular modelling has come, and will continue to come, from computational chemists who are deeply embedded in the drug discovery process as full partners to the project teams. Despite the advances that have been made in theory, and the increase in computer power, this human factor is still the key to successful modelling.

References

1. D. Lowe, *Nature*, 2010, doi:10.1038/news.2010.223.
2. A. R. Fersht, J.-P. Shi, J. Knill-Jones, D. M. Lowe, A. J. Wilkinson, D. M. Blow, P. Brick, P. Carter, M. M. Y. Waye and G. Winter, *Nature*, 1985, **314**, 235.
3. E. Freire, *Drug Discovery Today*, 2008, **13**, 869.
4. E. Freire, *Chem. Biol. Drug Des.*, 2009, **74**, 468.
5. Y. Kawasaki, E. E. Chufan, V. Lafont, K. Hidaka, Y. Kiso, L. M. Amzel and E. Freire, *Chem. Biol. Drug Des.*, 2010, **75**, 143.
6. C. R. Groom and F. H. Allen, *Future Med. Chem.* 2010, **2**, 933.
7. D. Mayer, C. B. Naylor, I. Motoc and G. R. Marshall, *J. Comput. -Aided Mol. Des.*, 1987, **1**, 3.
8. N. L. Allinger, X. F. Zhou and J. Bergsma, *J. Mol. Struc-Theochem*, 1994, **118**, 69.
9. N. L. Allinger, *J. Am. Chem. Soc.*, 2010, **99**, 8127.
10. H. A. Carlson, T. B. Nguyen, M. Orozco and W. L. Jorgensen, *J. Comp. Chem.*, 1993, **14**, 1240.
11. G. Klebe and T. Mietzner, *J. Comput. Aided Mol. Des.*, 1994, **8**, 583.
12. J. Sadowski and J. Bostrom, *J. Chem. Inf. Model.*, 2006, **46**, 2305.
13. C. Bissantz, B. Kuhn and M. Stahl, *J. Med. Chem.*, 2010, **53**, 5061.
14. A. Klamt, F. Eckert and M. Hornig, *J. Comput. Aided Mol. Des.*, 2001, **15**, 355.

15. A. Klamt and M. Diedenhofen, *J. Comput. Aided Mol. Des.*, 2010, **24**, 357.
16. J. W. Ponder, C. J. Wu, P. Y. Ren, V. S. Pande, J. D. Chodera, M. J. Schnieders, I. Haque, D. L. Mobley, D. S. Lambrecht, R. A. DiStasio, M. Head-Gordon, G. N. I. Clark, M. E. Johnson and T. Head-Gordon, *J. Phys. Chem. B*, 2010, **114**, 2549.
17. J. J. Stewart, *J. Comput. Aided Mol. Des.*, 1990, **4**, 1.
18. W. J. Hehre, *Phys. Chem. Chem. Phys.*, 2006, **8**, 3172.
19. M. P. Gleeson and D. Gleeson, *J. Chem Inf. Model.*, 2009, **49**, 670.
20. W. G. Zhong, J. P. Gallivan, Y. N. Zhang, L. T. Li, H. A. Lester and D. A. Dougherty, *Proc. Natl. Acad. Sci. U.S.A.*, 1998, **95**, 12088.
21. X. A. Xiu, N. L. Puskar, J. A. P. Shanata, H. A. Lester and D. A. Dougherty, *Nature*, 2009, **458**, 534.
22. V. P. Ananikov, D. G. Musaev and K. Morokuma, *J.Mol. Catal. A: Chem.*, 2010, **324**, 104.
23. A. R. Leach, V. J. Gillet, R. A. Lewis and R. Taylor, *J. Med. Chem.*, 2010, **53**, 539.
24. D. Barnum, J. Greene, A. Smellie and P. Sprague, *J. Chem. Inf. Comput. Sci.*, 1996, **36**, 563.
25. N. J. Richmond, C. A. Abrams, P. R. N. Wolohan, E. Abrahamian, P. Willett and R.D. Clark, *J. Comp. Aided Mol. Design*, 2006, **20**, 567.
26. G. Jones, P. Willett and R. C. Glen, *J. Comput. Aided Mol. Des.*, 1995, **9**, 532.
27. S. L. Dixon, A. M. Smondyrev, E. H. Knoll, S. N. Rao, D. E. Shaw and R. A. Friesner, *J. Comp. Aided Mol. Des.*, 2006, **20**, 647.
28. S. Cross and G. Cruciani, *Drug Discovery Today*, 2010, **15**, 23.
29. A. C. Good and W. G. Richards, *J. Chem. Inf. Comput. Sci.*, 1993, **33**, 112.
30. T. Cheeseright, M. Mackey, S. Rose and A. Vinter, *J. Chem. Inf. Model.*, 2006, **46**, 665.
31. I. J. Bruno, J. C. Cole, M. Kessler, J. Luo, W. D. S. Motherwell, L. H. Purkis, B. R. Smith, R. Taylor, R. I. Cooper, S. E. Harris and A. G. Orpen, *J. Chem. Inf. Comput. Sci.*, 2004, **44**, 2133.
32. K. A. Brameld, B. Kuhn, D. C. Reuter and M. Stahl, *J. Chem Inf. Model.*, 2008, **48**, 1.
33. M. Hendlich, A. Bergner, J. Gunther and G. Klebe, *J. Mol. Biol.*, 2003, **326**, 607.
34. P. J. Goodford, *J. Med. Chem.*, 1985, **28**, 849.
35. H. J. Bohm, *J. Comp. Aided Mol. Des.*, 1992, **6**, 61.
36. T. Halgren, *Chem. Biol. Drug Des.*, 2007, **69**, 146.
37. A. Caflisch, A. Miranker and M. Karplus, *J. Med. Chem.*, 1993, **36**, 2142.
38. R. D. Cramer and B. Wendt, *J. Comp. Aided Mol. Des.*, 2007, **21**, 23.
39. D. K. Agrafiotis, A. C. Gibbs, F. Zhu, S. Izrailev and E. Martin, *J. Chem. Inf. Model.*, 2007, **47**, 1067.
40. J. Sadowski and J. Gasteiger, *Chem. Rev.*, 1993, **93**, 2567.
41. F. Fogolari, A. Brigo and H. Molinari, *J. Mol. Recognit.*, 2002, **15**, 377.
42. P. A. Kollman, I. Massova, C. Reyes, B. Kuhn, S. H. Huo, L. Chong, M. Lee, T. Lee, Y. Duan, W. Wang, O. Donini, P. Cieplak, J. Srinivasan, D. A. Case and T. E. Cheatham, *Acc. Chem. Res.*, 2000, **33**, 889.

43. I. Massova and P. A. Kollman, *Perspect. Drug Discovery Des.*, 2000, **18**, 113.
44. R. W. Zwanzig, *J. Chem. Phys.*, 1954, **22**, 1427.
45. M. B. K. Smith, B. M. Hose, A. Hawkins, J. Lipchock, D. W. Farnsworth, R. C. Rizzo, J. Tirado-Rives, E. Arnold, W. Zhang, S. H. Hughes, W. L. Jorgensen, C. J. Michejda and R. H. Smith, *J. Med. Chem.*, 2003, **46**, 1940.
46. C. Oostenbrink and W. F. van Gunsteren, *Proteins: Struct., Funct., Genet.*, 2004, **54**, 237.
47. G. L. Warren, C. W. Andrews, A. M. Capelli, B. Clarke, J. LaLonde, M. H. Lambert, M. Lindvall, N. Nevins, S. F. Semus, S. Senger, G. Tedesco, I. D. Wall, J. M. Woolven, C. E. Peishoff and M.S. Head, *J. Med. Chem.*, 2006, **49**, 5912.
48. M. Vieth and D. J. Cummins, *J. Med. Chem.*, 2001, **43**, 3020.
49. E. J. Martin and D. C. Sullivan, *J. Chem. Inf. Model.*, 2008, **48**, 861.
50. S. B. A. de Beer, N. P. E. Vermeulen and C. Oostenbrink, *Curr. Top. Med. Chem.*, 2010, **10**, 55.
51. J. E. Ladbury, *Chem. Biol.*, 1996, **3**, 973.
52. P. Y. S. Lam, P. K. Jadhav, C. J. Eyermann, C. N. Hodge, Y. Ru, L. T. Bacheler and O. M. J. Meek, *Science*, 1994, **263**, 380.
53. W. R. Ewing, M. R. Becker, V. E. Manetta, R. S. Davis, H. W. Pauls, H. Mason, Y. M. Choi-Sledeski, D. Green, D. Cha, A. P. Spada, D. L. Cheney, J. S. Mason, S. Maignan, J. P. Guilloteau, K. Brown, D. Colussi, R. Bentley, J. Bostwick, C. J. Kasiewski, S. R. Morgan, R. J. Leadley, C. T. Dunwiddie, M. H. Perrone and V. Chu, *J. Med. Chem.*, 1999, **42**, 3557.
54. M. L. Verdonk, G. Chessari, J. C. Cole, M. J. Hartshorn, C. W. Murray, J. W. M. Nissink, R. D. Taylor and R. Taylor, *J. Med. Chem.*, 2005, **48**, 6504.
55. R. Abel, T. Young, R. Farid, B. J. Berne and R. A. Friesner, *J. Am. Chem. Soc.*, 2008, **130**, 2817.
56. R. A. Pearlstein, Q. Y. Hu, J. Zhou, D. Yowe, J. Levell, B. Dale, V. K. Kaushik, D. Daniels, S. Hanrahan, W. Sherman and R. Abel, *Proteins: Struct., Funct. Bioinf.*, 2010, **78**, 2571.
57. J. D. Watson and F. H. C. Crick, *Nature*, 1953, **171**, 737.
58. A. M. Davis, S. J. Teague and G. J. Kleywegt, *Angewe. Chem. Int. Ed.*, 2003, **42**, 2718.
59. M. Stahl, W. Guba and M. Kansy, *Drug Discovery Today*, 2006, **11**, 326.
60. R. Kneller, *Nat. Rev. Drug Discovery*, 2010, **9**, 867.
61. C. A. Lipinski, F. Lombardo, B. W. Dominy and P. J. Feeney, *Adv. Drug Delivery Rev.*, 1997, **23**, 3.
62. W. J. Egan, K. M. Merz and J. J. Baldwin, *J. Med. Chem.*, 2000, **43**, 3867.
63. P. Furet, G. Caravatti, N. Lydon, J. P. Priestle, J. M. Sowadski, U. Trinks and P. Traxler, *J. Comp. Aided Mol. Des.*, 1995, **9**, 465.
64. C. Bissantz, A. Logean and D. Rognan, *J. Chem. Inf. Comput. Sci.*, 2004, **44**, 1162.
65. P. C. Astles, T. J. Brown, C. M. Handscombe, M. F. Harper, N. V. Harris, R. A. Lewis, P. M. Lockey, C. McCarthy, I. M. Mclay, B. Porter, A. G. Roach, C. Smith and R. J. A. Walsh, *Eur. J. Med. Chem.*, 1997, **32**, 409.

66. P. C. Astles, C. Brealey, T. J. Brown, V. Facchini, C. Handscombe, N. V. Harris, C. McCarthy, I. M. Mclay, B. Porter, A. G. Roach, C. Sargent, C. Smith and R. J. A. Walsh, *J. Med. Chem.*, 1998, **41**, 2732.
67. P. Meindl, G. Bodo, J. Lindner and P. Palese, *Z. Naturforsch.*, 1971, **B 26**, 792.
68. P. Meindl, G. Bodo, P. Palese, J. Schulman and H. Tuppy, *Virology*, 1974, **58**, 457.
69. J. N. Varghese, W. G. Laver and P. M. Colman, *Nature*, 1983, **303**, 35.
70. M. Von Itzstein, W. Y. Wu, G. B. Kok, M. S. Pegg, J. C. Dyason, B. Jin, T. V. Phan, M. L. Smythe, H. F. White, S. W. Oliver, P. M. Colman, J. N. Varghese, D. M. Ryan, J. M. Woods, R. C. Bethell, V. J. Hotham, J. M. Cameron and C. R. Penn, *Nature*, 1993, **363**, 418.
71. J. V. Duncia, A. T. Chiu, D. J. Carini, G. B. Gregory, A. L. Johnson, W. A. Price, G. J. Wells, P. C. Wong, J. C. Calabrese and P. B. M. W. Timmermans, *J. Med. Chem.*, 1990, **33**, 1312.
72. D. J. Carini, J. V. Duncia, P. E. Aldrich, A. T. Chiu, A. L. Johnson, M. E. Pierce, W. A. Price, J. B. Santella and G. J. Wells, *J. Med. Chem.*, 1991, **34**, 2525.
73. M. L. Benson, R. D. Smith, N. A. Khazanov, B. Dimcheff, J. Beaver, P. Dresslar, J. Nerothin and H. A. Carlson, *Nucleic Acids Res.*, 2008, **36**, D674.
74. A. H. Goller, M. Hennemann, J. Keldenich and T. Clark, *J. Chem. Inf. Model.*, 2006, **46**, 648.
75. T. J. Ritchie, P. Ertl and R. A. Lewis, *Drug Discovery Today*, 2011, **16**, 65.
76. J. Bajorath, L. Peltason, M. Wawer, R. Guha, M. S. Lajiness and J. H. Van Drie, *Drug Discovery Today*, 2009, **14**, 698.
77. M. T. Sisay, L. Peltason and J. Bajorath, *J. Chem. Inf. Model.*, 2009, **49**, 2179.
78. R. P. Sheridan, P. Hunt and J. C. Culberson, *J. Chem. Inf. Model.*, 2006, **46**, 180.
79. J. P. Snyder, *Med. Res. Rev.*, 1991, **11**, 641.

CHAPTER 5

Contribution of Structure-Based Drug Design to the Discovery of Marketed drugs

ALEXANDER A. ALEX AND DAVID S. MILLAN

Pfizer Global Research and Development, Ramsgate Road, Sandwich CT13 9NJ, UK

5.1 Introduction

More than 100 years ago, aspirin was patented (in 1900) and barbiturates, as well as procain (in 1903), were launched as human therapeutics. It took another 70 years before the X-ray structure of a human protein, haemoglobin, could be used to design pharmacologically active compounds.[1] The realisation of structure-based drug design (SBDD) as a new approach to drug discovery can be attributed to Max Perutz, for solving the phase problem of imaging protein X-ray diffraction data at atomic resolution.[2] SBDD involves the integration of a number of independent sciences. X-ray crystallography, and to a lesser extent nuclear magnetic resonance (NMR), provide the data for detailed atomic-level description of the binding site.[3] Molecular modelling; synthetic organic chemistry; medicinal chemistry; molecular biology and knowledge about pharmacokinetics/pharmacodynamics (PK/PD), absorption, distribution, metabolism and excretion (ADME) are all required for successful SBDD.[3] The history and the development of SBDD over recent years has been reviewed extensively elsewhere.[3,4] There are a number of early examples of successful SBDD, several of which were highlighted by Greer *et al.* in a review in 1994,

RSC Drug Discovery Series No. 13
Drug Design Strategies: Quantitative Approaches
Edited by David J. Livingstone and Andrew M. Davis

Published by the Royal Society of Chemistry, www.rsc.org

highlighting examples of the use of X-ray structure information for HIV protease, carbonic anhydrase and thymidylate synthase.[5] Further examples of the use of X-ray structure information in the design of protease inhibitors were highlighted by Leung *et al.* in 2000.[6] Since then, structure-based drug discovery has become increasingly important as a key component of drug discovery[7] and is now an essential contributor to addressing the need to improve research and development productivity faced by the pharmaceutical industry.[3,8,9] Today, there are just over 67 000 structures in the Brookhaven Protein Data Bank (PDB) as of 16 Aug 2010.[10] Of these, 58 000 are X-ray structures and 8 500 are NMR structures. In a review of the impact of structural biology on drug discovery, Congreve *et al.* listed 10 drugs which were attributed to having come from structure-based approaches prior to 2003.[7] Since then, 145 more new chemical entities (NCEs) have been approved by the FDA.[11–19]

The purpose of this review is to highlight the impact of protein structures solved by X-ray and NMR techniques on structure-based discovery of approved drugs up till the end of 2009. Obviously there is an abundance of other excellent examples of structure-based drug design published in the literature that has not yet resulted in the discovery of a marketed drug. Many of these compounds are moving through the development pipelines of companies, but we will restrict this chapter to predominantly focus on marketed drugs, which in reality is what is required for improvements in human health. In addition to X-ray crystallography, we will also consider the role of biophysical methods, for example isothermal titration calorimetry (ITC) and NMR, in gaining a deeper understanding of the thermodynamic components of binding, free energy, enthalpy and entropy,[20–22] as well as the role of water.[23] In this chapter, the importance of these methods to drug discovery will be underscored by examples of marketed drugs, discovered utilising structural information about the target protein.

5.2 Experimental Methods in Structure-Based Drug Design

5.2.1 Opportunities and Limitations for X-Ray Crystallography

X-Ray crystallography has been in routine use in SBDD design only since the late 1980s. Although the capabilities of solving protein structures had been available since the solving of the phase problem by Perutz,[2] it took several decades before crystal structures of proteins could be solved in a reasonable timeframe, aligned with drug discovery workflows and processes. Since then there has been a number of significant technological advances in all areas of the gene to structure process. These important advances include improved molecular biology techniques, protein expression, purification and crystallisation techniques. The way diffraction data is collected has also changed from simple X-ray sources to state of the art synchrotron facilities.[24] Together with automation and computational advances, protein crystal structure generation is now at a stage where it can keep pace with medicinal chemistry design.

The applicability and impact of X-ray crystallography has massively increased with the development of high throughput X-ray crystallography approaches about 10 years ago.[25,26] With the onset of high throughput X-ray crystallography (as well as NMR), the new paradigm of fragment-based drug discovery has emerged in the last two decades.[27–32] In recent years, X-ray crystallography has been combined with high-resolution neutron analysis to reveal intricate details of enzyme mechanisms, and particularly the role of water molecules, at very high resolutions.[33] More recently, very high resolution neutron structures of carbonic anhydrase provided insights into the proton transfer underlying the catalytic activity of the enzyme.[34] The X-ray structures of the β_1, β_2 and A_{2A} receptors have also become available,[35–37] opening up the huge new area of structure-based drug design in the G-protein coupled receptor families.[38] In addition, protein–protein interactions have become more important for structure-based drug design over the last decade, particularly in the discovery of antibacterial drugs,[39] oncology[40] and antivirals.[41]

As discussed, the PDB currently contains close to 58 000 X-ray structures. With the availability of large numbers of protein–ligand X-ray structures in targets relevant to drug discovery efforts, elaborate molecular modelling techniques like molecular dynamics have been applied to rationalise binding events and the role of water molecules in binding at a molecular level.[42,43] However, there are potential limitations to the use of X-ray in drug discovery. Limitations and artefacts in protein–ligand X-ray structures and their implications for drug design have been extensively reviewed elsewhere.[44–46] A very interesting aspect, which is highlighted by a number of examples of the use of structure–based design in drug discovery, is that although the visualisation of ligands bound in their respective targets is an immensely powerful tool for the design of ligands with improved binding, only very few of these designed ligands have progressed over the years to become drugs.[44] In the early examples of structure-based drug discovery, compounds were designed with excellent binding affinities for their biological target, however, their physicochemical properties proved to be suboptimal for becoming a drug. One of these examples is the discovery of the HIV protease inhibitor, DMP323, a cyclic urea. This constitutes a very elegant structure-based design, where the resulting ligand was very different from the peptidomimetic inhibitors at the time (*e.g.* nelfinavir and saquinavir), with exquisite potency against HIV protease (IC_{50} 0.031 nM). However, the compound suffered from poor bioavailability due to low solubility and metabolic instability, issues that were probably not recognised early enough as potentially problematic in design at that time in the pharmaceutical industry. Since then, immense progress has been made in understanding the effect of physicochemical properties on the success and failure of compounds in development,[47–52] and structure-based ligand design has now become more like structure-based drug design. The application of this learning in structure-based drug design is now as important as designing for potency against a particular target. This was highlighted by the discovery of HIV protease drug, tipranavir a few years after DMP323, when excellent pharmacokinetic properties were achieved already as part of the hit-to-lead process.[44,53] Although the availability of X-ray derived structural information on protein–ligand complexes is increasing, the ambiguities associated with structural models derived from X-ray data may not be fully appreciated. The process of deriving an atomic model from X-ray electron density data disguises uncertainties in terms of ligand identity and position, as well as those of water and protein atoms in the complex.[44,45] The observed ligand and protein conformation can also be affected by experimental conditions (*e.g.* crystallisation). In particular, changes in the protonation states of ligands and proteins may lead to significant changes in the binding orientations and affinities.[20] Additionally, post-translational modifications, particularly the phosphorylation of kinases, can impact on topology and folding of the target protein. It can be difficult, even for the most skilled medicinal chemist, to avoid drawing potentially misleading or incomplete conclusions from these models. It is therefore imperative that collaboration with the X-ray crystallographers, NMR scientists and computational chemists takes place in order to ensure that the optimum value is extracted from the available experimental data. The ambiguities, both in experimental data and the models derived from this data, have important consequences for the application of structure-based design approaches in general. For example, calculation of binding energies is still not sufficiently accurate to guide design beyond the current chemical scaffold or series,[54] particularly when significant movement of areas of the target protein are a consequence of the binding event.[55,56] This induced fit, which can be described as the protein adjusting to the ligand in order to increase binding complementarity, also adds to the complexity of predicting binding modes and affinities. Despite these inadequacies, remarkable successes in structure-based docking and scoring (virtual screening) in a variety of targets has been reported in recent years.[57,58] Further work is needed in order to better understand the nature of the interactions and the relationship between structure and energetics, which will be discussed in more detail later.

5.2.2 Opportunities and Limitations for Nuclear Magnetic Resonance

The availability of NMR structure determination of proteins has added a second physical technique with which to image potential drug receptors. The first NMR structure determination was published in 1985.[59] NMR has been used in SBDD since the early 1990s. As discussed earlier, the PDB currently contains around 8500 NMR structures. NMR has two advantages over X-ray crystallography: the protein does not need to be crystallised and NMR provides real-time dynamic information, while crystal structures at present provide static and thermodynamically stable structures. A major disadvantage of NMR spectroscopy has been the molecular size limit of around 35 000 Da. Although there is little evidence in the literature for the use of NMR-determined structures in SBDD, NMR techniques have been used extensively in fragment-based lead discovery,[60] building on the seminal paper by Hajduk *et al.* on "SAR by NMR", published in 1997.[61] The use of NMR in drug discovery has been reviewed extensively elsewhere.[62]

5.2.3 Opportunities and Limitations for Isothermal Titration Calorimetry

Like any other spontaneous process, a non-covalent binding event takes place only when it is associated with a negative, favourable binding free energy (ΔG). It is the sum of an enthalpy term (ΔH) and an entropy term ($-T\Delta S$), which may be of equal or opposite sign, and therfore leads to various thermodynamic signatures of a binding event, ranging from enthalpy-driven to entropy-driven. An increasing body of data from isothermal titration calorimetry (ITC) is available on the thermodynamic profiles of many complexes.[63–75] Where crystal structure information exists as well, it is tempting to speculate about the link between thermodynamics and geometry of protein–ligand complexes. A rough correlation between the burial of apolar surface area and free energy could be derived,[67] but beyond that practically useful relationships between structure and the components of free energy remain elusive. This is not surprising as both entropy and enthalpy terms obtained from calorimetric experiments contain solute and solvent contributions and thus cannot be interpreted on the basis of structural data alone. ITC is an emerging tool for drug discovery, providing a differentiation between the thermodynamic components (free energy, enthalpy and entropy) during a binding event. The role of ITC in drug discovery has been highlighted in a number of applications to drug discovery problems, particularly the role of enthalpy–entropy balance, or compensation, and the relationship with drug properties.[68–69,75] A more detailed discussion of the merit of ITC in drug discovery has been published elsewhere.[76] Although the use of ITC has increased over recent years, like X-ray and NMR it does not enable the development of hypotheses by directly linking observed protein–ligand structures and the associated energetics of protein–ligand interactions. Any hypothesis can only be based on knowledge of other experimental structures and measurements. Therefore many drug design hypotheses are based on tacit knowledge of the drug discoverers involved and experiences in previous drug discovery programmes. However, applicability of this knowledge from analogous drug discovery examples to specific target or ligand classes is not always clear; for example, how any relationship between "best in class" and binding enthalpy, entropy or free energy could be used to direct molecular design.[68,75] Molecular modelling can sometimes rationalise and even predict protein–ligand binding modes and interaction energies, but confidence in modelling results are often moderate at best, and rigorous testing of any hypothesis using available techniques like X-ray, NMR and ITC is essential for guiding successful drug design efforts.

5.2.4 Energetics of Molecular Interactions in Protein–Ligand Complexes

Biophysical methods like X-ray crystallography, NMR and ITC are essential to improve our understanding of molecular interactions in principle, however, they provide little detailed insight into the relationship between structure and energetics beyond the components of free energy, enthalpy and entropy. In order to improve our ability to design drugs, a better understanding of binding interactions between small molecules and proteins is essential. However, this requires the use of molecular modelling techniques to support the interpretation of experimental results[76] and to analyse in more detail the observed interactions between individual atoms in ligands and proteins. Attempts to rationalise and understand the relationship between observed ligand-target structure and thermodynamics go back two decades to the pioneering work in this area of Williams *et al.* in the early 1990s.[77–79] Prediction of protein–ligand binding energy from structural data has been a focus of research for at least the last twenty years, and numerous attempts have been made to develop so-called empirical scoring functions based on available binding data and X-ray structures, with LUDI being the earliest example,[80] followed by many others.[81–83] There have also been attempts to rationalise protein–ligand binding energies on the basis of available structural data by developing potentials of mean force (PMF), utilising the frequency of occurrence of an interaction in a given dataset as an indication of its energy contribution to binding.[84,85] However, at that time protein–ligand binding and corresponding structural data was relatively sparse,[86,87] therefore this area could potentially be revisited with the massively increased amount of available structural data. Most known interactions of small molecules with proteins are non-covalent and reversible, and there is considerable interest in utilising covalent interactions for drug design due to their massive potency and selectivity potential.[88] However, for the purpose of this review, we will focus on non-covalent interactions. Based on the data and information available from various databases like the PDB or crystal structure database (CSD),[89] guiding principles have been established to help in particular the medicinal chemist in the drug discovery effort.[20,76] In addition, particular attention has been drawn to the importance of tautomeric equilibria in small molecules,[90] highlighting the preferences found in protein–ligand complexes in the PDB.[91] Furthermore, the geometric preferences of hydrogen bonds have been explored by mining of the PDB and CSD in order to derive guidelines for understanding of the role of hydrogen bonds in protein–ligand binding.[92] Binding interactions have been of interest to researchers for many decades, since they are the underlying fundamental physical process responsible for drugs binding to receptors.[77] In order to better understand the components and contributions which are important in binding, the complexes of peptides with vancomycin antibiotics have been used by Williams *et al.* to obtain the relative magnitudes of the interactions,[77]as shown in equation (5.1).

$$\Delta G = \Delta G_{r+t} + \Delta G_{rot} + \Delta G_{conf} + \Delta G_{p} + \Delta G_{vdW} + \Delta G_{h} \tag{5.1}$$

ΔG_{r+t} is the unfavourable free energy arising from the restriction of translational and rotational entropy upon converting the host and guest into a complex. ΔG_{rot} is the unfavourable free energy resulting from the restriction of internal rotations. ΔG_{conf} is the free energy cost of any conformational energy change required for binding. ΔG_{p} is the free energy of polar interactions such as hydrogen bonds. ΔG_{vdW} is the free energy associated with van der Waals interactions, and ΔG_h the hydrophobic binding free energy associated with solvation. This equation formed the basis for many attempts to develop empirical computational scoring functions based on X-ray structural information and binding energy data. The first such scoring function was LUDI, developed by Boehm in 1994,[80] which contains binding components that are very similar to equation (5.1). Many other scoring functions have followed since, but their usefulness in drug discovery has been very limited. To our knowledge, there are very few reports of their successful use, partly due to the fact that most scoring functions are not accurate enough to work in the small range of binding energies in a typical drug discovery programme of 10 μM to 10 nM, which equates to a range of binding energies of only about 5 kcal mol^{-1}. For comparison, the range of binding energies that was used for the development of LUDI was about 13 kcal mol^{-1}. Scoring functions also do not work very well for inactive compounds, since those are not part of the training set other than through random decoy sets. Therefore, scoring functions will produce large numbers of false positives by not sufficiently differentiating between active and inactive compounds. This also indicates that the impact of structural information from X-ray lies mainly in the visual representation of the binding orientation and interaction, without any reference to energetic considerations. The presence of a particular interaction, for example a hydrogen bond, in an X-ray structure gives no indication as to whether, or to what extent, this interaction is energetically favourable. Its presence is merely a consequence of an overall favourable physical process of intermolecular interactions between ligand, protein and water. The total free binding energy is also the sum of all contributions, where some observed interactions could be favourable, and others unfavourable. This makes the discovery of drugs through structure-based methods a very difficult endeavour, since in most cases the actual design and prediction of improvements has to be done *via* traditional medicinal chemistry optimisation approaches, relying entirely on compound SAR to guide the process. Even with the most advanced molecular modelling techniques, it is very difficult and time-consuming to accurately estimate the energetic contribution of any single binding interaction, for example a hydrogen bond or a hydrophobic contact. However, a simplified concept of the complexity of molecular interactions has been developed, based on the observation that binding free energies correlate with the contact surface area of the binding partners (0.018–0.025 kcal mol^{-1}).[93,94] This indicates that specific polar interactions may typically not have large favourable contributions to binding. This has also been exemplified by the work on binding energy contributions per non-hydrogen atom,[95,96] and the usefulness of this very simple measure (termed ligand efficiency) in drug discovery.[97] It has been shown that amide–amide hydrogen bonds in proteins, for example, form only in a cooperative process, which is a result of the hydrophobic collapse of the unfolded protein and the consequent shedding of waters from main-chain amides.[98] In addition, the binding interactions of very small molecules in proteins gives some indication that polar interactions may not be as significant in the overall contributions as is widely assumed. Formaldehyde (as the acetal) and acetic acid have binding energies of 4.7 and 3.1 kcal mol^{-1} respectively in hydrolase.[99] Hydroxamic acid binding to a zinc cation in the matrixmetalloprotease stromelysin (MMP-3) binding site has a slightly lower binding energy of only 2.4 kcal mol^{-1}, despite the charge–charge interaction between the presumably deprotonated hydroxamic acid and the doubly positive zinc ion.[61] Imidazole binding *via* the iron to the haem group in haemoglobin has about the same binding energy of 2.6 kcal mol^{-1}, despite the fact that only one binding partner is likely to be charged.[100] However, the uncharged, lipophilic molecules, benzene and toluene have binding energies of 2.2 and 2.6 kcal mol^{-1} respectively in chymotrypsin.[101] This indicates that polar and lipophilic interactions are perhaps on average of very similar strength, with perhaps no more than a factor of two, in terms of binding energy per heavy atom, between them. This equivalence of polar and non-polar interactions may be somewhat counterintuitive,[102] and it appears that there is historically a focus on hydrogen bonds and ionic interactions in chemistry,[103] particularly for visualisations in publications of X-ray structures or molecular modelling.[76] Perhaps this is because they are easier to identify since they are more defined in terms of distance and angles between atoms than van der Waals interactions, and have been analysed in great detail from X-ray structures of small molecules.[89] Additionally, increases in binding energy from introducing lipophilic substituents are often linearly correlated with the molecular weight increase and can be relatively small.[72] There are many examples in the literature where affinity is enhanced through hydrophobic interactions at the expense of hydrogen bonds.[73,103] An example of this is the binding of a series of inhibitors to thermolysin, for which X-ray structures are available (5tmn, 6tmn).[104] The general scaffold of the inhibitors is shown in Figure 5.1.

When X = NH, the compound is very potent (ZG^PLA, K_i = 16.5 nM, 5 tmn), in contrast to X = O [$ZG^P(O)LA$, K_i = 16 000 nM, 6 tmn]. This significant decrease in potency can be rationalised from the X-ray structures of both compounds (5 tmn and 6 tmn, respectively), with the NH forming a hydrogen bond to two backbone carbonyl groups in the enzyme. In the case of X = O, the compound binds in almost exactly the same orientation, resulting in a poor electrostatic interaction between the oxygen atom in the compound and the protein carbonyl oxygen atoms, with the binding energy decreasing 1000-fold.

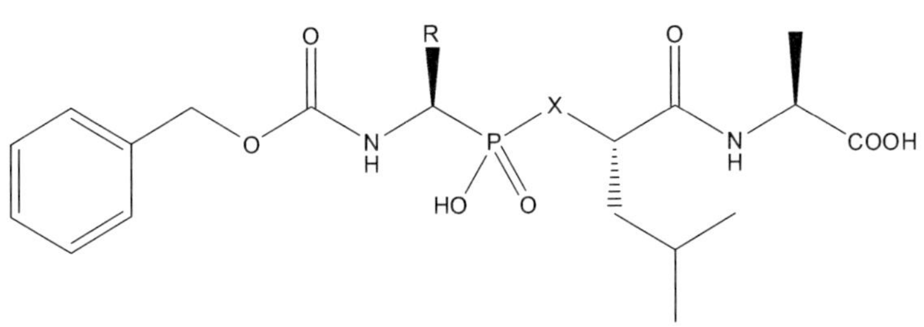

Figure 5.1 Scaffold of thermolysin inhibitor series.

However, adding hydrophobic groups [ZA^{P}(O)LA, R = methyl], [ZL^{P}(O)LA, R = isobutyl] and [ZF^{P}(O)LA, R = benzyl] increases the binding affinity of the compound to K_i = 1800, 680 and 45 nM, respectively. This indicates that even when starting with a poor scaffold, hydrophobicity can somewhat increase the binding energy to the desired level, at the expense of potentially poor molecular properties. This example is also evidence that even though an interaction is observed in an X-ray structure, there is no indication as to whether a particular interaction is attractive or repulsive. The fact that a compound can be crystallised in a receptor at all implies not only that it forms an attractive interaction with a receptor, but also that the scientists solving the structure were sufficiently interested in a particular compound, which in a drug discovery setting usually means that the compound is one of the best examples in a series and therefore potent enough for the structural information to be of value in successive design cycles. We therefore mostly observe attractive interactions in protein–ligand structure databases and are left guessing about the non-attractive interactions. This is one of the caveats for developing empirical scoring functions for molecular docking and binding energy prediction approaches, since these are exclusively built on observed, and therefore mostly attractive, interactions. Additionally, the strength of hydrogen bonds and ionic interactions are significantly influenced by the environment in which they are formed, for example they are stronger in a lipophilic environment than in water.[20,105,106] Although accurate rationalisation and prediction of any type of interaction between small molecules and receptors is very difficult and rarely successful,[43] structure-based drug design offers valuable opportunities for the development of testable design hypotheses that have significant advantages over traditional design, based on SAR or small molecule overlay hypotheses.[107] The examples of successes of structure-based drug design described in Section 5.3 are indicative of the significant advances not only in X-ray crystallography and NMR, but also in molecular modelling, which have contributed to new chemical entities reaching the market in the last 20 years.

5.3 Examples of Structure-Based Drug Design in the Discovery of Marketed Drugs and Advanced Clinical Candidates

To our knowledge, there are currently 31 examples of marketed drugs for which there is evidence in the literature of X-ray and NMR approaches making a significant contribution to their design and discovery. In a review in 2005 on the impact of structural biology on drug discovery, Congreve *et al.* described 10 examples of marketed drugs which were derived from structure-based approaches.[7] We have reviewed and extended this original list beyond 2005, to drugs approved up till the end of 2009, and revisited structure-based drug discovery efforts, particularly in HIV and cancer targets. This resulted a chronological list of 31 approved drugs for which there is evidence in the literature of contributions from structure-based drug design to their discovery (Table 5.1).

Table 5.1 Chronological list of FDA approved drugs (unless otherwise noted) for which a structure-based component in their discovery and development has been reported in the literature.

Name of compound	*Marketed name*	*Company*	*Year launched*	*Target*	*Indication*	*Reference(s)*
Captopril	Capoten	Bristol-Myers Squibb	1981	Angiotensin-converting enzyme	Hypertension	108
Dorzolamide	Trusopt	Merck	1995	Carbonic anhydrase	Glaucoma	109
Saquinavir	Invirase/Fortovase	Roche	1995	HIV protease	HIV/AIDS	110
Ritonavir	Norvir	Abbott	1996	HIV protease	HIV/AIDS	111
Indinavir	Crixivan	Merck	1996	HIV protease	HIV/AIDS	112
Brinzolamide	Azopt	Alcon	1999	Carbonic anhydrase	Glaucoma	113
Nelfinavir	Viracept	Agouron/Pfizer and Lilly	1999	HIV protease	HIV/AIDS	114
Amprenavir	Agenerase	Vertex and GlaxoSmithKline	1999	HIV protease	HIV/AIDS	115
Lopinavir	Aluviran	Abbott	1999	HIV protease	HIV/AIDS	116
Zanamivir	Relenza	Monash University and GlaxoSmithKline	1999	Neuraminidase	Influenza	117
Oseltamivir	Tamiflu	Gilead and Roche	1999	Neuraminidase	Influenza	118
Imatinib	Gleevec	Novartis	2001	Bcr-Abl kinase	Chronic myelogenous leukemia	119,120
Gefitinib	Iressa	AstraZenca	2003	EGFR	Oncology, non-small cell lung cancer	121
Atazanavir	Reyataz	Bristol-Myers Squibb	2003	HIV protease	HIV/AIDS	122
Fosamprenavir	Lexiva/Telzir	GlaxoSmithKline	2003	HIV protease	HIV/AIDS	123
Erlotinib	Tarceva	OSI and Genentech	2004	EGFR	Oncology, non-small cell lung cancer	124
Ximelagatran, melagatran	Exanta	AstraZeneca	2004 (Europe only), withdrawn 2006	Thrombin	Venous thromboembolic events	125

Table 5.1 (*Continued*)

Name of compound	*Marketed name*	*Company*	*Year launched*	*Target*	*Indication*	*Reference(s)*
Sorafenib	Nexavar	Bayer and Onyx	2005	VEGFR	Renal cell carcinoma	126
Tipranavir	Aptivus	Boehringer Ingelheim	2005	HIV protease	HIV/AIDS	53
Udenafil	Zydena	Dong-A	2005	Phosphodiesterase-5	Erectile dysfunction	127,128
Sunitinib	Sutent	Pfizer	2006	Multitargeted kinase inhibitor	Imatinib-resistent gastrointestinal stromal tumour and advanced renal cell carcinoma	129
Darunavir	Prezista	Tibotec	2006	HIV protease	HIV/AIDS	130
Vorinostat	Zolinza	Merck	2006	Histone deacetylase inhibitor	Cutaneous T-cell lymphoma	131
Dasatinib	Sprycel	Bristol-Myers Squibb	2006	Tyrosine kinase	Chronic myeloid leukemia	132
Nilotinib	Tasigna	Novartis	2006	BCR-ABL kinase	Chronic myeloid leukemia	133
Aliskiren	Tekturna	Novartis	2007	Renin inhibitor	Hypertension	134
Lapatinib	Tykerb	GlaxoSmithKline	2007	Kinase inhibitor with targets including HER2	Breast cancer	135
Rivaroxaban	Xarelto	Bayer, Johnson & Johnson	2008 Europe and Canada	Factor X	Venous thrombolic events	136
Dabigatran	Pradaxa	Boehringer Ingelheim	2008	Thrombin	Venous thrombolic events	137
Etravirine	Intelence	Janssen	2008	NNRT	HIV/AIDS	138
Pazopanib	Votrient	GlaxoSmithKline	2009	Multikinase inhibitor with targets that include vascular endothelial growth factor receptors	Ovarian cancer, soft tissue sarcoma	139

If a drug is not listed in the Table 5.1, this indicates that we could not find any evidence in the literature that structure-based drug design played a role in the discovery of that compound. In some cases, X-ray information may have been available during the lifetime of a particular drug discovery programme that eventually led to a drug, but the information may not have been utilised in compound design or may have become available later. Examples where the impact of structure-based design is unclear are Sitagliptin (MK-0431)[140,141] and saxagliptin (BMS-477118),[142] both inhibitors of dipeptidyl peptidase IV for the treatment of diabetes, as well as febuxostat (TEI-6720),[143] an inhibitor of xanthine oxidase for the treatment of hyperuricemia or gout. It would go beyond the scope of this chapter to undertake a detailed analysis and explore the reasons why in those programmes structure-based information did not impact on design.

Throughout this chapter, the picture representations of X-ray structures of drugs bound to their targets are intended as illustrations, rather than detailed analyses of the specific binding interactions between ligand and protein. We refer to the primary literature for this purpose, and the pdb code is included for every illustration, in case the reader wishes to undertake a more in depth analysis.

5.3.1 Angiotensin-Converting Enzyme Inhibitor as a Drug for the Treatment of Hypertension: Captopril

Although the discovery of captopril has been attributed to structure-based drug design,[7] the evidence from the primary literature is inconclusive.[108] It appears that a hypothesis was developed based on the sequence similarity between angiotensin-converting enzyme (ACE) as a zinc-containing exopeptidase and the pancreatic carboxypeptidases. The mechanism of action was believed to be similar to that of bovine pancreatic carboxypeptidase A (CPA), which was considered to be well understood at the time (in 1977).[108] It is not clear from the literature whether the structure-based design of captopril went beyond the use of a schematic representation of the binding pocket similarity between ACE and bovine CPA. Therefore, our conclusion is that although protein sequence and structure considerations were taken into account, the discovery of captopril was not based on a three-dimensional assessment of the target structure, or that of a related target structure, by means of homology modelling.

5.3.2 Carbonic Anhydrase Inhibitors Dorzolamide and Brinzolamide as Drugs for the Treatment of Glaucoma

Dorzolamide was launched in 1995 as a treatment for glaucoma. It inhibits human carbonic anhydrase II (HCA II) and lowers ocular pressure. It is based on a thienothiopyran-2-sulfonamide scaffold, which was first reported in 1987 as a class of carbonic anhydrase inhibitor with good aqueous solubility compared to earlier agents for the treatment of glaucoma, such as acetazolamide.[144] Carbonic anhydrase inhibitors and their therapeutic potential in a number of indications have been reviewed extensively elsewhere and will therefore not be discussed in detail.[145–147] The general strategy for the design of a topically active HCA II inhibitor is highlighted by the development of the initial prototype lead, MK-927.[5,109,144] It is water-soluble and rapidly penetrates ocular tissue, lowering intraocular pressure in animal models. X-Ray structures of the two enantiomers in HCA II revealed very similar binding modes, which did not rationalise the 100-fold difference in potency between the *S* and *R* enantiomers [K_i of 0.61 (*S*, MK-417) and 71 nM, (*R*), respectively]. However, quantum mechanics calculations showed that the conformation of the *R* enantiomer appeared to be non-optimal with respect to the N–S–C–S dihedral angle, with calculated strain energy of about 1 kcal mol^{-1}. A second conformational difference between the two enantiomers involved the geometry of the 4-isobutylamino substituent. The side chain is *trans* in the *S* enantiomer and *gauche* in the *R* enantiomer. Again, *ab initio* calculations suggested that the *trans* geometry should be preferred by 1 kcal mol^{-1}. These two conformational features can account for most of the potency difference found between the two enantiomers (2.8 kcal mol^{-1}).[109] Further structure-based optimisation of the lead compound was achieved by incorporation of a methyl group at the 6-position of the thienothiopyran. This was thought to reduce the penalty for the higher energy pseudo-axial conformation of the isobutylamino group, which was required to fit the enzyme active site. A methyl group was introduced at the 6-position of the thienothiopyran ring system, which modelling predicted to eliminate the pseudo-equatorial preference.[148] To counter the enhanced lipophilicity introduced by the methyl group, the 4-isobutylamino group was modified to an ethylamino group. All of the four possible optical isomers were prepared and their X-ray structures in HCA II were determined.[109] Again, molecular modelling was applied to rationalise the observed potency differences and to guide further compound design, eventually leading to the discovery of dorzolamide (Figures 5.2 and 5.3).[109,144,148,149]

Brinzolamide, a close analogue of dorzolamide, was launched in 1999.[113] It is structurally very similar to dorzolamide, differing only in the replacement of the

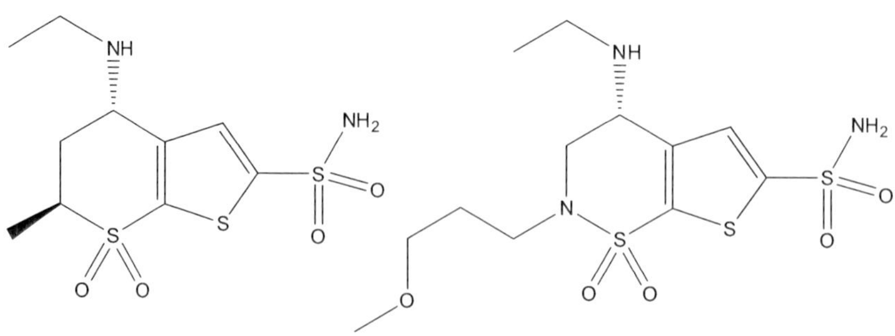

Figure 5.2 Structures of dorzolamide (left) and brinzolamide.

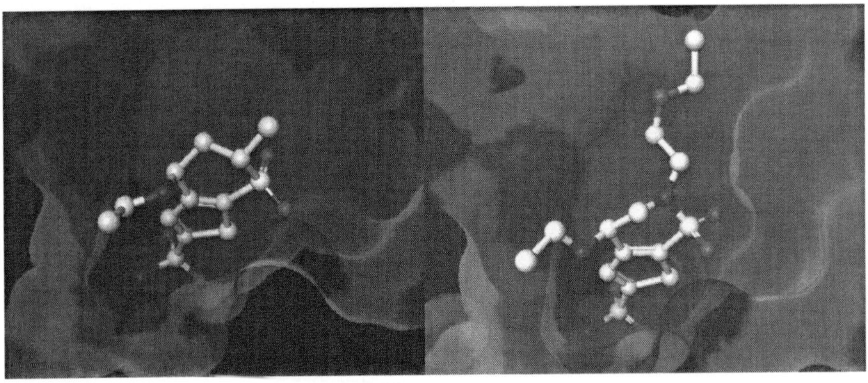

Figure 5.3 X-Ray structure of dorzolamide in human carbonic anhydrase IV (pdb 3fw3, left) and X-ray structure of brinzolamide in human carbonic anhydrase II (pdb 1a42).

stereocentre at the 6-position by introducing a nitrogen atom into the ring and extending the 6-substituent to a methoxypropyl group.[150] Further examples of carbonic anhydrase inhibitors discovered by structure-based design have been reported, but so far none of these drug candidates have made it to market.[151,152]

5.3.3 HIV Protease Inhibitors Saquinavir, Ritonavir, Indinavir, Nelfinavir, Amprenavir, Lopinavir, Atazanavir, Fosamprenavir, Tipranavir and Darunavir as Drugs against HIV Infection

Saquinavir (Ro 31-8959),[153] an inhibitor of HIV protease as a treatment for HIV infection, was launched in 1995 and was amongst the first wave of HIV protease inhibitors to be approved for the treatment of AIDS. It was followed by ritonavir (ABT-538)[111] and indinavir (L-735,524)[112] in 1996, along with nelfinavir (AG1343),[114] amprenavir (VX-478)[115] and lopinavir (ABT-378)[116] in 1999. Atazanavir (BMS-232632)[122] and fosamprenavir (phosphate prodrug of amprenavir)[123] followed in 2003, as well as tipranavir (PNU-140690)[53] and darunavir (TMC-114)[130,154] in 2005 and 2006 respectively. Therefore, there are currently 10 FDA approved HIV protease inhibitors on the market. After the X-ray structure of HIV protease was published in 1989 by several laboratories,[155–158] the race was on to find inhibitors that would be suitable as a therapy for AIDS. HIV research has been going on for almost 30 years now, and progress in the field has been reviewed recently.[159] For the design of the first generation HIV protease inhibitors, several hypotheses were developed, yielding diamino alcohols and diamino diols as transition state mimic HIV inhibitors. With the exception of tipranavir, all currently marketed inhibitors of HIV protease are peptidomimetic transition-state analogues that contain a non-hydrolysable transition state isostere. These drugs have the problem of poor bioavailability typical for peptide-based molecules. This class of drugs also suffers from resistance and cross-resistance, as well as specific toxicity profiles (*i.e.* lipodystrophy).[159] One approach to overcoming resistance is combination therapy, which has shown remarkable success in treating HIV infection.[160] Low bioavailability and resistance issues have prompted the search for new, non-peptidic inhibitors of HIV protease, with cyclic urea, 4-hydroxycoumarin, mannaric acid or 4-hydroxy-5,4-dihydro-2-pyrone as the central scaffold, instead of peptidomimetic hydroxyethylene. This approach eventually led to the discovery of tipranavir, which shows little cross-resistance with the peptidomimetic inhibitors.[159] The discovery of tipranavir will be discussed in more detail in Section 5.3.4. All currently marketed HIV protease inhibitors have been discovered with the aid of structure-based drug design, a remarkable achievement and clear evidence of the potential of the structure-based approach in the discovery of new medicines. For example, the discovery of indinavir has been described as utilising X-ray structures of known inhibitors, as well as docking and energy minimisation of compounds in the HIV active site.[112]

5.3.4 Non-Peptidic HIV Protease Inhibitor, Tipranavir as a Drug for the Treatment of HIV

Tipranavir (PNU-140690),[53] the first and so far only non-peptidic HIV protease inhibitor, was approved by the FDA in 2005 for the treatment of AIDS. At that time, there were already eight HIV protease inhibitors on the market and viral resistance had emerged for some therapies.[161] However, synthetic efforts towards peptidomimetic inhibitors of HIV protease faced challenges regarding low oral bioavailability and high clearance; non-peptidic inhibitors were potentially able to overcome those challenges. The discovery of tipranavir started with the identification of procoumon, a weak, non-peptidic HIV protease inhibitor, *via* broad screening of the Pharmacia & Upjohn compound library. Therefore, compounds based on procoumon would constitute an alternative scaffold which could potentially avoid issues for peptidomimetic inhibitors mentioned above.[161] Procoumon inhibited HIV protease with a K_i of 1 μM, and showed weak antiviral activity (ED_{50} 100–300 μM).[53] This provided the starting point for several iterations of structure-based design, leading to enhancements in terms of potency. Replacing the 4-hydroxycoumarin ring system with the corresponding 4-hydroxypyrone system resulted in compound PNU-96988, with an improved K_i of 38 nM. Further introduction of carboxamide functionality at the *meta* position of the phenyl side chain led to a compound with a K_i of 68 nM. Improvements in potency were achieved by combining these changes and modifying the amide into a sulfonamide functionality, eventually leading to PNU-140690 (Figures 5.5 and 5.6).[53] Representative compounds are shown in Figure 5.4.

Despite the fact that there are already 10 marketed HIV protease inhibitors, discovery and development of HIV protease inhibitors is still a very active field of pharmaceutical research.[162] This is largely due to the deficiencies of current therapies with regards to bioavailability and resistance, as well as risks in combination therapy relating to P450 inhibition.[159]

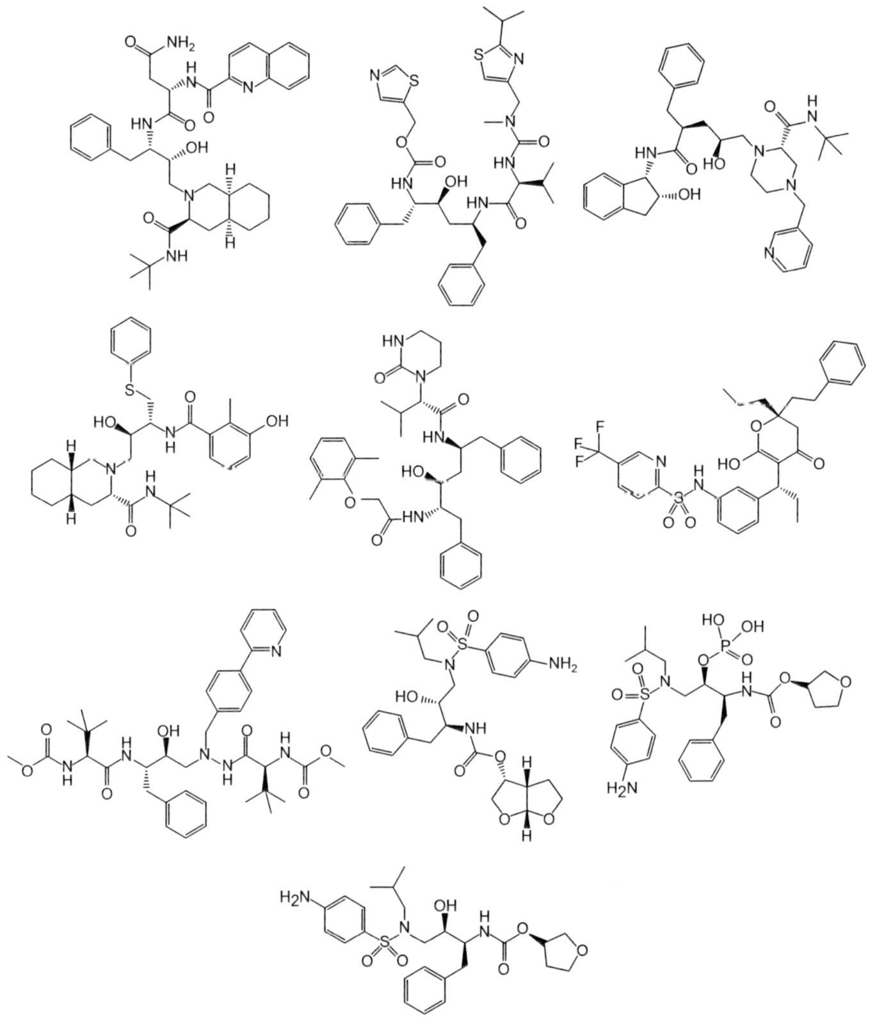

Figure 5.4 Structures of marketed HIV protease inhibitors. Saquinivir (top left), ritonavir (top middle), indinavir (top right), nelfinavir (second row left), lopinavir (second row middle), tipranavir (second row right), atazanavir (third row left), darunavir (third row middle), fosamprenavir (third row right) and amprenavir (bottom).

5.3.5 Nucleoside Analogue HIV Reverse Transcriptase Inhibitors as Drugs for the Treatment of HIV and Hepatitis B: Zidovudine, Didanosine, Stavudine, Lamivudine, Abacavir, Emtricitabine and Entecavir

Nucleoside analogues have been used as treatments for HIV and have been the first line of defence against acquired immunodeficiency syndrome (AIDS) since

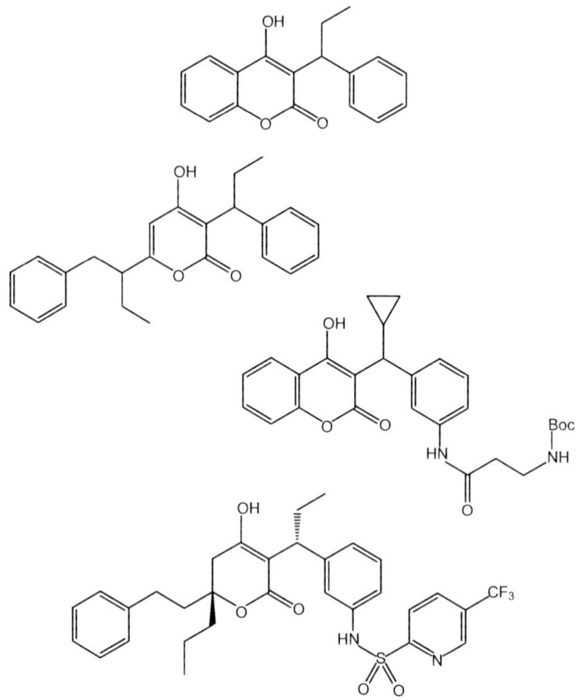

Figure 5.5 Structures of procoumon (top), PNU-96988 (middle left), carboxamide analogue of procoumon (middle right) and tipranavir (PNU-140690, bottom).

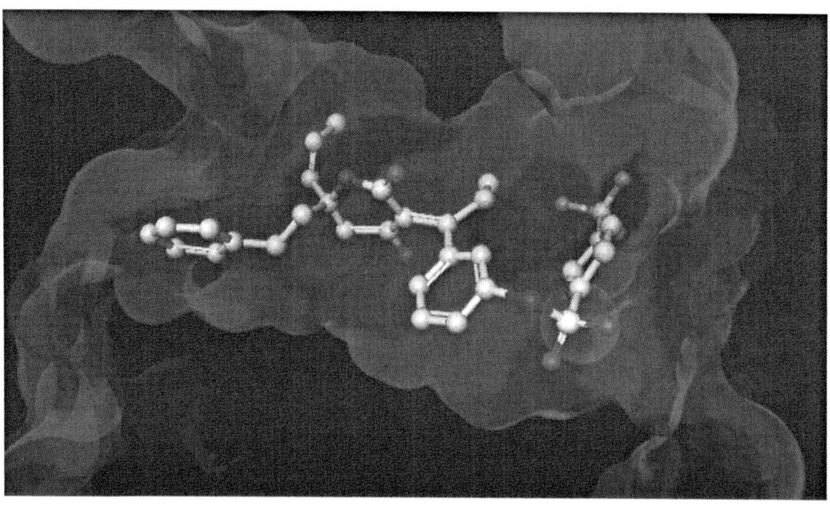

Figure 5.6 X-Ray structure of tipranavir with surface representation of the active site of HIV-1 protease (pdb 1d4s).

the approval of zidovudine (AZT, Retrovir, Retrovis)[163] in 1987. Since then, didanosine (ddI, 1991), stavudine (d4T, 1994),[164] lamivudine (3TC, BCH-189, 1995),[165,166] abacavir (ABC, GX623, 1592U89, 1998),[167,168] emtricitabine (FTC, 2003)[165,169] and tenofovir (TDF, PMPA, 2006)[170] have been approved by the FDA for the treatment of AIDS. Two of these retroviral drugs, lamivudine and tenofovir, have also been approved for the treatment of hepatitis B. Adefovir (2002)[171] and entecavir (2005),[172] which are in the same class of nucleoside analogue drugs, have been developed and approved for the treatment of hepatitis B only. One of the drawbacks of nucleoside reverse transcriptase inhibitors (NRTIs), as well as other classes of HIV agents, is the emergence of drug-resistant variants.[173,174] X-Ray structures of HIV-1 reverse transcriptase have been available since the early 1990s (*e.g.* pdb 1hrh, 3hvt).[175,176] The discovery of zidovudine (in 1964),[177] didanosine (in 1966),[178] stavudine (in 1966)[179] and lamivudine (in 1989) preceded the availability of HIV reverse transcriptase crystal structures and, based on the relevant publications, it can be assumed that structure-based design played no role in their discovery. Interestingly, even for later antiretroviral drugs like abacavir, adefovir and enticavir, there is no evidence in the literature for the use of structure-based drug design of nucleoside analogues. However, it should be noted that generating protein crystal structures on NRTIs is an extremely challenging task, due to the multicomponent complex that is required.[180] Another reason could be the close structural relationship to nucleosides and the fact that there appears to be a large number of known nucleoside analogues, which have been synthesised since at least the early 1960s. So it appears that the discovery of NRTIs has been driven more by screening of nucleoside analogues for antiviral activity than by the design of new scaffolds. However, this seems to have changed recently since new classes of NRTIs with less cross-resistance with existing treatments are highly desirable, and design efforts in this area are ongoing, utilising structure-based design in the process.[181]

5.3.6 Non-Nucleoside Reverse Transcriptase Inhibitor as Drugs for the Treatment of HIV: Efavirenz, Nevirapine, Delavirdine, Apricitabine and Etravirine

One of the targets for structure-based drug design over the last twenty years has been the non-nucleoside binding site of HIV reverse transcriptase (RT).[182,183] Drugs that bind the nucleoside binding site of RT—the nucleoside reverse transciptase inhibitors (NRTIs)—have been known for many years, but they all suffer from the effects of viral resistance through binding site mutations. In contrast to the nucleoside binding site which NRTIs bind, the non-nucleoside binding site of HIV RT only exists as a cavity when ligands are bound.[183] There are currently 4 non-nucleoside reverse transcriptase inhibitor (NNRTI) drugs on the market. Nevirapine[184] was approved by the FDA in 1996, followed by delavirdine (U-90152S) in 1997,[185] efavirenz (L-743,726, DMP 266) in 1998[186] and etravirine (TMC125-R165335, Figures 5.7 and 5.8) in 2008.[187,188] The discovery of nevirapine, delavirdine and efavirenz preceded the availability of

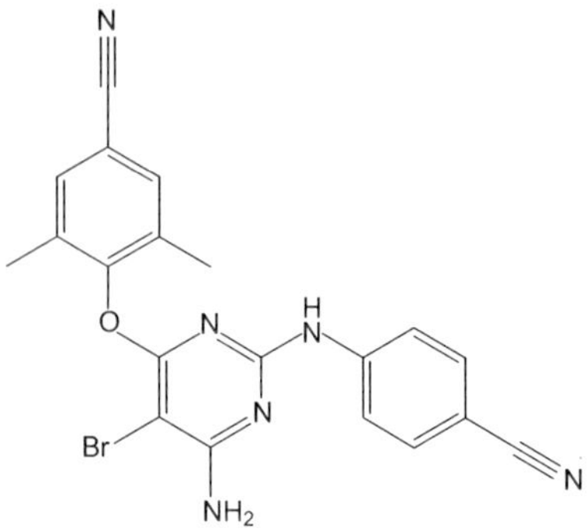

Figure 5.7 Structure of etravirine.

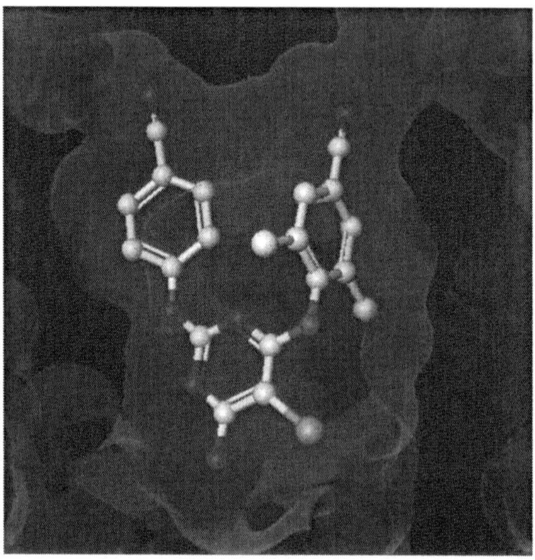

Figure 5.8 X-ray structure of etravirine in HIV RT (pdb3mec).

X-ray structures of HIV RT, so structure-based design did not seem to play a role in their discovery. The discovery of NNRTIs appears to have started with selective screening of the compound library in cell culture at Janssen Pharmaceutica in 1987, which led to the discovery of the first generation NNRTIs, the TIBO (tetrahydro-imidazo[4,5,l-jk][1,4]-benzodiazepin-2(1H)one) compounds. Two of these first generation NNRTIs, tivirapine, a TIBO derivative,[189] and loviride, an α-APA (α-anilinophenylacetamide) derivative,[190] were effective against wild-type HIV-1 but had significantly lower potency when tested against common NNRTI-resistant mutants. As chemical modifications were introduced in these TIBO and α-APA derivatives, a systematic structure-based molecular modelling study played a key role in understanding the three-dimensional structure-activity relationships in these two chemically distinct series.[138] This information was used for predicting NNRTI potency against both wild-type and NNRTI-resistant HIV-1 variants and in designing new NNRTIs, effective against a wide range of HIV-1 variants. Crystal structure analysis of HIV-1 RT has revealed important features of the enzyme's structure and function, including details of NNRTI binding,[138] which were utilised in the discovery of NNRTI drugs. With the increased number of X-ray structures available, the impact of structure-based drug design in the discovery of novel NNRTIs is very likely to increase in future,[191] and novel inhibitors (for example lersivirine) discovered by utilising structure-based drug design have recently been published.[192] Although it is not entirely clear whether structure-based design has played a role in any of the currently marketed HIV NNRTI drugs, with the exception of etravirine, it is likely that due to the availability of X-ray structures of RT since the early 1990s, there have been some structure-based design efforts in many discovery programmes since then (for example GW678248)[193] which have not yet resulted in marketed drugs.

5.3.7 Thrombin Inhibitors Melagatran/Ximelagatran and Dabigatran as Drugs for the Treatment of Thromboembolytic Disease

Oral anticoagulants for the treatment of thromboembolytic disease have been used in clinical practice for more than 60 years. The most commonly prescribed oral anticoagulant has been warfarin, or longer acting coumarin. Although warfarin is an effective anticoagulant, it has a narrow therapeutic window with significant risks of hemorrhage at therapeutic drug concentrations. This unpredictable and variable pharmacological response necessitates frequent monitoring of prothrombin time, raising the need for alternative oral treatments with wider therapeutic windows and fewer safety implications.[194] The first oral, direct thrombin inhibitor for the prevention of thrombosis, ximelagatran, was launched in Europe in 2004. However, marketing and development was halted in 2006 because of potential hepatotoxicity and the need for monitoring liver biochemistry for at least the first 6 months of treatment.[194] Despite its withdrawal from the market, ximelagatran provided proof of concept for the efficacy of oral, direct thrombin inhibition.[195] Ximelagatran is the prodrug of melagatran, the actual inhibitor of thrombin, since melagatran has low membrane permeability and is poorly absorbed after oral dosing. However, the oral bioavailability of ximelagatran is approximately 20%.[194] Ximelagatran undergoes rapid enzymatic conversion to melagatran *via* hydroxy-melagatran, formed by hydrolysis of the ethyl ester.[194] The structures of ximelagatran and melagatran are shown in Figures 5.9 and 5.10.

In 2008, dabigatran (BIBR-953) had also been approved by the FDA for use as an anticoagulant for the treatment of thromboembolytic diseases.[137] It is marketed as a double prodrug, dabigatran etexilate.

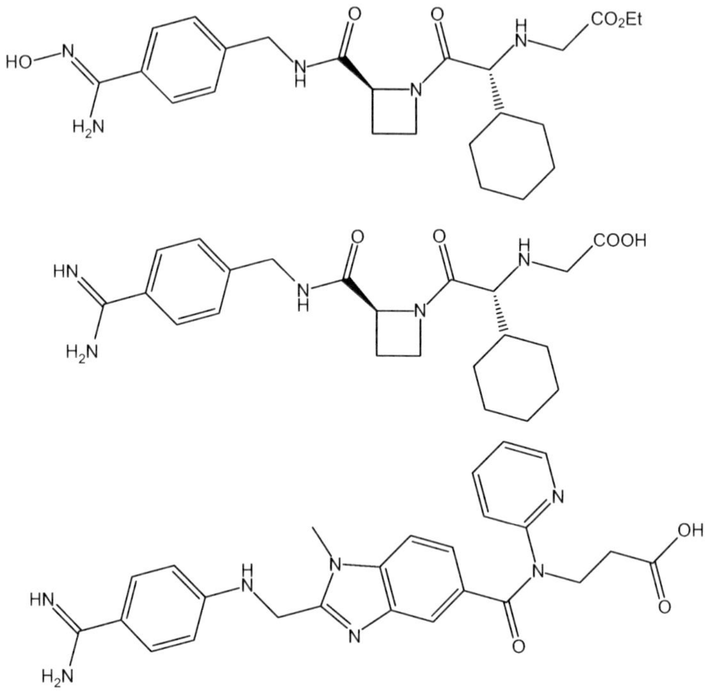

Figure 5.9 Structures of ximelagatran (top, prodrug of melagatran), melagatran (middle) and dabigatran (bottom).

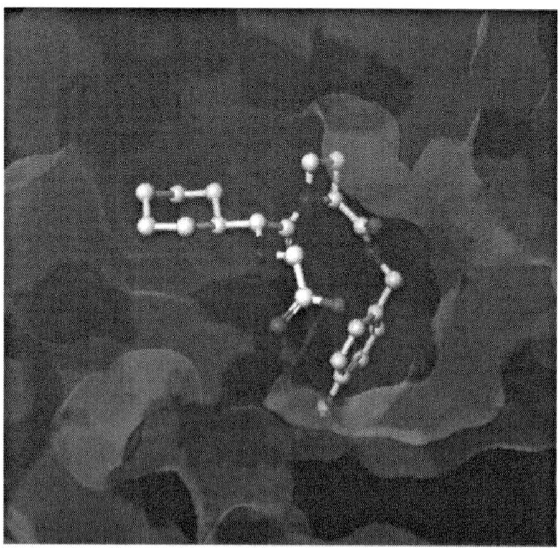

Figure 5.10 X-Ray structure of melagatran in thrombin (pdb 1k22).

5.3.8 Inhibitors of Factor X as Drugs for the Treatment of Thromboembolytic Disease: Rivaroxaban and Apixaban

One of the most important recent advances in drug development for thromboembolytic disease has been the recent introduction of oral direct factor X (FXa) inhibitors into clinical practice as a new class of anticoagulants.[195,196] The discovery and development of FXa inhibitors in recent years has been reviewed extensively.[195] One of the first potent FXa inhibitors was DX-9065a, an intravenous agent due to its low bioavailability of 2–3% in humans, which had a K_i of 41 nM against FXa and significant selectivity over thrombin (K_i > 2000 μM) and trypsin ($K_i = 620$ nM).[197] Since then, a large number of different scaffolds have been discovered to inhibit FXa while having more favourable molecular properties for potential oral absorption.[197] One of these was apixaban (BMS-562247, Figures 5.11 and 5.12),[198] which although not yet launched, clearly highlights the value of structure-based drug design in the discovery of small molecule oral agents and the progress that has been made over recent decades in the use of X-ray structures of pharmaceutically relevant biological targets. Clinical trials on apixaban for thrombosis and other indications are ongoing.[199]

More recently (2008), rivaroxaban (BAY 59-7939) has been approved in both the EU and Canada for use in the prevention of venous thromboembolism.[136] The discovery of rivoxaban started with the optimisation of a lead compound from high throughput screening (HTS), which yielded a potent FXa inhibitor (IC_{50} 120 nM) leading to isoindolinones with IC_{50} values in the low nanomolar range.[136] In this series however, the target pharmacological profile could not be achieved as compounds with sufficient potency generally showed low bioavailabilities due to low absorption. During the lead optimisation process, it was learned that the 5-chlorothiophene-2-carboxamide moiety was essential for potent FXa inhibition,[136] although it is not clear whether structure-based drug

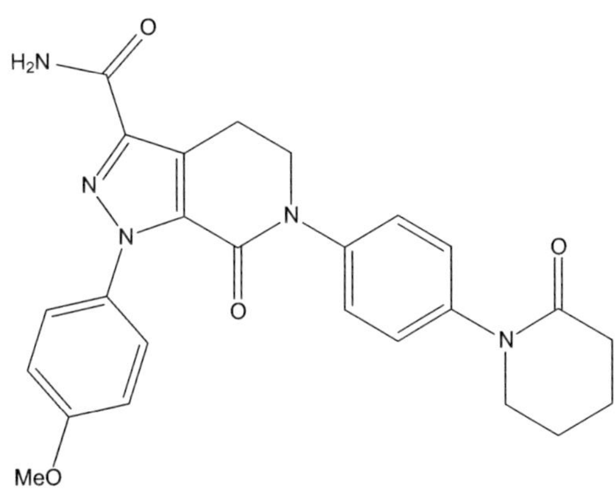

Figure 5.11 Structure of apixaban.

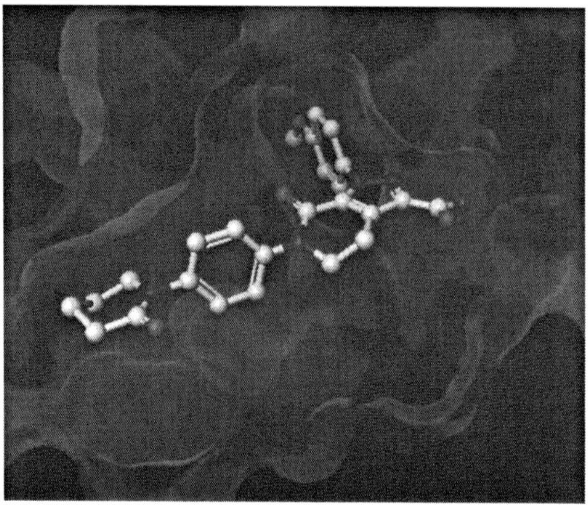

Figure 5.12 X-Ray structure of apixaban in factor X (pdb 2p16).

design was involved at this stage. After re-evaluation of the HTS hits by similarity considerations, oxazolidinone type compounds were identified as very weak inhibitors of FXa (IC_{50} 20 μM). Applying knowledge from the isoindoline class and replacing the thiophene with a 3-chlorothiophene resulted in a new lead compound with more than a 200-fold improved potency (IC_{50} 90nM). A clear preference for the *S*-configuration at the oxazolidinone core was observed, indicating a specific interaction with FXa.[136] Again, it is not clear from the publication whether structure-based considerations played a role up until this point, although similarity considerations are specifically mentioned. Therefore it is probably fair to assume that structure-based design was not playing a major role in the identification of the lead compounds. The lead compound derived from the aforementioned efforts offered an excellent starting point for optimisation to an orally active FXa inhibitor. It did not have a highly basic group, such as amidine, or indeed any other basic groups, which were for some time believed to be essential for interaction with the S1 pocket but which usually contribute only to poor oral absorption.[200,201] Based on this promising lead, a medicinal chemistry programme was initiated in order to evaluate the SAR and to further improve potency and the pharmacokinetic profile. It appears that X-ray crystallography was used to clarify the binding mode and the stringent requirements for high affinity observed within the oxazolidinone series. This implies the use of structure-based design and binding mode hypotheses, which were then confirmed by X-ray structure analysis. A schematic of the optimisation steps from the early HTS hit to rivaroxaban is shown in Figure 5.13. The X-ray structure of rivaroxaban in factor Xa is shown in Figure 5.14.

Encouraged by these successes in developing orally active therapies for thromboembolytic diseases, research in the area of FXa and thrombin inhibitors remains very active, with very recent examples of the discovery of more orally active compounds.[202]

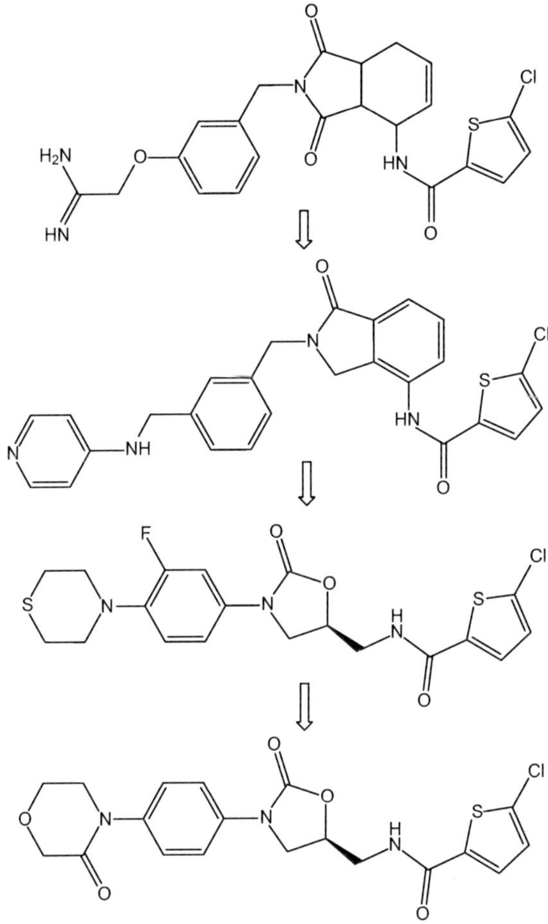

Figure 5.13 Lead optimisation of FXa inhibitors from an early HTS lead (top) to rivaroxaban (bottom).

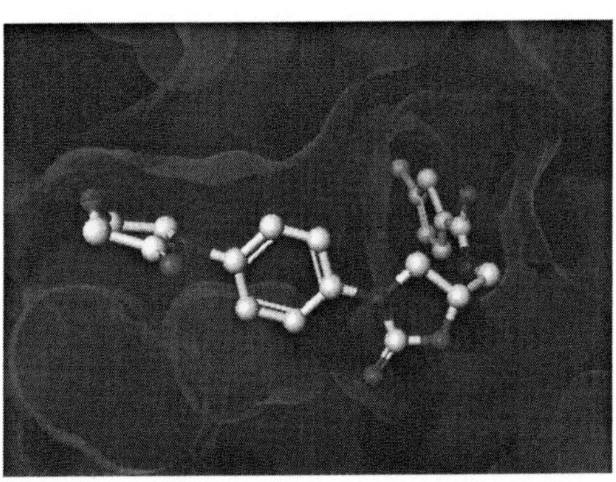

Figure 5.14 X-Ray structure of rivaroxaban in factor Xa (pdb 2w26).

5.3.9 Renin Inhibitor Aliskiren as Drug for the Treatment of Hypertension

Renin is a member of the family of aspartic proteases, and it controls the first and rate-limiting step of the renin–angiotensin system, catalysing the cleavage of the Leu10-Val11 peptide bond of angiotensinogen and releasing decapeptide angiotensin I.[203] Angiotensinogen is the only known physiological substrate for renin; therefore renin is an absolutely essential and very specific enzyme. A large variety of peptide inhibitors of human renin, with different stable transition-state analogues of the scissile peptide bond, have been developed. Because none of these compounds succeeded in drug development, there was a need for new classes of non-peptidic renin inhibitors that would fulfil all criteria for becoming drugs.[134] Three-dimensional structures of a number of mammalian and viral aspartic proteases have been determined, and crystal structures of human and mouse renin in complex with peptide and peptidomimetic inhibitors have been reported.[204] Peptidic inhibitors of renin have been known for many years and have suffered from low bioavailability.[134] The search for lower molecular weight, non-peptidic compounds was therefore initiated, to achieve good oral bioavailability and efficacy in animal models and stability against metabolic degradation. After analysis of the shape and chemical properties of the active site of human renin, initially using a protein homology model, it became apparent that the S1 and S3 pockets together form a large hydrophobic cavity. New compounds were designed to complement the extended hydrophobic surface corresponding to the large S3-S1 cavity, potentially leading to an increase in binding energy through improved van der Waals contacts. This would also allow elimination of the peptide main chain from P1 to P4, leading to the non-peptide inhibitors shown in Figures 5.15 and 5.16. Variations and extensions of the P3 moitety were explored synthetically and analysed by X-ray crystallography in an iterative design process.[134]

Aliskiren was launched in 2007 as a treatment for hypertension and is one of the most recent examples of the impact of structure-based drug design and the use of X-ray structural information, by molecular modelling techniques, for improving compound design with regards to potency and bioavailability.

5.3.10 Abl Kinase Inhibitors Imatinib, Nilotinib and Dasatinib as Drugs for the Treatment of Chronic Myeloid Leukemia

The impact of structure-based design on the discovery of kinase inhibitors has been very significant in the last decade, and there are numerous examples of the design of drug candidates.[205] This has been highlighted particularly for chronic myeloid leukemia (CML). The disease is characterised by increased and unregulated proliferation of predominantly myeloid cells in the bone marrow. It occurs most commonly in the middle-aged and elderly and accounts for 15–20% of all cases of adult leukemia in Western populations.[206] The underlying cause of CML is a characteristic reciprocal translocation between

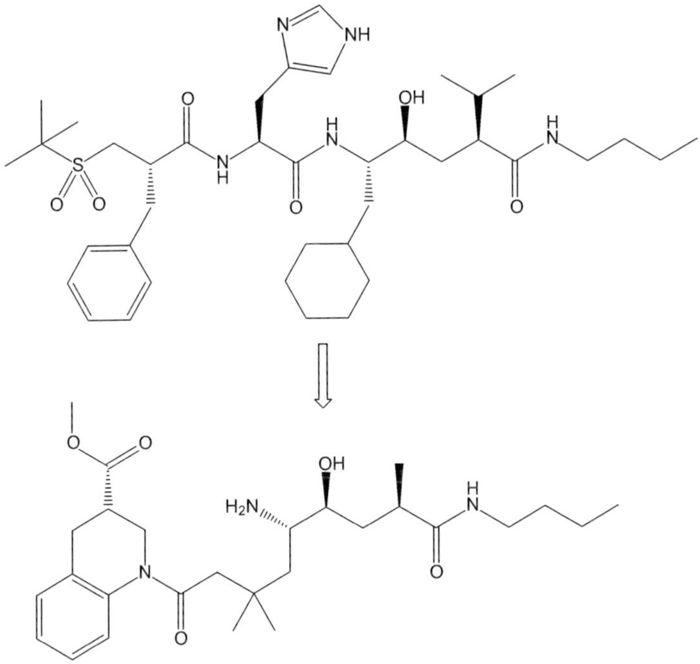

Figure 5.15 Structure of initial peptidic lead compound CGP 38'560 (top), which led to the design of non-peptidic aliskiren (bottom).

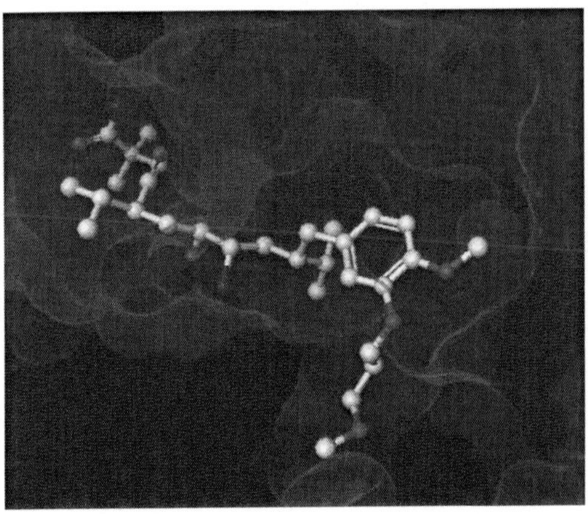

Figure 5.16 X-Ray structure of aliskiren in human rennin (pdb 2v0z).

chromosomes 9 and 22, which cytogenetically results in the Philadelphia chromosome and molecularly gives rise to the chimeric BCR-ABL gene. The protein product of this hybrid gene is an active protein kinase called the BCR-ABL kinase. The BCR-ABL kinase drives the pathogenesis of CML through the phosphorylation and activation of a broad range of downstream substrates that play critical roles in cellular signal transduction and transformation, thus representing potential therapeutic targets. In this respect, the remarkable clinical success of the ABL tyrosine kinase inhibitor imatinib (Gleevec, formerly STI571 or CGP 5148B) in the treatment of patients with CML has highlighted the potential of molecularly targeted anticancer therapies and has sparked the extensive development and use of such agents in cancers in general.[207,208] Protein kinases such as ABL have evolved highly specialised mechanisms for transitioning between active and inactive states, and crystal structures of inactive kinases have revealed a remarkable plasticity in the kinase domain, facilitating the adoption of distinct conformations. The highly conserved kinase domain of protein kinases consists of a bilobed structure in which Mg-ATP is located in a deep cleft between the N-terminal and C-terminal lobes.[209] Kinase conformations and binding motifs have been reviewed extensively elsewhere.[210] Most tyrosine kinase inhibitors (TKIs) have been designed to target the ATP site of the kinase in the active conformation. TKIs acting in this way have been termed type I kinase inhibitors. In contrast, type II kinase inhibitors bind preferentially to an inactive conformation of the kinase, locking it in that state and preventing its activation.[209] In addition to the ATP binding cleft, type II inhibitors bind in an adjacent hydrophobic pocket created by the activation loop, in which the phenylalanine of the conserved motif DFG (Asp-Phe-Gly) swings more than 10 Å away from its position (DFG-out) in the kinase active conformation.[209] TKIs targeting this hydrophobic pocket (for example imatinib) bind a group of amino acids that are less conserved than those surrounding the ATP binding site, giving rise to potential issues with mutations and resistance.[211]

However, resistance to imatinib is being increasingly encountered in the clinic, primarily a result of mutations within the kinase domain of BCR-ABL. To date, mutations at 17 different amino acid positions within the BCR-ABL kinase domain have been associated with the clinical resistance of imatinib in CML patients. Imatinib (formerly STI571 or CGP5148B, Figures 5.17 and 5.18) was approved by the FDA in 2001 and is now a frontline therapy for CML. It was initially discovered from a phenylaminopyrimidine (PAP) screening hit. Introduction of the so-called 'flag' methyl and a *meta*-benzamide functionality into imatinib converted the compound, which was initially a scaffold that targeted protein kinase C (PKC) and cyclin-dependent kinases (CDKs), into a compound that preferentially inhibited ABL kinase. Imatinib was obtained through addition of a methylpiperazine group to the *para* position of the benzamide, which greatly enhanced the water solubility and physicochemical properties of the compound. Although medicinal chemistry was based on a molecular model that incorrectly predicted that the PAP[119] motif would form the expected hydrogen bonding pair to the kinase hinge, the enzyme-selectivity assays still guided optimisation correctly. Once the crystal structure of imatinib in complex with ABL was solved, it became clear that the flag methyl and *meta*-benzamide substitutions had introduced conformational preferences and hydrogen bonding interactions that gave the PAP scaffold the

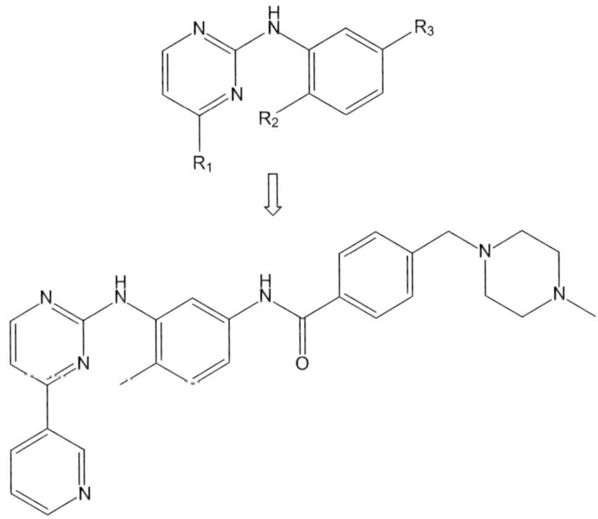

Figure 5.17 Structure of the initial PKC lead (top) and imatinib (bottom).

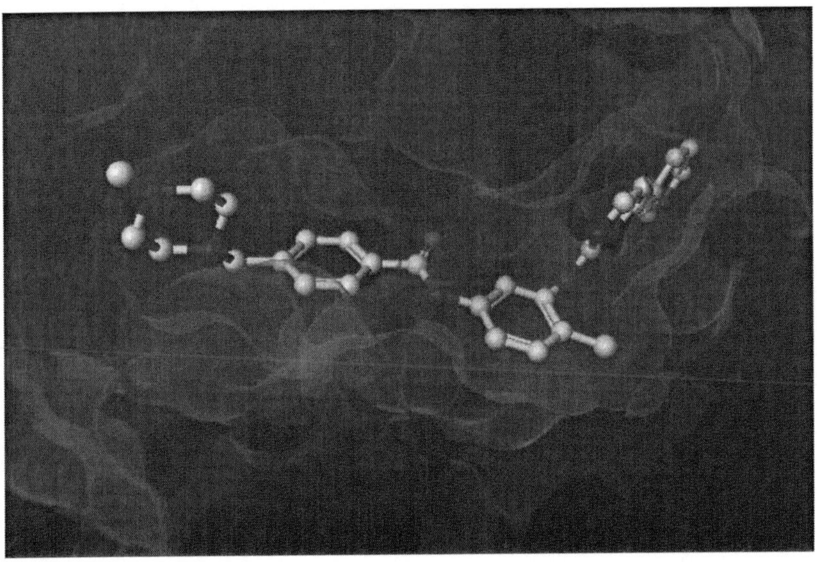

Figure 5.18 X-Ray structure of imatinib bound to human ABL kinase (pdb 1iep).

ability to stabilise the unexpected DFG-out conformation of ABL. Consistent with the observed binding mode, imatinib preferentially inhibits the enzymatic activity of the unphosporylated ABL ($K_i = 37$ nM) over the phosphorylated form ($K_i = 7$ mM).[212] The inability of PKCs and CDKs to assume the DFG-out comformation, combined with the energetic penalty resulting from imatinib's assumption of the pyrimidine *cis*-conformation (required for hinge binding), may provide an explanation for the selectivity of imatinib over other kinases. However, imatinib has recently been crystallographically demonstrated to bind to Syk, a target that it inhibits only very weakly (IC_{50} > 10 mM), by using the pyrimidine *cis*-conformation.[213] It appears that the discovery of imatinib has not been aided by structure-based drug design.[214] The second generation BCR-ABL inhibitors nilotinib and dasatinib have proved highly efficacious in patients with CML, following failure of imatinib therapy.[214] Both agents have shown remarkable activity against most imatinib-resistant ABL mutants, with the exception of the T315I mutation.[214] Although less frequent than ABL kinase domain mutations, other mechanisms linked to ABL TKI resistance have been identified *in vivo*; examples include BCR-ABL1 gene amplification,[215] overexpression of BCR-ABL protein, activation of SRC family kinases (SFKs) and transporters involved in drug efflux, suggesting that targeting this group of proteins could be a useful characteristic of agents to treat CML.[206] The c-Src proto-oncogene in particular plays a major role in the development, growth, progression and metastasis of a wide variety of human cancers.[216] Src activation, in the form of elevated kinase activity and/or expression levels, has been demonstrated in several major cancer types, including colon, breast, pancreatic, lung and brain carcinomas. Src kinase modulates signal transduction through multiple oncogenic pathways, including EGFR, Her2/neu, PDGFR, FGFR and VEGFR. It is therefore anticipated that blocking signalling through the inhibition of Src will be an effective means to modulating aberrant pathways that drive oncologic transformation of cells. Although imatinib was not discovered by structure-based design, the fact that molecular modelling was involved shows that efforts were made to rationalise the SAR and predict the feasibility of new designs. These efforts also provided the basis and learnings for subsequent work in the discovery of drugs that address the resistance issues with imatinib, particularly the increased awareness of kinase flexibility and the existence of alternative binding modes for compounds.

Nilotinib (AMN107) is a PAP derivative developed by the reconciliation of the crystal structures of imatinib and ABL kinase in complex.[133] Replacing the *N*-methylpiperazine ring in imatinib, which participates in hydrogen bond interactions with both Ile360 and His361, increases affinity for the inactive conformation of wild-type BCR-ABL 20-30 fold, while a similar activity against mast/stem-cell growth factor receptor (KIT; IC_{50} 60nM) and platelet-derived growth factor receptor (PDGFR; IC_{50} 57 nM) is maintained.[133] It appears that X-ray structure data of imatinib bound to ABL kinase has been used to develop hypotheses around the incorporation of alternative binding groups for the *N*-methylpiperazine group, while retaining an amide pharmacophore to maintain the H-bond interactions to Glu286 and Asp381. This structure-based approach resulted in the discovery of nilotinib (Figures 5.19 and 5.20).[133]

Dasatinib (BMS-354825, Figures 5.21 and 5.22) is a multikinase inhibitor with potent activity against BCR-ABL kinase (IC_{50}< 1 nM) and SFKs (IC_{50} 0.2–1.1 nM)[132] and has been approved by the FDA for the treatment of

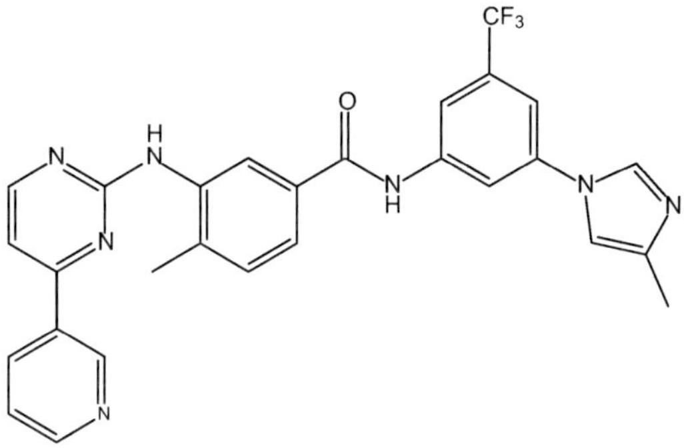

Figure 5.19 Structure of nilotinib.

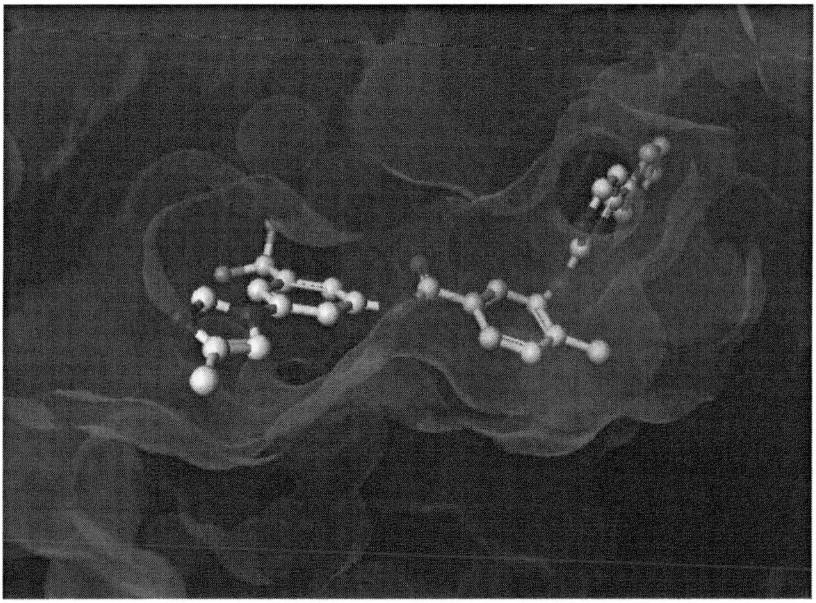

Figure 5.20 X-Ray structure of nilotinib bound to human ABL kinase (pdb 3cs9).

patients with CML, following failure with or intolerance to imatinib therapy. Unlike other ABL or SFK inhibitors, dasatinib was originally designed as an immunosuppressant. A potentially advantageous feature of dasatinib over imatinib and nilotinib is its ability to bind ABL with greater affinity, due, at least in part, to it recognizing multiple states of the enzyme.[132,214] The resolution of the co-crystal structure of dasatinib in complex with the active conformation of ABL appears to support this hypothesis.[217] Structure-based drug design based on the X-ray structure of the SRC kinase (pdb 2src) and molecular modelling appeared to be an important component in the discovery of dasatinib.[132]

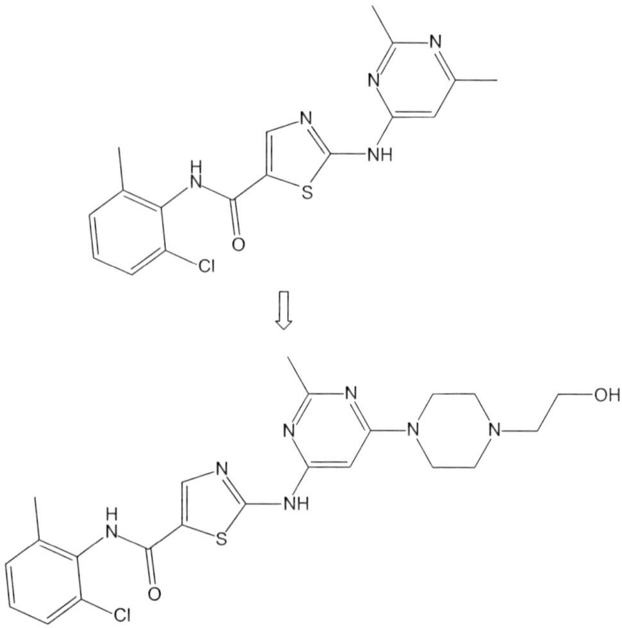

Figure 5.21 Structure of dasatinib.

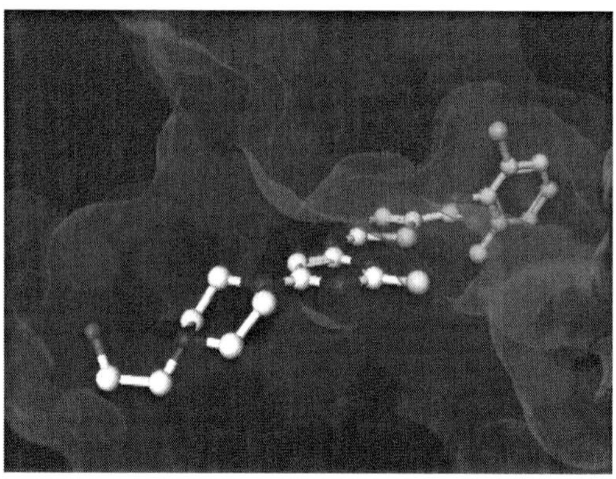

Figure 5.22 X-Ray structure of dasatinib in BCR-ABL kinase (pdb 2gqg). The binding orientation of dasatinib in cSrc kinase (pdb 3g5d) is virtually identical.

Unfortunately, the T315I mutation poses a formidable therapeutic challenge as it mediates complete resistance not only to imatinib, but also to many of the next generation of ABL kinase inhibitors, including both dasatinib and nilotinib.[206] Efforts to find compounds that overcome the issues caused by resistance to imatinib, nilotinib and dasatinib are therefore ongoing,[218] for example by combining allosteric with ATP site inhibitors,[219,220] and there have been recent reports in the literature of success in finding inhibitors for the T315I mutant in ABL kinase.[221,222]

5.3.11 Influenza Neuraminidase Inhibitors Zanamivir and Oseltamivir as Drugs for the Treatment of Influenza

Zanamivir[117] and oseltamivir (GS 4071, the ethyl ester prodrug is GS 4104)[118] were approved by the FDA in 1999 as treatments for influenza. Both compounds inhibit influenza neuraminidase, an enzyme that catalyses the cleavage of the terminal sialic acid attached to glycoproteins and glycolipids. This process is believed to be necessary for the release of the newly formed virus from infected cells, and for efficient spread of the virus in the respiratory tract.[223] For both zanamivir and oseltamivir (Figures 5.23 and 5.24), structure-based drug design seems to have been essential for the rationalisation of binding orientations in neuraminidase and SAR. The use of structural information from X-ray structures of sialic acid and its analogues complexed with influenza neuraminidase, as well as the use of molecular modelling techniques, is explicitly mentioned in publications regarding their discovery, specifically claiming rational, computer-assisted drug design.[117,118] Both drugs were based

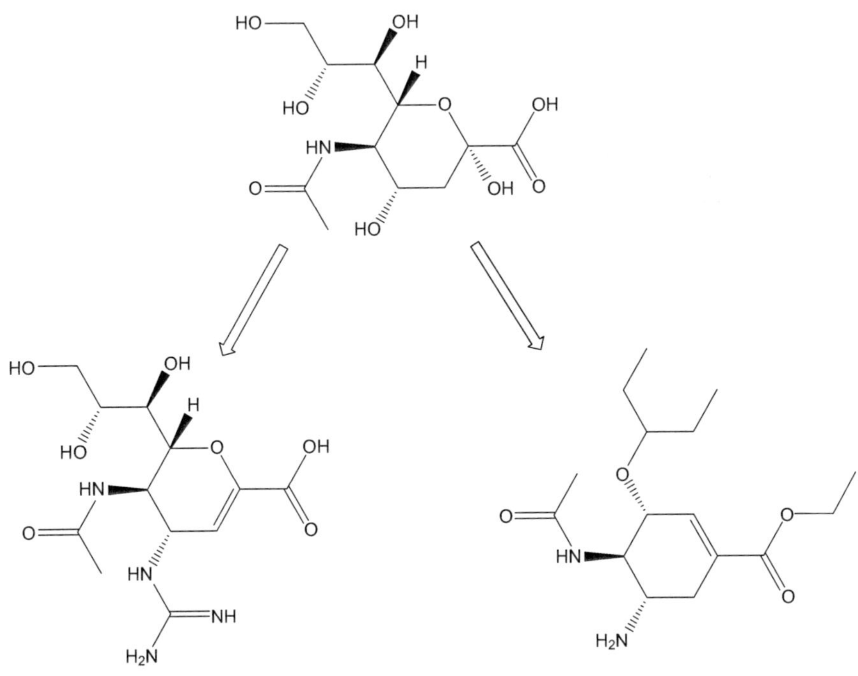

Figure 5.23 Structures of sialic acid (top), zanamivir (left) and oseltamivir.

Figure 5.24 X-Ray structures of zanamivir (pdb 3b7e, left) and oseltamivir (pdb 117f) in neuraminidase.

on the binding orientation of sialic acid in neuramidase, followed by modifications of the core ring as well as the side chains.

5.4 Has Structure-Based Drug Design Increased the Speed and Productvity of Drug Discovery? A Medicinal Chemist's Perspective

The previously mentioned examples of drugs discovered with the aid of structure-based drug design can give the impression that this technology has impacted on the productivity and speed of the research-based pharmaceutical industry. It is a tempting conclusion to make, but of course the control experiment has not been conducted. There are numerous success stories, particularly in the HIV protease and protein kinase area, but there are also structure-based drug discovery efforts which have been far less productive, such as the search for inhibitors of renin and thrombin. It could be argued however, that these failures are perhaps more associated with the physicochemical properties of the chemical leads and not the technology *per se*. This review seeks to give a frank perspective on the benefits and advantages of structure-based drug design, but also to highlight the pitfalls and hype surrounding this technology. In particular, it will draw on examples to highlight its impact and limitations on assessing druggability, as well as factors to consider when using structure-based drug design in lead generation, and subsequent optimisation of leads for potency, selectivity and pharmacokinetic properties.

Understanding the druggability of a novel protein target with an orally bioavailable, small molecule is one of the first questions a drug hunter should try to understand. In the absence of lead matter or structural knowledge of a novel target this is a very difficult question to answer without prior knowledge of the gene family. Hopkins *et al.* used a sequence similarity approach to understand the druggability of the human genome based on the druggability of known gene families,[224] suggesting that somewhere between 600–1500 human drug targets that would bind rule of 5 (Ro5) compliant small molecules are in existence. Rather than using the amino acid sequence of a given protein, and making assumptions about similarities in tertiary structure within a gene family, it would be more desirable to understand its tractability through data driven knowledge of topology, folding and the nature of the residues that make up the binding site of the specific protein. This can only be gained through structure determination.

The research-based pharmaceutical industry is aimed at discovering drugs, not ligands, therefore druggability should be considered not only in terms of potency but also selectivity, safety, pharmacokinetic and pharmaceutical properties, to highlight a few. With a structural understanding of the binding site of a novel target, considerable information can be deduced with regard to likely ligand properties. Having this information at the outset of a drug discovery programme could dissuade, or encourage, the drug designer to target a particular protein. This valuable information could point to the resources and investment that will be required to mitigate potential risks.

If one were to consider the druggability of kinases, as a gene family for example, it appears tractable with regard to generating potent leads, as generally the ATP binding site contains buried pockets which could conceivably deliver high affinity. However, with structural genomic knowledge in mind, it appears a daunting gene family to achieve selectivity within, if targeting the ATP binding site. Furthermore, if one considers the shape of the binding site, it is likely that relatively flat and lipophilic ligands will make up part of the pharmacophore, therefore low solubility and metabolic instability may be undesirable properties that will be encountered. Of course these observations are largely based on hindsight, but could conceivably have been recognised at the outset of the first kinase drug discovery programme if structural knowledge were available. These observations may have triggered an emphasis on risk mitigation strategies around, for example, selectivity and solubility. If this information was available early on in a chemistry programme, it may help triage chemical series.

The strong biological rationale for targets such as renin, thrombin and factor X have led to them being been pursued aggressively within the pharmaceutical industry, with many compounds entering clinical development, but most failing for pharmacokinetic and/or safety reasons.[134,195,225] Although drugs that are providing clinical benefit have emerged from these targets, significant investment has been made in their discovery and it is far from certain that this investment will yield a significant return. With suitable structural knowledge in hand at the outset of the drug discovery programme, coupled with a critical analysis of the binding site and likely lead matter, could the decision not to work on such difficult targets have been taken? Could it have changed the strategy and pointed to investment in other areas such as allosteric inhibitors, or other targets on the same pathway? Alternatively, could more have been invested in improving the understanding and predictability of the pharmacokinetics with respect to the lead chemical matter? Ultimately extremely valuable insights, such as Ro5,[47] arose from such issues but did so in a more 'knee-jerk' response, rather than a prospective manner. High quality structural information is sometimes not available at the outset of a drug discovery programme against a novel target, but the aforementioned observations and thoughts could point towards investing more heavily in obtaining this information before embarking on a costly HTS campaign or committing chemistry resources to equally expensive design–synthesis–screening cycles.

The previous discussion considers druggability of protein targets from a binding site perspective, but arguably the true power of structure-based drug design is not revealed until co-crystal structures of proteins complexed with ligand molecules are generated. Before considering the co-crystal structure of a ligand–protein complex, it is worth highlighting many of the pitfalls and caveats that should be considered, and in turn the crucial questions that must be asked by the medicinal chemist and structural biologist before embarking on a design strategy based on structural information. Before a co-crystal structure is generated, the protein must be cloned, expressed, purified and then co-crystallised with a ligand. Each of these steps is far from trivial and can require extensive and very time-consuming experimentation. Often the full length protein, with post-translational modifications, cannot be obtained, leading the biochemists to develop truncated or modified protein which becomes the substrate for generating co-crystals. Consequently, modified protein could potentially result in changes to the topology and folding of the protein and co-crystals may therefore not be representative of the native state. Co-crystallisation conditions may also play a part in producing structures that are inconsistent with native structures, with solvent and pH at times influencing the tertiary structure. Ultimately the drug designer will want to draw conclusions from structural data and relate this to the potency in a biochemical assay. Therefore the drug designer must be cognisant of the nature of the protein in the biochemical assay as well as the protein construct and conditions used to generate the co-crystal, to ensure 'apples are compared with apples, and not pears'.

Once a diffraction pattern from a co-crystal structure is obtained, it must be processed by *in silico* methods and the resulting electron density maps interpreted by the structural biologist. This involves expressing an electron density map in terms of a chemical structure. Interpreting this data can therefore lead to some degree of subjectivity, particularly when lower resolution data is obtained. Allied to this, a co-crystal structure should not be considered as a representation of a static ligand bound to a static protein. In reality, the protein and the ligand are constantly flexing, twisting and moving many thousands of times a second, which is of course akin to the ligand bound to protein in its native environment. For this reason what one actually observes from a co-crystal structure is a time averaged snapshot that has been inferred from an electron density map, aiming to describe a dynamic system.[44,103] The subjectivity involved in interpreting and refining this data can mean that differing representations of the data could be derived from one structural biologist to another. There have been several cited retractions of published structures where this data has been carefully peer reviewed.[44] Therefore constructive challenge and seeking a second opinion is sometimes a sensible measure.

When one obtains a crystal structure it is also important to know at what resolution the structure has been determined, in order to get some idea of what useful information can be derived. Typically, resolutions of about a single bond length, or ≤ 2Å, are considered high resolution. Therefore using structures with lower resolution (≥ 2Å) in design where geometry and distance are key considerations (as in optimising hydrogen bonding), can provide often misleading information and could result in a lot of head-scratching and potentially wasted time and resources. A well cited example is the position of flexible side chains such as lysine or glutamate which can sometimes be poorly resolved; therefore targeting interactions with the side chains of such residues must be done with input from the structural biologist.[226,227] For macromolecular type resolutions, the ligand itself can sometimes be difficult to accurately model from the electron density map, when compared to a small molecule X-ray crystal structure. Therefore vectors can be misinterpreted. Atoms such as nitrogen and carbon are very difficult to differentiate, and spatial arrangement of some nitrogen-containing heterocyclic ring systems must be inferred from local hydrogen bonding networks or is simply not known. The same can be said for ligands that can undergo tautomerisation.[91] In summary, to confidently use structural information in a rational manner, whether it is targeting an interaction with a side chain or drawing conclusions on ligand conformation and/or tautomerism, the drug designer must use the structural information in close collaboration with structural biologists and molecular modellers.

A protein–ligand co-crystal structure also contains a significant amount of water molecules. Some crystals may contain up to 70% water.[45] Pockets within the protein structure, both lipophilic and polar, will contain water molecules.[67,69] These water molecules may make up the bulk solvent, while others could be structural water molecules which are integral to the tertiary structure of the protein. Again, the position of water molecules must be inferred from the diffraction pattern and therefore low resolution structures may not be suitable for mapping the local hydrogen bonding network. Structural water molecules can potentially lead to a larger gain in binding energy if displaced, compared to bulk water molecules, and it is therefore tempting to target these for displacement with a suitable isostere. This tactic is more often than not met with a decrease in binding affinity, for example in thrombin.[73] Despite major efforts, replacement of the basic S1 group in thrombin inhibitors (argatroban and melagatran) with a lipophilic residue, thereby displacing a conserved water molecule in the recognition pocket, has not yet led to a marketed drug. Marketed inhibitors contain the basic substituents of the earliest known thrombin inhibitors, which are essentially mimics of arginine in the natural substrate.[225] This relatively high likelihood of failure of replacing structural water molecules could be due to an unsuitable isostere that does not satisfy all of the hydrogen bonding contacts exhibited by the targeted water molecule; or perhaps due to the fact that the local hydrogen bond network of the targeted water molecule was not well defined from the structural information, hence the designer was not well informed; or it could be a combination of the two. A sensible conclusion therefore might be not to target structural water molecules for displacement, unless this strategy has advantages in terms of achieving better drug properties and a significant loss in potency can be tolerated. Instead those water molecules could be considered part of the protein structure, and it might be more successful to design ligands to form hydrogen bonds to them directly, or simply leave them alone.

Lead generation and evolution are fundamental steps in early drug discovery, but identifying chemical starting points for novel targets is not trivial.[224] Understanding the binding mode of an endogenous ligand offers the opportunity to design novel leads *de novo*. Two such examples are the neuraminidase inhibitors oseltamivir and zanamivir.[117,118] Through understanding the binding mode of the the natural ligand, sialic acid, the respective medicinal chemistry teams were able to hydridise the chemical structure of sialic acid towards a more potent molecule. In the case of zanamivir, the introduction of a polar basic group led to a potent compound with properties suitable for topical application to the lung *via* inhalation. Structure-based design enabled the Gilead group to optimise for potency whist maintaining physicochemical properties to enable oral bioavailability. The endogenous substrate of the enzyme renin, angiotensinogen, also inspired a variety of peptide-based inhibitors as chemical starting points, with an understanding of the binding mode and catalytic activity of the enzyme being a key factor in the design of lead compounds. These few examples are amongst many that highlight the power of structural knowledge of endogenous binding modes and how this knowledge can be used in idea generation towards identifying novel potent leads. It is worth noting, however, that even though structure-based design has enabled lead generation and evolution of targets such as renin, through knowledge of binding modes, it has not translated into rapid discovery of multiple drugs in this class.[134]

An alternative to lead generation through structural knowledge of endogenous substrates is HTS of compound collections.[228,229] Screening for low molecular weight (MW <250 Da) fragment-based leads has seen an explosion in popularity in recent years.[27,29,31,230] Although this strategy cannot claim to have led to the discovery of many marketed drugs, with the exception of Tipranavir perhaps,[53,161] its impact is yet to be realised. Fragment screening claims to cover more chemical space due to the reduced number of heavy atoms and in essence, a reduced complexity. This tactic often results in the identification of leads with high ligand efficiency[97] and if coupled with good synthetic tractability offers powerful starting points for drug discovery. Moreover, when used in combination with ITC, it can be used to identify leads with binding affinities enriched in enthalpic contribution and hence higher lipoidal efficiences.[68,75] To make rapid advances in the evolution of a millimolor or micromolar fragment lead into a nanomolar lead compound relies heavily on structure-based drug design. Predicting the binding mode of a fragment within a protein is often very difficult to achieve by virtue of its structural simplicity, and probably the currently available docking algorithms more often than not fail.[231,232] The 'growing' of fragments by addition of further pharmacophores requires careful consideration of geometry and distances which can really only be understood through structural knowledge. The linking of fragments is an even harder endeavour and more often than not leads to a loss of potency and therefore ligand efficiency.[61,233,234] In reality, fragment-based evolution involves the same structure-based design principles as conventional higher molecular weight lead evolution but offers the opportunity to optimise the ligand and lipoidal efficiencies of the starting point.

As alluded to, conformational understanding of binding modes of ligands in proteins is arguably one of the most powerful pieces of information that can be garnered from a co-crystal structure. Knowledge of binding modes is probably best used in idea generation. As highlighted, most proteins are flexible, even in the solid state, therefore binding pockets and protein loops should not be considered as rigid barriers. A structural analysis of the protein kinase gene family reveals many varied binding modes of ligands, despite the fact that all bind a common endogenous ligand, adenosine triphosphate (ATP). Examples of this include the BCR-ABL kinase inhibitor imatinib,[119,120] in which the drug binds in the ATP pocket, as well as a pocket adjacent to the ATP binding site that is created by a large (~10Å) movement of the activation loop, leading to a binding mode known as the DFG-out mode. Lapatinib, another marketed kinase inhibitor,[135] also occupies a region of the active site, remote from the ATP binding site created by a ligand induced movement of the C-helix, which also results in a modulation of the enzyme kinetics. These examples highlight the fact that that as ligands become larger and occupy more of the binding site of a protein, they can sometimes influence the native structure of the protein through inducing protein movement, emphasising that some proteins are inherently flexible. In addition, the same compound can bind to two kinases in very different orientations, as exemplified by imatinib.[213] Without the protein co-crystal structures, these binding modes could probably never have been rationalised through screening in biochemical assays, again highlighting the power of structure-based design in progressing and understanding SAR in a drug discovery programme. It could be argued that structure-based drug design has been hugely enabling with respect to understanding interactions at the molecular level, but a possible counterargument could be that it stifles innovation and serendipity. If one were to restrict the design of new inhibitors for the BCR-ABL kinase, for example, to fit the native DFG-IN binding mode of the kinase, then drugs such as imatinib may never have been discovered. Dedicating some design to challenging the model creates the opportunity to get lucky and make unexpected leaps forward.

As mentioned, once a binding mode of a compound is understood from high resolution co-crystal data, it can be an immensely powerful tool in idea generation. This knowledge drives lead hopping through hybridisation of existing chemical leads, guided by crystallographic overlays. This is often evident in the field of kinase research, where many different aromatic heterocyclic cores may be used to mimic the adenine heterocyclic structure of ATP. It offers the opportunity to modulate physicochemical properties and synthetic tractability of kinase leads. Another good example is in the field of non-nucleoside reverse transcriptase inhibitors, whereby researchers hybridised the marketed NNRTIs efavirenz and capravirine through co-crystal overlays to produce a novel series with improved metabolic stability and resistance profiles.[235] An increasingly important application of hydridisation and lead hopping is in the identification of novel chemical space. In kinase research, for example, the patent space is becoming increasingly congested and therefore the identification of novel intellectual property space through co-crystal inspired lead hopping is an important application.

Driving down primary potency against the desired biological target is paramount to the discovery of a drug molecule. In non-structural enabled drug discovery programmes this is largely an empirical process, based on design–synthesis–screening cycles to produce SAR and inform future design. Structure-based drug design, however, offers the opportunity to prospectively target regions of a binding site with the potential to improve potency, as well as optimise existing interactions. Whether potency is driven by lipophilic interactions, hydrogen bonding or various types of charged interactions, it is important to keep in mind the thermodynamics of the system as a whole. As mentioned at the beginning of this chapter, the pioneering work of Williams *et al.* on the vancomycin antibiotic began to unravel many of the thermodynamic contributors to an overall binding energy.[77] These, of course, involve many entropy and enthalpy contributions from both ligand and protein. It is impossible to measure all of these individually, although ITC will provide some insight. Therefore it should not come as a surprise that often our hypotheses around improving ligand potency, when reduced to practice, actually lead to no increase in potency, or a significant loss of potency when screening data is obtained. A co-crystal structure contains an abundance of information, however, and it is often subtle effects that go un-noticed, which may have a profound effect on a ligand's overall binding affinity.

A case study published by AstraZeneca on the discovery of iNOS inhibitors exemplifies many of the issues discussed here and highlights how sometimes structural knowledge can be a hindrance rather than a help.[45] This study involved a misinterpretation of an electron density map derived from low resolution data. From this misinterpretation, compounds were designed and tested. This resulted in a loss of activity that could not be explained but was eventually attributed to the designed ligands not satisfying all the interactions of the water molecule that the ligand had displaced. In such cases, a lot of time can be spent trying to rationalise negative data, with time and money wasted on medicinal chemistry design and synthesis. It could be argued that in such cases it may actually be beneficial to be blinded to the structural data, to prevent this type of 'wheel spin'. It would be an interesting experiment to take two design teams, one using structural data, the other blinded, and observe what strategies are adopted and how data is rationalised. What paradigm would deliver the most rapid and high quality advances in lead generation and optimisation?

Perhaps the best philosophy to adopt when investigating a hypothesis to improve potency from co-crystal data is to not 'over-design' ligands but instead design and synthesise a small group of structurally similar ligands that subtely probe geometry and volume. Moreover, one should consider the structural data from multiple protein–ligand structures, if available, when building hypotheses, as this will highlight subtle differences in binding mode and ligand induced changes that should be incorporated into design ideas. Using this type of approach, ideas can be triaged based on structural information and could mean chemical libraries are smaller, relative to those of non-structure-based programmes potentially moving the project forward more rapidly.

For many of the reasons highlighted, predicting potency from a co-crystal structure is fraught with danger. This is simply due to a lack of understanding of the thermodynamics of the system as a whole. It is therefore not surprising that *in silico* methods to predict binding modes, and in particular potency, are met with mixed support in the literature. Virtual screening tools such as docking programs should be used with caution when triaging 'hits' for progression to biochemical assays.[236] Many assumptions are made within these algorithms, particularly with regard to protein flexibility, therefore ligand induced protein movements may not be accounted for and important molecules and learnings may not be disovered.[55,237] As virtual screening is cheap and relatively quick, it offers the potential to be a powerful method for identifying leads based on structural data. However, as scoring functions probably do not have the required sophistication to accurately rank relative predicted potencies from a virtual screen, it is something that should be the focus of software designers working in this area.[238]

Achieving selectivity for a target in drug discovery programmes is an extremely important property but can often be a major hurdle, depending on the gene family being targeted. Selectivity strategies could involve increasing affinity for the desired target and decreasing affinity for closely related targets, or a combination of the two. The field of kinase research is a well cited area with regard to difficulties in achieving selectivity over the 500 or so present within the human genome. Structure-based drug design has played an important role in achieving this goal and is particularly useful for understanding the structural consequences of residue differences and whether or not these may be exploited. Occasionally, the size of lipophilic pockets differs between kinases, even though the residues making up the pocket are similar. Therefore one should not necessarily rely solely on sequence homology information to determine binding site differences, but should use it in combination with structure.

The kinase gene family has now succumbed to several relatively selective drugs that can be rationalised from a structural point of view. As exemplified previously, imatinib achieves selectivity for BCR-ABL over PKCs and CDKs by the introduction of the 'flag' methyl.[119,120] Although research and development efforts have not yet produced a marketed p38 kinase inhibitor, it has been an interesting story with regard to selectivity. Several opportunities to tune the selectivity profile have been identified. One of these is the relatively unique 'gate keeper' residue (Thr106) within the ATP binding site. It was discovered that this side chain may be moved, through introduction of an appropriate ligand, to open a pocket in p38 that was subsequently filled with the ligand, but sterically not tolerated in other kinases with larger 'gate keepers'. Several other p38 inhibitor classes have been shown to take advantage of a subtle amino acid flip in the hinge binding region (Gly110) of the kinase, resulting in a tolerated change for p38 but a decrease in affinity against a whole host of other kinases. Another strategy has involved the activation loop movement to produce the so called DFG-OUT mode that kinases such as BCR- ABL and p38 are able to accommodate but which is not possible in many other kinases, allowing selectivity to be achieved on this basis.[239,240]

In gene families where selectivity is one of the key issues, utilising one or more of these strategies may well improve a compound's selectivity, profile leading to a truly selective inhibitor. All of these examples highlight the need for a small or large ligand induced change to the protein, which is tolerated in the target protein but not in closely related selectivity targets. To reiterate a point made earlier, these structural differences between closely related proteins may never have been realised if it were not for the design, synthesis and screening of a molecule that challenged the binding site model accepted at that time. It is therefore crucial that this philosophy is adopted in all structure-based programmes.

There are examples, however, where a compound's selectivity profile cannot be rationalised from a structural perspective. Again, this relates to the point made earlier that small, ligand induced changes that go un-noticed actually contribute to profound selectivity differences. An interesting example of such a phenomenon is the JAK inhibitor CP-690,550. This compound exhibits remarkable *pan*-kinase selectivity which cannot easily be rationalised by structure.[241,242] When one analyses a co-crystal structure, we tend to focus on the residues that make up the binding site. But of course there are 'second shell' residues, remote from the binding site, that may play a role in binding affinity that is not usually considered. Learning more about the global residue differences between closely related proteins and the impact of these on structural changes, and therefore selectivity potential, would be worth investigating further.

Alongside optimising compounds for potency and selectivity, it is also crucial to consider physicochemistry and pharmacokinetic properties. Ro5 properties are usually a good guideline to keep in mind. As mentioned previously, many peptidic renin inhibitors have been described in the literature, with most suffering from poor oral bioavailability due to high clearance and poor permeability. Structural knowledge of the binding modes of these peptidic inhibitors inspired design teams to hydridise the peptidic structure into more metabolically stable and permeable peptidomimetic compounds, with aliskiren being the only drug to reach the market. The discovery of aliskiren is a remarkable success story given the difficulties in renin research. However, it has poor oral bioavailability and therefore significant improvements could be made. Could structure-based design enable these improvements? Have previous screening and design cycle campaigns challenged the accepted binding mode? Could a ligand induce protein movement that opens up hydrophobic pockets capable of binding high affinity, more moderately polar ligands?

The discovery of the neuraminidase inhibitor drugs zanamivir and oseltamivir highlights the potential for greater commercial success if physicochemistry is optimised carefully alongside potency.[117,118] As mentioned, the more polar drug zanamivir has little oral bioavailability and must be dosed *via* inhalation, while oseltamivir has more balanced physicochemistry and is orally bioavailable. The oral bioavailability of oseltamivir has led to it dominating the influenza market and experiencing significant commercial success. Clearly, improving the oral pharmacokinetic characteristics of drug compounds through the use of protein–ligand crystallography has been hugely impactful, paricularly in the field of peptidomimetic drugs.

Solubility is also key to achieving good oral bioavailability of drug compounds.[243] Certain drug classes, such as kinase inhibitors, are often plagued by poor solubility due to the flat aromatic heterocycles that frequent the kinase pharmacophore. Through an understanding of the binding mode of these compounds it is clear where the solvent exposed regions of the binding site are located and as a result where solubilising groups may be placed to improve aqueous solubility. This approach has been cited in the discovery in imatinib,[119,120] where poorly soluble leads were improved by addition of the piperazine basic centre. The addition of a basic centre also opened up the possibility of salt forms to improve solubility and pharmaceutical properties. Access to solvent-exposed regions also offers the potential to tune the physicochemical properties of leads.

5.5 Outlook

Over the last 25 years, structure-based drug discovery has become an intergral part of the research-based pharmaceutical industry. Its value and contributions have been highlighted by the fact that this technology has contributed at least in part to the discovery of the drugs listed in Table 5.1. The opportunities for structure-based drug design are ever increasing, with new structures of biological targets becoming available at an ever increasing rate. For example, the structural genomics consortium is producing more than a third of published structures against medically relevant targets, year on year. As well as individual novel targets, whole target classes, such as G-protein coupled receptors,[35,36] are succumbing to structural determination. As a result, huge areas of druggable target space[244–246] can potentially be guided by structure-based design. In addition, the discovery of new antibacterial drugs could be enabled by the availability of protein structures,[39] as well as recent successes in the discovery of new drug candidates using molecular modelling techniques.[247] Will other target classes, such as ion channels, yield high resolution structure determination? Advances in such areas will no doubt have impact on the discovery of important new medicines.

As discussed earlier, it can be very difficult even for the most skilled and experienced medicinal chemist to avoid drawing misleading conclusions from structural information; therefore it is crucial that data is assessed in consultation with the X-ray crystallographers, NMR scientists and computational chemists. Moreover, compound design should take place as a collaborative effort in order to get the maximum value from the available experimental data. As mentioned previously, there is much information that can be garnered from a co-crystal structure that goes beyond the obvious hydrogen bonding contacts, for example. When this more subtle information is used in combination with other biophysical techniques, such as ITC or surface plasmon resonance, together with potency derived from biochemical assays, it can be extremely powerful. Should this become the way structural information is routinely analysed? Collectively, these points cannot be emphasised enough and highlights that perhaps a cultural change is required with respect to the way data is extracted and utilised within a design environment. More constructive challenge is required, both from medicinal chemists' perspective as well as the perspective of structural biologists and computational chemists, with regard to what information is extracted and how it is used.

As discussed earlier, predicting potency, and therefore selectivity over closely related protein targets, from structural information is an extremely difficult task. This is probably one of the greatest shortcomings of structure-based drug design. Therefore utilising *in silico* methods to conduct virtual screening must be done with extreme care and caution. It could be argued that we do not currently have suitable *in silico* tools to reliably rank a virtual screen in terms of potency, and therefore its usefulness in hit identification is questionable. This is therefore an area with huge scope for impact on drug discovery productivity. Of course advances in ranking and/or predicting potency *via* scoring functions can only be made when we are better able to characterise the interactions present in co-crystal structures. This further emphasises the need for biophysical techniques to complement structural biology. The alternative of course is the current paradigm of HTS, which involves producing and maintaining large screening files. These are both resource intensive and expensive, therefore reducing and focussing screening efforts through *in silico* techniques would be hugely enabling, particularly within the biotechnology sector, which operates with limited funds.

References

1. C. R. Beddell, *Br. J. Pharmacol.*, 1976, **57**, 201.
2. M. F. Perutz, *Science*, 1963, **140**, 863.
3. D. J. Abraham, in *Comprehensive Medicinal Chemistry II*, ed. J. B. Taylor and D. J. Triggle, Elsevier, Amsterdam, 1st Edition, 2007, vol. 4, *Computer-Assisted Drug Design*, p. 65.
4. H. Jhoti and A. G. Leach, ed., *Structure-Based Drug Discovery*, Springer, Dordrecht, 2007.
5. J. Greer, J. W. Erickson, J. J. Baldwin and M. D. Varney, *J. Med. Chem.*, 1994, **37**, 1035.
6. D. Leung, G. Abbenante and D. P. Fairlie, *J. Med. Chem.*, 2000, **43**, 305.
7. M. Congreve, C. W. Murray and T. L. Blundell, *Drug Discovery Today*, 2005, **10**, 895.

8. G. Lange, in *Comprehensive Medicinal Chemistry II*, ed. J. B. Taylor and D. J. Triggle, Elsevier, Amsterdam, 1st Edition, 2007, vol. 4, *Computer-Assisted Drug Design*, p. 597.
9. S. M. Paul, D. S. Mytelka, C. T. Dunwiddie, C. C. Persinger, B. H. Munos, S. R. Lindborg and A. L. Schacht, *Nat. Rev. Drug Discovery*, 2010, **9**, 203.
10. http://www.pdb.org/pdb/home/home.do.
11. J. Owens, *Nat. Rev. Drug Discovery*, 2007, **6**, 99.
12. B. Hughes, *Nat. Rev. Drug Discovery*, 2008, **7**, 107.
13. B. Hughes, *Nat. Rev. Drug Discovery*, 2009, **8**, 93.
14. B. Hughes, *Nat. Rev. Drug Discovery*, 2010, **9**, 89.
15. S. Hegde, M. Schmidt and W. Anthony, *Annu. Rep. Med. Chem.*, 2006, **41**, 439.
16. S. Hegde, M. Schmidt and E. M. John, *Annu. Rep. Med. Chem.*, 2007, **42**, 505.
17. S. Hegde, M. Schmidt and E. M. John, *Annu. Rep. Med. Chem.*, 2008, **43**, 455.
18. S. Hegde, M. Schmidt and E. M. John, *Annu. Rep. Med. Chem.*, 2009, **44**, 577.
19. S. Hegde and M. Schmidt, *Annu. Rep. Med. Chem.*, 2010, **45**, 467.
20. H. Gohlke and G. Klebe, *Angew. Chem. Int. Ed.*, 2002, **40**, 2644.
21. K. N. Houk, A. G. Leach, S. P. Kim and X. Zhang, *Angew. Chem. Int. Ed.*, 2003, **42**, 4872.
22. V. M. Krishnamurthy, G. K. Kaufman, A. R. Urbach, I. Gitlin, K. L. Gudiksen, D. B. Weibel and G. M. Whitesides, *Chem. Rev.*, 2008, **108**, 946.
23. S. W. Homans, *Drug Discov. Today*, 2007, **12**, 534.
24. M. Chruszcz, D. Borek, M. Domagalski, Z. Otwinowski, W. Minor and J. Andrzej, in *Advances in Protein Chemistry and Structural Biology*, Academic Press, 1st Edition, 2009, vol. 77, p. 23.
25. A. Sharff and H. Jhoti, *Curr. Opin. Chem. Biol.*, 2003, **7**, 340.
26. T. L. Blundell and S. Patel, *Curr. Opin. Pharmacol.*, 2004, **4**, 490.
27. M. J. Hartshorn, C. W. Murray, A. Cleasby, M. Frederickson, I. J. Tickle and H. Jhoti, *J. Med. Chem.*, 2005, **48**, 403.
28. A. R. Leach, M. M. Hann, J. N. Burrows and E. J. Griffen, *Mol. BioSyst.*, 2006, **2**, 429.
29. W. Jahnke and D. A. Erlanson, *Fragment-Based Approaches in Drug Discovery*, Wiley-VCH, Weinheim, 2006.
30. J. S. Albert, N. Blomberg, A. L. Breeze, A. J. H. Brown, J. N. Burrows, P. D. Edwards, R. H. A. Folmer, S. Geschwindner, E. J. Griffen, P. W. Kenny, T. Nowak, L.-L. Olsson, H. Sanganee and A. B. Shapiro, *Curr. Top. Med. Chem.*, 2007, **7**, 1600.
31. A. A. Alex and M. M. Flocco, *Curr. Top. Med. Chem.*, 2007, **7**, 1544.
32. R. E. Hubbard, *J. Synchrotron Rad.*, 2008, **15**, 227.
33. T. Tamada, T. Kinoshita, K. Kurihara, M. Adachi, T. Ohhara, K. Imai, R. Kuroki and T. Tada, *J. Am. Chem. Soc.*, 2009, **131**, 11033.
34. S. Z. Fisher, A. Y. Kovalevsky, J. F. Domsic, M. Mustyakimov, R. McKenna, D. N. Silverman and P. A. Langan, *Biochemistry*, 2009, **49**, 415.

35. T. Warne, M. J. Serrano-Vega, J. G. Baker, R. Moukhametzianov, P. C. Edwards, R. Henderson, A. G. W. Leslie, C. G. Tate and G. F. X. Schertler, *Nature*, 2008, **454**, 486.
36. S. G. F. Rasmussen, H.-J. Choi, D. M. Rosenbaum, T. S. Kobilka, F. S. Thian, P. C. Edwards, M. Burghammer, V. R. P. Ratnala, R. Sanishvili, R. F. Fischetti, G. F. X. Schertler, W. I. Weis and B. K. Kobilka, *Nature*, 2007, **450**, 383.
37. V. P. Jaakola, M. T. Griffith, M. A. Hanson, V. Cherezov, E. Y. Chien, J. R. Lane, A. P. IJzerman and R. C. Stevens, *Science*, 2008, **322**, 1211.
38. P. Kolb, D. M. Rosenbaum, J. J. Irwin, J. J. Fung, B. K. Kobilka and B. K. Shoichet, *Proc. Nat. Acad. Sci. U. S. A.*, 2009, **106**, 6843.
39. F. Franceschi and E. M. Duffy, *Biochem. Pharmacol.*, 2006, **71**, 1016.
40. D. Fry and L. Vassilev, *J. Mol. Med.*, 2005, **83**, 955.
41. H.-G. Kräusslich, R. Bartenschlager, T. Melby and M. Westby, in *Antiviral Strategies*, ed. K. Starke, G. V. R. Born, S. Duckles, M. Eichelbaum, D. Ganten, F. Hofmann, W. Rosenthal and G. Rubanyi, Springer, Berlin Heidelberg, 1st Edition, 2009, vol. 189, p. 177.
42. J. W. Park and W. H. Jo, *J. Chem. Inf. Model.*, 2009, **49**, 2735.
43. S. Alcaro, A. Artese, F. Ceccherini-Silberstein, F. Ortuso, C. F. Perno, T. Sing and V. Svicher, *J. Chem. Inf. Model.*, 2009, **49**, 1751.
44. A. M. Davis, S. J. Teague and G. J. Kleywegt, *Angew. Chem. Int. Ed.*, 2003, **42**, 2718.
45. A. M. Davis, S. A. St-Gallay and G. J. Kleywegt, *Drug Discoery Today*, 2008, **13**, 831.
46. C. R. Sondergaard, A. E. Garrett, T. Carstensen, G. Pollastri and J. E. Nielsen, *J. Med. Chem.*, 2009, **52**, 5673.
47. C. A. Lipinski, F. Lombardo, W. B. Dominy and P. J. Feeney, *Adv. Drug Del. Rev.*, 1997, **23**, 3.
48. T. I. Oprea, A. M. Davis, S. J. Teague and P. D. Leeson, *J. Chem. Inf. Comput. Sci.*, 2001, **41**, 1308.
49. C. A. Lipinski, *Drug Discovery Today: Technologies*, 2004, **1**, 337.
50. P. D. Leeson, A. M. Davis and J. Steele, *Drug Discovery Today: Technologies*, 2004, **1**, 189.
51. M. S. Lajiness, M. Vieth and J. Erickson, *Curr. Opin. Drug Discovery Dev.*, 2004, **7**, 470.
52. P. D. Leeson and B. Springthorpe, *Nat. Rev. Drug Discovery*, 2007, **6**, 881.
53. S. R. Turner, J. W. Strohbach, R. A. Tommasi, P. A. Aristoff, P. D. Johnson, H. I. Skulnick, L. A. Dolak, E. P. Seest, P. K. Tomich, M. J. Bohanon, M.-M. Horng, J. C. Lynn, K.-T. Chong, R. R. Hinshaw, K. D. Watenpaugh, M. N. Janakiraman and S. Thaisrivongs, *J. Med. Chem.*, 1998, **41**, 3467.
54. G. Klebe, *Drug Discovery Today*, 2006, **11**, 580.
55. C. B-Rao, J. Subramaniana and S. D. Sharma, *Drug Discovery Today*, 2009, **14**, 394.
56. S. Grosdidier and J. Fernández-Recio, *Expert Opin. Drug Discovery*, 2009, **4**, 673.

57. S. Grüneberg, B. Wendt and G. Klebe, *Angew. Chem. Int. Ed.*, 2001, **40**, 389.
58. S. Grüneberg, M. T. Stubbs and G. Klebe, *J. Med. Chem.*, 2002, **45**, 3588.
59. M. P. Williamson, T. F. Havel and K. Wuthrich, *J. Mol. Biol.*, 1985, **182**, 295.
60. M. Schade, in *Frontiers in Drug Design & Discovery,* ed. G. W. Caldwell, Atta-ur-Rahman, M. R. Player and M. I. Choudhary, Bentham, Hilversum, 1st Edition, 2007, vol. 3, p. 105.
61. P. J. Hajduk, G. Sheppard, D. G. Nettesheim, E. T. Olejniczak, S. B. Shuker, R. P. Meadows, D. H. Steinman, G. M. J. Carrera, P. A. Marcotte, J. Severin, K. Walter, H. Smith, E. Gubbins, R. Simmer, T. F. Holzman, D. W. Morgan, S. K. Davidsen, J. B. Summers and S. W. Fesik, *J. Am. Chem. Soc.*, 1997, **119**, 5818.
62. D. J. Craig and R. J. Clark, in *Burger's Medicinal Chemistry and Drug Discovery*, ed. D. J. Abraham, John Wiley, Hoboken, 1st Edition, 2003, vol. 1, p. 507.
63. R. Talhout, A. Villa, A. E. Mark and J. B. F. N. Engberts, *J. Am. Chem. Soc.*, 2003, **125**, 10570.
64. A. J. Ruben, Y. Kiso and E. Freire, *Chem. Biol. Drug Des.*, 2006, **67**, 2.
65. C. Gerlach, M. Smolinski, H. Steuber, C. A. Sotriffer, A. Heine, D. Hangauer and G. Klebe, *Angew. Chem. Int. Ed.*, 2007, **46**, 8511.
66. V. Lafont, A. A. Armstrong, H. Ohtaka, Y. Kiso, L. M. Amzel and E. Freire, *Chem. Biol. Drug Des.*, 2007, **69**, 413.
67. T. S. G. Olsson, M. A. Williams, W. R. Pitt and J. E. Ladbury, *J. Mol. Biol.*, 2008, **384**, 1002–1017.
68. E. Freire, *Drug Discovery Today*, 2008, **13**, 869.
69. J. E. Ladbury, G. Klebe and E. Freire, *Nat. Rev. Drug Discovery*, 2009, **9**, 23.
70. B. Baum, L. Muley, A. Heine, M. Smolinski, D. Hangauer and G. Klebe, *J. Mol. Biol.*, 2009, **391**, 552.
71. A. D. Scott, C. Phillips, A. Alex, M. Flocco, A. Bent, A. Randall, R. O'Brien, L. Damian and L. H. Jones, *ChemMedChem*, 2009, **4**, 1985.
72. Y. Kawasaki, E. E. Chufan, V. Lafont, K. Hidaka, K. Yoshiaki, L. M. Amzel and E. Freire, *Chem. Biol. Drug Des.*, 2009, **75**, 143.
73. B. Baum, M. Mohamed, M. Zayed, C. Gerlach, A. Heine, D. Hangauer and G. Klebe, *J. Mol. Biol.*, 2009, **390**, 56.
74. J. Key, T. H. Scheuermann, P. C. Anderson, V. Daggett and K. H. Gardner, *J. Am. Chem. Soc.*, 2009, **131**, 17647.
75. E. Freire, *Chem. Biol. Drug Des.*, 2009, **74**, 468.
76. C. Bissantz, B. Kuhn and M. Stahl, *J. Med. Chem.*, 2010, **53**, 5061.
77. M. S. Searle, D. H. Williams and U. Gerhard, *J. Am. Chem. Soc.*, 1992, **114**, 10697.
78. M. S. Searle and D. H. Williams, *J. Am. Chem. Soc.*, 1992, **114**, 10690.
79. M. S. Westwell, M. S. Searle, D. J. Wales and D. H. Williams, *J. Am. Chem. Soc.*, 1995, **117**, 5013.
80. H.-J. Boehm, *J. Comput. Aided Mol. Des.*, 1994, **8**, 243.
81. R. S. DeWitte and E. I. Shakhnovich, *J. Am. Chem. Soc.*, 1996, **118**, 11733.

82. R. Wang, L. Liu, L. Lai and Y. Tang, *J. Mol. Model*, 1998, **4**, 379.
83. S. Raub, A. Steffen, A. Kaemper and C. M. Marian, *J. Chem. Inf. Model.*, 2008, **48**, 1492.
84. I. Muegge and Y. C. Martin, *J. Med. Chem.*, 1999, **42**, 791.
85. J. B. O. Mitchell, R. A. Laskowski, A. Alex and J. M. Thornton, *J. Comput. Chem.*, 1999, **20**, 1165.
86. J. B. O. Mitchell, R. A. Laskowski, A. Alex, M. J. Forster and J. M. Thornton, *J. Comput. Chem.*, 1999, **20**, 1177.
87. I. Nobeli, J. B. O. Mitchell, A. Alex and J. M. Thornton, *J. Comput. Chem.*, 2001, **22**, 673.
88. A. J. T. Smith, X. Zhang, A. G. Leach and K. N. Houk, *J. Med. Chem.*, 2009, **52**, 225–233.
89. F. Allen, *Acta Cryst. Sec. B*, 2002, **58**, 380.
90. A. R. Katritzky, D. C. Hall, B. E.-D. M. El-Gendy and B. Draghici, *J. Comput. Aided Mol. Des.*, 2010, **24**, 475.
91. F. Milletti and A. Vulpetti, *J. Chem. Inf. Model.*, 2010, **50**, 1062.
92. L. Simon and J. M. Goodman, *J. Org. Chem.*, 2009, **75**, 1831.
93. M. D. Cummings, T. N. Hart and R. J. Read, *Protein Sci.*, 1995, **4**, 2087.
94. V. J. Hilser, J. Gómez and E. Freire, *Proteins: Struct., Funct., Genet.*, 1996, **26**, 123.
95. P. R. Andrews, D. J. Craik and J. L. Martin, *J. Med. Chem.*, 1984, **27**, 1648.
96. I. D. Kuntz, K. Chen, K. A. Sharp and P. A. Kollman, *Proc. Nat. Acad. Sci. U. S. A.*, 1999, **96**, 9997.
97. A. L. Hopkins, C. R. Groom and A. Alex, *Drug Discovery Today*, 2004, **9**, 430.
98. E. S. Eberhardt and R. T. Raines, *J. Am. Chem. Soc.*, 1994, **116**, 2149.
99. J. D. Findlater and B. A. Orsi, *FEBS Lett.*, 1973, **35**, 109.
100. M. Bolognesi, E. Cannillo, P. Ascenzi, G. M. Giacometti, A. Merli and M. Brunori, *J. Mol. Biol.*, 1982, **158**, 305.
101. R. A. Wallace, A. N. Kurtz and C. Niemann, *Biochemistry*, 1963, **2**, 824.
102. K. A. Dill, *Biochemistry*, 1990, **29**, 7133.
103. A. M. Davis and S. J. Teague, *Angew. Chem. Int. Ed.*, 1999, **38**, 736.
104. P. A. Bartlett and C. K. Marlowe, *Biochemistry*, 1987, **26**, 8553.
105. S. Cotesta and M. Stahl, *J. Mol. Model*, 2007, **12**, 436.
106. P. Gilli, L. Pretto, V. Bertolasi and G. Gilli, *Acc. Chem. Res.*, 2009, **42**, 33.
107. K. H. Kim, *J. Comput. Aided Mol. Design*, 2007, **21**, 63.
108. D. W. Cushman, H. S. Cheung, E. F. Sabo and M. A. Ondetti, *Biochemistry*, 1977, **16**, 5484.
109. J. J. Baldwin, G. S. Ponticello, P. S. Anderson, M. E. Christy, M. A. Murcko, W. C. Randall, H. Schwam, M. F. Sugrue and P. Gautheron, *J. Med. Chem.*, 1989, **32**, 2510.
110. N. A. Roberts, J. A. Martin, D. Kinchington, A. V. Broadhurst, J. C. Craig, I. B. Duncan, S. A. Galpin, B. K. Handa, J. Kay, A. Kröhn, R. W. Lambert, J. H. Merrett, J. S. Mills, K. E. B. Parkes, S. Redshaw, A. J. Ritchie, D. L. Taylor, G. J. Thomas and P. J. Machin, *Science*, 1990, **248**, 358.

111. D. J. Kempf, H. L. Sham, K. C. Marsh, C. A. Flentge, D. Betebenner, B. E. Green, E. McDonald, S. Vasavanonda, A. Saldivar, N. E. Wideburg, W. M. Kati, L. Ruiz, C. Zhao, L. Fino, J. Patterson, A. Molla, J. J. Plattner and D. W. Norbeck, *J. Med. Chem.*, 1998, **41**, 602.
112. B. D. Dorsey, R. B. Levin, S. L. McDaniel, J. P. Vacca, J. P. Guare, P. L. Darke, J. A. Zugay, E. A. Emini and W. A. Schleif, *J. Med. Chem.*, 1994, **37**, 3443.
113. L. H. Silver, *Surv. Ophthalmol.*, 2000, **44**, S147.
114. S. W. Kaldor, V. J. Kalish, J. F. Davies, B. V. Shetty, J. E. Fritz, K. Appelt, J. A. Burgess, K. M. Campanale, N. Y. Chirgadze, D. K. Clawson, B. A. Dressman, S. D. Hatch, D. A. Khalil, M. B. Kosa, P. P. Lubbehusen, M. A. Muesing, A. K. Patick, S. H. Reich, K. S. Su and J. H. Tatlock, *J. Med. Chem.*, 1997, **40**, 3979.
115. E. E. Kim, C. T. Baker, M. D. Dwyer, M. A. Murcko, B. G. Rao, R. D. Tung and M. A. Navia, *J. Am. Chem. Soc.*, 1995, **117**, 1181.
116. H. L. Sham, D. J. Kempf, A. Molla, K. C. Marsh, G. N. Kumar, C.-M. Chen, W. Kati, K. Stewart, R. Lal, A. Hsu, D. Betebenner, M. Korneyeva, S. Vasavanonda, E. McDonald, A. Saldivar, N. Wideburg, X. Chen, P. Niu, C. Park, V. Jayanti, B. Grabowski, G. R. Granneman, E. Sun, A. J. Japour, J. M. Leonard, J. J. Plattner and D. W. Norbeck, *Antimicrob. Agents Chemother.*, 1998, **42**, 3218.
117. M. von Itzstein, W.-Y. Wu, G. B. Kok, M. S. Pegg, J. C. Dyason, B. Jin, T. V. Phan, M. L. Smythe, H. F. White, S. W. Oliver, P. M. Colman, J. N. Varghese, D. M. Ryan, J. M. Woods, R. C. Bethell, V. J. Hotham, J. M. Cameron and C. R. Penn, *Nature*, 1993, **363**, 418.
118. C. U. Kim, W. Lew, M. A. Williams, H. Wu, L. Zhang, X. Chen, P. A. Escarpe, D. B. Mendel, W. G. Laver and R. C. Stevens, *J. Med. Chem.*, 1998, **41**, 2451.
119. J. Zimmermann, E. Buchdunger, H. Mett, T. Meyer, N. B. Lydon and P. Traxler, *Bioorg. Med. Chem. Lett.*, 1996, **6**, 1221.
120. J. Zimmermann, E. Buchdunger, H. Mett, T. Meyer and N. B. Lydon, *Bioorg. Med. Chem. Lett.*, 1997, **7**, 187.
121. L. F. Hennequin, A. P. Thomas, C. Johnstone, E. S. E. Stokes, P. A. Ple, J.-J. M. Lohmann, D. J. Ogilvie, M. Dukes, S. R. Wedge, J. O. Curwen, J. Kendrew and C. Lambert-van der Brempt, *J. Med. Chem.*, 1999, **42**, 5369.
122. B. S. Robinson, K. A. Riccardi, Y.-F. Gong, Q. Guo, D. A. Stock, W. S. Blair, B. J. Terry, C. A. Deminie, F. Djang, R. J. Colonno and P.-F. Lin, *Antimicrob. Agents Chemother.*, 2000, **44**, 2093.
123. S. Becker and L. Thornton, *Expert Opin. Pharmacother.*, 2004, **5**, 1995.
124. V. A. Pollack, D. M. Savage, D. A. Baker, K. E. Tsaparikos, D. E. Sloan, J. D. Moyer, E. G. Barbacci, L. R. Pustilnik, T. A. Smolarek, J. A. Davis, M. P. Vaidya, L. D. Arnold, J. L. Doty, K. K. Iwata and M. J. Morin, *J. Pharmacol. Exp. Ther.*, 1999, **291**, 739.
125. D. Gustafsson, J.-E. Nyström, S. Carlsson, U. Bredberg, U. Eriksson, E. Gyzander, M. Elg, T. Antonsson, K.-J. Hoffmann, A.-L. Ungell, H. Sörensen, S. Någård, A. Abrahamsson and R. Bylund, *Thromb. Res.*, 2001, **101**, 171.

126. S. M. Wilhelm, C. Carter, L. Tang, D. Wilkie, A. McNabola, H. Rong, C. Chen, X. Zhang, P. Vincent, M. McHugh, Y. Cao, J. Shujath, S. Gawlak, D. Eveleigh, B. Rowley, L. Liu, L. Adnane, M. Lynch, D. Auclair, I. Taylor, R. Gedrich, A. Voznesensky, B. Riedl, L. E. Post, G. Bollag and P. A. Trail, *Cancer Res.*, 2004, **64**, 7099.
127. T. Oh, K. Kang, B. Ahn, M. Yoo and W. Kim, *Arch. Pharmacal. Res.*, 2000, **23**, 471.
128. B. O. Ahn, K. K. Kang, G. J. Ahn, J. W. Kwon, W. B. Kim, K. S. Kang and Y. S. Lee, *Int. J. Impot. Res.*, **15**, 405.
129. D. B. Mendel, A. D. Laird, X. Xin, S. G. Louie, J. G. Christensen, G. Li, R. E. Schreck, T. J. Abrams, T. J. Ngai, L. B. Lee, L. J. Murray, J. Carver, E. Chan, K. G. Moss, J. Ã. Haznedar, J. Sukbuntherng, R. A. Blake, L. Sun, C. Tang, T. Miller, S. Shirazian, G. McMahon and J. M. Cherrington, *Clin. Cancer Res.*, 2003, **9**, 327.
130. A. K. Ghosh, Z. L. Dawson and H. Mitsuya, *Bioorg. Med. Chem.*, 2007, **15**, 7576.
131. W. K. Kelly, O. A. O' Connor, L. M. Krug, J. H. Chiao, M. Heaney, T. Curley, B. MacGregore-Cortelli, W. Tong, J. P. Secrist, L. Schwartz, S. Richardson, E. Chu, S. Olgac, P. A. Marks, H. Scher and V. M. Richon, *J. Clin. Oncol.*, 2005, **23**, 3923.
132. L. J. Lombardo, F. Y. Lee, P. Chen, D. Norris, J. C. Barrish, K. Behnia, S. Castaneda, L. A. M. Cornelius, J. Das, A. M. Doweyko, C. Fairchild, J. T. Hunt, I. Inigo, K. Johnston, A. Kamath, D. Kan, H. Klei, P. Marathe, S. Pang, R. Peterson, S. Pitt, G. L. Schieven, R. J. Schmidt, J. Tokarski, M.-L. Wen, J. Wityak and R. M. Borzilleri, *J. Med. Chem.*, 2004, **47**, 6658.
133. E. Weisberg, P. W. Manley, W. Breitenstein, J. Brüggen, S. W. Cowan-Jacob, A. Ray, B. Huntly, D. Fabbro, G. Fendrich, E. Hall-Meyers, A. L. Kung, J. Mestan, G. Q. Daley, L. Callahan, L. Catley, C. Cavazza, A. Mohammed, D. Neuberg, R. D. Wright, D. G. Gilliland and J. D. Griffin, *Cancer Cell*, 2005, **7**, 129.
134. J. Rahuel, V. Rasetti, J. Maibaum, H. Rüeger, R. Göschke, N. C. Cohen, S. Stutz, F. Cumin, W. Fuhrer, J. M. Wood and M. G. Grütter, *Chem. Biol.*, 2000, **7**, 493.
135. D. W. Rusnak, K. Lackey, K. Affleck, E. R. Wood, K. J. Alligood, N. Rhodes, B. R. Keith, D. M. Murray, W. B. Knight, R. J. Mullin and T. M. Gilmer, *Mol. Cancer Ther.*, 2001, **1**, 85.
136. S. Roehrig, A. Straub, J. Pohlmann, T. Lampe, J. Pernerstorfer, K.-H. Schlemmer, P. Reinemer and E. Perzborn, *J. Med. Chem.*, 2005, **48**, 5900.
137. L. A. Sorbera, J. Bozzo and J. Castaner, *Drugs Future*, 2005, **30**, 877.
138. K. Das, A. D. Clark, P. J. Lewi, J. Heeres, M. R. de Jonge, L. M. H. Koymans, H. M. Vinkers, F. Daeyaert, D. W. Ludovici, M. J. Kukla, B. De Corte, R. W. Kavash, C. Y. Ho, H. Ye, M. A. Lichtenstein, K. Andries, R. Pauwels, M.-P. de Béthune, P. L. Boyer, P. Clark, S. H. Hughes, P. A. J. Janssen and E. Arnold, *J. Med. Chem.*, 2004, **47**, 2550.

139. S. Sleijfer, I. Ray-Coquard, Z. Papai, A. Le Cesne, M. Scurr, P. Schöffski, F. Collin, L. Pandite, S. Marreaud, A. De Brauwer, M. van Glabbeke, J. Verweij and J.-Y. Blay, *J. Clin. Oncol.*, 2009, **27**, 3126.
140. D. Kim, L. Wang, M. Beconi, G. J. Eiermann, M. H. Fisher, H. He, G. J. Hickey, J. E. Kowalchick, B. Leiting, K. Lyons, F. Marsilio, M. E. McCann, R. A. Patel, A. Petrov, G. Scapin, S. B. Patel, R. S. Roy, J. K. Wu, M. J. Wyvratt, B. B. Zhang, L. Zhu, N. A. Thornberry and A. E. Weber, *J. Med. Chem.*, 2005, **48**, 141.
141. D. Kim, J. E. Kowalchick, L. L. Brockunier, E. R. Parmee, G. J. Eiermann, M. H. Fisher, H. He, B. Leiting, K. Lyons, G. Scapin, S. B. Patel, A. Petrov, K. D. Pryor, R. S. Roy, J. K. Wu, X. Zhang, M. J. Wyvratt, B. B. Zhang, L. Zhu, N. A. Thornberry and A. E. Weber, *J. Med. Chem.*, 2008, **51**, 589.
142. D. J. Augeri, J. A. Robl, D. A. Betebenner, D. R. Magnin, A. Khanna, J. G. Robertson, A. Wang, L. M. Simpkins, P. Taunk, Q. Huang, S.-P. Han, B. Abboa-Offei, M. Cap, L. Xin, L. Tao, E. Tozzo, G. E. Welzel, D. M. Egan, J. Marcinkeviciene, S. Y. Chang, S. A. Biller, M. S. Kirby, R. A. Parker and L. G. Hamann, *J. Med. Chem.*, 2005, **48**, 5025.
143. K. Okamoto, B. T. Eger, T. Nishino, S. Kondo, E. F. Pai and T. Nishino, *J. Biol. Chem.*, 2003, **278**, 1848.
144. G. S. Ponticello, M. B. Freedman, C. N. Habecker, P. A. Lyle, H. Schwam, S. L. Varga, M. E. Christy, W. C. Randall and J. J. Baldwin, *J. Med. Chem.*, 1987, **30**, 591.
145. C. T. Supuran and A. Scozzafava, *Expert Opin. Ther. Pat.*, 2000, **10**, 575.
146. C. T. Supuran and A. Scozzafava, *Curr. Med. Chem.: Immunol. Endoc. Metab. Agents*, 2001, **1**, 61.
147. C. T. Supuran and A. Scozzafava, *Expert Opin. Ther. Pat.*, 2002, **12**, 217.
148. C. A. Hunt, P. J. Mallorga, S. R. Michelson, H. Schwam, J. M. Sondey, R. L. Smith, M. F. Sugrue and K. L. Shepard, *J. Med. Chem.*, 1994, **37**, 240.
149. A. Scozzafava, L. Menabuoni, F. Mincione, F. Briganti, G. Mincione and C. T. Supuran, *J. Med. Chem.*, 1999, **42**, 2641.
150. H.-H. Chen, S. Gross, J. Liao, M. McLaughlin, T. Dean, W. S. Sly and J. A. May, *Bioorg. Med. Chem.*, 2000, **8**, 957.
151. X. de Leval, M. Ilies, A. Casini, J.-M. Dogné, A. Scozzafava, E. Masini, F. Mincione, M. Starnotti and C. T. Supuran, *J. Med. Chem.*, 2004, **47**, 2796.
152. A. Scozzafava, L. Menabuoni, F. Mincione, F. Briganti, G. Mincione and C. T. Supuran, *J. Med. Chem.*, 2000, **43**, 4542.
153. N. A. Roberts, J. A. Martin, D. Kinchington, A. V. Broadhurst, J. C. Craig, I. B. Duncan, S. A. Galpin, B. K. Handa, J. Kay, A. Kröhn, R. W. Lambert, J. H. Merrett, J. S. Mills, K. E. B. Parkes, S. Redshaw, A. J. Ritchie, D. L. Taylor, G. J. Thomas and P. J. Machin, *Science*, 1990, **248**, 358.
154. M. Rittweger and K. Arastéh, *Clin. Pharmacokinet.*, 2007, **46**, 739.
155. B. M. McKeever, C. T. Leu, J. C. Heimbach, W. K. Herber, I. S. Sigal, P. L. Darke and J. P. Springer, *Nature*, 1989, **337**, 615.

156. A. Wlodawer, M. Miller, M. Jaskolinski, B. K. Sathyanarayana, E. Baldwin, I. T. Weber, L. M. Selk, L. Clawson, J. Schneider and S. B. M. Kent, *Science*, 1989, **245**, 616.
157. R. Lapatto, T. Blundell, A. Hennings, J. Overington, A. Wilderspin, S. Wood, J. R. Merson, P. J. Whittle, D. E. Danley, K. F. Geoghegan, S. J. Hawrylik, S. E. Lee, K. G. Scheld and P. M. Hobart, *Nature*, 1989, **342**, 299.
158. S. Spinelli, Q. Z. Liu, P. M. Alzari, P. H. Hirel and R. J. Poljak, *Biochimie*, 1989, **73**, 1391.
159. Y. Mehellou and E. De Clercq, *J. Med. Chem.*, 2009, **53**, 521.
160. M. Boffito, A. Winston, A. Jackson, C. Fletcher, A. Pozniak, M. Nelson, G. Moyle, I. Tolowinska, R. Hoetelmans, D. Miralles and B. Gazzard, *AIDS (London)*, 2007, **21**, 1449.
161. S. Thaisrivongs and J. W. Strohbach, *Biopolymers*, 1999, **51**, 51.
162. A. Ali, G. S. K. K. Reddy, M. N. L. Nalam, S. G. Anjum, H. Cao, C. A. Schiffer and T. M. Rana, *J. Med. Chem.*, 2010, **53**, 7699.
163. M. A. Fischl, D. D. Richman, M. H. Grieco, M. S. Gottlieb, P. A. Volberding, O. L. Laskin, J. M. Leedom, J. E. Groopman, D. Mildvan, R. T. Schooley, G. G. Jackson, D. T. Durack and D. King, *N. Engl. J. Med.*, 1987, **317**, 185.
164. M. M. Mansuri, J. E. Starrett, I. Ghazzouli, M. J. M. Hitchcock, R. Z. Sterzycki, V. Brankovan, T. S. Lin, E. M. August and W. H. Prusoff, *J. Med. Chem.*, 1989, **32**, 461.
165. T. S. Mansour, H. Jin, W. Wang, E. U. Hooker, C. Ashman, N. Cammack, H. Salomon, A. R. Belmonte and M. A. Wainberg, *J. Med. Chem.*, 1995, **38**, 1.
166. J. M. Cameron, P. Collis, M. Daniel, R. Storer and P. Wilcox, *Drugs Future.*, 1993, **18**, 319.
167. S. M. Daluge, *U.S. Pat.* 5 034 394, 1991.
168. M. T. Crimmins and B. W. King, *J. Org. Chem.*, 1996, **61**, 4192.
169. L. W. Frick, L. St John, L. C. Taylor, G. R. Painter, P. A. Furman, D. C. Liotta, E. S. Furfine and D. J. Nelson, *Antimicrob. Agents Chemother.*, 1993, **37**, 2285.
170. L. Naesens, N. Bischofberger, P. Augustijns, P. Annaert, G. Van den Mooter, M. N. Arimilli, C. U. Kim and E. De Clercq, *Antimicrob. Agents Chemother.*, 1998, **42**, 1568.
171. K. R. Reddy, M. C. Matelich, B. G. Ugarkar, J. E. Gómez-Galeno, J. DaRe, K. Ollis, Z. Sun, W. Craigo, T. J. Colby, J. M. Fujitaki, S. H. Boyer, P. D. van Poelje and M. D. Erion, *J. Med. Chem.*, 2008, **51**, 666.
172. K. A. Sims and A. M. Woodland, *Pharmacotherapy*, 2006, **26**, 1745.
173. B. A. Larder, D. G. and D. D. Richman, *Science*, 1989, **243**, 452.
174. M. H. St Clair, J. L. Martin, G. Tudor-Williams, M. C. Bach, C. L. Vavro, D. M. King, P. Kelham, S. D. Kemp and B. A. Larder, *Science*, 1991, **253**, 1557.
175. J. F. Davies 2nd, Z. Hostomska, Z. Hostomsky, S. R. Jordan and D. A. Matthews, *Science*, 1991, **252**, 88.

176. S. J. Smerdon, J. Jager, J. Wang, L. A. Kohlstaedt, A. J. Chirino, J. M. Friedman, P. A. Rice and T. A. Steitz, *Proc. Natl. Acad. Sci. U. S. A.*, 1994, **91**, 3911.
177. J. P. Horwitz, J. Chua and M. J. Noel, *Org. Chem. Ser. Monogr.*, 1964, **29**, 2076.
178. M. J. Robins, J. R. J. McCarthy and R. J. Robins, *Biochemistry*, 1966, **5**, 224.
179. J. P. Horwitz, J. Chua, M. A. Da Rooge, M. Noel and I. L. Klundt, *J. Org. Chem.*, 1966, **31**, 205.
180. S. Tuske, S. G. Sarafianos, A. D. Clark, J. Ding, L. K. Naeger, K. L. White, M. D. Miller, C. S. Gibbs, P. L. Boyer, P. Clark, G. Wang, B. L. Gaffney, R. A. Jones, D. M. Jerina, S. H. Hughes and E. Arnold, *Nat. Struct. Mol. Biol.*, 2004, **11**, 469.
181. J. Wang, Y. Jin, K. L. Rapp, M. Bennett, R. F. Schinazi and C. K. Chu, *J. Med. Chem.*, 2005, **48**, 3736.
182. E. De Clercq, *Antiviral Res.*, 1998, **38**, 153.
183. J. Ren, L. E. Bird, P. P. Chamberlain, G. B. Stewart-Jones, D. I. Stuart and D. K. Stammers, *Proc. Nat. Acad. Sci. U. S. A.*, 2002, **99**, 14410.
184. K. D. Hargrave, J. R. Proudfoot, K. G. Grozinger, E. Cullen, S. R. Kapadia, U. R. Patel, V. U. Fuchs, S. C. Mauldin and J. Vitous, *J. Med. Chem.*, 1991, **34**, 2231.
185. N. Fan, D. B. Evans, K. B. Rank, R. C. Thomas, W. G. Tarpley and S. K. Sharma, *FEBS Lett.*, 1995, **359**, 233.
186. S. D. Young, S. F. Britcher, L. O. Tran, L. S. Payne, W. C. Lumma, T. A. Lyle, J. R. Huff, P. S. Anderson, D. B. Olsen, S. S. Carroll, D. J. Pettibone, J. A. O'Brien, R. G. Ball, S. K. Balani, J. H. Lin, I.-W. Chen, W. A. Schleif, V. V. Sardana, W. J. Long, V. W. Byrnes and E. A. Emini, *Antimicrob. Agents Chemother.*, 1995, **39**, 2602.
187. K. Andries, H. Azijn, T. Thielemans, D. Ludovici, M. Kukla, J. Heeres, P. Janssen, B. De Corte, J. Vingerhoets, R. Pauwels and M.-P. de Béthune, *Antimicrob. Agents Chemother.*, 2004, **48**, 4680.
188. D. S. Schiller and M. Youssef-Bessler, *Clin. Therapeut.*, 2009, **31**, 692.
189. R. Pauwels, K. Andries, Z. Debyser, M. J. Kukla, D. Schols, H. J. Breslin, R. Woestenborghs, J. Desmyter, M. A. Janssen and E. De Clercq, *Antimicrob. Agents Chemother.*, 1994, **38**, 2863.
190. R. Pauwels, K. Andries, Z. Debyser, P. Van Daele, D. Schols, P. Stoffels, K. De Vreese, R. Woestenborghs, A. M. Vandamme and C. G. Janssen, *Proc. Nat. Acad. Sci. U. S. A.*, 1993, **90**, 1711.
191. E. B. Lansdon, K. M. Brendza, M. Hung, R. Wang, S. Mukund, D. Jin, G. Birkus, N. Kutty and X. Liu, *J. Med. Chem.*, **53**, 4295.
192. R. Corbau, J. Mori, C. Phillips, L. Fishburn, A. Martin, C. Mowbray, W. Panton, C. Smith-Burchnell, A. Thornberry, H. Ringrose, T. Knöchel, S. Irving, M. Westby, A. Wood and M. Perros, *Antimicrob. Agents Chemother.*, 2010, **45**, 4451.
193. K. R. Romines, G. A. Freeman, L. T. Schaller, J. R. Cowan, S. S. Gonzales, J. H. Tidwell, C. W. Andrews, D. K. Stammers,

R. J. Hazen, R. G. Ferris, S. A. Short, J. H. Chan and L. R. Boone, *J. Med. Chem.*, 2005, **49**, 727.

194. S.-J. Ho and T. A. Brighton, *Vasc. Health Risk Manag.*, 2006, **2**, 49.
195. D. J. P. Pinto, J. M. Smallheer, D. L. Cheney, R. M. Knabb and R. R. Wexler, *J. Med. Chem.*, 2010, **53**, 6243.
196. Y. K. Lee and M. R. Player, *Med. Res. Rev.*, 2009, DOI: 10.1002/med.20183.
197. T. Nagahara, Y. Yokoyama, K. Inamura, S.-I. Katakura, S. Komoriya, H. Yamaguchi, T. Hara and M. Iwamoto, *J. Med. Chem.*, 1994, **37**, 1200.
198. D. J. P. Pinto, M. J. Orwat, S. Koch, K. A. Rossi, R. S. Alexander, A. Smallwood, P. C. Wong, A. R. Rendina, J. M. Luettgen, R. M. Knabb, K. He, B. Xin, R. R. Wexler and P. Y. S. Lam, *J. Med. Chem.*, 2007, **50**, 5339.
199. http://clinicaltrials.gov/ct2/show/NCT00457002.
200. D. J. P. Pinto, M. J. Orwat, S. Wang, J. M. Fevig, M. L. Quan, E. Amparo, J. Cacciola, K. A. Rossi, R. S. Alexander, A. M. Smallwood, J. M. Luettgen, L. Liang, B. J. Aungst, M. R. Wright, R. M. Knabb, P. C. Wong, R. R. Wexler and P. Y. S. Lam, *J. Med. Chem.*, 2001, **44**, 566.
201. P. Y. S. Lam, C. G. Clark, R. Li, D. J. P. Pinto, M. J. Orwat, R. A. Galemmo, J. M. Fevig, C. A. Teleha, R. S. Alexander, A. M. Smallwood, K. A. Rossi, M. R. Wright, S. A. Bai, K. He, J. M. Luettgen, P. C. Wong, R. M. Knabb and R. R. Wexler, *J. Med. Chem.*, 2003, **46**, 4405.
202. T. Fujimoto, Y. Imaeda, N. Konishi, K. Hiroe, M. Kawamura, G. P. Textor, K. Aertgeerts and K. Kubo, *J. Med. Chem.*, **53**, 3517.
203. I. A. Reid, R. J. Morris and W. F. Ganong, *Annu. Rev. Physiol.*, 1978, **40**, 377.
204. L. Tong, S. Pav, D. Lamarre, B. Simoneau, P. Lavallée and G. Jung, *J. Biol. Chem.*, 1995, **270**, 29520.
205. C. McInnes, in *Frontiers in Drug Design & Discovery,* ed. G. W. Caldwell, Atta-ur-Rahman, M. R. Player and M. I. Choudhary, Bentham, Hilversum, 1st Edition, 2007, vol. 3, p. 105.
206. A. Quintas-Cardama, H. Kantarjian and J. Cortes, *Nat. Rev. Drug Discovery*, 2007, **6**, 834.
207. G. Q. Daley, R. A. Van Etten and D. Baltimore, *Science*, 1990, **247**, 824.
208. T. G. Lugo, A. M. Pendergast, A. J. Muller and O. N. Witte, *Science*, 1990, **247**, 1079.
209. Y. Liu and N. S. Gray, *Nat. Chem. Biol.*, 2006, **2**, 358.
210. F. Zuccotto, E. Ardini, E. Casale and M. Angiolini, *J. Med. Chem.*, 2009, **53**, 2681.
211. J. A. Bikker, N. Brooijmans, A. Wissner and T. S. Mansour, *J. Med. Chem.*, 2009, **52**, 1493.
212. T. Schindler, W. Bornmann, P. Pellicena, W. T. Miller, B. Clarkson and J. Kuriyan, *Science*, 2000, **289**, 1938.
213. S. Atwell, J. M. Adams, J. Badger, M. D. Buchanan, I. K. Feil, K. J. Froning, X. Gao, J. Hendle, K. Keegan, B. C. Leon, H. J. Müller-Dieckmann, V. L. Nienaber, B. W. Noland, K. Post, K. R. Rajashankar,

A. Ramos, M. Russell, S. K. Burley and S. G. Buchanan, *J. Biol. Chem.*, 2004, **279**, 55827.
214. N. P. Shah, C. Tran, F. Y. Lee, P. Chen, D. Norris and C. L. Sawyers, *Science*, 2004, **305**, 399.
215. M. E. Gorre, M. Mohammed, K. Ellwood, N. Hsu, R. Paquette, P. N. Rao and C. L. Sawyers, *Science*, 2001, **293**, 876.
216. M. C. Frame, *Biochim. Biophys. Acta*, 2002, **1602**, 114.
217. J. S. Tokarski, J. A. Newitt, C. Y. J. Chang, J. D. Cheng, M. Wittekind, S. E. Kiefer, K. Kish, F. Y. F. Lee, R. Borzillerri, L. J. Lombardo, D. Xie, Y. Zhang and H. E. Klei, *Cancer Res.*, 2006, **66**, 5790.
218. M. Getlik, C. Grütter, J. R. Simard, S. Klüter, M. Rabiller, H. B. Rode, A. Robubi and D. Rauh, *J. Med. Chem.*, 2009, **52**, 3915.
219. J. Zhang, F. J. Adrian, W. Jahnke, S. W. Cowan-Jacob, A. G. Li, R. E. Iacob, T. Sim, J. Powers, C. Dierks, F. Sun, G.-R. Guo, Q. Ding, B. Okram, Y. Choi, A. Wojciechowski, X. Deng, G. Liu, G. Fendrich, A. Strauss, N. Vajpai, S. Grzesiek, T. Tuntland, Y. Liu, B. Bursulaya, M. Azam, P. W. Manley, J. R. Engen, G. Q. Daley, M. Warmuth and N. S. Gray, *Nature*, **463**, 501.
220. W.-S. Huang, X. Zhu, Y. Wang, M. Azam, D. Wen, R. Sundaramoorthi, R. M. Thomas, S. Liu, G. Banda, S. P. Lentini, S. Das, Q. Xu, J. Keats, F. Wang, S. Wardwell, Y. Ning, J. T. Snodgrass, M. I. Broudy, K. Russian, G. Q. Daley, J. Iuliucci, D. C. Dalgarno, T. Clackson, T. K. Sawyer and W. C. Shakespeare, *J. Med. Chem.*, 2009, **52**, 4743.
221. H. G. Choi, P. Ren, F. Adrian, F. Sun, H. S. Lee, X. Wang, Q. Ding, G. Zhang, Y. Xie, J. Zhang, Y. Liu, T. Tuntland, M. Warmuth, P. W. Manley, J. Mestan, N. S. Gray and T. Sim, *J. Med. Chem.*, 2010, **53**, 5439.
222. W.-S. Huang, C. A. Metcalf, R. Sundaramoorthi, Y. Wang, D. Zou, R. M. Thomas, X. Zhu, L. Cai, D. Wen, S. Liu, J. Romero, J. Qi, I. Chen, G. Banda, S. P. Lentini, S. Das, Q. Xu, J. Keats, F. Wang, S. Wardwell, Y. Ning, J. T. Snodgrass, M. I. Broudy, K. Russian, T. Zhou, L. Commodore, N. I. Narasimhan, Q. K. Mohemmad, J. Iuliucci, V. M. Rivera, D. C. Dalgarno, T. K. Sawyer, T. Clackson and W. C. Shakespeare, *J. Med. Chem.*, 2010, **53**, 4701.
223. P. M. Colman, *Protein Sci.*, 1994, **3**, 1687.
224. A. L. Hopkins and C. R. Groom, *Ernst Schering Res. Found. Workshop*, 2003, **42**, 11.
225. W. C. J. Lumma, K. M. Witherup, T. J. Tucker, S. F. Brady, J. T. Sisko, A. M. Naylor-Olsen, S. D. Lewis, B. J. Lucas and J. P. Vacca, *J. Med. Chem.*, 1998, **41**, 1011.
226. C. M. N. Allerton, C. G. Barber, K. C. Beaumont, D. G. Brown, S. M. Cole, D. Ellis, C. A. L. Lane, G. N. Maw, N. M. Mount, D. J. Rawson, C. M. Robinson, S. D. A. Street and N. W. Summerhill, *J. Med. Chem.*, 2006, **49**, 3581.
227. M. J. Palmer, A. S. Bell, D. N. A. Fox and D. G. Brown, *Curr. Top. Med. Chem.*, 2007, **7**, 405.

228. A. Alanine, M. Nettekoven, E. Roberts and A. W. Thomas, *Comb. Chem. High Throughput Screening*, 2003, **6**, 51.
229. J. W. Davies, M. Glick and J. L. Jenkins, *Curr. Opin. Chem. Biol.*, 2006, **10**, 343.
230. M. Congreve, D. Aharony, J. Albert, O. Callaghan, J. Campbell, R. A. E. Carr, G. Chessari, S. Cowan, P. D. Edwards, M. Frederickson, R. McMenamin, C. W. Murray, S. Patel and N. Wallis, *J. Med. Chem.*, 2007, **50**, 1124.
231. G. Marcou and D. Rognan, *J. Chem. Inf. Model.*, 2006, **47**, 195.
232. C. P. Mpamhanga, D. Spinks, L. B. Tulloch, E. J. Shanks, D. A. Robinson, I. T. Collie, A. H. Fairlamb, P. G. Wyatt, J. A. Frearson, W. N. Hunter, I. H. Gilbert and R. Brenk, *J. Med. Chem.*, 2009, **52**, 4454.
233. E. T. Olejniczak, P. J. Hajduk, P. A. Marcotte, D. G. Nettesheim, R. P. Meadows, R. Edalji, T. F. Holzman and S. W. Fesik, *J. Am. Chem. Soc.*, 1997, **119**, 5828.
234. B. G. Szczepankiewicz, G. Liu, P. J. Hajduk, C. Abad-Zapatero, Z. Pei, Z. Xin, T. H. Lubben, J. M. Trevillyan, M. A. Stashko, S. J. Ballaron, H. Liang, F. Huang, C. W. Hutchins, S. W. Fesik and M. R. Jirousek, *J. Am. Chem. Soc.*, 2003, **125**, 4087.
235. L. H. Jones, G. Allan, O. Barba, C. Burt, R. Corbau, T. Dupont, T. Knöchel, S. Irving, D. S. Middleton, C. E. Mowbray, M. Perros, H. Ringrose, N. A. Swain, R. Webster, M. Westby and C. Phillips, *J. Med. Chem.*, 2009, **52**, 1219.
236. D. Plewczynski, M. Łaźniewski, R. Augustyniak and K. Ginalski, *J. Comput. Chem.*, 2010, **32**, 742.
237. H. V. Namboodiri, M. Bukhtiyarova, J. Ramcharan, M. Karpusas, Y. Lee and E. B. Springman, *Biochemistry*, **49**, 3611.
238. A. D. Andricopulo, R. V. C. Guido and G. Oliva, *Curr. Med. Chem.*, 2008, **15**, 37.
239. A. L. Gill, M. Frederickson, A. Cleasby, S. J. Woodhead, M. G. Carr, A. J. Woodhead, M. T. Walker, M. S. Congreve, L. A. Devine, D. Tisi, M. O'Reilly, L. C. A. Seavers, D. J. Davis, J. Curry, R. Anthony, A. Padova, C. W. Murray, R. A. E. Carr and H. Jhoti, *J. Med. Chem.*, 2005, **48**, 414.
240. D. R. Goldberg, M.-H. Hao, K. C. Qian, A. D. Swinamer, D. A. Gao, Z. Xiong, C. Sarko, A. Berry, J. Lord, R. L. Magolda, T. Fadra, R. R. Kroe, A. Kukulka, J. B. Madwed, L. Martin, C. Pargellis, D. Skow, J. J. Song, Z. Tan, C. A. Torcellini, C. S. Zimmitti, N. K. Yee and N. Moss, *J. Med. Chem.*, 2007, **50**, 4016.
241. J.-K. Jiang, K. Ghoreschi, F. Deflorian, Z. Chen, M. Perreira, M. Pesu, J. Smith, D.-T. Nguyen, E. H. Liu, W. Leister, S. Costanzi, J. J. O'Shea and C. J. Thomas, *J. Med. Chem.*, 2008, **51**, 8012.
242. M. E. Flanagan, T. A. Blumenkopf, W. H. Brissette, M. F. Brown, J. M. Casavant, C. Shang-Poa, J. L. Doty, E. A. Elliott, M. B. Fisher, M. Hines, C. Kent, E. M. Kudlacz, B. M. Lillie, K. S. Magnuson, S. P. McCurdy, M. J. Munchhof, B. D. Perry, P. S. Sawyer, T. J. Strelevitz,

C. Subramanyam, J. Sun, D. A. Whipple and P. S. Changelian, *J. Med. Chem.*, 2010, **53**, 8468.
243. C. A. S. Bergström, C. M. Wassvik, K. Johansson and I. Hubatsch, *J. Med. Chem.*, 2007, **50**, 5858.
244. A. L. Hopkins and C. R. Groom, *Nat. Rev. Drug Discovery*, 2002, **1**, 727.
245. J. P. Overington, B. Al-Lazikani and A. L. Hopkins, *Nat. Rev. Drug Discovery*, 2006, **5**, 993.
246. G. V. Paolini, R. H. B. Shapland, W. P. van Hoorn, J. S. Mason and A. L. Hopkins, *Nature Biotechnol.*, 2006, **24**, 805.
247. K. J. Simmons, I. Chopra and C. W. G. Fishwick, *Nat. Rev. Microbiol.*, 2010, **8**, 501.

CHAPTER 6

Representing Chemical Structures in Databases for Drug Design

JOHN M. BARNARD,[a]* PETER W. KENNY[b] AND PAUL N. WALLACE[c]

[a] Digital Chemistry Ltd., 30 Kiveton Lane, Todwick, Sheffield, S26 1HL, UK; [b] UK; [c] GlaxoSmithKline, Gunnels Wood Road, Stevenage, Herts, SG1 2NY, UK

6.1 Introduction

Clearly, any sort of computer-based strategy for drug design requires a computer representation of the chemical structures being considered as potential drugs. Various types of representation have been used throughout the history of chemical information systems, and these are discussed in Section 6.2. Very often the appropriate representation depends on the purpose for which it is required, thus different types of representation are used for different aspects of drug design.

Two important characteristics of a chemical structure representation are the extent to which it provides an unambiguous representation of the structure (*i.e.* whether or not the representation corresponds to a single chemical entity), and whether or not it is a unique, or "canonical", representation of the structure (*i.e.* if there is only one correct representation of this type for a particular chemical entity). The question of unambiguity depends on the definition of a

RSC Drug Discovery Series No. 13
Drug Design Strategies: Quantitative Approaches
Edited by David J. Livingstone and Andrew M. Davis

Published by the Royal Society of Chemistry, www.rsc.org

single chemical entity, for example, are different conformers of a molecule, or molecules with differently isotope-labelled atoms, to be regarded as different chemical entities with different representations? For some purposes such distinctions may not matter, while for others they may be vital. The procedures required to achieve unique and appropriately unambiguous representations are discussed in Sections 6.3 and 6.5.

A further issue is the extent to which the representation corresponds to chemical reality. In many situations (sometimes to deal with issues such as uniqueness and unambiguity) representations may modify a chemical structure by neutralising opposing charges on a molecule, or by preferring a particular tautomeric form or ionisation state, thus giving a representation that does not correspond to the way in which the molecule exists in reality (at least under certain conditions, such as pH, solvation *etc.*). Calculated physicochemical or other properties that are based on this representation may then give values that are significantly different from those measured experimentally. This issue is discussed in further detail in Section 6.4.

Some representations are deliberately ambiguous, in other words a single representation is used for a class of related molecules which may, for example, have similar biological activities. Much work has been done in recent years in developing such representations to assist in drug design. These types of representations, as well as other processes used to bring related molecules together, are discussed in Section 6.6.

6.2 Computer Representation of Chemical Structures

Computer representations of chemical structures have evolved over several decades, and continue to do so as new and possibly, though not necessarily, improved formats achieve popularity or market penetration. There are a number of general discussions on chemical structure representations, both in textbooks[1–4] and in review articles[5,6]. This section attempts to provide an up-to-date summary of those currently considered most important, in an historical context.

6.2.1 Topological Graphs

Most computer representations of chemical structures treat the molecule as a topological graph. This is an abstract mathematical structure in which *nodes* or *vertices* (corresponding to atoms) are linked in pairs by *edges* (corresponding to bonds). Different properties (atom type, charge, bond order, *etc.*) can be applied to the nodes and edges. The theory of topological graphs is very well understood,[7,8] and has been applied in a wide range of disciplines, such as computer science. The development of many useful algorithms has allowed graphs to be processed in a variety of ways, for example to identify cycles (rings) in a graph, or to determine whether or not two graphs are identical (isomorphic), or whether or not the nodes and edges of one graph can be mapped onto a subset of the nodes and edges of another (subgraph isomorphism). Though extremely useful, it is important to recognise that the analogy between chemical structures and topological graphs is not a perfect one, for example graph theory says nothing about the spatial arrangement of the nodes and edges (3D structure representations are discussed in Section 6.2.8), and the occurrence of chemical phenomena such as resonance and tautomerism challenges the rigid node/edge structure of graphs. Indeed, almost all of the problem areas of chemical structure representation derive from deficiencies in this analogy.

6.2.2 Structure Diagrams

The chemist's "natural language" is the two-dimensional chemical structure diagram,—whether printed in a learned journal or scribbled on the back of a restaurant menu—and this is, of course, a form of topological graph. It can be, and sometimes is, represented simply as a graphical image on the computer. But this conveys little or no direct information about the way the atoms and bonds in the molecule are connected. More often, there is some internal topological representation (normally a connection table, discussed in Section 6.2.5) underlying a graphical representation. Several free and commercial programs allow structure diagrams to be sketched using a keyboard and mouse, and converted to other representations; market leaders include Accelrys Draw (formerly ISIS/Draw and SymyxDraw),[9] ChemDraw,[10] ChemSketch[11] and Marvin.[12] An up-to-date review of chemical drawing programs is provided by Gunda.[13] Much work has also been done on the automatic analysis of chemical structure diagrams, and their conversion to structure-searchable records.[14]

6.2.3 Chemical Nomenclature

Chemical nomenclature, like the structure diagram, pre-dates the computer age, but has been adapted to it, although arguably made obsolete by it. Systematic names have the advantage, from the computer's point of view, of being entirely text-based—though this is somewhat hampered by the use of special font faces, superscripts, subscripts, and occasional non-Roman letters—yet they convey substantial amounts of structural information. However, they are unwieldy and error-prone and it has been shown that the quality of computer-generated names now substantially exceeds that of manually-generated systematic names.[15] Several excellent programs are now available to both convert existing chemical names to other forms of structural representation, and to generate correct systematic names from chemical structures.[16]

6.2.4 Line Notations

Line notations represent chemical structures as linear strings of alphanumeric symbols. Though the earliest ones were developed for use as manual ciphering tools, more convenient than systematic nomenclature, their compactness made them especially suitable for use in computer systems at a time when computer storage was much more limited than it is now. They were the predominant form for computer representation of chemical structures in the 1960s and 1970s, and one or two notations (in particular, SMILES) remain widely used today. Figure 6.1 shows a simple chemical structure and its representation in several different line notations.

Among the now superseded line notations, is the Wiswesser Line Notation (WLN), in which the more common elements were represented by several different symbols, indicating different bonding patterns (*e.g.* X and Y for branching carbon atoms), while a single symbol was used for commonly-occurring groups of atoms (*e.g.* "R" for a benzene ring). WLN had complex rules to ensure that there was only one correct order for the symbols, thus giving canonical representations, which could be sorted alphanumerically, and was thus well suited to archival storage of compounds. Although it was time-consuming to learn, expert coders could readily recognise structural features (especially ring systems) by inspection of the notation, something which cannot be said of more recent notations.

The Simplified Molecular Input Line Entry System (SMILES)[17] is the most popular line notation in use today, and though it technically remains a proprietary product of Daylight Chemical Information Systems Inc.,[18] it has been widely implemented by other vendors. Unfortunately this has led to some divergence of "dialects" of the notation, especially with respect to extensions to

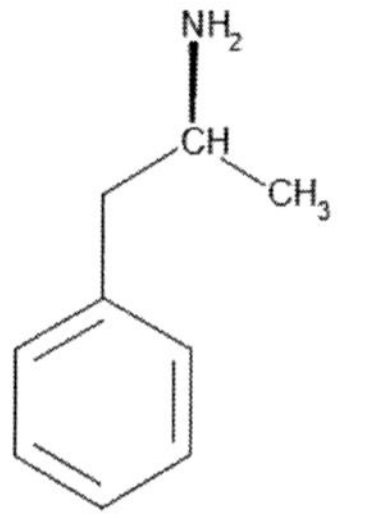

Trivial name:	(S)-amphetamine
IUPAC Name:	(2S)-1-phenylpropan-2-amine
WLN:	ZY1&1R
SMILES:	C[C@H](N)Cc1ccccc1
SLN:	CH3C[s=IE]H(NH2)CH2C[1]:CH:CH:CH:CH:CH:@1
InChI:	InChI=1S/C9H13N/c1-8(10)7-9-5-3-2-4-6-9/h2-6,8H,7,10H2,1H3/t8-/m0/s1
InChIKey:	KWTSXDURSIMDCE-QMMMGPOBSA-N

Figure 6.1 A simple chemical structure and its representation in several different line notations.

the original format, though Daylight's implementation can be regarded as definitive. SMILES uses a few simple rules to encode chemical structures:

- atoms are represented by standard atomic symbols (in square brackets for less common elements, or where additional properties are specified)
- bonds are represented by symbols separating the atoms ("-" for single bonds, "=" for double bonds, "#" for triple bonds and ":" for bonds in aromatic rings), though the single and aromatic bond symbols are usually omitted
- branches are represented by nested parentheses
- rings are represented by matching "ring-closure" numerals immediately after two non-adjacent atom symbols, to represent the bond which joins them and therefore completes the ring

A particular feature of SMILES is its use of lower-case atomic symbols to represent atoms in aromatic rings. In general, the atoms in SMILES can occur in any order, but if a canonicalisation algorithm (see Section 6.3.1) is applied, then a unique ordering is obtained, and molecules can be compared for identity simply by comparing their canonical SMILES strings. Though Daylight have developed a canonicalisation algorithm,[19] other vendors have their own, thus canonical SMILES from different vendors are not necessarily comparable. Though SMILES can be readily encoded manually, recognition of gross structural features, such as complex ring systems, by inspection of the notation is rather more difficult.

SMILES is one of a number of related notations, designed for special purposes. Especially important among these is SMARTS (SMILES Arbitrary Target Specification),[20] which is a sophisticated means of specifying query patterns that can be used to search for substructures in a database of chemical structures. Figure 6.2 shows an example, and a web-based program has recently been described for convenient visualization of SMARTS patterns.[21] The power of SMARTS notation is that it can be used to impose precisely defined views of chemistry on databases of chemical structures in a transparent and automated manner. Again, Daylight's own implementation can be regarded as definitive, and other vendors use subtly-different variants, with their own extensions and interpretations. SMIRKS[22] (SMIles ReaKtion Specification) is a hybrid of SMILES and SMARTS, which specifies how the atoms and bonds in the matching substructure should be modified—for example in the course of a chemical reaction—although it can also be used to specify structural modifications to be made during structure normalisation (see Section 6.3.2 and Figure 6.3).

Sybyl Line Notation (SLN)[23] acknowledges its debt to SMILES, and integrates the capabilities of the various Daylight notations into a single system, which is able to represent molecules, queries, reactions and some types of Markush structure (see Section 6.2.10). It remains a proprietary product of Tripos L.P. Representation of Structure Diagram Arranged Linearly (ROSDAL),[24] developed by the Beilstein Institute is also an "integrated" notation, capable of representing several different types of chemical entity. Though

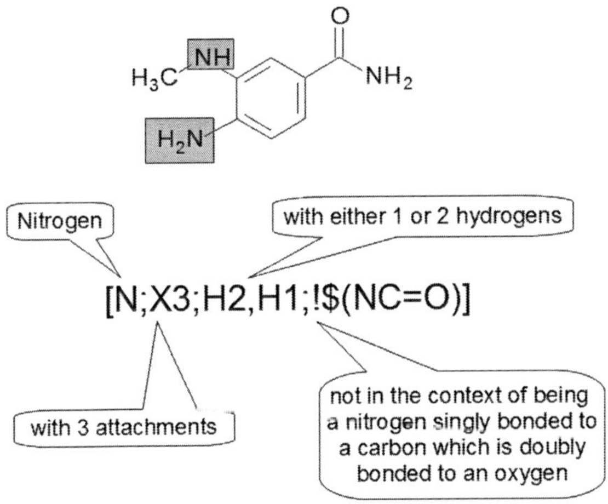

Figure 6.2 A SMARTS notation to specify a primary or secondary amine that is not an amide. This substructure pattern will match the shaded nitrogen atoms in the structure diagram, but not the unshaded one.

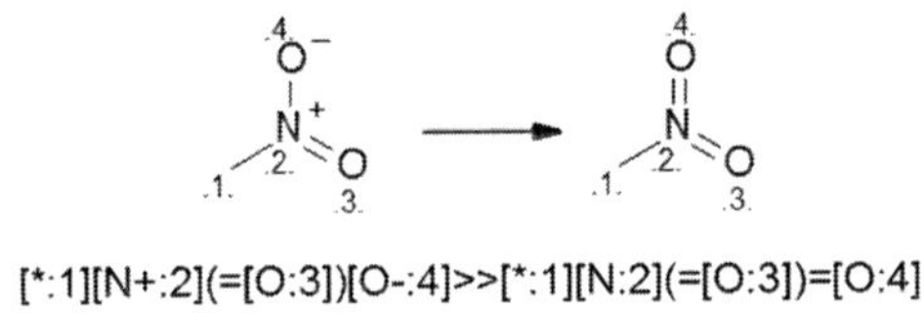

Figure 6.3 A SMIRKS notation specifying the conversion of a charge-separated nitro group to a representation with two double bonds and a 5-valent nitrogen atom. Corresponding atoms on each side have the same number (shown after the colon in the SMIRKS), and the atom attached to the nitro group is shown in the SMIRKS as * indicating "any atom".

both these notations can be read and written by a number of other vendors' programs, they have not established the popularity of SMILES and SMARTS.

6.2.5 Connection Tables

Connection tables are the predominant form of "internal" representation of chemical structures in computer memory. That is, they form the data structures to which various processing algorithms can be applied. Many "external" connection table file formats also exist for disk storage and exchange of structure representations.

At its simplest, a connection table is a table of records, one row for each atom, with columns to capture properties such as element type, charge, *etc.*, and a list of the row numbers to which the atom in question has bonds (with the appropriate bond order). There are many variants on this general principle. In some cases the atoms and bonds may be shown in separate tables; in some cases each bond is shown twice—once in for each of the atoms it to which it is joined, a so called *redundant* connection table—while in others, each bond is shown only once.

Sometimes hydrogen atoms are shown separately in designated rows of the table, though in *hydrogen-suppressed* connection tables a "hydrogen count" is shown as a property of each heavy atom. The latter approach not only has the advantage of storage compactness (probably less important now than it once was), but can also significantly improve processing times when applying algorithms for which time requirement increases exponentially with the number of atoms, as many graph-processing algorithms do.

Some authors[25,26] have proposed more radical departures from the conventional organisation of the connection table, especially with regard to the representation of bonds, putting emphasis on the electronic structure of the atoms involved. These approaches have distinct advantages in the handling of certain chemical species, such as carbenes and those involving multi-centre bonds.

6.2.5.1 *Connection Table File Formats*

Many file formats have been designed for connection tables, usually by software developers who need some means to save a chemical structure to a disk file. None of these can be considered ideal, though the *de facto* industry standard[27] has for many years been the Molfile[28,29] and its associated formats and variants, developed by MDL (subsequently acquired by Symyx Inc., and now part of Accelrys[9]); Figure 6.4 shows an example. Most formats are

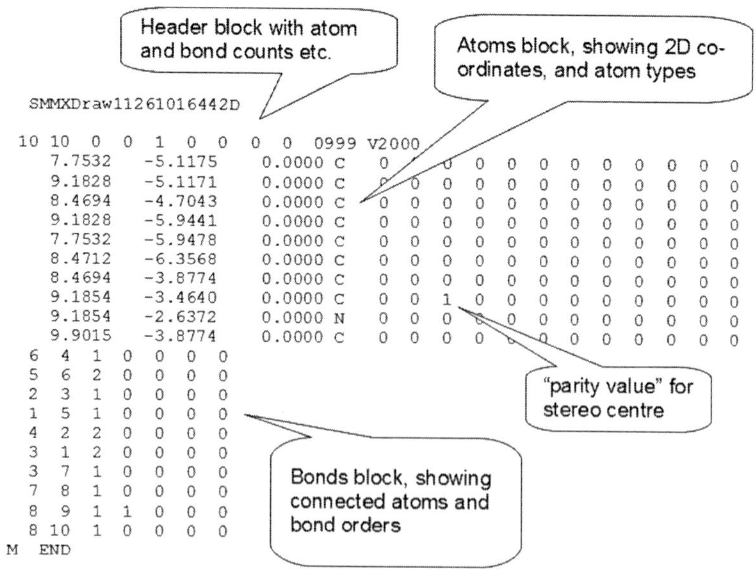

Figure 6.4 The Molfile representation of the structure shown in Figure 6.1.

commercially proprietary and may not be publicly documented or readable other than by the originator's own software.

There have been numerous attempts to develop "standard" connection table formats for data exchange, almost all of which have sunk rapidly and without trace. An XML-based connection table format, Chemical Markup Language (CML) was proposed in the late 1990s[30] and has now reached version 2.4[31] which incorporates a number of extensions for representation of features such as spectra, reactions and polymers. Though there is now an extensive CML literature, its adoption by commercial software vendors has thus far been very limited. Greater penetration has been achieved by the IUPAC Chemical Identifier, which is essentially a highly-compacted connection table, discussed in the next section.

6.2.6 IUPAC International Chemical Identifier (InChI)

The IUPAC International Chemical Identifier (InChI)[32] is a relatively recent arrival on the chemical structure representation scene, and combines some of the characteristics of connection table, line notation and registry number identifier. A comprehensive technical description has yet to be published, though substantial details are given in the documentation which accompanies the open-source software provided by IUPAC,[33] and a number of authors have provided good overviews.[6,34]

The InChI is a canonical representation of a chemical structure, and was explicitly designed to be used as a unique, algorithmically generated identifier, which could be used in place of arbitrarily-assigned registry numbers. In order to achieve this, several normalisation and canonicalisation procedures (of the type discussed in Section 6.3) are applied to the structure during the process of InChI generation. The InChI itself consists of a number of separate "layers", representing different aspects of the structure, such as Hill formula, non-hydrogen atom connectivity, hydrogen atoms, stereochemistry, isotopic composition *etc.* (see Figure 6.1 for an example). This has the advantage that the relevant substrings of full InChIs can be compared if features such as stereochemistry or isotope labelling are not of interest. InChIs have the additional advantage over registry numbers that the original structure can, within the limitations of the normalisation that is imposed, be recreated directly from the InChI string.

Because full InChI strings can be quite long, especially where large molecules are concerned, and incorporate many punctuation symbols, they have proved unsuitable for use as index terms in internet search engines. For this reason, a fixed-length, hash-coded version of the full InChI, known as the InChlKey, has been developed. This uses only uppercase Roman letters, divided into a group of 14 letters (encoding the main connection layer of the InChI), a group of 10 letters (encoding the stereochemistry and isotopic layers), and a single-character checksum. Of course, because it is a hash-coded version of the full InChI, it is not possible to regenerate the InChI from it. A number of websites[35,36] now provide "InChIKey Resolvers", which are able to perform the regeneration by looking keys up in a database of known InChIKeys. There is also a small but finite probability that two different structures with different InChIs will coincidentally hash to the same InChI key. So far this has only been observed in what might be called a pathological case, involving a structure with huge numbers of stereoisomers, among which hash collisions are observed at the theoretically-predicted frequency.

A number of scientific publishers are in the process of adopting InChIs and InChIKeys as a means of indexing the chemical structures mentioned in the text of academic papers *etc.*, and allowing them to be searched for in the same way as words and phrases in the full text article. InChIKeys are also being added to many chemical databases, and to internet resources such as Wikipedia. Several publishers and other interested bodies have formed a new organisation, the InChI Trust,[37] which will support the development and extension of the InChI standard, and the maintenance of definitive open–source software. Though the InChI is not above criticism,[38] it is showing signs of becoming the elusive internationally-accepted standard for canonical representation of chemical structures.

6.2.7 Stereochemistry

Topological graphs are concerned only with which nodes (atoms) are joined by which edges (bonds), thus have difficulty in representing stereochemistry directly. A number of approaches have been used to incorporate information about stereochemical configuration into chemical structure representations.[39–41]

An obvious choice, when using a connection table, is simply to record additional columns of data for each atom, specifying their 3-dimensional coordinates, though this has the disadvantage of conflating configuration and conformation; the use of 3-dimensional coordinates is discussed in Section 6.2.8. The conventional use of wedged and hatched bonds in chemical structure diagrams is frequently extended into computer systems, where connection tables mark bonds as "up", "down" *etc.*, though this approach is beset with the potential danger of showing ambiguous or geometrically impossible structures; the best software is able to identify such ambiguities as input errors.

A more commonly used approach, especially for simple tetrahedral stereochemistry around chiral carbon atoms, is to assign a "parity" label to the stereocentre. This essentially involves assigning a priority order to the atoms attached to the stereocentre, conceptually viewing the centre from the highest (or lowest) priority atom, and determining whether the remaining atoms show increasing priority in a clockwise or anticlockwise direction. Many different means are available for assigning the priorities, which can be grouped into those which are based globally on the molecule as a whole, and those that are defined purely locally. The well known Cahn-Ingold-Prelog (CIP) rules[42] use the global approach, in which the priorities are determined by a complex sequence of rules based on the relevant element types, though some circumstances have been identified in which they break down.[43] Another example of the global approach is provided by the Stereochemically-Extended Morgan Algorithm (SEMA) Name,[44] in which the priority order at each stereocentre is determined by the algorithm assigning a canonical ordering to the atoms in the connection table (see Section 6.3.1). The local approach is exemplified by SMILES, in which the priority order is simply the sequence of the relevant atom symbols in the SMILES string; the atom stereo parity value used in Molfiles[28–29] is similarly determined by the arbitrary sequence of atoms in the connection table. In SMILES, the parity is indicated by a symbol appended to the chiral atom indicating if its neighbours, in the order they occur in the string and viewed from the earliest of them, appear anticlockwise (@) or clockwise (@@). The advantage of the local approach is that it can be applied to substructures and remains unchanged if more atoms are added, provided the order of the atoms involved in the stereocentre is not changed.

Another approach, which lends itself particularly to forms of stereochemistry involving non-tetrahedral geometries (*e.g.* in coordination compounds) is the *stereovertex list*. Here, the atom numbers around the stereocentre are mapped as an n-tuple onto the corners of a standard polyhedron representing the geometry in question, and stored separately from the main connection table.[40]

Many cheminformatics programs are able to interconvert different forms of stereochemistry representation, for example to calculate appropriate parity values from wedged and hatched bonds input using a structure drawing program.

6.2.8 Three-Dimensional Structure Representation

Databases of 3D structures are widely used in directed screening, which can be defined as the process of prioritising compounds for biological evaluation. In virtual screening, 3D structures for potential ligands are docked into protein structures and the resulting poses are scored to predict the likelihood of observing activity.[45] In the absence of a protein structure, 3D structures can be matched geometrically to pharmacophore models[46] or shape-matched[47] against 3D structures of known active compounds. A pharmacophore can be defined in general terms as a set of molecular recognition features (*e.g.* anionic oxygen, aromatic ring centroid) and the geometric relationships between them, which is hypothesised to be necessary but not sufficient for biological activity. Pharmacophore matching was the first directed screening application to use databases of 3D structures, and pharmacophoric elements may be encountered in both virtual screening and shape matching. Quantitative structure activity relationships (QSARs) frequently use descriptors calculated from 3D structures, and there may be advantages in extracting these from a database rather than generating them on-the-fly.

In principle, extension of the topological graph representations discussed in the preceding sections of this chapter requires little more than the addition of 3D coordinates for the atoms, for example as additional columns in a connection table. Typically these coordinates are generated using a program such as CONCORD or CORINA to automatically build low energy conformations from topological input.[48] The issue of conformational flexibility can be addressed by storing multiple conformations for each molecule, although docking and pharmacophore matching programs can often explore conformational space during the search process if provided with a single low energy conformation. Although lack of availability limits the use of small molecule crystal structures in directed screening applications, analysis of databases of experimentally determined structures provides valuable insight into the nature of intermolecular interactions and the conformational preferences of molecules.[49]

6.2.9 Fragment Codes and Structure Fingerprints

Before the development of unambiguous computer representations of full chemical structures (and in particular, of software capable of performing substructure searches on them), structure searching was based primarily on so-called *chemical fragment codes*. These involved the identification (originally manually) of small groups of connected atoms (often functional groups and rings) present in the molecule, and their use as "index terms" for the structure. Searches, which could be implemented using punched card sorting equipment, involved identifying those molecules that contained the required fragments in appropriate Boolean combinations. Such coding systems (albeit updated from the punched card era, and with automatic identification of the fragments) still retain some use in specialised areas such as Markush structure retrieval.

Automatically-identified structural fragments are also widely used in the generation of chemical structure *fingerprints*, which are binary strings (usually containing a few hundred or a few thousand bits) in which a 'one' indicates that a fragment is present in the structure, and a zero indicates that it is absent. Such fingerprints can be regarded as the electronic equivalents of punched cards. Rather than chemically meaningful functional groups however, the fragments usually represent topological patterns within the chemical structure, such as linear paths of atoms and bonds, or an atom and its immediate neighbours. Such fragments can easily be identified algorithmically and either looked up in a predefined fragment dictionary (which specifies the bit position to be set in the fingerprint) or "hashed" automatically to generate a bit position in the appropriate range. The former approach means that some fragments present in the molecule (if not found in the dictionary) will not be represented in the fingerprint, while the latter approach means that structurally unrelated fragments may finish up setting the same bit position. Another type of fingerprint, the extended connectivity fingerprint (ECFP)[50] combines the fragment identification and hash-coding stages by iteratively refining integers (*e.g.* atomic number) initially assigned to each atom, at each stage combining the values assigned to its neighbours, thus taking into account larger and larger circles round each atom.

Structural fingerprints are used for two primary purposes. Firstly, as an initial screening stage in substructure searches,[51] to rapidly eliminate structures that do not contain the required fragments before a more time consuming

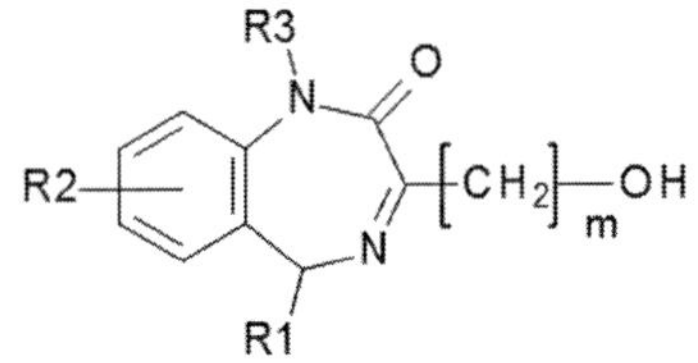

R1 = propyl, $CH_3CH(Br)CH_3$, OMe

R2 = F

R3 = C1-4 alkyl

m = 1-3

Figure 6.5 An example of a Markush structure.

atom-by-atom search stage. Secondly, for the calculation of numerical measures of "similarity" between structures, based on the fragments they have in common.[52] The two purposes have somewhat different requirements, and fingerprints that are suitable for one may not be appropriate for the other. In particular, "redundancy" in the fingerprint (where different bits effectively represent the same structural feature, perhaps as part of a larger feature) can introduce bias into similarity calculations, whereas they will not have an adverse effect on substructure search screening other than a very slight degradation in efficiency.

6.2.10 Markush Structures

Generic or "Markush" structures are a form of structural representation that can be used to encompass many individual molecules having common structural features. Figure 6.5 shows a simple example. They are particularly associated with chemical patents,[53] but may also be used (often with a table of values for the R-groups) to represent sets of related molecules involved in structure – activity relationship studies, or large combinatorial libraries.[54] The RGfile variant of the Molfile format[29] can represent restricted forms of Markush structure, as can Sybyl line notation,[23] and a variety of vendors have implemented *ad hoc* extensions (which have some Markush capabilities) to SMILES.

6.3 Canonicalisation and Normalisation

There are two related aspects to creating a unique representation of a chemical structure. One is to provide a unique encoding for a chosen representation, which primarily involves determining an order in which the atoms should be listed, while the other is to determine which of a number of possible variations in the way that equivalent structures can be represented (positions of double bonds, charges, mobile hydrogens, *etc.*) should be chosen. The former process is usually referred to as *canonicalisation* or *canonisation* and the latter as *normalisation* or *standardisation*. Comparison of the canonical forms of a number of different possible representations may be required as part of the normalisation process.

6.3.1 Canonicalisation

This is essentially a graph theoretical procedure. Initially the numbering of the nodes in the graph (the atoms) is quite arbitrary, and canonicalisation is a procedure for renumbering them in a unique way, which will be reproducible whatever initial numbering has been assigned. This enables two chemical structure graphs to be compared for identity by simple sequential matching (or string matching in the case of SMILES), rather than using much more time-consuming graph isomorphism algorithms.

Most cheminformatics literature cites the algorithm proposed by Morgan in 1965[55] and implemented at Chemical Abstracts Service. Figueras[56] provides a theoretical analysis of this and the various extensions and modifications subsequently proposed for it, which includes one[44] taking stereochemistry into account. The graph theory literature prefers to cite McKay's algorithm,[57] which has been explicitly acknowledged as the algorithm used in generating InChIs (see Section 6.2.6). Both algorithms are based on the principle of assigning initial numerical labels to each node (based on some characteristic of the node, such as the atom type it represents or the number of connected nodes it has) which is iteratively refined by considering the labels of its neighbours. Once all nodes (other than those that are symmetrically equivalent) have different labels, they can be cited in a priority order based on their labels.

6.3.2 Normalisation and Business Rules

Different types of chemical structure equivalence may be encountered in pharmaceutical compound collections and it is important to be able to recognise these in an automated manner. Different topological graphs can be written for a particular molecule and an individual compound may be represented by different molecular structures. Even when the problem of duplicate structures has been addressed by defining "business rules" for in house databases, there is no guarantee that compounds from an external supplier or merger partner will have been registered according to the same conventions. A molecular editor capable of modifying chemical structures in a controlled and automated manner can be used to apply business rules. The Leatherface program[58] applies structural transformations defined as SMARTS,[20] but newer tools[59–61] exploit SMIRKS[22] notation which has been designed specifically for this purpose and allows each transformation to be specified as a single string. Chemical scripting languages such as Cheshire[9] and Standardizer[12] are often used to apply business rules.

Definition of unambiguous business rules is not always straightforward and in these situations it is useful to generate an ensemble of equivalent structures.[58,60,61] If rules to rank the structures in the ensemble can be defined, then it becomes possible to select one of these to represent the ensemble. For example, enumeration of tautomers using Leatherface creates a sorted list of canonical SMILES, the first of which is used to represent the ensemble.[58] This effectively extends the concept of canonicalisation to an ensemble of structures, and the term 'canonical tautomer' may be encountered in the literature.[61–62]

When a molecule is presented as more than one topological graph, the individual representations are termed mesomers or, in the context of structure registration, mesomeric duplicates.[61] Mesomers differ only in bond order and placement of formal charges and are used to represent bond delocalisation in a molecule.

Examples include the Kekulé forms that can be written for benzene (Scheme 6.1) and the different topological graphs used to represent the hypervalency of nitrogen in amine N-oxides (Scheme 6.2) and azides (Scheme 6.3).

Different topological graphs may be written for a charged species (Scheme 6.4) to represent the delocalisation of the charge in the molecule. It is customary to link such forms by a double headed arrow to indicate that they are different representations of the same molecule, rather than distinct molecular species.

Scheme 6.1

Scheme 6.2

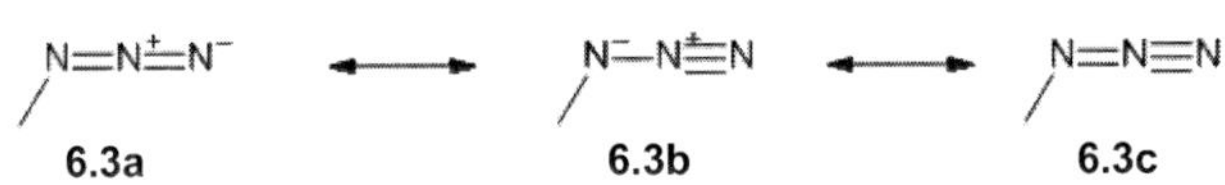

Scheme 6.3

Scheme 6.4

The difficulties posed by Kekulé forms of benzene can be addressed by defining aromatic atom and bond types. However, this can introduce new problems in that the connectivity of an imidazole can no longer be used to assign implicit hydrogen atoms. In cases like amine *N*-oxides (Scheme 6.2), business rules can be defined to specify one form as preferred, and a molecular editing tool can be used to convert inappropriate forms. Using opposite charges to represent hypervalency usually adds complexity (*e.g.* Scheme 6.3) to chemical database systems. A strong case can be made for assigning a valency of five to hypervalent nitrogen and restricting the use of formal charges to charged and zwitterionic species.

Delocalised cations such as that shown in Scheme 6.4 are less easily handled using business rules. The substructural pattern used to define the transformation between forms matches both forms, so the transformation can be described as 'degenerate'. Although business rules might be defined for imidazolium cations like that in Scheme 6.4, these will need to encode properties of substituents at all positions in order to be comprehensive. This is a good example of a situation in which it is best to enumerate the different forms and select the highest ranking one as preferred.

Isomers such as tautomers that can readily inter-convert represent another type of structural equivalence that is encountered frequently in databases of chemical structures. Although the different forms are distinct molecular species, techniques used for processing mesomers are equally applicable to tautomers.[58]

For example, a business rule can be defined to select the typically more stable 2-pyridone (**6.5b**) in preference to its 2-hydroxypyridine tautomer (**6.5a**).

6.5a 6.5b

Scheme 6.5

6.6a 6.6b

Scheme 6.6

The 'degenerate' tautomerism of **6.6** is analogous to the structural equivalence of the mesomeric duplicates **6.4a** and **6.4b**. Handling tautomerism and mesomerism becomes more difficult when the hydrogen atoms and charges move over long distances, and Sayle has discussed global approaches to the problem.[61] The treatment of tautomerism in commercial chemical information management systems has been reviewed by Warr.[62]

Acids and bases are often treated as equivalent to their salts in pharmaceutical databases because under assay conditions the presence of counter-ions is usually irrelevant. However, there are situations where it is useful to know when an acid or base is in a salt or neutral form. For example, fragment screening libraries are stored as high concentration stock solutions in dimethylsulfoxide, and using neutral forms of acids and bases appears to lead to fewer problems with precipitation on storage.

6.4 Physiologically Relevant Forms

Drugs interact with their targets in buffered aqueous media and computational tools for directed screening (*e.g.* docking, scoring and pharmacophore matching) usually require that molecules be input in physiologically relevant forms. The processes of ionisation and tautomerism typically invert polarity of contact points in molecules and are particularly relevant to molecular recognition.[58] Computational tools for directed screening vary widely in their ability to generate physiologically relevant forms so there are advantages in generating these beforehand and storing them in a database.

Many measured values of pK_a have been reported in the literature for acids and bases. Ionisation typically represents a local structural change, so it is usually possible to use structural analogy to estimate pK_a and to derive substructural rules for identifying the most abundant protonation state(s) at a given pH. Although the tautomerism literature is more fragmented than that for pK_a, it is still sufficiently extensive to allow derivation of useful rules for the most probable tautomers under physiological conditions.[63] These rules must encode both biased equilibria, where one form predominates, and balanced equilibria, where more than one form must be used to represent the state of the compound in assay buffer. For example, docking carboxylic acids exclusively as their anions represents a low risk option for a virtual screen at a pH of 7.4 because the pK_a of these species rarely exceeds 5. In contrast, a realistic representation of imidazole requires one protonated form (**6.7c**) and two tautomeric neutral forms (**6.7a** and **6.7b**) because the pK_a is close to physiological pH.

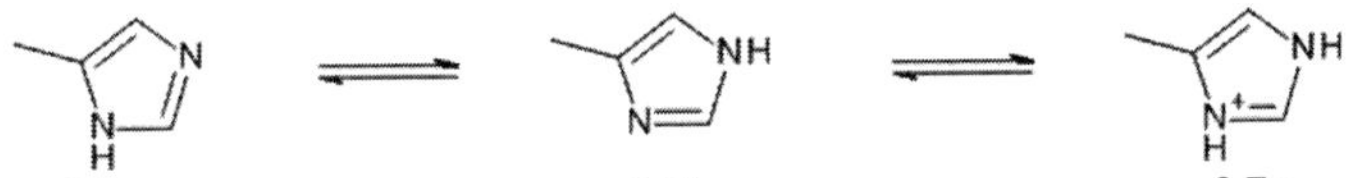

6.7a **6.7b** **6.7c**

Scheme 6.7

A comprehensive treatment of tautomerism and ionisation in structural databases requires storage of the different forms in which a compound can exist, as well as values of pK_a (some proteases function in the acidic interior of the lysosome) and tautomeric ratios. Automatic prediction of pK_a and tautomer ratio is an active area of research,[60,61,64] therefore it is still useful to store the forms which are believed to be physiologically relevant without explicit reference to stability.

6.5 Registration Systems

Registration systems are maintained by many organisations, including pharmaceutical companies, in order to control the addition of chemical structures to a central database. A key requirement for a registration system is the determination of structural novelty, both to unambiguously record a structure for intellectual property reasons and to prevent a wasteful and possibly expensive re-synthesis when a sample already exists. This implies that structures should be normalised to a consistent representation. There is no need for the "best" normalised form here, or even the physiologically active form; any form will suffice as long as it is consistently derived. As different companies have their own specialist needs, normalisation is typically driven by a set of business rules written in a chemical scripting language (such as Cheshire[9] or Standardizer[12]). These scripts can then be customised as required. Care must be taken when initially determining these rules, as any changes implemented at a later date may affect how structures are grouped in the registry.

Typical rules include converting nitro groups to the uncharged form (**6.8**), expanding shortcuts (**6.9**), replacing dative bonds in coordination compounds with separated charges (**6.10**), and neutralising zwitterions (**6.11**).

6.8a **6.8b**

Scheme 6.8

6.9a **6.9b**

Scheme 6.9

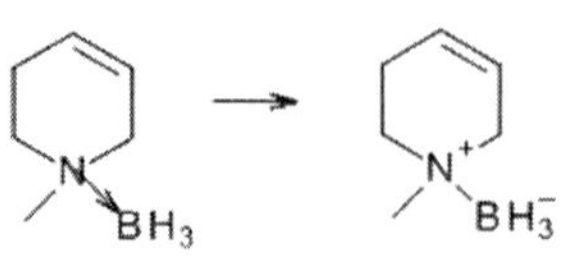

6.10a **6.10b**

Scheme 6.10

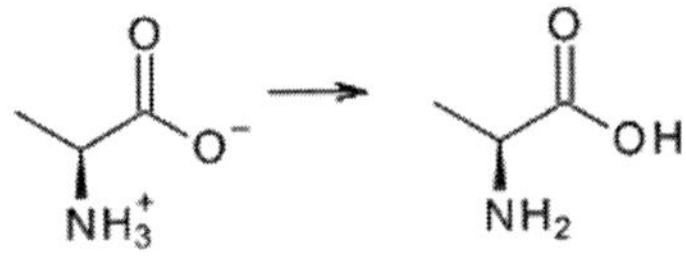

6.11a **6.11b**

Scheme 6.11

For tautomers, the ideal situation is to store one preferred tautomeric form and register other forms as synonyms of this preferred one. However, there is no universally accepted algorithm for tautomer generation, and each registration system will have its own implementation. There will be cases where the system perceives two structures to be tautomers but the chemist knows better and wants to store them separately. Conversely, due to limitations in the tautomer perception algorithm, two tautomeric forms may not be recognised as such. A practical registration system must therefore allow chemists to override the system-provided default.

6.5.1 Parent/Version/Preparation Hierarchy

In a pharmaceutical context, the biologically active part of a structure is of the most importance, and it is convenient to group biologically equivalent forms together. A technique which has worked for many years in GSK is the concept of a parent structure.[65] The parent structure is derived from the version structure (as supplied by the chemist) by stripping off any salts and solvates and converting any unnatural isotopes to their normally abundant form. Isotopically labelled compounds are typically used for metabolism studies and are therefore normalised and grouped under the same parent as the non-isotope form.

Other typical transformations to make the parent include breaking any covalent bonds to group I metals and neutralising charges as far as possible by adding or removing protons subject to maintenenance of the correct valency.

It is important to realise that the parent structure may not actually exist; it is simply used for the grouping of biologically equivalent structures. Parent structures are also beloved of modellers who don't want salts and isotopes complicating their SAR work.

Many batches (preparations) of a version structure may be prepared, so the parent version hierarchy can be extended to the preparation level. Some registration systems further extend the hierarchy to the sample level, with each preparation having many samples.

6.5.2 Stereochemistry

Biological targets are stereospecific, so any registration system needs to clearly specify and distinguish different stereoisomeric forms. There is still some disagreement about the best way to represent stereochemistry in a 2D structural diagram.[66–67] Examples **6.12** to **6.16** use MDL[9] Enhanced Representation.[29,68]

6.12a shows a single resolved enantiomer and **6.12b** is its mirror image. **6.12c** has the chiral centre in an "AND" group and represents a racemic mixture of **6.12a** and **6.12b**. All three structures will have different biological activity and must be registered separately.

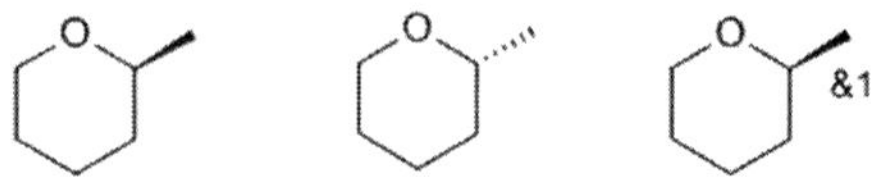

6.12a 6.12b 6.12c

Scheme 6.12

As the number of chiral centres grows, so does the number of permutations. **6.13** has two chiral centres and can exist as one of the four enantiomers, a–d.

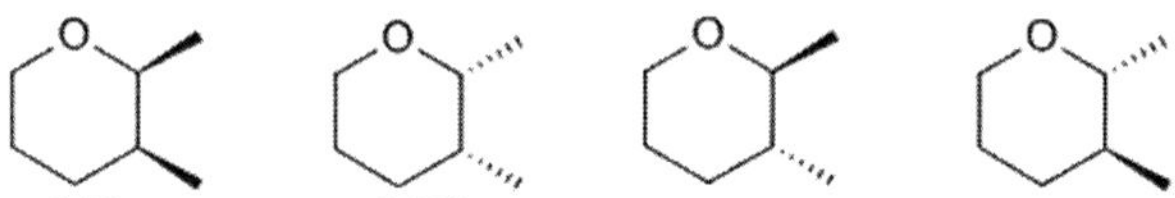

6.13a 6.13b 6.13c 6.13d

Scheme 6.13

Compound **6.14** is an example of relative stereochemistry; as the two chiral centres are in the same "AND" group, it represents a racemic mixture where the two methyl groups are relatively *cis*. Compound **6.15** similarly represents a racemic mixture of the *trans* isomers. Compound **6.16** has each chiral centre in a different "AND" group and therefore represents a mixture of all four diastereoisomers. Again, all of these will have different properties and must be registered as distinct compounds.

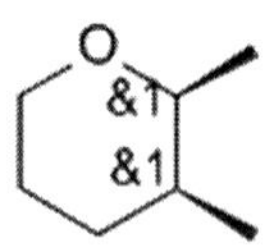

Scheme 6.14

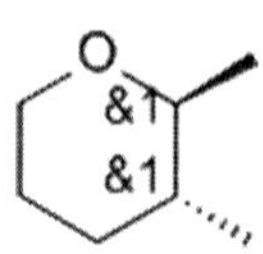

Scheme 6.15

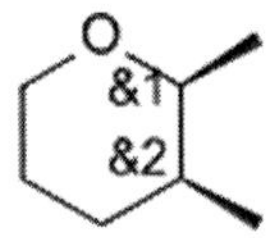

Scheme 6.16

Scheme 6.17

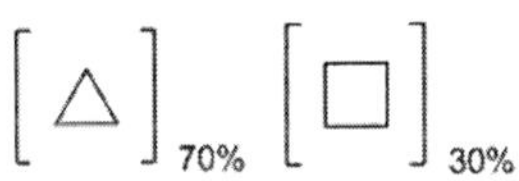

Scheme 6.18

6.5.3 Mixtures

Mixtures arise either as products of a reaction that cannot be separated, or as pure components that have been deliberately mixed together. A common approach is to register all structures combined into one Molfile, and also to register each component separately.

The precise composition of a mixture will be recorded in fields in the registration database, but it is convenient also to display this information along with the structure. There is no established convention for such mark-up and it depends on the abilities of the underlying software. For Molfile-based systems, SGroup brackets can be used for this purpose.[29] For stoichiometric mixtures with integral multipliers, the multiple SGroup bracket (MUL) can be used (Scheme 6.17).

This has the advantage that many chemical editors and toolkits will correctly calculate the molecular formula and molecular weight. Non-integral multipliers can use the SRU(n) bracket (Scheme 6.18).

The use of "n" as the label is recommended by IUPAC for repeating groups of unknown size, but any text is allowed here. Racemic mixtures are treated as a special case; as the chemical structure of each component is essentially the same, it is clearer to use the same skeleton with annotations. Racemic mixtures can be represented using the aforementioned "AND" group enhanced mark-up.

6.5.4 Structural Uncertainty

6.5.4.1 Stereochemical Uncertainty

A particular complication for registration systems is the requirement to register compounds whose structure is not completely known. For example, a chemist

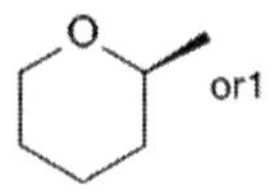

Scheme 6.19

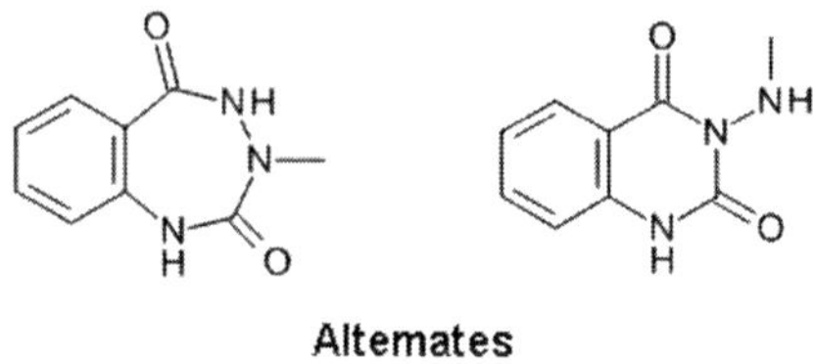

Scheme 6.20

may have registered and tested a racemic mixture and found it to be active. The next step might be to resolve the enantiomers on a chiral column so they can be independently tested. If the chromatography is successful, the chemist will have two separated, pure enantiomers but will not know which is which. It is useful to be able to register these enantiomers separately so that biological data can be accumulated against each isomer until the structure is fully elucidated. MDL enhanced stereochemistry[68] allows the "OR" tag to be applied to the chiral centre, which represents a single unknown enantiomer (Scheme 6.19).

6.5.4.2 Chemical Uncertainty

This includes cases where the specific isomer is unknown. In this case, all the possible structures can be combined into one Molfile and registered as a single entity, together with a flag indicating that these are alternative structures and not a mixture (Scheme 6.20).

In all such cases of uncertainty, both isomers will have the same structural representation and will need to be marked as distinct in the registry.

6.5.5 Unknown Structures

There is a need to register and test batches of material whose structure is not yet known or is partially known, *e.g.* natural products, polymers and resin bound fragments.

If the structure is partially known, it is valuable to record that part of the structure which is known, together with some descriptive "chemically significant text" (CST). Where the structure is attached to a resin or biomolecule, then the point of attachment can be indicated. The CST is attached to the whole structure as a labelled SGroup.

If the structure is completely unknown then there is no structure to register! In these cases, it is common to register a placeholder structure consisting of a single "star" atom to which the CST can be attached.

Because the structure is not fully specified, it will be uncertain whether two batches are the same; is one batch of green slime the same as another? Determining whether they are the same (and if not, changing the CST to avoid a match) will require manual intervention by a registrar.

6.6 Analysis of Structural Databases

Once a number of structures have been accumulated in a database, a great deal of useful information can be obtained by comparing them and identifying commonalities and differences between them.

6.6.1 Clustering and Subsetting

The use of structural fingerprints for calculating numerical measures of similarity between individual molecules was discussed in Section 6.2.9. Such pairwise similarities can also be used as the basis for *clustering* large numbers of molecules into subgroups whose members are similar to each other but different from those of other subgroups. Many different procedures have been proposed for this,[69] and they are widely used in industry for purposes such as selection of compounds for acquisition for high-throughput screening, or to assist in the development of quantitative structure-activity relationship models in homogeneous datasets. Other structural similarity-based approaches have also been used to identify maximally diverse subsets.[70]

6.6.2 R-group Decomposition

R-group decomposition, or R-group analysis, is the process of building a Markush structure (see Section 6.2.10) which can be used to visualise the structural commonalities in a set of individual molecules, for example a common scaffold surrounded by variable R-groups. This can be achieved either by pre-specifying the required scaffold and performing a substructure search to find those molecules that contain it,[71] or by performing some form of automated *maximum common subgraph (MCS)* detection to find the largest connected groups of atoms and bonds common to the molecules being analysed.[72]

6.6.3 Reduced Representations

In recent years, a number of ambiguous representations of chemical structures have been developed in which different but closely related molecules have identical representations, and can thus in some sense be regarded as "equivalent". In feature trees,[73–74] single nodes replace fragments of the full molecule, each representing either a ring or an acyclic functional group or hydrophobic fragment, appropriately classified. A similar concept is also used in reduced graphs,[75–77] though here each node is more explicitly selected to represent some feature such as a hydrogen bond donor group or aromatic ring system.

Molecular equivalence numbers[78] are derived by breaking the molecule into separate components (*e.g.* rings and acyclic groups), and sequentially blurring detailed distinctions within each (*e.g.* precise atom and bond types, or ring sizes) to form "equivalence classes". A unique naming function (essentially a canonicalisation process) is then used to generate alphanumeric identifiers for these classes.

All three of these representations can be used for purposes such as similarity searching, giving results that might be missed by conventional structure-fingerprint-based similarity (thus providing opportunities for so-called "scaffold-hopping" in drug discovery) and for assessing the structural diversity of compound collections.

6.6.4 Matched Molecular Pair Analysis

An alternative approach to analysis of structural databases is to focus on relationships between structures. A relationship between structures can be a simple substitution (**6.21a** is mono-chloro analogue of **6.21b**) or something more abstract (**6.22a** and **6.22b** are 'reversed' amides).

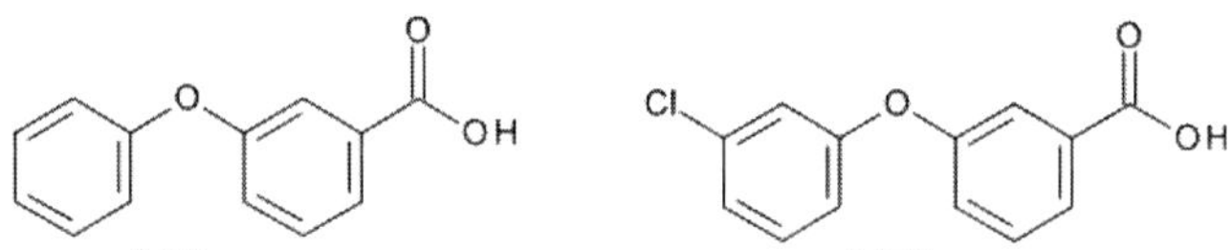

6.21a **6.21b**

Scheme 6.21

6.22a **6.22b**

Scheme 6.22

The process of identifying pairs of structures that have a specific relationship and quantifying the effects of the structural change has been termed matched molecular pair analysis (MMPA).[58] MMPA allows the effect of a specific structural change to be averaged over diverse structures and can be thought of as special type of local QSAR model. The underlying assumption is that the effect of a small structural change on a property can be predicted more accurately than the value of the property. Although superficially similar to Free-Wilson Analysis,[79] MMPA is actually a data-analytic equivalent of the free energy perturbation (FEP) method.[80]

MMPA has proven useful as a database mining tool for quantifying the effects of specific structural changes on a number of physicochemical properties such as aqueous solubility.[81] Substituting tetrazole for a carboxylic acid is perhaps the prototypical bioisosteric replacement, and MMPA has been used to show that this is likely to lead to a reduction in the fraction of free drug in the plasma.[82] MMPA revealed that *N*-methylation of secondary amides typically leads to an increase in aqueous solubility and this was interpreted as evidence for the importance of crystal packing as a determinant of this physicochemical property.[81] Support for this view was provided by the observation that *N*-methylation of cyclic secondary amides did not result in a significant increase in aqueous solubility.[82] One area in which the concept of matched molecular pairs has proven useful is in exploring the link between chemical substitutions and activity cliffs.[83] Both increasing the efficiency and increasing the degree of automation for matched pair identification remain active areas of research.[84]

6.7 Conclusions and Future Directions

The basic forms of representation for chemical structures are well established, and although some changes to the details of exchange formats, and the popularity of different ones, remains likely, the principles on which they are constructed will probably remain the same. The changes that are most likely to occur will concern matters such as stereochemistry and aromaticity (and other delocalised bonding patterns) and the point at which differently represented structures are to be regarded as distinct. The problem here is less that new techniques are required and more that none of the established techniques provide the perfect solution.

Standardisation of representations has not had a happy history, partly as a result of the differing requirements of different groups of users. Though the InChI does seem to be establishing itself in some quarters as a widely used standard, under international control, different flavours of and extensions to established formats (such as SMILES) continue to appear.

There is a trend towards storing increasing amounts of information (*e.g.* tautomers, conformers and pK_a) in chemical structure databases. Some of this information (*e.g.* energy of a tautomer or molecular surface area of a conformer) is not associated with individual atoms and bonds . New techniques and data structures will be required to ensure that the databases can be mined, maintained efficiently and interfaced with an ever-growing number of applications which feed off the data.

In the future, it is likely that information about relationships between structures will be stored in chemical structure databases to an increasing extent, so defining and encoding these when multiple forms are used to represent compounds will require careful attention. Registration of new compounds will inevitably reveal structural relationships that have not previously been encountered, and database systems will need sufficient flexibility to process new information of this nature.

Commercial registration systems are available from a variety of vendors, each with their own unique ways of representing not only individual chemical structures but also the grouping of these structures into related classes. As a consequence, it is hard to convert from one vendor's system to another without losing information in the process. This is leading to the formation of pre-competitive collaborative groups such as the Pistoia Alliance,[85] one aim of which is to standardise the information models and interfaces used by registration systems, with the eventual goal of increasing data quality and interoperability.

In the near future, registration systems will need to move beyond their traditional domain of small molecules into the realm of larger "biological" entities such as proteins, plasmids and vaccines. Although technically still chemical structures, their size and complexity will require the development of new representations and algorithms that permit rapid searching and novelty checking.

References

1. J. Gasteiger and T. Engel, *Chemoinformatics: a Textbook*, Wiley-VCH, Weinheim, 2003.
2. J. Bajorath, *Chemoinformatics: Concepts, Methods, and Tools for Drug Discovery*, Humana Press, Totowa, 2004.
3. T. I. Oprea, *Chemoinformatics in Drug Discovery*, Wiley-VCH, Weinheim, 2005.
4. A. R. Leach and V. J. Gillet, *An Introduction to Chemoinformatics*, Springer Verlag, Dordrecht, 2007.
5. J. M. Barnard, in *Handbook of Chemoinformatics*, ed. J. Gasteiger, Wiley-VCH, Weinheim, 2003, 27.
6. B. Kosata, in *Chemical Information Mining*, ed. D. L. Banville, CRC Press, Boca Raton, 2009, 77.
7. D. R. J. Wilson, *Introduction to Graph Theory*, Prentice Hall, Harlow, 2010.
8. O. Ivanciuc, in *Handbook of Chemoinformatics*, ed. J. Gasteiger, Wiley-VCH, Weinheim, 2003, 103.
9. Accelrys Inc., San Diego, CA, USA, *http://accelrys.com*.
10. CambridgeSoft Corporation, Cambridge, MA, USA, *http://www.cambridgesoft.com/*.
11. Advanced Chemistry Development Inc., Toronto, Canada, *http://www.acdlabs.com/*.
12. ChemAxon Kft, Budapest, Hungary, *http://www.chemaxon.com*.
13. T. E. Gunda at *http://dragon.unideb.hu/~gundat/rajzprogramok/dprog.html*.
14. A. P. Johnson and A. T. Valko, in *Chemical Information Mining*, ed. D.L. Banville, CRC Press, Boca Raton, 2009, 45.
15. G. A. Eller, *Molecules*, 2006, **11**, 915.
16. A. J. Williams and A. Yerin, in *Chemical Information Mining*, ed. D.L. Banville, CRC Press, Boca Raton, 2009, 21.

17. D. Weininger, *J. Chem. Inf. Comput. Sci.*, 1988, **28**, 31.
18. Daylight Chemical Information Systems Inc., Laguna Niguel, CA, USA, *www.daylight.com*.
19. D. Weininger, A. Weininger and J. L. Weininger, *J. Chem. Inf. Comput. Sci.*, 1989, **29**, 97.
20. Daylight Theory Manual, Chapter 4, *http://www.daylight.com/dayhtml/doc/theory/theory.smarts.html*.
21. K. Schomburg, H. Ehrlich, K. Stierand and M. Rarey, *J. Chem. Inf. Model.*, 2010, **50**, 1529.
22. Daylight Theory Manual, Chapter 5, *http://www.daylight.com/dayhtml/doc/theory/theory.smirks.html*.
23. R. W. Homer, J. Swanson, R. J. Jilek, T. Hurst and R. D. Clark, *J. Chem. Inf. Model.*, 2008, **48**, 2294.
24. J. M. Barnard, C. J. Jochum and S. M. Welford, in *Chemical Structure Information Systems. Interfaces, Communications, and Standards, ACS Symposium Series*, 1989, **400,** 76.
25. A. Dietz, *J. Chem. Inf. Comput. Sci.*, 1995, **35**, 787.
26. S. Bauerschmidt and J. Gasteiger, *J. Chem. Inf. Comput. Sci.*, 1997, **37**, 705.
27. J. L. Baumgras and A. E. Rogers, *Journal of the American Society for Information Science*, 1995, **46**, 623.
28. A. Dalby, J. G. Nourse, W. D. Hounshell, A. K. I. Gushurst, D. L. Grier, B. A. Leland and J. Laufer, *J. Chem. Inf. Comput. Sci.*, 1992, **32**, 244.
29. Symyx CTfile Formats, June 2010, *http://www.symyx.com/downloads/public/ctfile/ctfile.pdf*.
30. P. Murray-Rust and H. S. Rzepa, *J. Chem. Inf. Comput. Sci.*, 1999, **39**, 928.
31. Chemical Markup Language *http://xml-cml.org/index.php*.
32. A. McNaught, *Chem. Int.* 2006, **28**, 12.
33. IUPAC International Chemical Identifier Software Download Site. *http://www.iupac.org/inchi/download/index.html*.
34. W. A. Warr, *QSAR World, 2008, http://www.qsarworld.com/INCHI1.php*.
35. Chemspider InChI Resolver. *http://inchis.chemspider.com/Default.aspx*.
36. NCI/CADD Chemical Identifier Resolver. *http://cactus.nci.nih.gov/chemical/structure/documentation*.
37. The InChI Trust, Berlin, Germany, *http://www.inchi-trust.org/*.
38. K. Boda, presented at the 5th Joint Sheffield Conference on Chemoinformatics, Sheffield, 2010. *http://cisrg.shef.ac.uk/shef2010/talks/52.pdf*.
39. H. W. Davis, *Computer Representation of the Stereochemistry of Organic Molecules*, Birkhäuser, Basel, 1976.
40. J. M. Barnard, A. P. F. Cook and B. Rohde, in *Chemical Information Systems - Beyond the Structure Diagram*, ed. D. Bawden and E. M. Mitchell, Ellis Horwood, Chichester, 1990, 29.
41. B. Rohde, in *Handbook of Chemoinformatics*, ed. J. Gasteiger, Wiley-VCH, Weinheim, 2003, 206.
42. R. S. Cahn, C. Ingold and V. Prelog, *Angew. Chem. Int. Ed.* 1966, **5**, 385.
43. P. Mata and A. M. Lobo, *Tetrahedr. Asymmetry*, 2005, **16**, 2215.

44. W. T. Wipke and T. M. Dyott, *J. Am. Chem. Soc.*, 1974, **96**, 4834.
45. G. Warren, S. Warren, in *Drug Design Strategies: Quantitative Approaches,* RSC Pub, Cambridge, 2011.
46. A. R. Leach, V. J. Gillet, R. A. Lewis and R. Taylor, *J. Med. Chem.*, 2010, **53**, 539.
47. A. Nicholls, G. B. McGaughey, R. P. Sheridan, A. C. Good, G. Warren, M. Mathieu, S. W. Muchmore, S. P. Brown, J. A. Grant, J. A. Haigh, N. Nevins, A. N. Jain and B. Kelley, *J. Med. Chem.*, 2010, **53**, 3862.
48. D. V. S. Green, in *Designing Bioactive Molecules: Three-Dimensional Techniques and Applications*, ch. 3, Oxford University Press, New York, 1998.
49. F. H. Allen and R. Taylor, *Chem. Soc. Rev.*, 2004, **33**, 463.
50. D. Rogers and M. Hahn, *J. Chem. Inf. Model.* 2010, **50**, 742.
51. J. M. Barnard, *J. Chem. Inf. Comput. Sci.*, 1993, **33**, 532.
52. P. Willett, J. M. Barnard and G. M. Downs, *J. Chem. Inf. Comput. Sci.*, 1998, **38**, 983.
53. E. S. Simmons, *World Patent Inform.*, 2003, **25**, 195.
54. J. M. Barnard and G. M. Downs, *Perspect. Drug Discovery Des.*, 1997, **7-8**, 13.
55. H. L. Morgan, *J. Chem. Doc.*, 1965, **5**, 107.
56. J. Figueras, *J. Chem. Inf. Comput. Sci.*, 1993, **33**, 717.
57. B. D. McKay, *Congessus Numerantium*, 1981, **30**, 45.
58. P. W. Kenny and J. Sadowski, in *Chemoinformatics in Drug Discovery*, ed. T. I. Oprea, Wiley-VCH, Weinheim, 2005, 271.
59. F. Oellien, J. Cramer, C. Beyer, W.-D. Ihlenfeldt and P. M. Selz, *J. Chem. Inf. Model.*, 2006, **46**, 2342.
60. M. Sitzmann, W.-D. Ihlenfeldt and M. C. Nicklaus, *J. Comput. Aided. Mol. Des.*, 2010, **24**, 521.
61. R. A. Sayle, *J. Comput. Aided Mol. Des.*, 2010, **24**, 485.
62. W. A. Warr, *J. Comput. Aided Mol. Des.*, 2010, **24**, 497.
63. J. Elguero, C. Marzin, A. R. Katritzky and P. Linda, *The Tautomerism of Heterocycles*, Academic Press, London, 1976.
64. Y. C. Martin, *J. Comput. Aided Mol. Des.*, 2010, **23**, 693.
65. G. Lock, EuroMUG, 2004, *http://www.daylight.com/meetings/emug04/Lock/mistry.pdf.*
66. H. Maehr, *J. Chem. Inf. Comput. Sci.*, 2002, **42**, 894.
67. J. Brecher, *Pure Appl. Chem.*, 2006, **78**, 1897.
68. MDL's Enhanced Stereochemistry Representation, *http://www.symyx.com/products/pdfs/Enhanced_Stereochemical_Representation.pdf.*
69. G. M. Downs and J. M. Barnard, in *Reviews in Computational Chemistry*, ed. K. B. Lipkowitz and D. Boyd, Wiley-VCH, Weinheim, 2002, vol. 18, pp. 1–40.
70. M. Snarey, N. K. Terrett, P. Willett and D. J. Wilton, *J. Mol. Graphics Modell.*, 1997, **15**, 372.
71. C. Kibbey and A. Calvet, *J. Chem. Inf. Model.*, 2005, **45**, 523.
72. E. J. Gardiner, V. J. Gillet, P. Willett and D. A. Cosgrove, *J. Chem. Inf. Model.*, 2007, **47**, 354.

73. M. Rarey and J. S. Dixon, *J. Comp. Aided Mol. Des.*, 1998, **12**, 471.
74. FTrees, Version 2.3, BioSolveIT GmbH: Sankt Augustin, Germany, *http://www.biosolveit.de/FTrees/*.
75. V. J. Gillet, P. Willett and J. Bradshaw, *J. Chem. Inf. Comput. Sci.*, 2003, **43**, 338.
76. G. Harper, G. S. Bravi, S. D. Pickett, J. Hussain and D. V. S. Green, *J. Chem. Inf. Comput. Sci.*, 2004, **44**, 2145.
77. K. Birchall and V. J. Gillet, in *Chemoinformatics and Computational Chemical Biology*, ed. J. Bajorath, Springer, Clifton, 2011, 197.
78. Y. Xu and M. Johnson, *J. Chem. Inf. Comput. Sci.*, 2002, **42**, 912.
79. S. M. Free and J. W. Wilson, *J. Med. Chem.*, 1964, **7**, 395.
80. R.W. Zwanzig, *J. Chem. Phys.* 1954, **22**, 1420.
81. A. G. Leach, H. D. Jones, D. A. Cosgrove, P. W. Kenny, L. Ruston, P. MacFaul, J. M. Wood, N. Colclough and B. Law, *J. Med. Chem.* 2006, **49**, 6672.
82. A. M. Birch, P. W. Kenny, I. Simpson and P. R. O. Whittamore, *Bioorg Med. Chem. Lett.*, 2009, **19**, 850.
83. A. M. Wassermann and J. Bajorath, *J. Chem. Inf. Model.*, 2010, **50**, 1248.
84. J. Hussain and C. Rea, *J. Chem. Inf. Model.*, 2010, **50**, 339.
85. The Pistoia Alliance, *http://www.pistoiaalliance.org/*.

CHAPTER 7

Modeling Chemical Structure-Log P

RAJESHWAR P. VERMA,[a] CYNTHIA D. SELASSIE[a]* AND ALBERT J. LEO[b]

[a] Pomona College, Department of Chemistry, 645 North College Avenue, Claremont, California 91711, USA; [b] BioByte Corporation, 201 West 4th St., #204, Claremont, California 91711, USA

7.1 Introduction

Many of the major advances in drug discovery over the last one hundred years have been buoyed mostly by serendipity, as in the case of the benzodiazepines, and sometimes by critical observation (Digitalis) or by chemical exploitation (antimetabolites). A rational approach to drug discovery/design emerged only in the last fifty years when it was recognized that the behavior of a drug could be related to its intrinsic physicochemical properties. Among the critical attributes of a physicochemical nature that drug candidates possess, two in particular stand out: their dissociation constants and hydrophobicities.

These two parameters are intimately involved with each other particularly in the charged state. Hydrophobicity, as measured by log P, refers to the partitioning of a solute that remains neutral over the pH range in question, while log D (distribution coefficient or apparent log P) constitutes a measure of the partitioning of both the neutral and charged species of a compound at a particular pH. Thus log D is related to log P *via* the ionization constant, pK_a.

RSC Drug Discovery Series No. 13
Drug Design Strategies: Quantitative Approaches
Edited by David J. Livingstone and Andrew M. Davis

Published by the Royal Society of Chemistry, www.rsc.org

$$\text{For acids:}\quad \log D_{pH} = \log P - \log[1 + 10^{(pH-pKa)}] \tag{7.1}$$

$$\text{For bases:}\quad \log D_{pH} = \log P - \log[1 + 10^{(pKa-pH)}] \tag{7.2}$$

It must be noted that permeability and aqueous solubility are also pKa dependent. Thus in the last twenty years, the need for accurate and precise pK_a values in the drug discovery process has fueled the development of methods to measure pK_a values and to predict them from structure. Measurements of log D and pK_a have been well addressed by Comer.[1] For further detailed discussion of pK_a values and their prediction methodologies, readers are referred to an important publication of Fraczkiewicz.[2] A few common methods for the pK_a values prediction are summarized in Table 7.1. The most commonly used programs for the pK_a calculation are listed in Table 7.2.

In a very recent publication,[10] a new quantitative structure–activity relationship (QSAR) methodology "multi-pH QSAR" was proposed. This QSAR approach may be utilized to differentiate the activity of neutral and ionized species by considering two species-specific terms, log D^N (neutral) and log D^I (ionized).

Table 7.1 Some common methods for the predicton of pK_a values.

No.	*Methods*	*Descriptions*					*Ref.*
1	Hammett-Taft Method	$(pK_a)_x = pK_a - \rho\Sigma\sigma_x$ where pK_a and $(pK_a)_x$ are ionization constants for the parent molecule and their X-substituted derivative respectively. ρ is the reaction constant and σ_x is the substituent constant.					3, 4
2	Multi-CASE Method	$pK_a = \text{constant} + \Sigma r_i\, n_i\, M_i$ where M_i is a binary {0,1} indicator for the presence of modulator i and n_i is the number of modulator in a compound. r_i is the regression coefficient.					5, 6
3	QSPR Method (MLR)	*Compd.*	*Descriptors*	*n*	r^2	*s*	
		Phenols	6 molecular descriptors	106	0.913	0.523	7
		Carboxylic acids	21 operational atomic contributions	826	0.941	0.104[a]	8
		Protonated amines	19 operational atomic contributions	802	0.933	0.182[a]	8
		Neutral and basic drugs	5 constututional, topological, geometrical, electrostatic, and quantum chemical	59	0.78	0.48	9

[a]rms in log unit.

Table 7.2 Available programs for pK_a calculations, with their methods and web addresses.

No.	Name of the programs	Methods	Web addresses
1	ACD/pK_a™	LFER	http://www.acdlabs.com
2	pKalc™	LFER	http://www.compudrug.com
3	ADMET Predictor™	QSPR	http://www.simulations-plus.com
4	Marvin™	QSPR	http://www.chemaxon.com
5	CSpK_a™	QSPR	http://www.chemsilico.com

Hydrophobicity plays such an important role in drug absorption and distribution that it will be addressed in some depth. In combination with other relevant electronic, topological, steric and hydrogen bond descriptors, hydrophobicity wields an important role in absorption, distribution, metabolism, elimination and toxicity (ADMET) phenomena.

Before there was any understanding of organic chemical structure, attempts were made to relate the properties of 'folk remedies' to their desired physiological activity as initially expressed by Crum-Brown and Fraser.[11] The history of these early attempts at structure activity relationships (SAR) are adequately treated elsewhere.[12,13]

7.2 General Hammett Approach

In 1935, Louis P. Hammett initiated a successful method of predicting reaction rates and equilibrium constants in *non*-biological systems in a *quantitative* manner using an "extra-thermodynamic" approach that employed substituent parameters based on model systems.[14] For example, benzoic acid served as a model 'parent', and the change in its ionization constant, pK_a, brought about by each substituent in the meta or para position would reflect the electronic influence of that substituent (electronic withdrawing or donating) on other 'parent features' on diverse aromatic systems. These were termed "sigma-meta" and "sigma-para" parameters. This empirical relationship was well represented by his now classical equation [equation (7.3)].

$$\log K_X/K_H \text{ or } \log k_X/k_H = \rho\sigma \tag{7.3}$$

The limitations of this approach, including substituents in the ortho position to the reaction center, which have been addressed elsewhere,[12] led to many refinements of Hammett sigma constants.[15–17]

7.3 Steric Constants

Despite the availability of a significant number of sigma constants, some model systems for evaluating substituent effects failed when steric effects were intensive or complex. One of the first measures of steric effects in physical organic chemistry was subsequently determined and defined by Taft by equation (7.4) that bears his name.[18]

$$E_S = \log\{k_X/k_H\}_A \tag{7.4}$$

where, k_X and k_H are rate constants for the acid hydrolysis of esters XCH_2COOR and CH_3COOR, respectively. Taft's work laid the basis for further developments in steric parameterization with modifications to E_S as well as the defining of other steric parameters that were based on average and minimum van der Waals radii.[15] Increased emphasis on intermolecular interactions led to the development of the STERIMOL parameters: L, B_1, and B_5; L represents a slightly modified length of the substituent while B_1 and B_5 correspond to the minimum and maximum widths orthogonal to L, respectively.[19,20]

Yet, in spite of their success in ordinary solution chemistry, where electronic interactions (and to a lesser degree steric) dominate, Hammett constants and their numerous, extended variants were found to be somewhat lacking in predicting rates and equilibria in biological systems. In all these systems, an interaction mediated by water and termed the "hydrophobic effect" caused clustering of hydrophobic moieties as first noted by Kauzmann.[21] Earlier, it was Langmuir who suggested that hydrophobicity was of critical importance in the folding of proteins into globular structures.[22] Frank and Evans were strong proponents of "hydrophobic hydration" and suggested that the significant loss of entropy on dissolution of a small apolar molecule in water was due to the formation of an ordered water structure around the solute akin to "iceberg" formation.[23] Ever since then, protein chemists, physical chemists and molecular dynamics specialists have been actively engaged in trying to explore hydration dynamics, describe the hydrophobic force, measure it experimentally and quantify it *via* calculations and simulations. It was clear that there was a strong need for a parameter to relate a chemical's affinity to lipid environments relative to an aqueous environment.

7.4 Hydrophobicity and Biological Activity

Following early leads provided by Meyer,[24] Overton,[25] and later by Collander,[26] Hansch *et al.*[27] established that the partition coefficient, the ratio of the partitioning of a solute between octanol and water (taken as $\log_{10}$ values to normalize it with free energy terms), provided an attractive way to extend the Hammett approach to biological reaction rates and equilibria. They demonstrated that log P between octanol and water was a facile and practical parameter that related to the transport of a chemical through various cellular compartments, from its point of entry to its site of action. At this site (usually the receptor site on an enzyme), this log P parameter could also reflect the binding energy exerted by the chemical (drug) exposed to a mixture of lipid-like and polar areas of the active site. The common prefix for this parameter is 'hydrophobic', but one should keep in mind that a hydrophobic solute (*i.e.* one with log $P_{oct} > 0$) still has an attraction to water—just a greater attraction to a lipid phase.

It may seem baffling that, with today's knowledge of biophysical chemistry, molecular mechanics and molecular dynamics, the properties of water—the most vital chemical to Life and its modeling in the simplest biological systems, cannot be adequately predicted from first principles.[28] Nevertheless, the importance of hydrophobicity in diverse phenomena from the formation of micro-emulsions to protein–protein and protein–ligand interactions has led to the development of numerous methodologies for the measurement of partition coefficients, beginning with the ubiquitous shake-flask method.[29,30] A number of other methodologies that are widely used are also summarized.[1,31]

7.5 Direct Measurement of Partition Coefficients

7.5.1 Shake-Flask Method

The traditional shake-flask technique remains the method of reference for the direct measurement of log P ranging from −3 to +6 with great accuracy and high precision if temperature control is maintained.[29,32,33] In this method the pure solute is partitioned between two immiscible but mutually pre-saturated liquid phases such as distilled/deionized/Milli-Q water and highly pure 1-octanol (note: when the hydrophobic parameter at physiological pH is desired, the aqueous phase is usually 0.1 M phosphate buffer). The quantification of solute is carried out using various methods such as UV-based procedures, GC, and HPLC.[34] Although this is the most popular and commonly used method for log P_{oct} measurement, it is time consuming and requires an appreciable amount of solute in some cases. To overcome well-known limitations of this method, several other direct log P_{oct} measurement methods have been developed, such as; slow stirring and filter-probe extractor methods *etc.*[35–37]

7.5.2 Potentiometric Titration Method

In the case of an *ionizable* organic compound, potentiometric titration may be used to determine log P_{oct} values over a wide range from −0.8 to +5.3. In this method, log P_{oct} of an organic compound is calculated by using equation (7.5)[38]

$$\mathrm{Log}\, P_{oct} = \log\left(\frac{10^{[\Delta pK_a]} - 1}{r}\right) \tag{7.5}$$

In equation (7.5), ΔpK_a is the difference in pK_a in the presence and absence of octanol $\left(pK_a = pK_a^{oct} - pK_a\right)$, and r is the octanol–water volume ratio.

7.5.3 High Throughput Screening (HTS) Method

In order to increase throughput, the traditional shake-flask methodology has been transferred onto 96-well plate technology and the sample preparation is carried out using a robotic liquid handler. After equilibration of the solute in

1-octanol and water, an aliquot is analyzed from each phase by applying fast gradient elution liquid chromatography with UV absorption,[39] or single ion monitoring mass spectrometry.[40] The compound concentrations in 1-octanol and water phases are determined by the peak areas obtained from their respective phases and that can directly provide partition coefficients by considering the volumes of the two partitioning solvents. It should be noted that this method is limited to the direct determination of log P_{oct} ranging from −2 to +4.[40–42] Parallel methods using the scan of 96/384-well plate by UV spectrophotometers are rapid and become 50-fold faster with the introduction of diode-array plate readers.[42]

7.6 Indirect Methods for Determination of Partition Coefficients

7.6.1 Reversed Phase Thin-Layer Chromatography

Reversed-phase thin layer chromatography (RP-TLC) is a simple and rapid method for the estimation of a hydrophobic parameter (R_m) without the need of expensive equipment. In this method, the compound is adsorbed on a solid support (usually silica or alumina) and an aqueous phase (with varying amounts of water and 1-octanol mixture) is used as the mobile phase. It has already been established that R_m values are linearly related to the log P_{oct} values and, therefore, R_m values are analogous to the log P_{oct} values.[43] R_m values are determined from the measured R_f values and defined by equation (7.6):[44,45]

$$R_m = \log 1/(R_f - 1) \tag{7.6}$$

7.6.2 Reversed Phase High-Performance Liquid Chromatography

Reversed phase high-performance liquid chromatography (RP-HPLC) is an indirect, rapid, and high throughput method for the determination of partition coefficients. In HPLC, the most frequently used retention factor, k, is defined as:

$$\log k = \log(t_R - t_0)/t_0 \tag{7.7}$$

where t_R and t_0 are the retention time and dead time, respectively.

In the isocratic mode of RP-HPLC, the use of retention factor extrapolated to pure water (log k_W) is very common in practice instead of retention factor itself, to obtain a more comparative hydrophobic index and avoiding the effect of changing the organic modifier. The extrapolation is based on the following Soczewiński-Snyder relationship:[46,47]

$$\log k = -S\varphi + \log k_W \tag{7.8}$$

where S is constant for a given solute and a given HPLC system, ϕ is the volume fraction of the organic modifier in eluent, and k_W corresponds to k in pure water as the mobile phase. There have been a number of publications confirming excellent correlations between log k_W and log P_{oct} values:[48–51]

$$\log P_{oct} = a \log k_W + c \tag{7.9}$$

where a and c are constants. In RP-HPLC, the most widely used stationary phases are the silanized silica gel and the polymer-based octadecyl-poly(vinyl alcohol) (ODP), while the mobile phases are methanol, acetonitrile and tetrahydrofuran.[52,53] For fast gradient RP-HPLC, a chromatographic hydrophobicity index (CHI) was proposed by Valkó *et al.*,[54] that can be calculated using equation (7.10):

$$\mathrm{CHI} = At_R + C \tag{7.10}$$

where A and C are the constants and derived from the calibration of analytes. t_R is the retention time and is directly related to the organic modifier concentration in eluent. A linear correlation was established between *C*log P and CHIN (CHI value of uncharged form of the molecule):

$$C \log P = 0.057(\pm 0.005)\ \mathrm{CHIN} - 1.107 \tag{7.11}$$

$$n = 52 \quad r^2 = 0.724 \quad s = 0.82 \quad F = 131$$

It has recently been demonstrated,[55] that RP-HPLC coupled with a 96-well plate auto-injector can be utilized for the simultaneous determination of log D, log P and pK_a of drug molecules. This method is associated with the following advantages: low sample requirement, accommodation of low solubility compounds, less restriction on compound purity, higher throughput, precise data and multiple results in one assay. However, the range in log P_{oct} of the test compounds in the dataset was limited from −1.7 to 4.4.

7.6.3 Artificial Membrane Chromatographic Method

Immobilized artificial membranes (IAMs) have successfully been introduced as HPLC column packing materials. IAM HPLC columns consist of phosphatidylcholine residues (most common phospholipids) covalently attached to a propylamino-silica support and consequently mimic the partitioning into a phospholipid bilayer.[56] In isocratic IAM HPLC method, log P_{oct} is expressed by the retention capacity (log k_{IAM}) when the volume ratio of the mobile and stationary phases remains constant. It must be noted that log P_{oct} and log k_{IAM} show a good correlation, despite the fact that the log P_{oct} is a measure of hydrophobicity while the log k_{IAM} is a measure of phospholipophilicity.[57,58] A gradient method has also been developed using an acetonitrile gradient on an IAM HPLC column that provides a new chromatographic hydrophobicity index (CHI IAM). The gradient retention times are usually calibrated against those obtained from usage of the isocratic method.[41] A number of extensive reviews on IAM chromatographic methods for determining partition coefficients may be found in the literature.[59,60]

7.6.4 Electrokinetic Chromatographic Methods

Electrokinetic chromatography (EKC), such as micellar electrokinetic chromatography (MEKC), microemulsion electrokinetic chromatography (MEEKC), and multiplexed microemulsion electrokinetic chromatography (MMEEKC) provides an attractive, expedient alternative to the indirect methods for log P_{oct} determination, because of its adequate accuracy and precision, as well as high degree of automation.[61] In addition, the advantages of employing 96-capillary MEEKC for its high speed and high throughput reliable estimation of log P_{oct}, appears to be very promising.[62,63] These methods are based on a chromatographic mechanism and thus analogous to RP-HPLC. They maintain all their advantages and are also able to eliminate some inherent problems of RP-HPLC.[64] The MEEKC method has been the system of choice for log P_{oct} determination due to its improved ability to handle insoluble compounds and minimize hydrogen bonding interference due to the presence of 1-butanol co-surfactant.[65,66] MEEKC, when performed in full pH range, can be utilized to determine the log P_{oct} values of solutes ranging from -1.0 to $+6.6$, and its modified system (MMEEKC and μMEEKC) can serve as a high throughput, rapid, accurate and reproducible method for log P_{oct} determination. MEEKC is, however, not suitable for determining log P_{oct} values of charged compounds due to the difficulties encountered by solute electrophoretic migration and ionic interactions with the charged microemulsions.[62] Solute capacity factors, log k', in MEEKC can be determined from migration times of the solute, DMSO and the dodecylbenzene microemulsion and used to calculate log P_{oct} values.

$$\log P_{oct} = A \times \log k' + C \tag{7.12}$$

There are obviously many methods to determine partition coefficients, directly and indirectly. Many of the newer high throughput procedures provide a rapid, easy to use, reproducible cost-effective way to assay the hydrophobicity of large numbers of diverse compounds; however, there are a few caveats. In some cases, the range of partition coefficients that has been validated is limited. In HT screens with 1-octanol, a 10,000 to 1 distribution ratio would be difficult to access and assess; thus the reliability of log P_{oct} values at the ends of the spectrum (≤ 3 or ≥ 5) would be much lower than shake-flask values when dealing with minute volumes such as those present in a 96-well plate.

7.7 Hydrophobicity and its Calculation

The earliest successful biological QSAR correlating plant growth regulating activity of X-phenoxyacetic acids utilized Hammett sigma constants for electronic effects and pi-constants for hydrophobicity.[30] Pi (designated by lower case π) for the most common substituents was taken as the difference in the *measured* log P_{oct} values between the aromatic parent, benzene, and the substituted analog. This QSAR highlighted the importance of the parabolic relationship between biological activity and hydrophobicity in addition to the electron-withdrawing attributes of the X substituents,

$$\log 1/C = 3.24\pi - 1.97\pi^2 + 1.86\sigma_p + 4.16 \quad (7.13)$$
$$n = 23 \quad r^2 = 0.776$$

where log $1/C$ represents the plant growth regulating activity of X-phenoxyacetic acids. It is important to note that the π value was considered constant for all aromatic systems that had little or no electronic interaction between the studied substituent and any others present on or in the aromatic system. Alternative π values must be employed when an electronically active parent, such as nitro-aromatics or pyridines, is being studied.

With the realization that log P_{oct} was a useful and critical tool in understanding bio-molecular interactions, and that the measurement of that parameter for every drug candidate was costly as well as impossible for those yet to be synthesized, a demand was created for a reliable method for calculating it from structure. Given the difficulties in calculating the properties of water itself, it is obvious that log P_{oct} calculations must be, for the most part, empirically based.

It was also clear that calculation of log P_{oct} for the immense variety of structures that might be considered in computer-assisted drug design (CADD) would be impractical based on the π-system, and thus 'fragmental' and 'whole molecule' approaches have dominated these calculations.[67–69] Since it plays such a pivotal role in the Hammett-Hansch QSAR methodology, the basis for log P_{oct} calculation will be examined in some detail because, even though they lack a firm, theoretical basis, the *calculation details* can often be as useful in CADD as the 'bottom line' number.

In 1993, an extensive review summarized the five major approaches to predicting/calculating log P_{oct} based on structure.[68] They included the π-substituent method, fragment-based methods, atomic contributions-based and surface area-based approaches, molecular-properties-method and one delineated by solvatochromic parameters.[70–74] A newer method includes a hybrid approach (AUTOQSAR) that marries molecular descriptors and atomic fragment values.[75] Many software programs are now available for use and the more commonly used are listed in Table 7.3. A number of comparative studies have also been carried out on the various programs and a recent study by Sakuratani *et al.*, indicates that CLOGP, KOWWIN and ACD/LOGP, which are all fragment-based methods, show excellent accuracy.[76]

7.7.1 CLOGP- Rules of Engagement

In attempting to calculate log P_{oct} by a 'whole molecule' approach, several factors must be considered:

(i) Since it takes more energy to create a cavity for the solute in water than in 1-octanol (difference in surface tensions), log P_{oct} should increase with

Table 7.3 Popular programs for log P calculations, with their methods and web addresses.

No.	*Name of the programs*	*Methods*	*Web addresses*
1[a]	CLOGP	Fragment-based	http://www.biobyte.com
2	ACD/Log*P*	Fragment-based	http://www.acdlabs.com
3	miLogP	Fragment-based	http://www.molinspiration.com
4	XLOGP3	Atom based	http://www.sioc-ccbg.ac.cn
5	KOWWIN	Atom/fragment based	http://epa.gov/oppt/exposure/pubs/episuite.htm
6	SLIPPER	Based on structural similarity and physicochemical properties	http://www.chemdbsoft.com/SLIPPER.html
7	SciLogP Ultra	Based on 2D E-State descriptors and neural net approach	http://www.twcbiosearch.com/DOCS/1/scLgPltr.htm
8	ALOGPs	Descriptor-based	www.vcclab.org

[a]BioLoom incorporates CLOGP.

the solute's size. Solvent-accessible surface area (SASA) might seem preferable over solute volume for the size effect, but in practice, the simple calculation of McGowan volume,[77] seems to perform as well as SASA calculated using a 1.5 Å water radius.

(ii) Since *solvent* water has two hydrogens to donate, compared to octanol's one, *solutes* with *H-bond acceptors* are more readily accommodated in the aqueous phase; *i.e.*, the sum of the H-bond acceptor strength of the solute reduces log P_{oct}.

(iii) Since both water and octanol have only one H-acceptor oxygen, the H-donor strength of the *solute* should have only a weak effect on log P_{oct}.

(iv) Water has one of the highest dipole moments of any solvent, and so a solute with a permanent dipole, or one that can be easily polarized, should have a lowered log P.

In a series of publications using this approach,[78–80] a number of equations were developed that increased our understanding of the solvation forces acting in different media. Quantifying the polarity/polarizability contribution solely from solute structural input proved the most difficult. The Kamlet-Taft group,[78] used dipole-moment squared and the symbol Π*, while Abraham *et al.*, used polarity (S) and added excess molar refractivity, E.

$$\log P_{oct} = -0.96\Pi^{*} + 0.0\varepsilon\alpha - 3.38\varepsilon\beta + 2.66\ V_x + 0.24 \qquad (7.14)$$

$$\log P_{oct} = +0.56E - 1.05S + 0.03A - 3.46B + 3.81V + 0.09 \qquad (7.15)$$

where, Π* = polarity/polarizability = S (Abraham *et al.*[80]), V_x = McGowan volume × 0.01, εα = summation of H-bond donor strength = A (for acidity,

Abraham *et al.*[80]), $\varepsilon\beta$ = summation of H-bond acceptor strength = B (for basicity, Abraham *et al.*[80]), and excess molar refractivity = E (Abraham *et al.*[80]).

The coefficients in equation (7.15) developed by Abraham *et al.*,[80] that also add a term in excess molar refractivity E, are very similar to those in (7.14), but (7.15) has a greatly reduced intercept. Although they give us a better ***understanding*** of the mechanics of the partitioning process, ***neither*** equation (7.14) nor (7.15) provides a ***practical*** basis for a computerized calculation of thousands of complex drug structures. Therefore, either 'atom-centered' or 'fragment-based' algorithms have gained the widest acceptance for that task.[67] Although these calculations are semi-empirical in nature and thus must be validated using the largest possible set of reliably measured values, they can still give deeper insights into the operation of the four points mentioned in the previous paragraph. Indeed, the *details* of many calculations (which have been termed 'latent information') can be surprisingly useful, as a few examples will soon illustrate.

Solvent water's ability to accommodate a solute's *dipolarity* appears to be *more localized* than one might expect. This is somewhat disappointing, because the simple dipole moment is easily available by measurement or calculation. This 'failure' of simple dipole moment is evidenced by the fact that measured log P values for *o*-dichlorobenzene and *p*-dichlorobenzene are indistinguishable experimentally (3.43 *versus* 3.44) while their dipole moments are quite different (2.3D and 0D). Note that the difference in solvent-accessible-surface-area (SASA), if pertinent, should also make log P for the *ortho* isomer appreciably lower (SASA *ortho* = 282.52; *para* = 285.60). Furthermore, the sequential chlorination of methane gives evidence that it is the ***localized*** bond dipoles that lower log P, and that in multiples they may shield each other. See Figure 7.1.

For the CLOGP algorithm,[68] a halogen'clustering' factor was developed to allow for this shielding of localized bond dipoles. For simplicity, it is given the same value for all halogens in a cluster (F, Cl, Br, or I) whether mixed or all the same. This seems to work for fluorine and chlorine (*e.g.* CHF_3; M = 0.64, C = 0.63 and $CHFCl_2$; M = 1.55, C = 1.51) but the larger halogens could use a larger correction ($CHBr_3$; M = 2.67, C = 2.37).

The electron attractive power of halogens is well characterized in Hammett-Taft methodology, either as σ_m or σ_p in aromatic systems and as σ^* in aliphatic. Thus halogens would be expected to reduce the electron density of nearby fragments with H-bond acceptor features and thus reduce their apparent hydro*philicity* (*i.e.* raise log P). Again, treating all halogens alike in this respect and allowing only for distance of separation and differences between aliphatic *versus* aromatic systems, the resultant average deviation is reasonably low. Although these Hammett sigma values (derived chiefly from the pK_a values of carboxylic acids) could be utilized directly in log P calculations, slightly better results were obtained if the correction factors were based on just a few very carefully measured partition coefficients of each type.

The methodology of the CLOGP program is somewhat unique in that fragments are determined by the computer—not preset by the designer. As will become evident, this has advantages in how much valuable 'latent information'

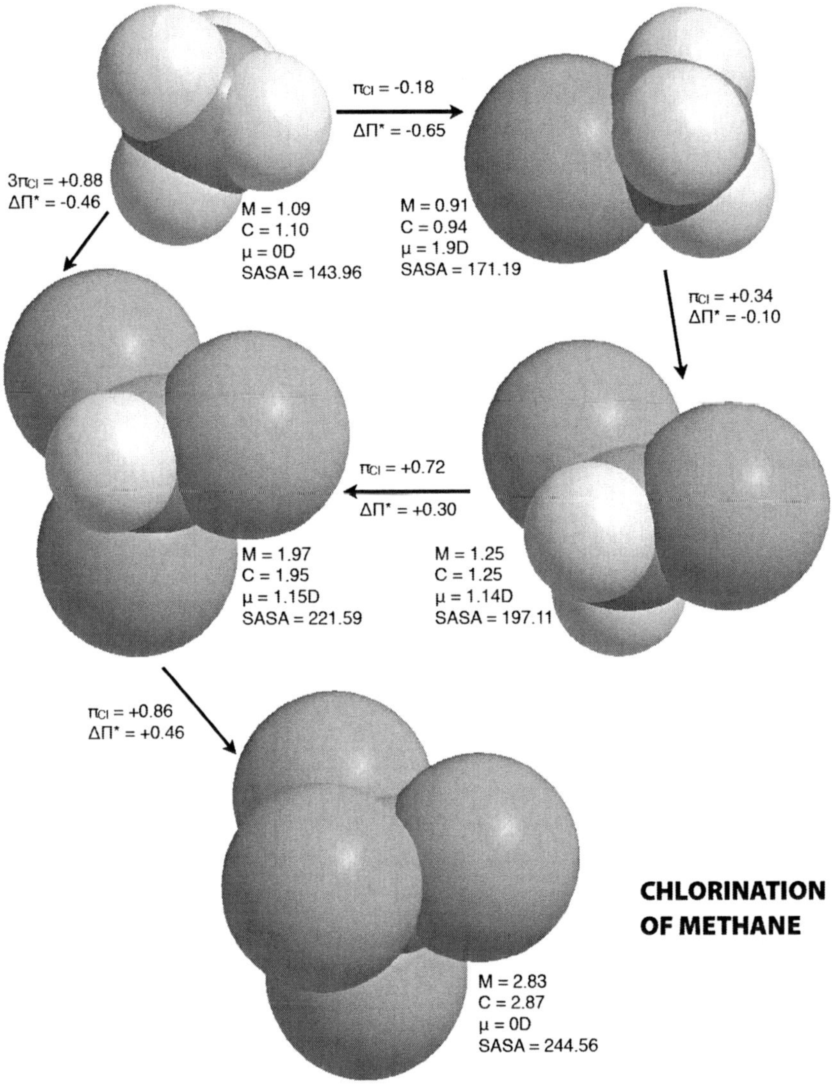

Figure 7.1 Sequential chlorination of methane and its effects on log P.

is contained in the 'calculation details'. Hydrogen bonding fragments interact in ways that have yet to be adequately explained in terms of molecular mechanics theory. These interacting polar fragments almost always lose polarity (have higher log P) than they do in isolation. In some cases, the formation of an internal H-bond seems a likely explanation, since that would reduce the sum of H-acceptor strength as seen by solvent water; that is, it reduces εβ in Equation (14) which carries a coefficient of −3.38. Consider the case of Aspirin where molecular mechanics predicts that in a vacuum the carboxylic OH could serve as a donor to either oxygen in the ortho ester group.[81] Partitioning data provides no evidence for such an internal H-bond, but rather suggests a steric interaction that requires a *negative* correction to calculated log P. It is the phenolic OH in salicylic acid that serves as a strong H-bond donor to the carbonyl in either o-acetylphenol or salicylic acid (Table 7.4).

There is long-standing evidence that water is more structured at the interface with a hydrocarbon. X-ray evidence does not support it being truly 'ice-like' but the first water layer does appear to be less dense and more structured than bulk water as indicated by NMR studies and time resolved infrared (IR) studies.[82,83] Frank and Evans,[23] likened it to a tight-fitting sweater and the free energy in the partitioning equilibrium of an alkane between water and octanol might have an appreciable entropy component—the entropy loss helping to drive the alkane from water into octanol (or other non-polar solvent). However, more recent 2D-IR studies suggest that the hydrogen bond network around hydrophobic groups may not be more rigid in comparison to bulk water but its translational mobility may be drastically compromised.[84] This may be the basis for the "proximity corrections" used in the CLOGP program. For example, in *n*-butyl butyrate, the di-aliphatic-attached ester fragment

Table 7.4 Phenolic OH serves as a strong H-bond donor to the *ortho* carbonyl in either *o*-acetylphenol or salicylic acid.

No.	*Compound*	Ortho	*HBD*	*Conclusions*
1	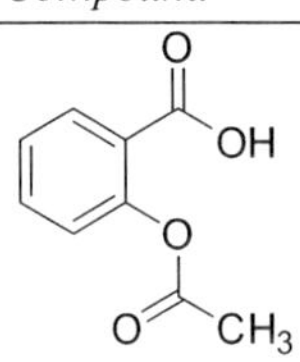	−0.43	0.69	No intra-molecular H-bond; fragments twisted from ring plane
2	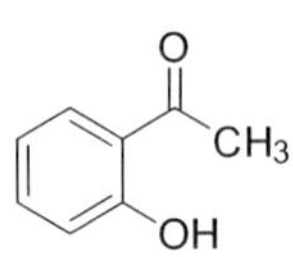	+0.46	0.0	Interaction reduces HBA; phenolic HBD lost thru IMHB
3	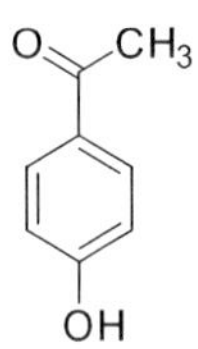	—	0.77	Phenolic HBD enhanced thru σ/ρ
4	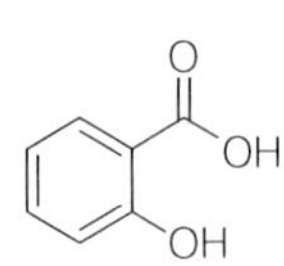	+0.63	0.88	Most of phenolic HBD lost thru IMHB; benzoic HBD enhanced by σ/ρ

(in isolation) is given a fragment value of -1.45. We can postulate that this negative value has two components: Firstly, it is an indication of how solvent water accommodates the solute by acting as an H-donor to the solute's two oxygen atoms. Secondly, the solute-to-solvent H-bond appears to disturb the structured water 'sweater' to some distance along each alkyl chain (probably a minimum of three alky carbons). Left undisturbed, this 'structured sweater' would encourage transfer into octanol, and so *its loss* is a *negative* contribution to the -1.45 fragment value. See Figure 7.2.

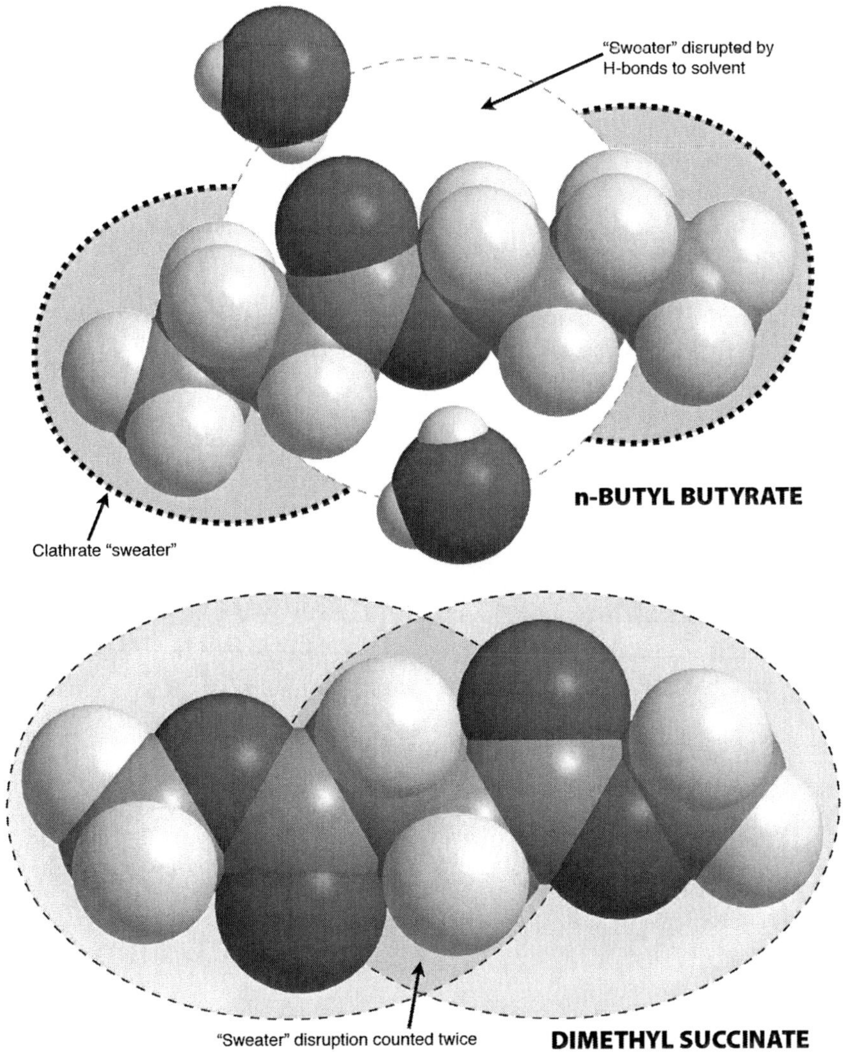

Figure 7.2 Lack of hydration of *n*-butyl butyrate.

In dimethyl succinate, the negative contribution from this water-structure component of each ester fragment overlaps, if it is considered to extend out to three alkyl carbons. In other words that portion of the fragment's negativity ($f = -1.45$) is, in effect, counted twice. Thus the H-bond acceptor contribution of each of the proximate ester fragments may remain constant, but the positive correction that is needed in dimethylsuccinate ($+0.75$) may result instead from the lower entropy change when the alkane portion is transferred from water to octanol.

7.8 Hydrophobicity and its Applications

At this point, it may be worthwhile to step back and ask a simple question: "What exactly is a log P_{oct} value telling us?" In the >13,000 biological QSARs constructed by the Hansch group at Pomona College, a hydrophobic parameter (expressed as a π-constant, a log P_{oct} or calculated *via* CLOGP) is a significant parameter in over 60% of them. It clearly plays a role in the CADD for antimicrobials, anti-virals, anti-neoplastics, as well as opioids, serotonin agonists, PPAR activators *etc.* It has a role in each of the components of ADMET. Since it obviously cannot serve as some sort of 'magic bullet', it must play a different role in many of these predictions. One role, where transport to an active site could be rate-controlling, was recognized early, and the ability to treat the hydrophobic parameter in parabolic or bi-linear fashion disclosed 'optimum values'.[15] This was not fully appreciated 'in the early days' of CADD (1960s and 1970s), when enzymatic assays indicated that greater activity could be achieved by steadily increasing log P_{oct} of the drug candidates. After progressing to *in vivo* animal studies, these hopes were soon dashed, as optimal values (log P_O) were considerably lower.

7.8.1 Hydrophobicity and Central Nervous System Toxicity

Blood-brain barrier (BBB) penetration is critical for CNS drugs. In order to minimize serious side effects in peripheral tissues in the brain, it is important that there be strict adherence to the principal of minimal hydrophobicity as delineated by Hansch *et al.*[85] Although hydrophobicity enhances BBB permeation, too much of it contributes to non-specific binding in the brain and subsequent loss of activity. In a seminal analysis of commonly used CNS drugs, Hansch *et al.* determined that most of these drugs had a log P (or log D) values in the range 2 ± 0.5. Other physicochemical descriptors have also been implicated in CNS activity and they include molecular weight, hydrogen bonds (donors especially), polar surface area, acidity and Δlog P. The latter variable is defined as:

$$\Delta \log P = \log P_{octanol} - \log P_{cyclohexane}$$

In a study of 20 H-2 histamine receptor antagonists, Young *et al.*,[86] used this descriptor to describe their blood-brain partitioning. See equation (7.16)

$$\log\{C_{brain}/C_{blood}\} = -0.485(\pm 0.160)\Delta\log P + 0.889(\pm 0.500)$$

$$n = 20 \quad r = 0.831 \quad s = 0.439 \quad F = 40.23 \qquad (7.16)$$

Individual correlations with either log $P_{octanol}$ ($r = 0.436$) or log $P_{cyclohexane}$ ($r = 0.732$) were not significant and it was suggested that these two independent variables represent two distinct processes involved in the distribution of a drug between peripheral blood and the brain; log $P_{cyclohexane}$ might reflect the partitioning process into non-polar regions of the brain. Log $P_{octanol}$ could be representative of significant protein binding in peripheral blood that could limit the availability of free drug for subsequent passage into the brain.

7.8.2 Hydrophobicity and Absorption

In recent years, there has been a great deal of emphasis on the importance of enhancing absorption, distribution, metabolism and excretion of potential new chemical entities early in the discovery process and not waiting until the drug development stage. Obviously physicochemical attributes play a key role in ultimately determining the efficacy of a drug *in vivo* and thus parameters such as hydrophobicity, pK_a and solubility are key factors to honing in on that critical balance between optimizing binding to a target receptor as well as maximizing accessibility to the therapeutic target *in vivo*. We will now consider a QSAR study of intestinal absorption of various drugs keeping in mind the dynamic balance of permeation (passive, active transport, efflux), solubility, enzymatic hydrolysis and dissolution in this process.

From data on human intestinal absorption of drug molecules (excluding compounds having absorptions either 0 or 100%), the following model was developed by Abraham *et al.*[87]

$$\log k = -0.025\ \mathrm{E} + 0.141\ \mathrm{S} - 0.409\ \mathrm{A} - 0.514\ \mathrm{B} + 0.204\ \mathrm{V} + 0.544 \qquad (7.17)$$

$$n = 127 \quad r^2 = 0.80 \quad s = 0.29 \quad F = 94$$

where k is the overall rate constant for human intestinal absorption, E is the solute excess molar refractivity, S is the solute dipolarity/polarizability, A is the solute hydrogen bond acidity, B is the solute hydrogen bond basicity, and V represents the McGowan volume of the solute. A well-established equation (7.15) from the same research group was delineated for partition coefficients of 613 solutes in the octanol–water system.[80] The coefficients of the independent variables in equations (7.15) and (7.17) differ considerably. On the other hand, the coefficients of equation (7.17) were found to be very similar to those of diffusion processes not involving transport through membrane. On the basis of these observations, the authors concluded that log P_{oct} is not a suitable predictor for human intestinal absorption. It was then suggested that intestinal absorption involved diffusion through a stagnant mucus layer, together with transfer across the mucus/membrane interface.[87]

This analysis raises a question. Does a comparison between the log k and log P_{oct} equations justify a conclusion that log P_{oct} is not a suitable predictor for human intestinal absorption? Although both log k and log P_{oct} constitute the identical dependent variables for equations (7.17) and (7.15), respectively, log k represents a biological activity while log P_{oct} is a physicochemical property. In order to clarify this confusion, we developed the following QSAR (7.18),[88] based on the same data used by Abraham *et al.*[87]

$$\log k = 0.31(\pm 0.03)C\log \mathrm{P} - 0.06(\pm 0.01)(C\log \mathrm{P})^2 + 0.05(\pm 0.06) \quad (7.18)$$

$$n = 119 \quad r^2 = 0.800 \quad s = 0.235 \quad F = 232$$

$$C\log \mathrm{P_O} = 2.52\ (2.22 - 2.91)$$

Outliers: ouabain, acarbose, lincomycin, sorivudine, lamivudine and amrinone.

Two compounds are not considered in the analysis due to their name/structure confusion. This is a parabolic correlation in terms of *c*log P with well-defined optimum *c*log P of 2.52 and suggests that log P_{oct} is a suitable predictor for the human intestinal absorption in a parabolic fashion. It has been shown that compounds with log P values between zero and three are ideal for passive transcellular absorption across intestinal epithelia.[89]

7.8.3 Hydrophobicity and Cytochrome P450 Inhibition

The administration of two or more drugs together can lead to unwanted drug–drug interactions mostly based on inhibition of Cytochrome P450 (CYP). Because these interactions can result in toxicity, it has become prudent to screen compound libraries early in the discovery process in order to minimize CYP inhibition.[90] The boom in assessment methodologies for CYP inhibition has led to the generation of a plethora of data, which in turn have fostered the development of QSAR models of CYP inhibition. Data published by Fu *et al.*,[91] on the inhibition by 4-X-phenoxyanilines of demethylation of aminopyrine catalyzed by rat CYP, led to the development of the following QSAR (7.19, unpublished results).[92]

$$\log 1/C = 1.10(\pm 0.49)\ c\log \mathrm{P} - 0.11(\pm 0.07)\ c\log \mathrm{P}^2 - 3.92(\pm 0.83) \quad (7.19)$$

$$n = 16 \quad r^2 = 0.914 \quad s = 0.106 \quad c\log \mathrm{P_o} \sim 5.23$$

Data on the inhibition of demethylation of aminopyrine in microsomes from Phenobarbital-treated rat liver CYP by 4-X-phenoxyanilines, resulted in the delineation of QSAR 7.20.

$$\log 1/C = 0.50(\pm 0.11)\ C\log \mathrm{P} - 0.33(\pm 0.25)\ \mathrm{CMR} + 4.92(\pm 1.31) \qquad (7.20)$$

$$n = 16 \quad r^2 = 0.917 \quad s = 0.103$$

In QSAR 7.19, there is a clear, strong parabolic dependence of inhibition of the enzyme by X-substituted anilines since *c*log P can explain 91% of the variance in the data. The optimal hydrophobicity is around 5.23. In QSAR 7.20 in the case of the microsomes, a size component comes into play, which indicates that inhibition of CYP in this environment is impacted by the larger shape/size and polarizability of the anilino compounds. These models provide guidance for substituent modification in order to minimize CYP inhibition: decrease hydrophobicity and increase steric bulk and polarizability.

7.8.4 Hydrophobicity and Toxicity: Mutagenicity and Carcinogenicity

Unanticipated toxicity is critical to the success of a new chemical entity, since it is a significant cause of attrition in drug development, and after the considerable expense of bringing it to the clinical phase. Because of the high costs and time outlays, toxicity studies have now joined ADME early in the discovery phase. In this section, we will examine the role of hydrophobicity in toxicity phenomena including mutagenicity, carcinogenicity, teratogenicity and blockage of the K^+ (hERG) channel.

Carcinogenicity and mutagenicity are two of the most important endpoints in toxicity assessment of new chemical entities. Thus the ability to predict and minimize human health and environmental burden has accelerated rigorous QSAR analysis of specific classes of molecules in these areas. Many excellent reviews on this subject are available in the literature.[93–95]

In a seminal study, Debnath *et al.*, analyzed the mutagenic activity of a large database of aromatic and heteroaromatic amines and developed QSARs 21 and 22 pertaining to their mutagenic potencies in *S. typhimurium* TA 98 and TA 100 (with S9 metabolic activation) strains.[96]

$$\begin{aligned}\log \mathrm{TA98} = {} & 1.08(\pm 0.26)\ C\log \mathrm{P} + 1.28(\pm 0.64)\ \mathrm{E_{HOMO}} \\ & - 0.73(\pm 0.41)\ \mathrm{E_{LUMO}} + 1.46(\pm 0.56)\ \mathrm{I_L} + 7.20(\pm 5.40)\end{aligned} \qquad (7.21)$$

$$n = 88 \quad r^2 = 0.806 \quad s = 0.860$$

The mutagenic potency (TA98) is expressed as number of revertants/nmol. $\mathrm{I_L}$ is an indicator variable that denotes amines with three or more fused rings and assumes a value of 1 for their presence. In this equation, most of the variance in the data (50%) is explained by hydrophobicity while the molecular orbital indices account for only 4% of the variance.

$$\begin{aligned}\log \mathrm{TA100} = {} & 0.92(\pm 0.23)\ C\log \mathrm{P} + 1.17(\pm 0.83)\mathrm{E_{HOMO}} \\ & - 1.18(\pm 0.44)\mathrm{E_{LUMO}} + 7.35(\pm 6.90)\end{aligned} \tag{7.22}$$

$$n = 67 \quad r^2 = 0.769 \quad s = 0.708$$

In both these cases, the most significant parameter was hydrophobicity. The positive coefficient with the $\mathrm{E_{HOMO}}$ descriptor suggests that amines with higher $\mathrm{E_{HOMO}}$ values are more easily oxidized. The lack of the indicator variable in QSAR 22 suggests that extended amines are more capable of inducing frame shift mutations; TA 98 is specific for frame-shift mutations while TA 100 is specific for base pair substitution mutations.

In a recent QSAR study of the genotoxic potential of quinolone antibacterials, Hu *et al.*, surmised from the mode of quinolone antibacterial binding with DNA, that genotoxic potential would be a factor of hydrophobicity and reactivity.[97] The following QSAR model was subsequently formulated for these interactions using *Salmonella typhimurium,* TA1535 with a plasmid bearing a *umu*C/lacZ gene fusion product:[98]

$$\begin{aligned}\log \mathrm{EC_{50}} = {} & -0.256(\pm 0.070)\ \log \mathrm{P} - 5.636(\pm 1.370)\ \mathrm{E_{HOMO}} \\ & - 47.57(\pm 12.13)\end{aligned} \tag{7.23}$$

$$n = 15 \quad r^2 = 0.935 \quad s = 0.230$$

In this equation, log P is the octanol–water partition coefficient at pH 6.98. In this case, the coefficient with the $\mathrm{E_{HOMO}}$ term is negative which implies that DNA accepts the electron from the quinolone antibacterial nucleus during binding. Thus genotoxicity increases with hydrophobicity and reactivity.

The carcinogenicity of a series of non-heterocyclic aromatic amines was examined by Benigni *et al.*, and the following models pertaining to the carcinogenic potency in mice (BRM) and rats (BRR) were formulated.[99] BRM and BRR were defined as log $\mathrm{MW/TD_{50}}$ where $\mathrm{TD_{50}}$ is the daily dose of the aromatic amine that halves the probability of an experimental animal remaining tumor-free to the end of its standard life span.

$$\begin{aligned}\mathrm{BRM} = {} & 0.88(\pm 0.27)\ \log \mathrm{P}^{*} + 0.29(\pm 0.20)\ \log \mathrm{P}^{**} + 1.38(\pm 0.76)\ \mathrm{E_{HOMO}} \\ & - 1,28(\pm 0.54)\ \mathrm{E_{LUMO}} - 1.06(\pm 0.34)\ \mathrm{MR_{2,6}} - 1.10(\pm 0.80)\ \mathrm{MR_3} \\ & - 0.20(\pm 0.15)\ \mathrm{E_S(R)} + 0.75(\pm 0.75)\ \mathrm{I(diNH2)} + 11.16(\pm 6.68)\end{aligned} \tag{7.24}$$

$$n = 37 \quad r^2 = 0.823 \quad s = 0.381 \quad F = 16.3,\ \mathrm{P} < 0.001$$

$$\text{BRR} = 0.35(\pm 0.18)\ \log \text{P} + 1.93(\pm 0.48)\ \text{I(Bi)} + 1.15(\pm 0.60)\ \text{I(F)} \\ - 1.06(\pm 0.53)\ \text{I(BiBr)} + 2.75(\pm 0.64)\ \text{I(RNNO)} - 0.48(\pm 0.30) \quad (7.25)$$

$$n = 41 \quad r^2 = 0.871 \quad s = 0.398 \quad \text{F} = 47.4,\ \text{P} < 0.001$$

Log P* and log P** represent the hydrophobicities of monoamines and diamines, respectively. $MR_{2,6}$ is the sum of MR (molar refraction) of substituents in the *ortho* positions and MR_3 is the MR of substituents in the *meta* position. $E_S(R)$ denotes Taft's E_S values for substituent groups on the amino terminus. All the indicator variables (I) pinpoint structural features of the amines that mostly enhance carcinogenicity in both rodent species. From these results, it is clear that hydrophobicity enhances carcinogenicity in both mice and rats. It is of interest to note that in contrast with mice, steric and electronic effects do not influence carcinogenic activity in rats. Recently, a new approach to discriminating between mutagens and non-mutagens and carcinogens and non-carcinogens has been developed and validated for the amine dataset; electronic reactivity factors (E_{HOMO}, E_{LUMO}) and steric factors represented by MR, dominate these models.[100]

7.8.5 Hydrophobicity and Toxicity: hERG Blockage

The protein product of the hERG (human Ether-à-go-go Related Gene) is the inner-pore forming portion of an important membrane bound potassium (K^+) channel in heart muscle. The K^+ channel is part of the ensemble of ion channels that generate the cardiac action potential. When a compound binds within the channel, it obstructs the outflow of K^+ ions, thus increasing the time for repolarization of the cell and lengthening the QT interval of the ECG.[101] Many drugs have been shown to block the hERG K^+ channel and induce arrhythmia. Clinical trials for potential hERG blockers are extensive, time-consuming and expensive, and so early detection of cardiac repolarization issues has shifted to the discovery phase. Several structural features that are incorporated in hERG blockers include a basic amine, which may be positively charged, hydrophobicity (*c*log P > 3.7), absence of negatively ionizable functionalities and minimum presence of oxygen-containing groups that act as hydrogen bond acceptors.[102]

A number of 3D-QSAR models have been developed for drugs that bind to the hERG channel in a similar way. However, the inner pore binding area of hERG is extensive and the great variability of hERG channel blockers indicates that there may be multiple sites of binding. Different attributes of inhibitors suggest that their binding affinities might vary depending on the channel state (activated/inactivated) and binding modes which would preempt the development of a single pharmacophore model.[103,104] Thus Keseru resorted to developing a 2D QSAR model with acceptable statistics that makes no explicit assumptions about the binding modes or binding sites.[105] The descriptors of importance were *C*log P, CMR, partial negative surface area (PNSA1), polarizability and hydrophobicity from Volsurf. Aptual and Cronin,[106] using high quality, reliable and consistent IC_{50} data culled from different sources, developed the following robust QSAR for QT prolonging drugs assayed in CHO cells:

$$pIC_{50} = 0.58(\pm 0.09)\ \log D + 0.30(\pm 0.06)\ D_{max} - 0.36(\pm 1.04) \qquad (7.26)$$

$$n = 19 \quad r^2 = 0.87 \quad s = 0.73 \quad q^2 = 0.73 \quad F = 54$$

Log D represents the hydrophobicity corrected for ionization while D_{max} is the maximum diameter of the molecules. Further analysis of the compounds indicate that the most active compounds in terms of their hERG –blocking ability have $D_{max} > 18$A.

A more recent analysis of 104 compounds tested in either CHO cells or human embryonic kidney (HEK) cells for inhibition of the hERG K^+ channel was carried out by Yoshida and Niwa.[107] They obtained the following QSAR:

$$pIC_{50} = 0.231(\pm 0.108)\ C\log P - 0.014(\pm 0.005)\ \text{TPSA} + 0.221(\pm 0.061)\ D + 0.029(\pm 0.017)\ \text{PEOE\_VSA-4} + 2.592(\pm 0.772)$$

$$n = 104 \quad r^2 = 0.636 \quad s = 0.847 \quad q^2 = 0.604 \quad F = 43.2 \qquad (7.27)$$

Increases in hydrophobicity and diameter of the compounds result in an increase in binding to the hERG channel. PEOE_VSA-4 represents the sum of the surface areas of atoms (O, F, N) with partial charges that range from –0.25 to –0.20 and it also favors an increase in blocking activity. TPSA is the topological polar surface area and it represents structural features that donate or accept hydrogen bonds. As TPSA decreases, hERG blocking is enhanced. The molecular features of this model are in close harmony with the structural characteristics of the pore region in a homology-modeled hERG channel. The inner pore is large and flexible, and mostly hydrophobic. A core of four polar residues with strong hydrogen bond donating capabilities in the inner pore can interact with the N-protonated regions of the drugs as well as other polar atoms present in their structures.[104]

7.8.6 Hydrophobicity and Bioaccumulation

Bioaccumulation is the critical process by which xenobiotics can influence the living species in the aquatic ecosystems. This process can be divided into two types: one involving bioconcentration, and the other biomagnification, which is more complex. Bioconcentration refers to an increase in the concentration of a chemical in the organism over a period of time compared with the chemical's concentration in the aqueous phase and is a net result of uptake, storage and elimination. The bioconcentration factor (BCF) is defined as the ratio of the concentration of a chemical in an aquatic organism to that in the aqueous phase under steady state conditions.[108] It is well recognized that bioconcentration is related to hydrophobicity and its well-established surrogate measurement; the octanol-water partition coefficient. Many studies in the last thirty years have established the strong correlation between log BCF and log P. In 1988, Isnard and Lambert derived the following model for several fish species[109] as shown in equation (7.28). In this model, considerable deviations from the linear relationship were observed for highly hydrophobic compounds for log P > 6.

$$\log \text{BCF} = 0.80 \log \text{P} - 0.52 \qquad (7.28)$$

$$n = 107 \quad r = 0.904 \quad s = 0.51$$

To circumvent this problem, Bintein, Devillers and Karcher developed one of the most comprehensive models for a wide range of chemicals including many pesticides for which bioconcentration data were acquired in five families of freshwater fish.[110] They utilized Kubinyi's bilinear model; the log P values of the compounds in the dataset ranged from 1 to 9.

$$\log \text{BCF} = 0.910 \log \text{P} - 1.975 \log (\beta.\text{P} + 1) - 0.784 \qquad (7.29)$$

$$n = 154 \quad r = 0.950 \quad s = 0.347 \quad F = 463.5 \quad \beta = 6.8 \times 10^{-7}$$

Biphasic functions with a linear ascending (slope = 0.91) and descending slope (–1.975 + 0.91 = –1.065), and a rounded apex are represented by this model. Highly hydrophobic compounds with log P values > 6 cover the apical and descending part of the curve. The uptake of chemicals from food and water has been compared and tissue concentrations are similar for water and food. Thus the uptake of pesticides from food is less critical than its uptake from water and only a fraction of the residue present in the lower level biota is transferred to the higher level of the food chain.[111]

The use of log P_{oct} in predicting environmental toxicity provides some useful insights into its value in CADD. It is the principal parameter that predicts bioaccumulation of PCBs and DDT analogs in the food chain,[112] but of even greater importance, log P establishes a "base line toxicity" of environmental hazards in aquatic and other systems as illustrated in QSAR 7.30.[113]

$$\text{pLC}_{50} = 0.90(0.04) \log \text{P} + 1.29(0.12) \qquad (7.30)$$

$$n = 150 \quad r = 0.959$$

This 4-day toxicity study conducted in juvenile fat-head minnows by the USEPA, encompassed a large series of stable, unreactive and non-ionizable compounds such as hydrocarbons, alcohols, esters, ketones and herbicides whose partition coefficients ranged from –1.30 to 6.

In view of the early observations of Meyer and Overton on the relationship of tadpole narcosis to hydrophobicity, and later the modern version of aquatic toxicity by Hansch and Dunn,[114] it is not surprising that simple narcosis is the main component of base-line toxicity and can be delineated by a single parameter, log P with a slope close to unity. When a certain structural class of potential toxicants such as anilines or acrylates, fit to a line with the same slope as alkanes, but with a greater intercept, the 'excess toxicity' of this structural feature can be expressed quantitatively. Alternatively, if the toxicity of a variety of structures is plotted against log P, those appearing above the 'background line' clearly exhibit this excess toxicity.

In a similar fashion, when the desired endpoint of a variety of drug candidates is plotted against log P, and the points all fall in line, not much useful information is gained. On the other hand, if the positive outliers can be 'brought on line' using an indicator variable that can be put into a structural context, this can be very insightful and lead to the next step in molecular design.[115]

7.9 Conclusion

The impact of hydrophobicity on most biological processes is of great significance as can be ascertained by the frequency of its usage in most QSAR studies. Despite the early usage by Meyer and Overton at the end of the nineteenth century, hydrophobicity eventually took its cues from the Hammett approach and finally crystallized fifty years later when the best known and now well-developed hydrophobic descriptor, log Po/w was delineated by Hansch and Fujita.

Log P values can be measured experimentally or predicted from molecular structure. Experimental approaches run the gamut from the highly reliable and traditional shake-flask method to newer methodologies such as electrokinetic chromatographic methods that are rapid, accurate and precise, and can be automated. However, for validation purposes, partition coefficient data of large, structurally diverse datasets obtained from such systems should be compared with that from the standard octanol/water system.

The expense and significant time that is generally required for log P measurements has led to the development of many robust software programs that are based on substructures or whole molecules. In whole molecule approaches, critical factors such as the decreased impact of hydrogen bond-rich solutes, the high-dipole moment of water, the sum of the hydrogen bond acceptor strength and the use of McGowan's volume as a quick representation of solvent accessible surface area, merits consideration. The chlorination of methane underscores the importance of localized bond dipoles that singularly lower log P but tend to shield each other when in clusters.

The complex contribution of hydrophobicity (log P) to various biological end-points is illustrated by examples drawn from the central nervous system (blood-brain penetration), absorption (intestinal absorption), metabolism (cytochrome P450), toxicity (mutagenicity, carcinogenicity and hERG channel inhibition) and aquatic bioaccumulation. Although the sign and magnitude of the log P term varies considerably in all these bio-systems, the significant dependence on hydrophobicity once again emphasizes the importance of elucidating and predicting behavior of drugs and organic molecules on isolated receptors, in mammalian cell culture systems, *in vivo* as well as in aquatic organisms.

Hydrophobicity as described by log P is particularly important in ADMET phenomena. Thus it strongly warrants a thorough understanding of its complexity as well as robust methods to predict and measure it. While trending toward a greater reliance on predicting partition coefficients, fewer actual measurements are being undertaken. Compounds bearing new or complex scaffolds need to be measured and some of the data acquired from more recent and cutting-edge technology should be validated using structurally diverse test sets.

References

1. J. E. A. Comer, in *Methods and Principles in Medicinal Chemistry*, ed. H. van de Waterbeemd, H. Lennernäs and P. Artursson, Wiley-VCH, Weinheim, 2003, vol. 18, p. 21.
2. R. Fraczkiewicz, in *Comprehensive Medicinal Chemistry: In Silico Prediction of Ionization*, ed. J.B. Taylor and D.B. Triggle, Elsevier, Oxford, UK, 2007, vol. 5, p. 603.
3. L. P. Hammett, *J. Am. Chem. Soc.*, 1937, **59**(1), 96.
4. R. W. Taft and I. C. Lewis, *J. Am. Chem. Soc.*, 1959, **81**(20), 5343.
5. G. Klopman, *Quant. Struct. Act. Relat.*, 1992, **11**(2), 176.
6. G. Klopman and D. Fercu, *J. Comput. Chem.*, 1994, **15**(9), 1041.
7. A. Habibi-Yangjeh, M. Danandeh-Jenagharad and M. Nooshyar, *J. Mol. Model,* 2006, **12**(3), 338.
8. A. Cherkasov, D. G. Sprous and R. Chen, *J. Phys. Chem. A,* 2003, **107**(45), 9695.
9. F. Luan, W. Ma, H. Zhang, X. Zhang, M. Liu, Z. Hu and B. Fan, *Pharm. Res.*, 2005, **22**(9), 1454.
10. R. A. Scherrer and A. J. Leo, *Mol. Inf.*, 2010, **29**(10), 687.
11. A. Crum-Brown and T. R. Fraser, *Trans. R. Soc. Edin.*, 1868–1869, **25**, 151.
12. C. Selassie and R. P. Verma, in *Burger's Medicinal Chemistry, Drug Discovery and Development*, ed. D. J. Abraham and D. P. Rotella, John Wiley & Sons, New York, USA, 7th edn, 2010, vol. 1, p. 1.
13. M. S. Tute, in *Comprehensive Medicinal Chemistry: The Rational Design, Mechanistic Study and Therapeutic Application of Chemical Compounds.*, ed. C. A. Ramsden, Pergamon, Oxford, UK, 1990, vol. 4, p. 1.
14. L. P. Hammett, *Chem. Rev.*, 1935, **17**(1), 125.
15. C. Hansch and A. Leo, in *Exploring QSAR. Fundamentals and Applications in Chemistry and Biology*, ed. S.R. Heller, American Chemical Society, Washington, DC, 1995.

16. O. Exner, in *Correlation Analysis of Chemical Data*, Plenum Publishing Corp., New York, 2nd edn, 1988, p. 128.
17. H. H. Jaffe, *Chem. Rev.*, 1953, **53**(2), 191.
18. R. W. Taft, in *Steric Effects in Organic Chemistry*, ed. M.S. Newman, Wiley, New York, 1956.
19. A. Verloop, *The STERIMOL Approach to Drug Design*, Marcel Dekker, New York, 1987.
20. A. Verloop, in *QSAR and Strategies in the Design of Bioactive Compounds, Proceedings of the 5th European Symposium on QSAR, Bad Segeberg 1984*, ed. J. K. Seydel, VCH Publishers, Deerfield Beach, FL, USA, 1986, p.98.
21. W. Kauzmann, *Adv. Protein Chem.*, 1959, **14**, 1.
22. I. Langmuir, *Proc. R. Soc. Lond., Ser. A*, 1939, **170**(940), 1.
23. H. S. Frank and M.W. Evans, *J. Chem. Phys.*, 1945, **13**(11), 507.
24. H. Meyer, *Arch. Exp. Pathol. Pharm.*, 1899, **42**, 109.
25. E. Overton, *Z. Physik. Chem.*, 1897, **22**, 189.
26. R. Collander, *Acta Chem., Scand.*, 1951, **5**, 774.
27. C. Hansch, P. P. Maloney, T. Fujita and R. M. Muir, *Nature*, 1962, **194**(4824), 178.
28. K. A. Dill, T. M. Truskett, V. Vlachy and B. Hribar-Lee, *Annu. Rev. Biophys. Biomol. Struct.*, 2005, **34**, 173.
29. A. J. Leo, *Methods Enzymol.*, 1991, **202**, 544.
30. C. Hansch, R. M. Muir, T. Fujita, P. P. Maloney, F. Geiger and M. Streich, *J. Am. Chem. Soc.*, 1963, **85**(18), 2817.
31. T. Fujita, K. Nishimura, C. Takayama, M. Yoshida and M. Uchida, in *Handbook of Pesticide Toxicology,*, ed. R. Krieger, Academic Press, London, 2001, vol. 1, p. 649.
32. A. Kristl and S. Pečar, *Eur. J. Med. Chem.*, 1997, **32**(1), 3.
33. A. Sarkar and G. E. Kellogg, *Curr. Top. Med. Chem.*, 2010, **10**(1), 67.
34. G. L. Grunewald, M. A. Pleiss, C. L. Gatchell, R. Pazhenchevsky and M. F. Rafferty, *J. Chromatogr.*, 1984, **292**(2), 319.
35. D. T. H. M. Sijm and T. L. Sinnige, *Chemosphere*, 1995, **31**(11-12), 4427.
36. E. Tomlinson, S. S. David, G. D. Parr, M. James, N. Farraj, J. F. M. Kinkel, D. Gaisser and H. J. Wynn, in *Partition Coefficient: Determination and Estimation*, ed. W.J. Dunn III, J.H. Block and R.S. Perlman, Pergamon, Oxford, UK, 1986, p. 83.
37. E. Tomlinson, *J. Pharm. Sci.*, 1982, **71**(5), 602.
38. C. Barzanti, R. Evans, J. Fouquet, L. Gouzin, N. M. Howarth, G. Kean, E. Levet, D. Wang, E. Wayemberg, A. A. Yeboah and A. Kraft, *Tetrahedron Lett.*, 2007, **48**(19), 3337.
39. L. Hitzel, A. P. Watt and K. L. Locker, *Pharm. Res.*, 2000, **17**(11), 1389.
40. D. M. Wilson, X. Wang, E. Walsh and R. A. Rourick, *Comb. Chem. High Throughput Screening*, 2001, **4**(6), 511.
41. K. Valkó, *J. Chromatogr. A*, 2004, **1037**(1-2), 299.
42. A. Avdeef and B. Testa, *Cell. Mol. Life Sci.*, 2002, **59**(10), 1681.
43. S. Marcinkiewicz and J. Green, *J. Chromatogr.*, 1963, **10**(5), 372.
44. E. C. B. Smith and R. G. Westall, *Biochim. Biophys. Acta,* 1950, **4**, 427.

45. A. Nasal, D. Siluk and R. Kaliszan, *Curr. Med. Chem.*, 2003, **10**(5), 381.
46. E. Soczewiński and C. A. Wachtmeister, *J. Chromatogr.*,1962, **7**, 311.
47. L. R. Snyder, J. W. Dolan and J. R. Gant, *J. Chromatogr.*,1979, **165**(1), 3.
48. F. Lombardo, M. Y. Shalaeva, K. A. Tupper, F. Gao and M. H. Abraham, *J. Med. Chem.*, 2000, **43**(15), 2922.
49. X. Liu, H. Tanaka, A. Yamauchi, B. Testa and H. Chuman, *J. Chromatogr. A*, 2005, **1091**(1-2), 51.
50. C. Giaginis, S. Theocharis and A. Tsantili-Kakoulidou, *J. Chromatogr. A*, 2007, **1166**(1-2), 116.
51. D. Benhaim and E. Grushka, *J. Chromatogr. A*, 2008, **1209**(1-2), 111.
52. S. F. Donovan and M. C. Pescatore, *J. Chromatogr. A*, 2002, **952**(1–2), 47.
53. X. Liu, H. Tanaka, A. Yamauchi, B. Testa and H. Chuman, *Helv. Chim. Acta*, 2004, **87**(11), 2866.
54. K. Valkó, C. Bevan and D. Reynolds, *Anal. Chem.*, 1997, **69**(11), 2022.
55. P.-C. Chiang and Y. Hu, *Comb. Chem. High Throughput Screening*, 2009, **12**(3), 250.
56. F. S. Sarr, Y. C. Guillaume and C. Andre, *J. Pharm. Biomed. Anal.*, 2008, **47**(3), 651.
57. T. Hartmann and J. Schmitt, *Drug Discovery Today: Technol.*, 2004, **1**(4), 431.
58. R.S. Ward, J. Davies, G. Hodges and D.W. Roberts, *J. Chromatogr. A*, 2003, **1007**(1-2), 67.
59. F. Barbato, *Curr. Comp.-Aided Drug Des.*, 2006, **2**(4), 341.
60. C. Giaginis and A. Tsantili-Kakoulidou, *J. Pharm. Sci.*, 2008, **97**(8), 2984.
61. C. W. Huie, *Electrophoresis*, 2006, **27**(1), 60.
62. K.-S. Wong, J. Kenseth and R. Strasburg, *J. Pharm. Sci.*, 2004, **93**(4), 916.
63. J. Tu, H. B. Halsall, C. J. Seliskar, P. A. Limbach, F. Arias, K. R. Wehmeyer and W. R. Heineman, *J. Pharm. Biomed. Anal.*, 2005, **38**(1), 1.
64. Y. S. Wu, H. K. Lee and S. F. Y. Liz, *Electrophoresis*, 1998, **19**(10), 1719.
65. W. L. Klotz, M. R. Schure and J. P. Foley, *J Chromatogr. A*, 2001, **930**(1-2), 145.
66. Y. Ishihama, Y. Oda, K. Uchikawa and N. Asakawa, *Anal. Chem.*, 1995, **67**, 1588.
67. R. Mannhold and H. van de Waterbeemd, *J. Comput.-Aided Mol. Des.*, 2001, **15**(4), 337.
68. A. J. Leo, *Chem. Rev.*, 1993, **93**(4), 1281.
69. R. F. Rekker, *The Hydrophobic Fragmental Constant*, Elsevier, Amsterdam, 1977.
70. T. Fujita, J. Iwasa and C. Hansch, *J. Am. Chem. Soc.*, 1964, **86**(23), 5175.
71. R. F. Rekker and R. Mannhold, *Calculation of Drug Lipophilicity*, VCH, Weinheim, 1992.
72. K. Iwase, K. Komatsu, S. Hirono, S. Nakagawa and I. Moriguchi, *Chem. Pharm. Bull.*, 1985, **33**(5), 2114.
73. N. Bodor, Z. Gabanyi and C. K. Wong, *J. Am. Chem. Soc.*, 1989, **111**(11), 3783.
74. M. J. Kamlet, J. L. M. Abboud, M. H. Abraham and R. W. Taft, *J. Org. Chem.*, 1983, **48**(17), 2877.

75. D. Eros, I. Kovesdi, L. Orfi, K. Takacs-Novak, G. Acsady and G. Keri, *Curr. Med. Chem.*, 2002, **9**(20), 1819.
76. Y. Sakuratani, K. Kasai, Y. Noguchi and J. Yamada, *QSAR Comb. Sci.*, 2007, **26**(1), 109.
77. J. C. McGowan, *Recl. Trav. Chim. Pays-Bas,* 1956, **75**, 193.
78. M. J. Kamlet, M.H. Abraham, R. M. Doherty, and R. W. Taft, *J. Am. Chem. Soc.*, 1984, **106**(2), 464.
79. R. W. Taft, M. Berthelot, C. Laurence and A. J. Leo, *Chem.-Tech.*, 1996, **26**(7), 20.
80. M. H. Abraham, H. S. Chadha, G. S. Whiting and R. C. Mitchell, *J. Pharm. Sci.*, 1994, **83**(8), 1085.
81. R. Glaser, *J. Org. Chem.*, 2001, **66**(3), 771.
82. W. Wachter, R. Buchner and G. Hefter, *J. Phys. Chem. B*, 2006, **110**(10), 5147.
83. Y. L. A. Rezus and H. J. Bakker, *J. Phys. Chem. A*, 2008, **112**(11), 2355.
84. A. A. Bakulin, C. Liang, T. La Cour Jansen, D.A. Wiersma, H. J. Bakker and M. S. Pshenichnikov, *Acc. Chem. Res.*, 2009, **42**(9), 1229.
85. C. Hansch, J. P. Bjoerkroth and A. Leo, *J. Pharm. Sci.*, 1987, **76**(9), 663.
86. R. C. Young, R. C. Mitchell, T. H. Brown, C. R. Ganellin, R. Griffiths, M. Jones, K. K. Rana, D. Saunders, I. R. Smith, N. E. Sore and T. J. Wilks, *J. Med. Chem.*, 1988, **31**(3), 656.
87. M. H. Abraham, Y. H. Zhao, J. Le, A. Hersey, C. N. Luscombe, D. P. Reynolds, G. Beck, B. Sherborne and I. Cooper, *Eur. J. Med. Chem.*, 2002, **37**(7), 595.
88. C. D. Selassie and R. P. Verma, unpublished results.
89. D. A. Smith, *Eur. J. Drug Metab. Pharmacokinet.*, 1994, **19**(3),193.
90. A. D. Rodrigues and J. H. Lin, *Curr. Opin. Chem. Biol.,* 2001, **5**(4), 396.
91. X. Fu, H. Jiang and Z. Liu, *Shengwu Huaxue Zazhi,* 1994, 10(1), 11.
92. C. D. Selassie and R. P. Verma, unpublished results.
93. R. Benigni, *Chem. Rev.*, 2005, **105**(5), 1767.
94. M. T. D. Cronin, J. C. Dearden, J. D. Walker and A. P. Worth, *Environ. Toxicol. Chem.*, 2003, **22**(8), 1829.
95. R. Benigni and A. Guiliani, *Med. Res. Rev.*, 1996, **16**(3), 267.
96. A. K. Debnath, R. L. Lopez de Compadre, G. Debnath, A. J. Shusterman and C. Hansch, *J. Med. Chem.*, 1991, **34**(2), 786.
97. J. Hu, W. Wang, Z. Zhu, H. Chang, F. Pan and B. Lin, *Environ. Sci. Technol.*, 2007, **41**(13), 4806.
98. Y. Oda, S. Nakamura, I. Oki, T. Kato and H. Shinagawa. *Mutat. Res.* 1985, **147**(5), 219.
99. R. Benigni, A. Giuliani, R. Franke and A. Gruska, *Chem. Rev.*, 2000, **100**(10), 3697.
100. R. Benigni, C. Bossa, T. Netzeva, A. Rodomonte and I. Tsakovska, *Environ. Mol. Mutagen.*, 2007, **48**(9), 754.
101. M. C. Sanguinetti and J. S. Mitcheson, *Trends. Pharmacol. Sci.*, 2005, **26**(3), 119.
102. R. J. Vaz, Y. Li and D. Rampe, *Prog. Med. Chem.*, 2005, **43**, 1.

103. E. Ficker, C. A. Obejero-Paz, S. Zhao and A. M. Brown, *J. Biol. Chem.*, 2002, **277**(7), 4989.
104. J. S. Mitcheson, J. Chen, M. Lin, C. Culberson, M. C. Sanguinetti, *Proc. Natl. Acad. Sci. U.S.A.*, 2000, **97**(22), 12329.
105. G. M. Keseru, *Bioorg. Med. Chem. Lett.* 2003, **13**(16), 2773.
106. A. O. Aptual and M. T. D. Cronin, *SAR QSAR Environ. Res.*, 2004, **15**(5-6), 399.
107. K. Yoshida and T. Niwa, *J. Chem. Inf. Model.*, 2006, **46**(3), 1371.
108. B. Streit, *Experientia*, 1992, **48**(10), 955.
109. P. Isnard and S. Lambert, *Chemosphere*, 1988, **17**(1), 21.
110. S. Bintein, J. Devillers and W. Karcher. *SAR QSAR Environ. Res.*, 1993, **1**(1), 29.
111. H. Ellgehausen, J. A. Guth and H.O. Esser, *Ecotox. Environ. Safety*, 1980, **4**(2), 134.
112. A. Leo, in *Environmental Health Chemistry*, ed. J.D. McKinney, Ann Arbor Science Publishing, Ann Arbor, MI, USA, 1981, ch. 16, p.323.
113. L. S. McCarty, D. Mackay, A. D. Smith, G. W. Ozburn and D. G. Dixon, *Environ. Toxicol. Chem.*, 1992, **11**(7), 917.
114. C. Hansch and W. J. Dunn, III, *J. Pharm. Sci.*, 1972, **61**(1), 1.
115. R. L. Lipnick, *Sci. Total Environ.*, 1991, **109–110**, 131.

CHAPTER 8

Characterising Chemical Structure Using Physicochemical Descriptors

DAVID J. LIVINGSTONE

ChemQuest, Sandown, Isle of Wight, UK and Centre for Molecular Design, University of Portsmouth, Portsmouth, UK

8.1 Introduction

We have already seen in Chapter 3 the dramatic effect of describing chemical structures in terms of properties. Yvonne Martin has shown how quantitative structure-activity relationships (QSAR) grew out of structure-activity relationships (SAR), how this enables the prediction of activity for quite diverse chemical structures and how it also allows us to draw mechanistic conclusions about drug actions. There are now thousands[1] of different molecular properties which can be more or less easily calculated or estimated for any given chemical structure,[2] but this hasn't always been the case. As shown in Chapter 7, at the start of what can be regarded as the beginning of systematic quantitative drug design, there was a very limited range of molecular descriptors available to the "drug designer". Forty years ago one would routinely use 3 types of molecular property to characterise a set of structures: hydrophobic, steric and electronic descriptors. Partition coefficients, either measured for whole molecules or tabulated as substituent constants, were used for hydrophobic effects, Taft's steric parameter (Es), later replaced by molar refraction (MR), was used for steric properties and Hammett σ values, again tabulated, were used for electronic effects. The way in which these properties were used was to look up

RSC Drug Discovery Series No. 13
Drug Design Strategies: Quantitative Approaches
Edited by David J. Livingstone and Andrew M. Davis

Published by the Royal Society of Chemistry, www.rsc.org

values in some sort of compilation. The major exception to this was topological parameters which could be rapidly calculated for any chemical structure.

So, how did we get to the situation we are in today? The following sections discuss the major categories of properties which can be used to describe chemical structures, showing how they were first used and how they have been developed, what works well, what works less well and, in some cases, what remains to be done.

8.2 Physical Organic Chemistry

The motivation behind the development of the branch of chemistry which has become known as physical organic chemistry was the desire to explain and/or predict changes in chemical reactivity with changes in chemical structure. Reactivity here can mean a variety of things, such as whether a reaction will happen at all, how fast a reaction will occur, the position of chemical equilibria and what will be the major products. The philosophy was predicated on the observation that a change in chemical structure, often a simple substituent change, in one chemical system can have a quantitatively predictable effect on chemical reactivity in other chemical systems, according to the nature of that structural change. The basis of the approach was to specify a simple chemical model system in which the changes in structure would result in changes in the monitored property (often an equilibrium constant). Changes in the monitored property could be assigned as a constant for that structure change/substituent. Different chemical systems would have different sensitivities to the effect of the structural change. The 3 main classes of molecular properties mentioned in the introduction are discussed in Sections 8.2.1–8.2.3.

8.2.1 Electronic Properties

Hammett, who is arguably the father of physical organic chemistry, proposed that the ionisation of benzoic acids could be used as a chemical model system to characterise the electronic properties of different substituents. In the Hammett equation (eqn (8.1)), the reaction is characterised by a reaction constant, ρ, which quantifies the sensitivity of that particular reaction to the electronic effect of the substituent, which itself is described by a substituent constant, σ:

$$\log K_X - \log K_H = \rho\sigma_X \tag{8.1}$$

In this equation, K_X is the equilibrium (or rate) constant for a process involving an X-substituted compound, and K_H is the corresponding constant for the unsubstituted parent. Hammett chose the ionisation of benzoic acids as a reference system and assigned a value of 1 as the reaction constant for this series. Substituent σ values were then computed from measured pK_a values of substituted benzoic acids, and quite extensive lists of σ values have been compiled. One feature of this approach to the modelling of electronic effects is that the substituent constant values are position dependent and thus there are (usually) *meta* and *para* values of σ for each substituent.

These σ values were shown to explain many different types of reactions and equilibria, but they were not universally applicable. One of the reasons for the failure is the nature of the model system itself, and how the electronic effect of the substituent "couples" with the developing change in electronic charge at the reaction centre relative to the calibrating reaction. For example, the negative charge on the oxygen atom in ionised phenols can delocalise its charge directly with the pi-system of the aromatic ring and potentially, depending on its substitution position, the pi-system of any substituent, which is not possible for the carboxylate anion in benzoic acids. This explains why the *p*-nitro substituent has a larger effect on the ionisation of phenols than is predicted from its effect on the ionisation of benzoic acids. Similarly, the electron donating *p*-methoxy substituent has a large effect on the development of positive charge at an anilinium centre or positive radical formation at a benzylic position. Thus many different σ scales based on alternative model systems (the ionisation of anilines, for example) have been proposed, to account for different types of resonance effects. There has been constant debate amongst physical organic chemists ever since, leading not only to a considerable number of different σ scales, but also to many extensions and enhancements of the original Hammett treatment[3] This multiplicity of σ scales makes the task of choosing the "right" one for a biochemical or biological application very difficult indeed. In an attempt to account for the many different σ scales and to produce more generally useful electronic descriptors, Swain and Lupton proposed[4] that any set of σ values could be expressed by a weighted linear combination of two components, $\mathfrak{F}$ and $\mathfrak{R}$, which they termed the field and resonance components respectively. As an example of this treatment, the standard *para* and *meta* σ values can be described as shown in eqn (8.2) and (8.3).

$$\sigma_p = 0.56\mathfrak{F} + 1.00\mathfrak{R} + 0.00 \tag{8.2}$$

$$\sigma_m = 0.60\mathfrak{F} + 0.27\mathfrak{R} + 0.00 \tag{8.3}$$

In these equations it can be seen that both σ scales contain approximately the same contribution from the field component but that the *para* scale has an increased contribution from the resonance component. This decomposition of electronic effects into the two components is surprisingly useful from the viewpoint of characterising structures for QSAR modelling but has not been generally accepted by the physical organic chemistry community. A revised treatment, based on a larger data set and using a more sophisticated statistical analysis, has been reported[5] but gave rise to considerable debate and criticism.[6]

An analogous descriptor for aliphatic systems, σ*, was proposed by Taft based on the ratio of acid catalysed to base catalysed hydrolysis reactions, as briefly discussed in Section 8.2.3.

8.2.2 Hydrophobic Properties

Octanol–water log P values and the substituent constant π values derived from them have been covered in detail in the previous chapter by Al Leo. It is of course undeniable that these descriptors have been hugely successful,† but they are not the only partition coefficients that can be useful in describing molecular properties. The difference between octanol–water and cyclohexane–water partition coefficients (Δlog P) has been shown to be useful in describing brain penetration,[7] and it was proposed that a set of partition coefficients from four different partitioning systems would provide useful, complementary information in order to describe the properties of real membranes.[8] The four partition systems involved water with the following second phases:

- An amphiprotic solvent (*e.g*. octanol).
- An inert solvent (*e.g*. any alkane, such as cyclohexane in the example cited).
- A pure proton donor solvent (*e.g*. chloroform).
- A pure proton acceptor solvent.

In addition to actual partition coefficient measurements, there are a number of other descriptors in which the main property involved is a measure of hydrophobicity. Examples of these are various chromatographic measurements, such as thin layer, paper and reverse phase high performance liquid chromatography. Another interesting treatment is the decomposition of partition coefficients into enthalpic and entropic components,[9] in an attempt to provide more "mechanism-based" parameters for hydrophobicity.

8.2.3 Steric Properties

Two fairly obvious descriptors of steric effects are molecular weight and some measure of molar volume, but these suffer from the disadvantage that they merely describe the whole molecule, not portions which might be subject to change. They also describe "size" or "bulk", not shape. Taft proposed[10,11] a free-energy related parameter, E_s [given by eqn (8.4)], to encode the steric effect of a substituent, based on the acid hydrolysis of substituted methyl esters.

$$E_s = \log(k/k_0)_A \tag{8.4}$$

Taft hypothesised that as the acid catalysed hydrolysis of methyl esters does not involve a change in overall charge between the ground state and the transition state, the reaction would be largely insensitive to the electronic nature of the substituent. By comparison, the base catalysed hydrolysis of the same series of methyl esters would involve the development of negative charge in the transition state relative to the ground state, hence the reaction would depend on both the steric and electronic effect of the substituent. Taft was also

† A survey of the QSAR journal in 1988 showed that 40% of the reported models involved a hydrophobicity descriptor, rising to more than 50% in a similar survey ten years later in 1998.

able to define a σ constant for aliphatic substituents, based on the ratio of the rate constants of the acid to based catalysed hydrolysis of methyl esters, as the very similar geometry of the transition state structures for the two reactions would experience similar steric influence of the substituent.

In eqn (8.4), k is the rate constant for the hydrolysis of an R substituted ester, and k_0 is the corresponding constant for the methyl substituted parent, thus all comparisons are made between the substituent and a methyl group. These substituent constant values are used in the same way as the electronic and hydrophobic substituent constants discussed earlier, that is to say they are found in tabulations of substituent constant values, and of course the same problems of missing values apply. In fact, the situation can be even worse for E_s as a number of substituents are themselves unstable under the conditions of acid hydrolysis. It has also been argued that this descriptor is not just a measure of steric effects, but that it also includes some electronic information. A number of more or less ingenious fixes were proposed to solve such problems, but a much more popular and generally useful measure of steric effects for both substituents and whole molecules was adopted in the form of molar refraction[12] (MR), as defined by the Lorentz-Lorenz equation [eqn (8.5)].

$$MR = \frac{n^2 - 1}{n^2 + 2} \cdot \frac{MW}{d} \tag{8.5}$$

In eqn (8.5), n is measured refractive index, MW is molecular weight and d is density. MR has been shown to be (mostly) additive, thus values can be obtained by summation of fragment values from look-up tables.

Both of these parameters suffer from the drawbacks associated with molecular weight and molar volume, as mentioned at the beginning of this section, in that they are a single measure of "size" or "bulk", not shape.

Verloop, in what was probably the forerunner of all the computational chemistry based descriptors, proposed[13] an ingenious system for characterising substituent shape, later extended to whole molecules. In this approach, a substituent is characterised using standard bond lengths and angles and van der Waals radii for the constituent atoms, as shown in Figure 8.1. This also assumes that the substituent adopts a "reasonable" conformation, as discussed further in Section 8.6.

A length parameter, L, is defined as the length of the substituent along the axis of the bond between the first atom of the substituent and the parent. Four width parameters, B_1 to B_4, were defined as the distance at the maximum point perpendicular to this bond axis and to each other. As shown in Figure 8.1, these parameters basically define a box around the substituent. There was debate of course about what constitutes a reasonable conformation, and the drawback of using 5 parameters to describe a single property was resolved to some extent by the adoption of just a minimum (B_1) and maximum (B_5) width parameter, along with L. These so called Sterimol parameters were tabulated for substituents in the same way as other substituent constants, such as π and σ, had been. A program was also available for their calculation, for both substituents

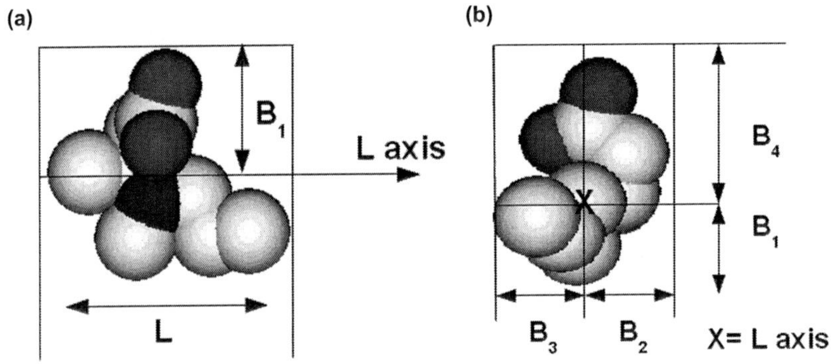

Figure 8.1 Definition of the original STERIMOL parameters, (a) Viewed orthogonal to the point of attachment of the substituent. A box is placed around the substituent and L is defined as the length of the box along the axis of attachment; (b) Viewed perpendicular to the axis of attachment (marked X). Four dimensions of the box, B1–B4 are defined as shown. Reproduced from reference 2, copyright 2000, American Chemical Society.

and whole molecules. Shape and size, of course, is very well covered in the properties calculated by molecular modelling systems.

8.3 Topological Descriptors

The first topological descriptors were developed in the late 1940s to explain some simple chemical properties but were brought to the attention of the drug design community in the 1970s by Kier and Hall,[14] who devised many different types of descriptors, which they termed molecular connectivity indices. Connectivity indices in their simplest form are computed from the hydrogen suppressed skeleton of a compound by the assignment of a degree of connectivity, δ_i, to each atom (i) representing the number of atoms connected to it. Figure 8.2 shows the degree of connectivity for each of the 6 heavy atoms in *iso*-butanoic acid.

For each bond in the structure, a bond connectivity (Ck) can be calculated by taking the reciprocal of the square root of the product of the connectivities of the atoms at either end of the bond. For example, the bond connectivity for the first carbon–carbon bond (from the left) in the structure is given by eqn (8.6).

$$C_1 = 1\sqrt{(1 \times 3)} \tag{8.6}$$

More generally, the bond connectivity of the kth bond is given by eqn (8.7).

$$C_k = 1\sqrt{(\delta_i \delta_j)} \tag{8.7}$$

In eqn (8.7), the subscripts i and j refer to the atoms at either end of the bond. The molecular connectivity index, χ, for a molecule is found by summation of

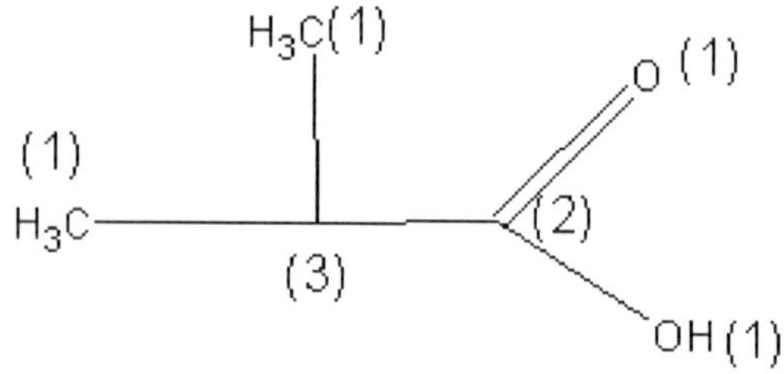

Figure 8.2 Illustration of the degree of connections of the heavy (non-hydrogen) atoms in *iso*-butanoic acid.

the bond connectivities over all of its N bonds.

$$\chi = \sum_{k=1}^{N} C_k \tag{8.8}$$

For the butanoic acid molecule shown in Figure 8.1, the five bond connectivities are the reciprocal square roots of (1 × 3), (1 × 3), (2 × 3), (2 × 1) and (2 × 1), which gives a molecular connectivity value of 2.977. This simple connectivity index is known as the first order index because it considers only individual bonds, in other words paths of two atoms in the structure. Higher order indices may be generated by the consideration of longer paths in a molecule, and other refinements—such as valence connectivity values, path, cluster and chain connectivities—have been introduced.[15]

These descriptors offered a number of advantages over the substituent constants that were in common use at that time. They could be easily and rapidly calculated for any structure and thus there were never any "missing values", as always seemed the case for substituent constants. There was no need for a "parent" structure and thus they could be applied to any collection of compounds, not just a congeneric series, and there was never any debate about which "position" a structural fragment occupied. Many of the substituent constants assumed different values depending on whether a substituent was *meta* or *para* to the reaction centre of the model system, as discussed earlier for σ. Recognition of substitution position is obvious for simple cyclic molecules but can be much less clear for complex, multi-ring structures.

So, topological descriptors have many advantages over the substituent constants described in the previous section, but they suffer from one major disadvantage and that is the question of interpretation. There is little doubt that they contain useful chemical information, given the variety of successful models that have been created using them, but for some researchers the difficulty or impossibility of interpretation of the physicochemical meaning underlying such descriptors has excluded their use. Attempts have been made to justify their significance[16] but this gave rise to a dialogue[17,18] reminiscent of the debate over the $\mathfrak{F}$ and $\mathfrak{R}$ parameters of Swain and Lupton, as described earlier. Charton has demonstrated[19] quite convincingly, by examining the properties of molecules with constant topology, that they are not "fundamental properties" but has also shown[20] that they are useful composite parameters. Thus they may not lend themselves to studies which are aimed at examining mechanism, but they have proved very useful in modelling environmental data that tends to consist of results for quite diverse sets of molecules, posing a problem for properties which require a "parent" structure. Another application area is where large numbers of compounds (*e.g.* chemical libraries) are involved, since they can be computed so rapidly from a 2D representation of chemical structure. An example of this is an artificial neural network (ANN) model of log P, using atom-type electrotopological-state (E-state) indices.[21] A training set of 1754 molecules, characterised by 38 E-state indices and molecular weight, was used to train[22] a 39-5-1 feed-forward neural network, giving a fit with $R^2 = 0.9$ and an RMSE value of 0.46. A test set of 116 diverse structures gave a predictive R^2 of 0.94 and an RMSE of 0.41. This performance is an improvement over the equivalent multiple linear regression (MLR) model, presumably because of the non-linearity built into the ANN modelling technique. Although construction, that is to say training, of an ANN model is time consuming compared with statistical models such as MLR, once built, an ANN model may be used to make predictions rapidly, since it consists only of a set of coefficients corresponding to the ANN weights. Thus this E-state ANN model can be rapidly used to make log P calculations for collections of thousands or hundreds of thousands of compounds.

Finally, there is one particular topological descriptor that should be mentioned because it has found considerable utility in a number of studies. This property, originally called a "chemically intuitive molecular index" by its inventor, Frank Burden[23], is based on a modified molecular adjacency matrix. The approach was extended by Pearlman[24] to take account of atomic charge, polarisability and hydrogen bonding ability, properties which are termed BCUT[‡] descriptors. BCUTs are available in a number of software packages and in addition to their utility in QSAR modelling, they have been shown to be useful measures of chemical similarity.

8.4 Experimental

Perhaps the most obvious way to characterise different chemical structures is to use experimental measurements of some characteristic properties, such as melting point, boiling point or refractive index for example. This approach has the advantage that these properties are unambiguously and easily defined and in many cases easily measured to any desired precision. Solubility is clearly an important factor in the biological activity of any compound, and indeed this was found to be one of the earliest successful molecular descriptors, where toxicity

[‡]The initials are derived as follows: B for Burden; C for chemical abstracts, where a study on similarity descriptors showed the original burden numbers to be useful and UT for University of Texas, where the modifications were made.

was shown to be inversely related to water solubility.[25] Other properties which have been found to be useful include: molecular weight,[26] NMR chemical shifts,[27] infrared and Raman stretching frequencies,[28] reactivity with 4-nitrophenol[29] and even rate of hydrolysis in mouse plasma![30] A disadvantage in the use of experimental properties is that they may be composed of a number of "fundamental" chemical properties, thus their inclusion in a mathematical model may make any mechanistic interpretation of the model difficult, if not impossible. Of course this doesn't matter if the model is only to be used for prediction, but perhaps a more fundamental problem in the use of experimental properties is the matter of their prediction for new structures. If a model is to be useful in prediction it should be possible to employ the model before a compound is synthesised, thus it is necessary to calculate or estimate any of the "independent variables" in the model from chemical structure alone. In fact, there is a surprising number of experimental properties which can be calculated[31,32,33] more or less accurately, although some of these calculations depend on knowledge (or an estimate) of some other experimental property. An example of the latter type of calculation is the relationship between melting point, water solubility and partition coefficient, shown by Yalkowsky and Valvani:[34,35]

$$\begin{gathered} log\ S_w = -1.05\ log\ P_{oct} - 0.012\ (Mp - 25) + 0.87 \\ \mathrm{n} = 155, \quad \mathrm{SD} = 0.308, \quad \mathrm{r}^2 = 0.98 \end{gathered} \tag{8.9}$$

In this equation $log\ S_w$ is the log of water solubility, $log\ P_{oct}$ is the octanol–water log P and Mp is melting point. This particular equation was generated for a set of 155 compounds and in a slightly modified form became known as the general solubility equation,[36] which of course can be very useful if the log P and melting point values are known or can be reliably estimated. Experimental properties are also sometimes needed in their own right, not to describe chemical structure but because their knowledge is useful, such as in a formulation, or because they are required for a registration document. It may also be that values of some experimental properties, such as solubility, will be used as a "virtual screen" to select compounds for further screening or progression. In such circumstances it is also very helpful to be able to make theoretical predictions of these experimental properties.[37]

8.5 Parameters from Molecular Modelling/ Computational Chemistry

In the late 1970s and early 1980s molecular modelling was in its infancy, as software was being developed and computing power/graphics ability was evolving. At this stage of development, molecular modelling or the construction of QSAR models were often seen as alternative methods of computer aided molecular design, although of course in reality they were complementary. This complementarity gradually dawned on the "drug designers" in industry, and systems were developed which would generate descriptors from the

Table 8.1 An example of a set of calculated properties (reproduced with permission from reference 41).

Calculated Property Set (81 parameters, 79 compounds)	
Whole Molecule Properties	
"Bulk" descriptors	Mol. Wt., van der Waals volume, dead space volume, collision diameter, approach diameter, surface area, molar refraction
"Shape" descriptors	Moment of inertia in X, Y and Z axes; principal ellipsoid axes in X, Y and Z directions
Electronic and energy descriptors	Dipole moment; X, Y and Z components of dipole moment; energies (total, core-core repulsion and electronic)
Hydrophobicity descriptors	Log P
Substituent Properties	
For 2 substituents	Coordinates (X, Y and Z) of the centre, ellipsoid axes (X, Y and Z) of the substituent
Atom-centred Properties	
Electronic	Atom charges, nucleophilic and electrophilic super-delocalisability, for atoms 1–14
Shape	Inter-atomic distances between 6 pairs of heteroatoms

computational models of molecules. At first these programs were developed using in-house molecular modelling packages[38,39] but gradually became available in the commercial computational chemistry software. Table 8.1 gives an example of the type of parameters routinely used in these early QSAR studies with theoretical parameters.

One particular type of descriptor missing from this table is a measure of hydrogen bonding ability, either as hydrogen bond donors or hydrogen bond acceptors. A simple approach to this is to use an indicator variable to show the presence or absence of donors/acceptors, or for compounds with multiple functional groups, a count of donor/acceptor groups. There is also a large number of descriptors, such as the energies of molecular orbitals (HOMO/LUMO), which can be obtained from molecular orbital calculations, either semi-empirical or *ab initio*, and these are often included in such data sets. Raevsky has recently reviewed the parameterisation of hydrogen bonding in QSAR.[40]

Of course the use of such descriptors also posed a problem, which can best be described as "an embarrassment of riches." Previously, the molecular designer would choose just 2 or 3 properties and then (usually) try to fit a linear regression model. There might have been some decisions to make as to which particular descriptors to choose, but the choice was fairly limited and the construction of regression models also involved little or no choice. All of this changed with the more general use of theoretical properties, and the process of QSAR model construction soon involved many more different types of mathematical modelling techniques,[41,42,43] as described in Chapters 9 and 10. There also arose the problem of variable selection. It was recognised even before the widespread use of theoretical descriptors that there was a danger of chance correlations when a very large number of variables were considered for inclusion in regression models.[44] In fact, as discussed in the next chapter, this is a problem for any "supervised learning method",[45] but apart from the question of chance effects there is also the matter of how to go about choosing variables. Consideration of every possible combination of variables is sometimes not a practical proposition. Kubinyi, for example, in an examination of a well known data set consisting of 53 descriptors for 31 compounds, has pointed out that if each combination of the possible regression models (from all 1 term equations up to all possible 29 term equations[§]) were computed at a speed of 1 second each, it would take 227 million years to evaluate them all.[46] An early example of an objective strategy for variable selection involved elimination of variables which were involved in the largest number of pairwise correlations with other variables in the set.[47] This results in a data set with a simplified correlation structure but tends to retain quite a large number of descriptors. An alternative strategy, which is still an "unsupervised learning method" and thus less prone to potential chance effects, involves selection of the variables with the lowest multiple correlations with other properties in the set.[48] A review in *Reviews in Computational Chemistry* discusses this problem in some depth.[49]

8.6 1D, 2D and 3D Descriptors and Conformation

A property such as molecular weight is sometimes called a 1 dimensional (1D) descriptor, since it only requires knowledge of the molecular formula. Other 1D descriptors are simple quantities such as the sum of van der Waals volumes, the sum of polarisabilities, mean electronegativity and so on. All of these descriptors only require knowledge of the molecular formula and probably, but not necessarily, contain only a little useful information. The next class of descriptors are 2D and require knowledge of the connections between the atoms, and here we begin to see some "chemistry" as we characterise functional groups, rings, chains and so on. The topological descriptors described in Section 8.3 are obviously 2D descriptors, as are calculated values of log P, polar surface area, principal ellipsoid axis (length!) and so on. All of these properties contain useful information, as can be seen from the many successful models generated using them. It might be thought, however, that 2D descriptors are lacking a vital piece of information, and that is the 3D structure of the molecules involved. We know from studies of biological macromolecules such as enzymes and receptors how sensitive natural systems are to 3D structure and thus we might expect that 3D descriptors—that is to say, properties where knowledge of 3D structure is required in their calculation—would show enhanced performance over 2D structures. Oddly enough, that doesn't appear to be the case,[50] although this may be a demonstration that we haven't yet devised a way to code chemical information effectively.

Another consideration with respect to 3D structure is the matter of conformation. There was a brief mention of conformation in the description of the

[§]There are 7.16×10^{15} possible combinations.

Sterimol parameters. Here, the operative term was "reasonable" (meaning of course "likely", or low energy) in terms of bond angles, lengths, torsion angles and van der Waals interactions for example. So a reasonable conformation is used in order to define the bounding box around the substituent. Similarly, when structures are modelled using a computational chemistry package (molecular mechanics or quantum mechanics), a low energy conformation is computed for each compound and this is then the basis for the calculation of a range of physicochemical descriptors. An alternative strategy is to use some experimentally determined conformation, usually (but not necessarily) based on an X-ray structure determination, as a template for the construction of analogues. Both of these approaches make good sense but there can also be arguments against them; a low energy conformation may not be the one that binds to a biological target, for example, or an X-ray structure may not be a good representation of the conformation in solution. What both of these objections mean of course, is that the template may not be a biologically relevant one. So, can we find or define or describe the biologically relevant conformation? The answer to this deceptively simple sounding question is a rather unsatisfactory "probably not". The biologically relevant conformation may be a relatively high energy structure, with the energy needed to achieve this coming from the binding energy, only seen in the presence of the biological target. Alternatively, there may be different conformations for different analogues in a series as they adopt slightly different poses when binding to the same target. In other cases of course, there may be an obvious conformational requirement to fulfil some binding interactions or other mechanistic need, and here we can define the biological conformation.

Another aspect of conformational flexibility in general though is the effect this may have on the properties which are computed in order to describe molecular structures. As an example of this, consider the pyrethroid structure shown in Figure 8.3. There are five flexible torsion angles in the two side chains of the cyclopropane ring, although torsion 3 is effectively fixed as it has some double bond character in resonance with the carbonyl oxygen.

A series of analogues of these compounds were modelled using a template derived from the X-ray crystal structure of deltamethrin, a potent pyrethroid insecticide. A large set of molecular properties ($\sim$90) were computed from these modelled structures and QSAR models were fitted to both *in vitro* and *in vivo* experimental measurements of insecticidal activity.[51] The QSAR modelling involved first carrying out a factor analysis of the molecular properties, where it was shown that 8 factors with eigenvalues greater than 1 (a common test of "significance" in factor analysis) could be generated from the set. The following equations were obtained, where NT is an *in vitro* measure of neurotoxicity and KA (kill) and KDA (knockdown) are *in vivo* measurements[52]

$$NT = -0.71\mathrm{F1} - 0.5\mathrm{F2} + 1.4\mathrm{F4} + 1.1\mathrm{F6} - 0.5\mathrm{F7} - 0.9 \quad (8.10)$$

$$KA = -0.61\mathrm{F1} - 0.3\mathrm{F2} + 0.6\mathrm{F4} - 0.5\mathrm{F7} - 0.6 \quad (8.11)$$

$$KDA = -0.3\mathrm{F1} - 0.2\mathrm{F2} + 0.8\mathrm{F4} - 0.4\mathrm{F5} - 0.9 \quad (8.12)$$

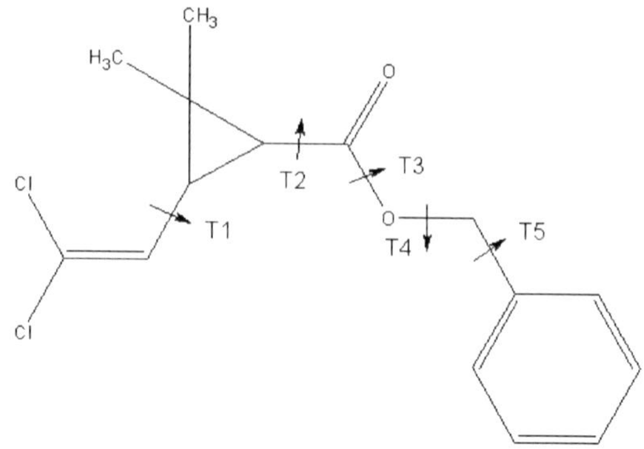

Figure 8.3 Pyrethroid parent structure showing flexible torsional angles marked by arrows.

Three factors (1, 2 and 4) are common in these three equations and it is tempting to imagine that these three factors are describing the interaction of the pyrethroid analogues with their molecular target, particularly since the signs of these terms in the equations are the same. Extending this sort of argument might then allow the identification of particularly important molecular properties (high loadings on factors) and other properties (loaded onto factors 5 and 7) which distinguish knockdown from kill. Caution, of course, must be used to avoid over-interpretation of empirical relationships. It is well known that correlation doesn't demonstrate causality, but it may inspire the design of experiments in order to test such notions.

The computer models of the compounds were analysed in a molecular dynamics simulation for 200 picoseconds, with values of the torsion angles calculated every 50 femtoseconds and a conformation recorded every 200 femtoseconds.[53] The resulting sets of 1000 conformations were analysed by cluster analysis, 25 representative conformations were selected and the molecular properties described were calculated for each conformation. These properties were averaged and a factor analysis conducted on the time averaged set. This yielded 9 "significant" factors, compared with 8 for the static set, and 3 of these corresponded closely to factors 1, 2 and 4 from that set. The molecular dynamics runs showed that the compounds existed mainly in one of two states, an extended conformation or a folded one. Regression analysis of the proportion of time spent in the folded state against the factors from the static property set showed a good correlation with two factors, F2 and F6. Repeating this analysis with the time averaged factors showed a better correlation with the same two factors, with very similar coefficients, and the "extra" factor 9. The conclusions from this study were that the choice of template structure (deltamethrin) was reasonable and that the calculation of properties based on a single conformation has general validity, but the incorporation of conformational flexibility information in this way is useful. Unfortunately, this process involves rather a lot of operator involvement but, as Lewis points out in his introduction to Chapter 4, proper use of molecular modelling still involves much interpretation.

8.7 Molecular Fields

Molecular field based descriptors are probably the ultimate 3D molecular properties since they are generated solely by consideration of the interactions of chemical probes with a 3D structure. The two most commonly used methods were devised at approximately the same time, albeit independently, and although they were intended to do quite different things, they are actually very similar. Comparative molecular field analysis (CoMFA) was designed at the outset to characterise small molecules,[54] whereas GRID was intended for the characterisation of macromolecular binding sites.[55] In each case, a grid of points is created in 3D space; for CoMFA, the grid forms a box which bounds the set of molecules under analysis; for GRID, the points overlay the region of the macromolecule being considered, usually (but not necessarily) a binding site. This is shown schematically in Figure 8.4, where a data matrix is built up from interactions on a 10X10X10 grid of points around a structure.

In the case of CoMFA, the steps involved in the procedure can be summarised as follows:

- Obtain a suitable 3D structure for each molecule in the training set.
- Derive partial atomic charges so that an electrostatic field can be generated.
- Align the molecules using some suitable alignment strategy (after conformational analysis, if required).

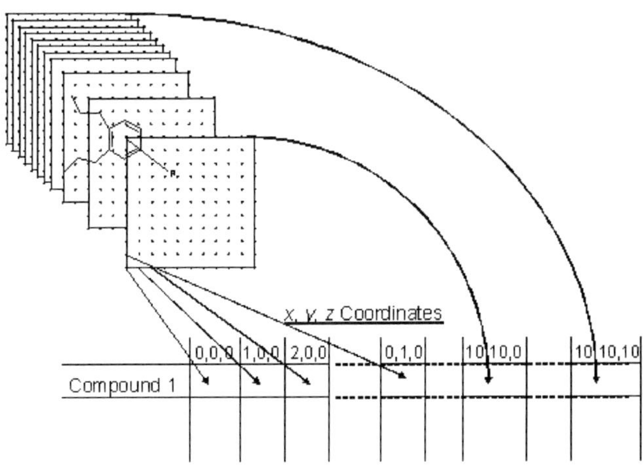

Figure 8.4 Illustration of the procedure for the production of a 10 × 10 × 10 matrix of molecular field values. Reproduced from reference 2, copyright 2000, American Chemical Society.

- Create a cubic lattice of points around the molecules (usually larger than the largest member of the set).
- Compute interaction energies using a probe, such as a pseudo methyl group, with a unit positive charge. This generates a steric interaction energy based on a Lennard-Jones potential and an electrostatic interaction energy based on a coulombic potential.
- Fit a PLS model to the biological response, using the interaction energies computed at individual grid points as the x-block variables.
- Make predictions for a test set and visualise the results by mapping the coefficient at each grid point variable back onto the 3-D grid, to create interaction contour plots.

This is a simplification of the whole process; a great deal of work has been involved in many of the steps listed above and debate, often quite vigorous, continues over a number of them. A review [56] of progress in CoMFA methodology lists four main areas where advances have been made:

1. Protocols for the alignment of compounds.
2. Introduction of new fields.
3. Variable selection techniques.
4. Statistical developments.

The most problematic of these areas is probably the alignment strategy, and it might even be argued that a common alignment is one reason why some CoMFA models give poor results since it has been shown that alternative binding modes exist for similar ligands.[57] Some published CoMFA studies which have used an X-ray conformation for alignment have been less successful than a reference conformation for the series.[58,59] A number of groups have tried to circumvent the "alignment problem" by using 3D fields in an alignment independent manner. For example, the ALMOND program uses GRID fields with autocorrelation transform to remove the need for alignment (GRID independent descriptors).[60] Topomer CoMFA decomposes the 3D structure into a series of superimposable fragments to remove the need for molecular alignment.[61]

Both CoMFA and GRID have proved popular, particularly as computer hardware has evolved and become more affordable, since they lend themselves well to visual analysis, which is particularly appealing to medicinal chemists. Indeed the progress of the CoMFA method in the early 1990s was remarkable; as Kubinyi points out, "the book *Quantitative Drug Design*,[62] comprising 766 pages and published in 1990, contains less than one page (!) on 3D QSAR methods related to CoMFA", and yet three years later he was the editor of an entire book devoted to the approach.[63]

These are not the only molecular field based methods; quite a large number of other variants have been developed[64–67] but there isn't space here to discuss them. There is one new development, however, which is qualitatively different to these approaches and that is a system called ParaSurf, which depends on a semi-empirical molecular orbital treatment of the problem, as opposed to these molecular mechanics force field approaches. The ParaSurf system calculates a variety of properties from the wavefunction at different positions on a surface, such as the van der Waals surface, and then uses these as molecular descriptors. Early applications of this approach in the derivation of QSAR models have shown some promise.[68]

Extracting molecular descriptors directly from observations from the drug-receptor interface determined from X-ray crystallographic information is the basis of the VALIDATE technique[69] This approach also forms the basis of empirical scoring functions as described in Chapter 16.

8.8 Fingerprints

Chemical fingerprints are not chemical properties as such, but are binary strings which code chemical structure information. The string consists of hundreds or, in some cases, thousands of bits, where a bit will be set to 1 to indicate the presence of some feature and 0 in the absence of this feature. The features may correspond to the sort of "natural" grouping that a chemist would recognise, such as a functional group or ring, or some other structural feature, or it may be a topological pattern such as a group of connected atoms or an atom and its neighbours. There are two main areas where chemical fingerprints are used: as a screening system in substructure searches in chemical databases (as described in Section 6.2.9 of Chapter 6) and as measures of chemical similarity. It is the latter application which is of interest in this section, since it is here that fingerprints most closely approximate to chemical properties. There are a variety of different schemes which may be used to generate the fingerprints, and the decision as to which scheme to choose will depend on their intended use. For example, Blomberg investigated the performance of 9 different types of fingerprints in retrieving active molecules from high throughput screening campaigns for 4 different protein classes. The selections of actives were complementary, guiding the authors to recommend a multi-fingerprint approach to balance the strengths and weaknesses of the different fingerprint methods.[70]

If the scheme includes redundancy, for example where small fragments are coded along with larger fragments that also contain them, then it is not so suitable for similarity measures since it will introduce bias, whereas this has little detrimental effect on substructure retrieval (Chapter 6, Section 6.2.9).

Since fingerprints are strings of binary bits, it can be argued that the "usual" measure of similarity, the euclidean distance, is not appropriate to compare molecules which are coded by fingerprints. Euclidean distance is our everyday measure of similarity, which we are accustomed to using when we compare objects in the 3D space of the physical world. In much the same way, two compounds characterised by any number of "regular" physicochemical properties (that is to say quantities expressed as continuous variables) can be compared using euclidean distance. In the case of fingerprints, where many of the bits may be set to "off", it is customary to use a measure of similarity called the Tanimoto coefficient. The chemometrics literature contains much discussion on this topic.

So, are fingerprints a better or more efficient way of expressing chemical similarity? Tiikainen and co-workers have compared the use of two different fingerprint techniques and two different molecular interaction field (MIF) methods in the characterisation of a set of small molecules tested in various cell lines.[71] The results require some consideration as this was a complex study, but in terms of the ability of the techniques to enrich biologically correlating compound pairs they found that the fingerprint methods outperformed the MIF techniques. Interestingly, they found that combining the two techniques always resulted in positive synergy. A similar study by Muchmore *et al.* using 10 different chemical similarity methods but using results based on protein targets reached similar conclusions.[72]

8.9 Which are the Best Properties to Use?

This is the $64 000 question. Unfortunately, there is no simple answer. It is not clear whether there is any "best" set and indeed it is very likely that the best properties to use are heavily dependent on the nature of the intended application. Surprisingly, little attention has been paid to this problem in the drug design community. Very often when a new way of describing molecules is invented or discovered, the authors will test this approach on one or two particular problems, but rarely is it compared with existing methods and even more rarely is this carried out in some systematic fashion. There are exceptions of course, as the examples of comparison of fingerprints showed at the end of the last section. Benigni and co-workers examined[73] a molecular eigenvalue descriptor in comparison with 5 other approaches for a set of 112 non-congeneric chemicals from a carcinogenicity database. This study showed, perhaps unsurprisingly, that the information coded by many types of descriptors is largely overlapping and thus they may be used more or less interchangeably. Stanton evaluated[74] an extension to the molecular eigenvalue descriptors (BCUT, see Section 8.3) and showed that they added extra information to a set of 105 topological, geometric and electronic parameters. This approach made use of an objective feature selection algorithm in order to produce an improved combined set. The logic behind this was that each set of properties would contain some useful information and that the use of the objective feature selection process would enable the extraction of informative subsets. In a similar approach, a study of 90 diverse mutagens reported[75] that superior classification functions could be built using descriptors from 3 different sets, as shown in Table 8.2. The classification results shown here are impressive (mutagen activity was taken as active/inactive) and it can be seen that different combinations of all three types of descriptor produced useful classification functions.

So, the only firm conclusions that can be drawn from these various investigations of different descriptor types is that combinations can be good and that the choice of descriptor, including the selection of subsets and possible combinations, is a vital part of the overall model building process.

Table 8.2 Summary of discriminant analysis results for a set of 90 mutagens (reproduced with permission from reference 75).

ID Analysis	*Pooled data set*[a]	*No. of variables*[b]	*Classification function (% correct)*	*Jackknifed validation (% correct)*
Pool 1	EVA 342 SMC + WHIM SMC	6	92.2	90.0
Pool 2	EVA 342 PDRv2.0 + WHIM SMC	9	96.7	96.7
Pool 3	EVA 288 CORCHOP + WHIM SMC	9	97.8	95.6
Pool 4	EVA 342 SMC + TSAR 58	9	92.2	90.0
Pool 5	EVA 342 PDRv2.0 + TSAR 58	9	93.3	91.1
Pool 6	EVA 288 CORCHOP + TSAR 58	5	90.0	86.7
Pool 7	EVA 342 SMC + TSAR stand.	9	92.2	90.0
Pool 8	EVA 342 PDRv2.0 + TSAR stand.	8	93.3	91.1
Pool 9	EVA 288 CORCHOP + TSAR stand.	5	90.0	86.7
Pool 10	WHIM SMC + TSAR 58	8	97.8	94.4
Pool 11	WHIM SMC + TSAR stand.	8	94.4	92.2
Pool 12	EVA 342 SMC + WHIM SMC + TSAR 58	9	97.8	94.4
Pool 13	EVA 342 PDRv2.0 + WHIM SMC + TSAR 58	9	97.8	94.4
Pool 14	EVA 288 CORCHOP + WHIM SMC + TSAR 58	9	96.7	93.3
Pool 15	EVA 342 SMC + WHIM SMC + TSAR stand.	10	96.7	92.2
Pool 16	EVA 342 PDRv2.0 + WHIM SMC + TSAR stand.	10	96.7	92.2
Pool 17	EVA 288 CORCHOP + WHIM SMC + TSAR stand.	8	95.6	92.2

[a]EVA, WHIM and TSAR refer to variables computed using the EVA and WHIM methods and the structure activity program TSAR. CORCHOP, SMC and PDRv2.0 refer to 3 different variable selection methods.
[b]The number of variables included in the discriminant function, see ref. 75 for details.

Another problem with the choice of descriptors is accessibility. Some types of proprietary parameters are only available through the licensing of commercial software. There are, however, some web-based resources, such as Chembench (http://chembench.mml.unc.edu) and the Virtual Computational Chemistry Laboratory[76] (www.vcclab.org) which both provide not only descriptor calculation facilities, but also access to statistical analysis routines. The molecular descriptors website (www.moleculardescriptors.eu) and QSAR world websites (www.qsarworld.com/qsar-web-based-programs.php) also provide useful links to resources such as databases and programs.

8.10 Conclusions

We have come a long way since the earliest use of quantitative descriptions of chemical structure using experimental measurements such as solubility, boiling point and olive oil–water partition coefficients. As mentioned in the introduction, we now have access to thousands of different molecular descriptors and Todeschini and Consonni's handbook of molecular descriptors, which describes them in detail, has now grown to 2 volumes.[77] In 1869, in what may be argued to be one of the earliest publications[78] on QSAR, it was stated that the reason it wasn't possible to obtain an accurate definition of the function relating "physiological action" to "chemical constitution" was that it wasn't possible to describe chemical constitution with a sufficient degree of "definiteness". It doesn't appear that the situation has changed greatly today, even though we have an embarrassment of riches in terms of properties. Attempts have been made to understand the relationships between different types of descriptors and to try to get a "feel" for the nature of the information they contain. Attempts have also been made to create more useful sets of parameters by extracting appropriate subsets from different classes of variables. It seems that there is still more to do in this area.

References

1. R. Todeschini and V. Consonni, *Handbook of Molecular Descriptors*, Wiley-VCH, Weinheim, 2000.
2. D. J. Livingstone, *J. Chem. Inf. Comput. Sci.*, 2000, **40**, 195.
3. J. Shorter, in *Similarity Models in Organic Chemistry, Biochemistry and Related Fields*, ed. R. I. Zalewski, T. M. Krygowski and J. Shorter, Elsevier, Amsterdam, 1991, pp. 77–147.
4. C. G. Swain and E. C. Lupton, *J. Am. Chem. Soc.*, 1968, **90**, 4328.
5. C. G. Swain, S. H. Unger, N. R. Rosenquist and M. S. Swain, *J. Am. Chem. Soc.*, 1983, **105**, 492.
6. See for example pages 1989–1992, 1992–1997 and 1997–2001 of *J. Org. Chem.*, 1984, **49**.
7. R. C. Young, R. C. Mitchell, T. H. Brown, C. R. Ganellin. R. Griffiths, M. Jones, Griffiths, M. Jones, K. K. Rana, D. Saunders, I. R. Smith, N. E. Sore and T. J. Wilks, *J. Med. Chem.*, 1988, **31**, 656.
8. D. E. Leahy, P. J. Taylor and A. R. Wait, *Quant. Struct.-Act. Relat.*, 1989, **8**, 17.
9. Y.-Z. Da, K. Ito and H. Fujiwara, *J. Med. Chem.*, 1992, **35**, 3382.
10. R.W. Taft, *J. Am. Chem. Soc.*, 1952, **74**, 3120.
11. R.W. Taft, *Steric Effects in Organic Chemistry*, ed. M. S. Newman, Wiley, New York, 1956, pp. 556–675.
12. C. Hansch and A. Leo, *Substituent Constants for Correlation Analysis in Chemistry and Biology*, Wiley, New York, 1979, pp. 44–47.
13. A. Verloop, W. Hoogenstraaten and J. Tipker, in *Drug Design*, ed. E. J. Ariens, Academic Press, New York, 1976, vol. III, pp. 165–207.
14. L. B. Kier and L. H. Hall, *Molecular Connectivity in Chemistry and Drug Research*, Academic Press, New York, 1976.
15. L. B. Kier and L. H. Hall, *Molecular Connectivity in Structure–Activity Analysis*, Wiley, New York, 1986.

16. A. K. Saxena, *Quant. Struct.-Act. Relat.*, 1995, **14**, 31 & 142.
17. H. Kubinyi, *Quant. Struct.-Act. Relat.*, 1995, **14**, 149.
18. A. K. Saxena, *Quant. Struct.-Act. Relat.*, 1995, **14**, 150.
19. M. Charton, *J. Comp. Aided Mol. Des.*, 2003, **17**, 197.
20. M. Charton and B. I. Charton, *J. Comp. Aided Mol. Des*, 2003, **17**, 211.
21. L. H. Hall, and L. B. Kier, *J. Chem. Inf. Comput. Sci.*, 1995, **35**, 1039.
22. J. J. Huuskonen, D. J. Livingstone and I. V. Tetko, *J. Chem. Inf. Comput. Sci.*, 2000, **40**, 947.
23. F. R. Burden, *Quant. Struct.-Act. Relat*, 1997, **16**, 309.
24. R. S. Pearlman and K. M. Smith, in *3D QSAR in Drug Design*, ed. H. Kubinyi, G. Folkers and Y. C. Martin, Kluwer/ESCOM, Dordrecht, 1998, vol. 2, pp. 339–354.
25. B. J. Richardson, *Medical Times and Gazette*, 1868, **2**, 703.
26. M. C. Richet, *Compt. Rend. Soc. Biol.*, 1893, **45**, 775.
27. J. K. Seydel, H. Ahrens and W. Losert, *J. Med. Chem.*, 1975, **18**, 234.
28. A. Rastelli, P. G. De Benedetti and G. G. Battistuzzi, *J. Med. Chem.*, 1975, **18**, 963.
29. N. J. Turner and R. D. Battershell, *Contrib. Boyce Thompson Inst.*, 1969, **24**, 139.
30. W. J. Wechter, M. A. Johnson, C. M. Hall, D. T. Warner, A. E. Berger, A. H. Wenzel, D. T. Gish and G. L. Neil, *J. Med. Chem.*, 1975, **18**, 339.
31. W. Lyman, W. Reehl and D. Rosenblatt, *Handbook of Chemical Property Estimation Methods*, McGraw-Hill, New York, 1982.
32. E. J. Baum, *Chemical Property Estimation: Theory and Application*, Lewis Publishers, Boca Raton, 1998.
33. R. S. Boethling and D. Mackay, *Handbook of Chemical Property Estimation Methods: Environmental and Health Sciences*, Lewis Publishers, Boca Raton, 1998.
34. S. H. Yalkovsky, and S. C. Valvani, *J. Pharm. Sci.*, 1980, **69**, 912.
35. S. H. Yalkovsky, S. C. Valvani, and T. J. Roseman, *J. Pharm. Sci.*, 1983, **72**, 866.
36. Y. Ran and S. H. Yalkowsky, *J. Chem. Inf. Comput. Sci.*, 2001, **41**, 354.
37. D. J. Livingstone, *Curr. Top. Med. Chem.*, 2003, **3**, 1171.
38. R. C. Glen and V. S. Rose, *J. Mol. Graphics*, 1987, **5**, 79.
39. D. J. Livingstone, D. A Evans and M. R. Saunders, *J. Chem. Soc.. Perkin Trans. 2*, 1992, 1545.
40. O. Raevsky, in *Methods and Principles in Medicinal Chemistry*, Wiley-VCH, Weinheim, 2008, vol. 37, pp. 127–154.
41. R. M. Hyde and D. J. Livingstone, *J. Comp.Aided Mol. Des.*, 1988, **2**, 145.
42. B. Hudson, D. J. Livingstone and E. Rahr, *J.Comp.Aided.Mol.Des.*, 1988, **3**, 55.
43. D. Livingstone, in *Molecular Design and Modeling: Concepts and Applications*, ed. J. J. Langone, Academic Press, New York, 1991, vol. 203, pp. 613–638.
44. J. G. Topliss and R. P. Edwards, *J. Med. Chem.*, 1979, **22**, 1238.

45. D. Livingstone, *A Practical Guide to Scientific Data Analysis*, Wiley, New York, 2009.
46. H. Kubinyi, *Quant. Struct.-Act. Relat.*, 1994, **13**, 285.
47. D. J. Livingstone and E. Rahr, *Quant.Struct- Act.Relat.*, 1989**, 8**, 103.
48. D. C. Whitley, M. G. Ford and D. J. Livingstone, *J. Chem. Inf. Comput. Sci.*, 2000, **40**, 1160.
49. D. J. Livingstone and D. W. Salt, in *Reviews in Computational Chemistry*, ed. K. Lipkowitz, R. Larter and T. R. Cundari, Wiley-VCH, Weinheim, 2005, vol. 21, pp. 287–348.
50. R. D. Brown and Y. C. Martin, *J. Chem. Inf. Comput. Sci.*, 1997, **37**, 1.
51. M. G. Ford, R. Greenwood, C.R. Turner, B. Hudson and D. J. Livingstone, *Pestic. Sci.*, 1989, **27**, 305.
52. D. J. Livingstone, B. D. Hudson, A. R. George and M. G. Ford, in *QSAR: Rational Approaches to the Design of Bioactive Compounds*, ed. C. Silipo and A. Vittoria, Elsevier, Amsterdam, 1991, pp. 557–560.
53. B. D. Hudson, A. R. George, M. G. Ford and D. J. Livingstone, *J. Comp.-Aid. Mol. Design*, 1992, **6**, 191.
54. R. D. Cramer, D. E. Patterson and J. D. Bunce, *J. Am. Chem. Soc.*, 1988, **110**, 5959.
55. P. J. Goodford, *J. Med. Chem.*, 1985**, 28**, 849.
56. U. Norinder, *Perspect. Drug Discovery Des.*, 1998, **12/13/14**, 25.
57. C. Mattos and D. Ringe, in *3D QSAR in Drug Design: Theory Methods and Applications*, ed. H. Kubinyi, ESCOM, Leiden, 1993, pp. 226–254.
58. A. J. Tervo, T. H. Nyroenen, T. Roenkkoe and A. Poso, *J. Chem. Inf. Comput. Sci.*, 2004, **44**, 807.
59. R. D. Clark, *J. Comp. Aided Mol. Des.*, 2007, **22**, 507.
60. M. Pastor, G. Cruciani, I. McLay, S. Pickett and S. Clementi, *J. Med. Chem.*, 2000, **43**, 3233.
61. R. D. Cramer, *J. Med. Chem.*, 2003, **46**, 374.
62. C. A. Ramsden, in *Comprehensive Medicinal Chemistry: The Rational Design, Mechanistic Study & Therapeutic Application of Chemical Compounds*, ed. C. Hansch, P. G. Sammes and J. B. Taylor, Pergamon Press, Oxford, 1990, vol. 4 (Quantitative Drug Design).
63. H. Kubinyi, Ed., *3D QSAR in Drug Design: Theory Methods and Applications*, ESCOM, Leiden, 1993.
64. G. Klebe, U. Abraham and T. Metzner, *J. Med. Chem.*, 1994, **37**, 4130.
65. B. D. Silverman and D. E. Platt, *J. Med. Chem.*, 1996, **39**, 2129.
66. R. Bursi, T. Dao, T. van Wijk, M. de Gooyer, E. Kellenbach and P. Verwer, *J. Chem. Inf. Comput. Sci.*, 1999, **39**, 861.
67. A. M. Doweyko, *J. Med. Chem.*, 1988, **31**, 1396.
68. D. J. Livingstone, T. Clark, M. G. Ford, B. D. Hudson and D. C. Whitley, *SAR QSAR Environ. Res.*, 2008, **19**, 285.
69. R. D. Head, M. L. Smythe, T. I. Oprea, C. L. Waller, S. M. Green and G. R. Marshall, *J. Am. Chem. Soc.*, 1996, **118**, 3959.
70. T. Kogej, O. Engkvist, N. Blomberg, and S. Muresan, *J. Chem. Inf. Model.* **2006,** *46*, 1201–1213.

71. P. Tiikkainen, A. Poso and O. Kallioniemi, *J. Comp.Aided Mol. Des.*, 2009, **23**, 227.
72. S.W. Muchmore, D. A. Debe, J. T. Metz, S. P. Brown, Y. C. Martin and P. J. Hajduk, *J. Chem. Inf. Model*, 2008, **48**, 941.
73. R. Benigni, L. Passerini, A. Pino and A Giuliani, *Quant. Struct- Act. Relat.*, 1999, **18**, 449.
74. D. T. Stanton, *J. Chem. Inf. Comput. Sci.*, 1999, **39**, 11.
75. D. J. Livingstone, R. Greenwood, R. Rees and M. D. Smith, *SAR QSAR Environ. Res.*, 2002, **13**, 21.
76. I. V. Tetko, J. Gasteiger, R. Todeschini, A. Mauri, D. J. Livingstone, P. Ertl, V. A. Palyulin, E.V. Radchenko, N. S. Zefirov, A. S. Makarenko, V. Y. Tanchuk, and V. V. Prokopenko, *J. Comp. Aided Mol. Des.*, 2005, **19**, 453.
77. R. Todeschini and V. Consonni, in *Methods and Principles in Medicinal Chemistry*, ed. R. Mannhold, H. Kubinyi and G. Folkers, Wiley-VCH, Weinheim, 2009, vol. 41.
78. A. Crum Brown and T. Frazer, *Trans. R. Soc. Edinburgh*, 1868-9, **25**, 151.

CHAPTER 9

Assessing Quantitative Model Quality and Performance

ANDREW M. DAVIS

AstraZeneca Respiratory and Inflammation Innovative Medicines, Pepparedsleden 1, 431 50 Mölndal, Sweden

9.1 Introduction

A random search through structure space is unlikely to yield a drug candidate. Conservative estimates of the potential number of synthesisable drug-like molecules are greater than the number of atoms in the universe. The use of quantitative predictive methods is necessary to reduce the search-space to a manageable number of synthesisable molecular targets within a drug discovery programme. A quantitative approach can be taken with many methods, to help answer the key question, which molecule should I make next?

Quantitative models used to predict chemical and biological properties can be divided into two fundamental types, physics-based models and empirical models.

9.1.1 Physics-Based and Empirical Models

Physics-based models are derived from first principles and attempt to capture the underlying physics of the system as a mathematically accurate description. Physics-based models are more or less precise, depending on the degree to which the systems are understood and coded within the mathematical description. If the physics of the property being modelled are completely

RSC Drug Discovery Series No. 13
Drug Design Strategies: Quantitative Approaches
Edited by David J. Livingstone and Andrew M. Davis

Published by the Royal Society of Chemistry, www.rsc.org

understood, all properties of the system should be predictable for all future molecules. Examples of physics-based models are quantum mechanical descriptions of a chemical system (or approximations thereof), estimates of affinity based on thermodynamically defined scoring functions, or mechanism-based mathematical models. The Schrodinger equation is a complete mathematical description of the hydrogen atom. Quantum mechanical approaches attempt to extend this complete solution to more complicated chemical systems by making approximations, which introduces error and uncertainty to the models and hence predictions.

Empirical models are based on observation. No prior assumption is made of the physics of the system. A set of data is used to train a model and a learner is used to find patterns within the data structure. The learner may be a human being or, more likely, a mathematical algorithm. The assumption is that the patterns identified in the training set can be extrapolated to a separate set of compounds which were not part of the training phase. The degree to which the model can describe the underlying physics of the property being modelled controls the ability of the model to generalise its predictions. Examples of empirical models would be quantitative structure activity (QSAR) models and expert systems.

The empirical model is limited by the amount of information encoded by the data in the training set. Hence the performance of most empirical models will depend upon the following:

- how much of the signal in the training set data can be modelled by the learner (fit of the model)
- the degree to which the training phase has modelled noise that cannot be extrapolated
- the degree to which the model has identified a signal within the training set that can be extrapolated to future compounds
- the similarity of the compounds being predicted to those in the training set

Although there has been extensive research into different machine learning algorithms to find the signal in the first instance, there is an even larger body of research devoted to understanding how to estimate the model's predictive performance.

The distinction between physics-based and empirically-based models concerns confidence. Physics-based models are characterised by our confidence in the mathematical description of the system. Empirical models are characterised by a lack of confidence in our ability to extrapolate a pattern observed in the training set to a relevant test set. The literature abounds with examples of over-fitted models, over-optimistic assessments of model quality, inappropriate use of statistical testing and models that are no more significant than random chance.[1,2] Even so, the promise of empirical models, such as QSAR models, to guide compound design and testing, means this is an increasingly important area of research and justifies continued effort and focus.

9.1.2 OECD Guidelines on QSAR Model Generation and Validation

The opportunity to use quantitative predictive models in the reduction, refinement and replacement of animal toxicity testing has stimulated regulatory authorities to issue guidance on QSAR model generation and validation. In 2004, the Organisation for Economic Cooperation and Development (OECD) agreed principles for model validation for QSARs used in assessing chemical safety. In 2007, the OECD published a "Guidance Document on the Validation of (Q)SAR Models", with the aim of providing guidance on how specific QSAR models can be evaluated with respect to the OECD principles. The guidance document defines qualitative structure activity relationships as derived from non-continious data, and quantitative structure activity relationships as derived from continuous activity data, and refer to them collectively as (Q)SARs. In this review we refer to them both as QSARs, and assign the "Q" to a quantitative description of chemical structure. The OECD has since developed a toolkit for the validation of QSAR models[3] to increase the regulatory acceptance of QSAR methods. The OECD has started the development of a "QSAR Toolbox" to make QSAR technology readily accessible and transparent, and less demanding in terms of infrastructure costs. The OECD guidance document recommends that QSAR models should be associated with the following:

1. A defined endpoint
2. An unambiguous algorithm
3. Appropriate measures of goodness of fit, robustness and predictivity
4. A defined domain of applicability
5. Ideally, a mechanistic interpretation

This document is recommended reading for all quantitative method builders and users. All guidance documents and materials generated by the OECD QSAR project to date are available free of charge on the internet.[4] The OECD principles for the Validation, for Regulatory Purposes, of (Q)SAR Models are described in more detail in Chapter 17.

In this chapter we will follow the structure suggested by the OECD recommendations. We will review the development of measures of goodness of fit, robustness, assessment of predictability for continuous models and models for classified endpoints, and the definition of domains of applicability. The aim is not to provide a comprehensive summary of all papers in this field, but to provide key references that have influenced the author's continuing journey to becoming a better QSAR scientist.

9.2 Goodness of fit

9.2.1 Pearsons Product Moment Correlation coefficient (r)

Francis Galton, in his empirical analysis of the dependence of sweetpea seed weights and plant heights on the properties of the parents, laid down the principles of correlation and regression. He coined the term "regression to the mean" to explain the observation that extreme heights of tall plants were not completely passed onto their offspring, which tended towards the average height of sweetpea plants. Inspired by Galton's observations, Karl Pearson developed a rigorous mathematical treatment of correlation and regression. Pearson's correlation coefficient (r) and its square, the coefficient of variation (r^2), are still the most widely used model quality indicators.

The r value is commonly employed to quantify the degree of association between predicted values (from either a physics-based or empirical model) and observed values from eqn (9.1). The endpoints could be as diverse as estimates of affinity from 3-dimensional protein–ligand complexes, to estimates of solubility from a quantitative structure–activity model. The coefficient of variation (r^2) expresses the fraction of the variation in the observed values that is explained by the predicted values, or more generally the fraction of the variation in the *y*-data that is explained by the *x*-data.

$$r = \frac{\sum x_i y_i - n\bar{x}\bar{y}}{n s_x s_y} \tag{9.1}$$

It is well recognised that a high r or r^2 value does not necessarily imply a linear relation between observed and predicted values. The magnitude of r or r^2 does not necessarily indicate a useful model for prediction, and a low r^2 does not necessarily indicate a model with no utility. Linear regression is a fundamental technique in analytical sciences. Sonnergaard pointed out in 2008 that "it appears every new generation of pharmaceutical scientists needs a brush-up regarding the pitfalls in statistical treatment of paired data and in evaluating results". He studied 23 papers published in *J. Pharm. Sci.* in 2005 and found that 8 of them used the correlation coefficient in a misleading, spurious or inappropriate way.[5] Common errors, originally highlighted by Anscome, include assuming that a linear model exists between the dependent and independent variables, and a single outlier driving a high correlation coefficient.[6] The International Conference on Harmonisation of Technical Requirements for Registration of Pharmaceuticals for Human Use (ICH) guideline on validation of analytical procedures recommends that evaluation of linearity is based on the correlation coefficient, *y*-intercept, slope of the regression line and residual sums of squares. Furthermore, a plot of the data and residuals should also be analysed. With problems in the application of simple linear models in the analytical sciences (still existing more than 80 years after the definition of linear regression) it may not be surprising that the complexity of multivariate regression methods also provides challenges.

9.2.2 The Problem of Using r/r^2 to Compare Models

The value of r or r^2 is often used to compare the performance of different models or modelling methods. However, it is not widely recognised that the performance of different models can only be directly compared using the r or r^2 value, if the training sets have identical variances. This is because the r or r^2 value is dependent upon the standard deviation of the x-data, relative to the unexplained variance of the model, and the slope of the regression line. The sample correlation r is an unbiased estimate of the population correlation coefficient only if the sample x standard deviation is identical (or at least comparable) to the population x standard deviation.

The problem was simply illustrated by Lee in 1992, reproduced here.[7] Lee generated a simple correlation using the following approach: from a set of 300 random 4 digit numbers, the first two digits were used to generate the x-values, and the y-values were generated as the difference between the first two digits and the last two digits of each random number, added to the x-values, plus a constant of 50. In this way, because y is part of x, the 300 pairs of numbers are correlated, with a theoretical r value of 0.7. Next, Lee generated two subsets of 150 pairs of points, each from of the population of 300 data pairs, based on low and high values of x. The correlations are reproduced in Figure 9.1, with regression statistics shown in Table 9.1.

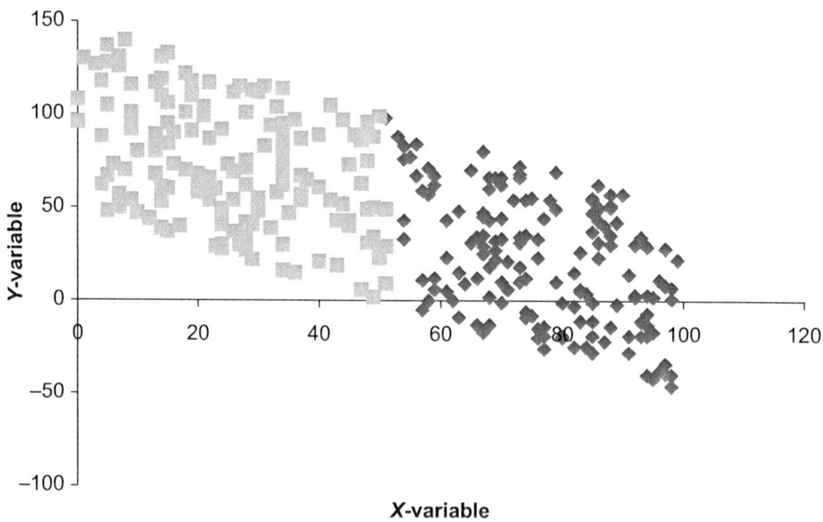

Figure 9.1 300 data-point population, with 150 data-point sub-populations, the diagnostics statistics of which are shown in Table 9.1.

Table 9.1 Diagnostic statistics of the 300 datapoint population, and the two 150 data point sub-populations from Figure 9.1.

n	*mean*	s	r
300	47.08	28.88	−0.712
150	73.39189	14.12	−0.417
150	21.4106	12.85	−0.476

Even though the two sub-datasets are drawn from the same overall population, the r values for the sub-correlations are lower than r value for the overall correlation, as the standard deviations for the sub-datasets are lower than the population standard deviation. The r value can be a useful comparative statistic if comparing the performance of different modelling approaches on a common dataset.

9.2.3 The Use of Root Mean Square Error in Fit and Prediction

A different metric used to compare models with different ranges/standard deviations of the y value, is the root mean square error in prediction (RMSE/RMSEP).

$$RMSE = \sqrt{\frac{\sum_{i=1}^{n}\left(y_{obs} - y_{pred}\right)^2}{n}} \tag{9.2}$$

A lower RMSE would indicate a better model. The advantage of RMSE is that it expresses the residual error of the model in the original units of the dependent variable, and it can be directly compared to the experimental error in determining the quality of the prediction. As a quality measure, it is independent of the variance of the test compound dataset. It is a measure that can be used to compare different models, even if based on different datasets with different variances, unlike r and r^2.

9.3 Model Robustness

Estimating the likelihood that the empirical model could not have arisen by chance is the primary concern of model builders and users, and describes model robustness. The central role of statistics in the derivation of empirical models is to quantify and provide confidence in the model. To provide an estimate of confidence that the observed model it is not a consequence of a chance relationship identified in the training set.

9.3.1 Parametric Statistics in Linear and Multiple Regression

Parametric statistics make assumptions about the type of probability distribution data is taken from, and make inferences about the parameters of that distribution. An example assumption is that data are random samples from a normally distributed population. With the sample mean and standard deviation known, the probability of any other value occurring can be calculated, assuming the new data comes from the same distribution. A second example assumption is that the ratio of variances, of two data samples randomly taken from two independent normally distributed populations with identical variances, follow a F-distribution. Knowing the variance of the two samples, the probability that the two samples come from distributions of different variance can be calculated. Linear regression and multiple regression methods rely on traditional parametric statistics. The application of these statistics methods to drug design can lead to over-optimistic estimates of model quality, as was first highlighted by Topliss in 1972.[8] For example, Livingstone and Salt demonstrated the weakness of tabulated F-statistics in a forward all subset multiple linear regression algorithm as the number of variables increase relative to the number of cases in the training set.[9] They demonstrated, that even for moderate sized datasets of 20 cases with 50 variables, where a 3-variable MLR model is selected from 50 variables, most MLR algorithms will almost certainly report a significant correlation if standard tabulated F-ratios are used. They proposed an adjustment of the F-statistic to prevent an over-optimistic assessment of model significance. An interpolation formula was presented to allow the standard F-tables to be used for datasets of varying numbers of cases and variables. Due to the computational limitations of the exhaustive search for all subset regression used, the analysis was constrained to small datasets. The analysis has now been extended to produce 90, 95 and 99% deciles of the F-distribution for stepping multiple linear regression, up to datasets of sizes commonly found in QSAR applications.[10]

The problem of over-optimistic estimates of model quality is a general one for all multivariate regression methods, and a number of model quality diagnostics have been developed that do not rely on parametric assumptions, to both limit the model fitting process and to assess the ability of the model to generalise beyond the training set.

9.3.2 q^2 and Q^2 Statistic

An alternative approach to using traditional parametric statistical methods to calculate the significance of fitted correlations, would be to directly assess the model based on its ability to predict, rather than merely to assess how well the model fits the training set. When the quality of the model is assessed by the prediction of a test set, rather than the fit of the model to its training set, a statistic related to r or r^2 can be defined, and denoted q or q^2,to indicate that the quality measure is assessed in prediction. A q^2 may be calculated by internal cross-validation techniques, or by the quality of predictions of an independent test set, in which case an upper-case Q^2 is used. The equation to calculate q^2 (or Q^2) is shown in equation 9.3.

Cross-validation is an internal validation technique, where a number of cases from the training set are left out during the model fitting process and then predicted from the resulting model. Multiple fitted models are generated so that each case is left out of one of the models and predicted. Leave-one-out cross-validation is the simplest and least challenging sampling validation and has been widely criticised.[11] Leaving out compounds in multiple groups is a tougher test for the model.

$$q^2 = 1 - \frac{\sum \left(y_{obs} - y_{pred}\right)^2}{\sum \left(y_{obs} - \bar{y}\right)^2} \tag{9.3}$$

Cross-validation can only give an unbiased estimate of future predictive ability if future compounds come from the same population as those in the cross-validation sample. This is not a safe assumption as the nature of drug design is such that compounds evolve with time.

Re-sampling techniques can also be applied to provide standard errors and confidence intervals of model parameters such as regression coefficients, and variable weights. Jack-knifing is related to leave-one-out cross-validation, while bootstrapping methods involve multiple re-sampling with replacement. Bootstrapping procedures can provide confidence in the importance and stability of overall variables in the model and, from a compound point of view, stability in the predictions.[12] For instance, the GOLPE procedure (Generating Optimal Linear PLS Estimations) uses the leave-n-out approach to assess the predictivity of a PLS model. As the n-groups can be formed in many ways, GOLPE re-samples the cross-validation procedure to provide confidence on the stability of the standard deviation error in prediction.[13] Bootstrapping, to assess the confidence in the prediction of a particular compound is related to using the dispersion of a prediction from ensemble-based methods, to assess prediction confidence, as discussed later.

9.3.3 *y*-Randomisation

While internal validation using cross-validation is an attempt to assess the ability of the model to generalise, additional confidence that the empirical model is not just a chance correlation found within the training set is often sought by using *y*-randomisation. The *y*-randomisation procedure compares the performance of models built for a randomly shuffled *y*-response. When repeated multiple times, the distribution of r^2 values obtained is a non-parametric method for assessing the significance of the original model. It has been referred to as "probably the most powerful validation procedure",[14] although only if the full model generation procedure, including variable selection, is repeated, rather than just permuting the *y*-variables of the final model.[15]

Rücker and co-workers have compared several variants of *y*-randomisation—including original response, permuted response and random pseudo-descriptor response—together with original descriptors and replacing original descriptors with random pseudo-descriptors for multiple linear regression models, using datasets taken from the literature. They consistently found that replacing the original descriptor set with random number descriptors (pseudo-descriptors) gave higher r^2 values than using the original descriptor set with random *y* or permuted *y*. This was attributed to the intercorrelation in the original descriptor set reducing the effective dimensionality of the descriptor set.[16] They suggest that a model r^2 should exceed both permuted *y* and random pseudo-descriptor r^2 in order to give confidence in the significance of the model. Where the observed r^2 falls between these two values, they also recommend that the significance of the original models should be questioned. They suggest that ignoring the higher r^2 achieved by using ransom pseudo-descriptors allots a decisive importance to the intercorrelation structure of the descriptor pool. Their explanation is that any set of descriptors are a more or less an arbitrary selection from a set of all possible descriptors, therefore we should not allot any fundamental importance to the intercorrelation structure within a given descriptor pool. This view could be challenged, as the size of the space generated by these random scales must be compatible with the size of the actual descriptor space[17] in order to represent a fair statistical test.

For example, the well-known Selwood dataset contained 53 highly intercorrelated descriptors, which were narrowed down in the original publication to 23 weakly intercorrelated descriptors by removing one descriptor from the highest intercorrelated pairwise comparisons within the original descriptor set.[18] From these, the 10 descriptors with the highest correlation with the response were selected and from these, the best 1, 2, and 3 descriptor models were selected. The best 2 variable model reported had an r^2 of 0.74. The randomisation experiments produced the best 2 variable model with the original y and 53 random pseudo-descriptors $r^2 = 0.62$, but with the randomised y and the original 53 descriptors this was only $r^2 = 0.53$. This difference between the reported model ($r^2 = 0.74$) and the randomised y model ($r^2 = 0.53$) was considered significant, but not compared to the model obtained with 53 random pseudo-descriptors. The best 2 variable model using only 23 random pseudo-descriptors gave an r^2 of 0.50, similar to the permuted y model with the original 53 descriptors.

9.4 Assessment of Predictivity

9.4.1 Calculation of Q^2 Using an External Test Set

An external test set is arguably a more robust test of the model than cross-validation. Protagonists of cross-validation maintain that an external test set, especially for small datasets, unnecessarily limits the information available for modelling.[19] A number of authors have shown that q^2 from internal validation is not correlated to the Q^2 for an external prediction set. Tropsha stated that a high cross-validated q^2 is a necessary but not sufficient condition for a model to have predictive ablility.[20,21] Model builders have used a number of different methods to select the training set and external test set, with ever-more esoteric methods from simple random selection, through statistical designs, sphere exclusions, k-means and clustering, to self organising and Kohonen maps, amongst others.[22]

The OECD guidelines stipulate that Q^2 be calculated according to the following equation:

$$Q^2 = 1 - \frac{\sum \left(y_{obs} - y_{pred}\right)^2}{\sum \left(y_{obs} - \bar{y}_{train\ set}\right)^2} \tag{9.4}$$

Where the Q^2 for external validation should be calculated with the sum of squared deviations referring to the training set mean. The use of the training set mean is pragmatic as it provides a fixed reference value, enabling comparison of Q^2 for differing test sets. But this formalism has been recently been questioned as it can lead to an overoptimistic estimate of the predictive ability of the model, caused by a difference in the training set and prediction set means. Schuurmann recommended using the test set mean:[23]

$$Q^2 = 1 - \frac{\sum (y_{obs} - y_{pred})^2}{\sum (y_{obs} - \bar{y}_{test\ set})^2} \tag{9.5}$$

However, using the test set mean can significantly underestimate the predictive ability of a model. Ambiguity occurs with the Q^2 statistic when the test set data is not evenly distributed over the range of the training set. As the variance of the external test set approaches the RMSE of the fitted model, the Q^2 measure would approach zero, even though it would appear that the predictions are in accordance with the model. Consonni defined a new Q^2 statistic that expresses the mean predicted error sum of squared deviations between the observed and predicted values for the test set, over the mean training set sum of squared deviations from the mean value:

$$Q^2 = 1 - \frac{\left[\sum_{i=1}^{n_{ext}} (\hat{y}_i - y_i)^2\right] / n_{ext}}{\left[\sum_{i=1}^{n_{TR}} (y_i - \bar{y}_{TR})^2\right] / n_{TR}} = 1 - \frac{PRESS/n_{ext}}{TSS/n_{TR}} \tag{9.6}$$

n_{ext} is the number of compounds in external set
y_i hat is the prediction estimate of y
y_i is the observed value,
n_{TR} is the number of compound in the training set
$\bar{y}_{TR}$ is the mean of the training set observed values
PRESS is predicted error sum of squares deviations between the observed and measured y values
TSS is the training set sum of squared deviations from the mean

Consonni demonstrated that this formulation of Q^2 is stable with test sets of different variances.[24]

Figure 9.2 shows predicted *vs.* observed values for a trained model with fitted $r^2 = 0.87$, with the predictions for a test set overlaid on the training set model. Diagnostic statistics for this model are shown in Table 9.2. By inspection, one can observe that the predictions of the test set are completely in accordance with the model and would be considered good predictions, but because of their small range they would provide a low Q^2 by the definition given in Equation 9.4, as recommended by OECD guidelines. Whereas the Consonni method demonstrates that the test set is predicted in line with the original model. The reader can use the diagnostics statistics shown in Table 9.2 for the data illustrated in Figure 9.2 to reproduce the Q^2s calculated using Equations 9.4–9.6.

The comparison of the test set RMSE with the training set RMSE provides a direct, unbiased relative assessment of the predictivity of the model. In assessing the predictivity of an external test set it is also useful to evaluate both the

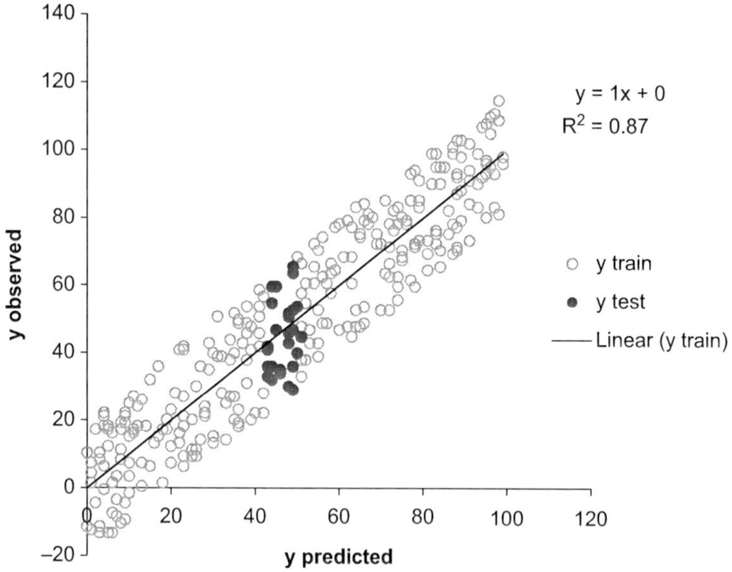

Figure 9.2 Simulated QSAR model with $n = 300$ data pairs fitted $r^2 = 0.87$. The model was used to make predictions for a test set of 25 compounds. The test set predictions are in accordance with the training set model performance, but are clustered over a small range of y.

mean absolute error along with the Q^2/ RMSE in the predictions. This indicates any systematic bias in the predictions. The mean absolute error is given by Equation (9.7).

$$mean\ absolute\ error = \frac{\sum_{i=1}^{n_{ext}} \left(y_{obs} - y_{pred}\right)}{n} \tag{9.7}$$

An external set of compounds may be predicted with a mean absolute error, which would inflate RMSE and decrease Q^2. A non-zero mean absolute error means the predictions may not be accurate, but may still be useful in ranking compounds. It may also indicate a structural or sub-structural feature present in the test set but not encoded in the training set or its description. Other authors recommend for accurate predictions that mean absolute error be low or zero.

9.4.2 Use of a Temporal Test Set

The ultimate test of an empirical model would be to test its ability to predict future data. After all, that is how we generally want to use the model, to predict future data. This simple philosophy makes the definition of test sets by random selection from training sets, or sampled from the training set using ever more sophisticated methods of clustering, rather inconsequential.

The temporal predictivity of 3 endpoints ($logD_{7.4}$, solubility in pH 7.4 buffer and human plasma protein binding) with 3 different statistical learners partial

Table 9.2 Diagnostic statistics for the data shown in Figure 9.2.

set	*n*	*Mean*	*sd*	r^2	Q^2 *equation 9.4*	Q^2 *equation 9.5*	Q^2 *equation 9.6*
training set	300.000	20.913	32.067	0.869			
test set	25.000	44.760	10.826	0.049	0.250	0.020	0.889

	$\sum_{i=1}^{n_{ext}} (\hat{y}_1 - y_i)^2$	$\sum (y_{obs} - \bar{y}_{tr\,set})^2$	$\sum (y_{obs} - \bar{y}_{test\,set})^2$
training set	297969.5246		
test set	2755.440323	3673.7478	2813.045513

least squares regression (PLS). Bayesian neural networks (BNN) and random forests][25] was studied with time. It was observed that predictivity of the models decreased as the test set grew more distant from the training set space in time, which is a surrogate for increasing chemical dissimilarity from the training set. Secondly, the assessment of which modelling method provided the best quality predictions changed with time. For example, with the human plasma protein binding endpoint, when predicting compounds close to the training set in time, the BNN model performed better than the PLS model—but as the test set became more distant in time from the training set, its performance decreased more rapidly than the PLS model—which tended to generalise better over longer periods of time. The random forests model maintained an improved predictivity over the other two methods throughout the timeframe of the study (Figure 9.3).

The quality of a model may change with time and different model learners may show different abilities to extrapolate forward with time. As David Clarke recently highlighted, *"There is a growing awareness that the validation of ADMET models should not be a 'one-off' process and that models need to be revalidated periodically to make sure their predictive power is not waning as new chemical classes emerge for prediction".*[26]

9.5 Classified Datasets

9.5.1 Cooper Statistics

Some endpoints cannot be measured on a continuous scale, or for scientific or regulatory purposes can only be divided into two or more qualitative classes, for instance active/inactive. Goodness of fit can be described by the Cooper statistics, introduced to assess the significance of carcinogenicity tests.[27] The goodness of fit parameters for a 2-level classification are shown in Table 9.3. This table is often, and very aptly, entitled a "confusion" matrix.

The Cooper statistics express the performance of a classification model, based on

- its ability to detect known active compounds (sensitivity)
- its ability to detect non-active compounds (specificity)
- its ability to correctly classify both classes (concordance or accuracy)
- the false positive and false negative rates can be calculated from the complement of specificity and sensitivity, respectively

The positive and negative "predictivities" focus on the effects on individual chemicals, since they act as conditional probabilities. Thus the positive "predictivity" is the probability that a chemical classified as active is really active, while the negative "predictivity" gives the probability that a classified non-active chemical is really non-active.

The focus on a particular Cooper statistic depends on how the model is to be used.

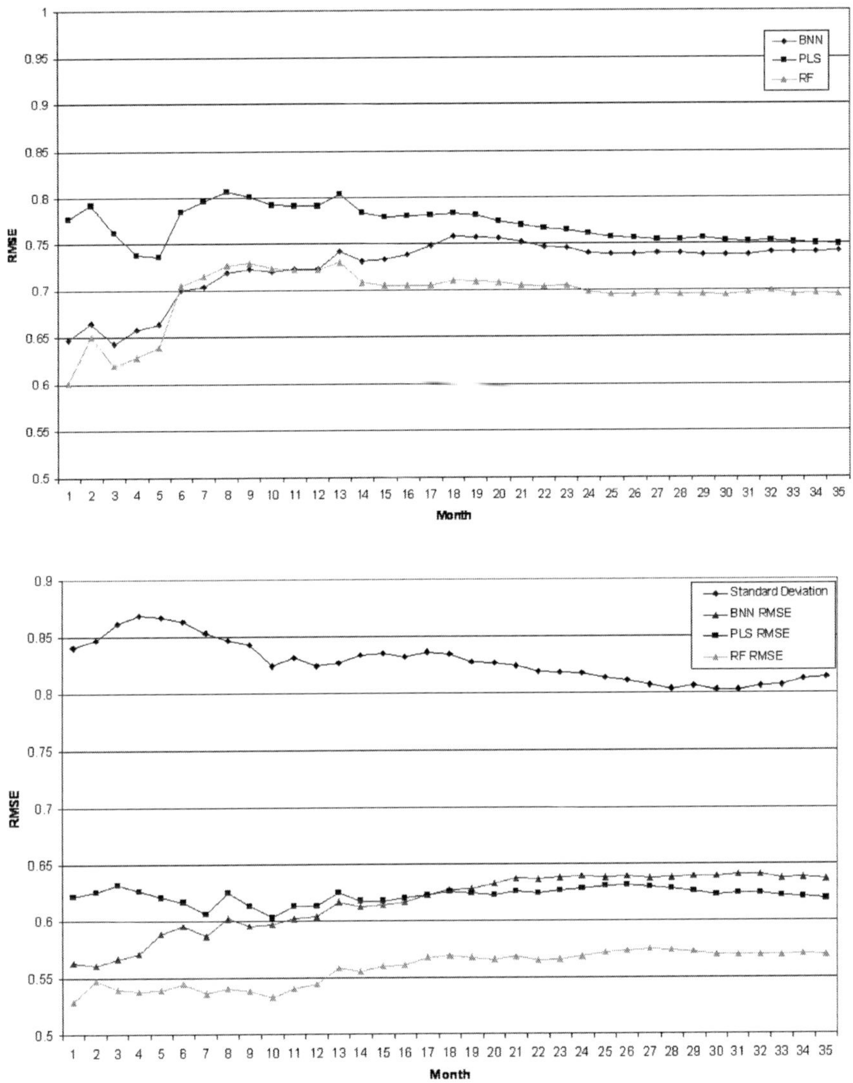

Figure 9.3 Initial model predictivity of cumulative monthly test sets. Human plasma protein binding (top), and lipophilicity log $D_{7.4}$ (bottom).

9.5.2 Prevalence in Classified Datasets

The Cooper statistics do not consider the prevalence within the training set, which will introduce a bias in the ability of the model to predict one or other class. For example, if the training set has 75% actives relative to inactives, the null probability will be three times as likely to predict a compound as active rather than an inactive compound. Cohen defined the kappa index to overcome the problem of prevalence when assessing the significance of classification

Table 9.3 2×2 contingency table and definition of the Cooper statistics.

		Predicted Class		
		Active	Not active	Totals
Observed class	Active	a	b	$a+b$
	Not active	c	d	$c+d$
	Totals	$a+c$	$b+d$	$a+b+c+d$

Statistic	*Formula*	*Definition*
Sensitivity(true positive rate)	$a/(a+b)$	Fractions of actives correctly assigned
Specificity (true negative rate)	$d/(c+d)$	Fraction of non-actives correctly assigned
Concordance or accuracy	$(a+d)/(a+b+c+d)$	Fraction of chemicals correctly assigned
Positive predictivity	$a/(a+c)$	Fraction of chemicals correctly assigned as active out all predicted actives
Negative predictivity	$d/(b+d)$	Fraction of compounds correctly assigned as not-active out all predicted not-actives
False positive (over classification rate)	$c/(c+d)$ *1-specificity*	Fraction of not-actives falsely classified as active
False positive (under classification rate)	$b/(a+b)$ *1-sensitivity*	Fraction of actives falsely classified as not-active

models, eqn (9.4). The kappa index is considered superior to overall correct classification as it considers the models improvement in prediction over chance alone.

$$\kappa = \frac{\textit{observed agreement} - \textit{chance agreement}}{\textit{total observed} - \textit{chance agreement}} \tag{9.4}$$

Koch and Landis have suggested ranges of the kappa value for significance,[28] where a kappa value of >0.2 is considered significant.

Randomisation procedures can be used to test significance in classified datasets, although this has not been widely investigated. Lusa and Blagus have investigated the effect of prevalence in high-dimensional datasets, such as microarray data, in classifying disease states, where classification methods are often employed.[29] They found that a number of common classifiers such as random forests and support vector machines are highly sensitive to the class imbalance of the training set, and that variable selection introduces an additional bias towards the prediction of the majority class. They found that using class-imbalance corrections, commonly embedded in some classifier methods such as random forests and support vector machines, does not correct the prediction accuracy for the minority class.

Downsizing, whereby the majority class from the training set is sampled to produce a balanced training set of actives and inactives, was found to worsen the predictive accuracy as it reduces the effective sample size considerably when the class imbalance is large. Multiple downsizing, whereby multiple models are produced with different samples of the majority class, was found to overcome the prediction variability and a modest improvement in prediction accuracy when compared to simple downsizing.

Multiple downsizing the training set to remove the class imbalance was found to be successful in improving the performance for the minority class, and the class specific prediction accuracy became similar for both classes, no matter what the class imbalance in the original training set. Multiple downsizing seems like a promising area for further research in developing predictive classification models.

9.5.3 Classification of Predictions from Continuous Models

Classification models are very popular in QSAR analyses, even when the original endpoint is measured on a continuous scale. This may be because it is feared that high variability in the endpoint may obscure any signal that the learner is hoping to find. A choice of separation value/cut value between the classes may be arbitrary (where an active/not active boundary may be placed), or conditioned by some selection or progression criteria (for instance a project progression criterion on Cytochrome P450 inhibition, or hERG inhibition). But the positioning of the cut-point imposes a degree of subjectivity upon the data and also removes any information on experimental variability in the *y*-data, which can be useful in assessing both the quality of the model and its predictions. In order to maintain the objectivity of the modelling approach and maintain information about the intrinsic variability of the endpoint, some researchers have preferred to build continuous models and then to classify the predictions made from those models.

Gleeson and co-workers used continuous regression methods to build Cytochrome-P450 models. In assessing the performance in an external prediction set, they used the RMSE of the fitted model to define an exclusion zone around the cut-off value where misclassification would be more likely, as illustrated in Figure 9.4.[30]

In using this fuzzy classification method, they were able to achieve very high classification success rates in both positives and negative classes, although at the expense of more compounds that were classified as indeterminate, as they were within the error of the model cut-point, Figure 9.5. The performance was similar in both high and low pIC50 classes, even though the prevalence of the datasets was biased towards inactive.

9.5.4 Receiver Operating Characteristic Curves

To compare the performance of a number of classification models, a receiver operating characteristic (ROC) curve can be used. These were first developed to correctly detect Japanese aircraft from their radar signatures after the attack on Pearl Harbour. ROC curves are widely used in evidence-based medicine. The *y*-axis is the sensitivity (true positive rate) and the *x*-axis is (1-specificity) the false positive rate. A perfect classification model would be a point in the top-left corner of the plot, and a random model would be a point on the positive diagonal line. In the case of continuous predictors, a ROC curve allows one to explore the relationship between sensitivity and specificity from different classifying thresholds, thus allowing the optimum threshold to be determined. An assessment of the goodness of the classification method is the area under the ROC curve.

ROC curves are commonly used in assessing the results of virtual screening experiments. Here, the x-axis effectively represents the rank of the active in the complete database. The area under the ROC curve can be used to compare the performance of different virtual screening approaches. The application of ROC curves to virtual screening has recently been questioned. Traditional ROC curves do not take into account the differences in inherent value in identifying early *vs.* late hits. It is suggested that using log(false positive rate) as the x-axis has a positive effect in emphasising the early part of the curve.[31]

9.6 Domain of Applicability

The OECD guidelines suggest the domain of applicability be assessed both in terms of the properties of the fitted model, and the structural similarity of the prediction compounds to the training set of the model.

A number of measures of different domains of applicability were noted in the OECD guidelines including:

- range-based methods (do the properties for the predicted compound fall within the range of the training set?)
- distance-based methods, either Euclidean or Mahalanobis distances from the centre of the training set, or Mahalanobis distance from the nearest neighbours in the training set. (The Euclidean distance to a fixed point in the training set space assumes that all variables contribute equally, *i.e.* the applicability domain is spherical. The Mahalanobis distance takes into consideration the intercorrelation structure of the predicting variables, hence the applicability domain is not spherical but ellipsoidal, driven by the intercorrelation structure.)
- parametric or non-parametric probability density based methods

No guidance was given on which should be applied and how they should be applied.

In order for the applicability domain to be prospectively useful in assessing the quality of the prediction, it should be demonstrated that the measure of distance of the prediction set to the training set is related to the error in prediction of the test set (see Figure 9.6). A lack of correlation may suggest an inappropriately defined applicability domain and distance measure.[32]

Gleeson,[32] based on work of Bruneau,[33] used an independent test set to define the relationship between the distance of test set compounds from the training set applicability domain and error in prediction. They investigated a number of different distance measures, including Euclidean and Mahalanobis

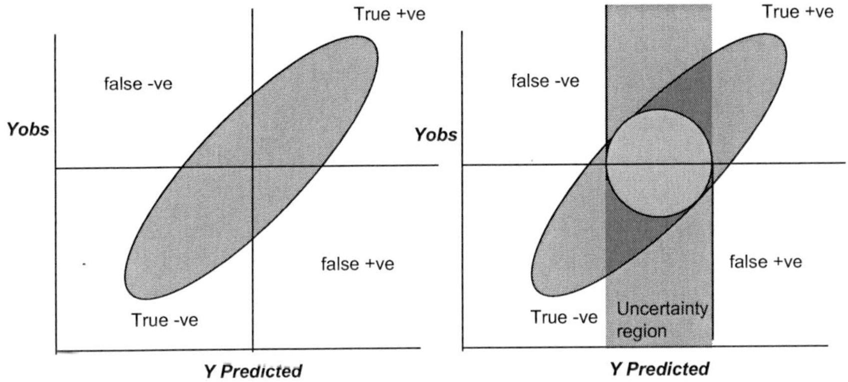

Figure 9.4 Illustration of how to use the error in the fitted model to define uncertainty regions for classification. The number of false positives and negatives are markedly reduced (portion of exposed elipse still visible in the top left and bottom right quadrant), increasing the confidence in a positive or negative prediction.

distances from training set centroid or near neighbours, and fingerprint distances defined by the Tanimoto index, but found that for a number of diverse datasets and statistical modelling methods the distance measures were correlated and gave similar relationships to error in prediction. Because of the often weak correlations between distance to the training set domain and the error in prediction, the test set distances were binned. To predict the likely error in prediction of a new compound, it was assigned the average error in prediction from the bin containing compounds at a similar distance. This was shown to be effective in both global and project specific models of plasma protein binding, and CYP P450 inhibition.

In a similar study, a group from Bayer Schering compared 4 different definitions of applicability domain for the prediction of aqueous solubility, based on an in-house dataset of 4000 compounds predicting a temporal test set of 632.[34] Two of the measures were based on fitting error assessments from ensemble methods (from random forests and Gaussian processes) and near neighbour based distance measures for ridge regression and support vector machines. The error-based assessment from ensemble models makes use of the information contained in the family of fitted models. The degree of confidence in different regions of the model space will be reflected in the dispersion of coefficients and predictions for compounds in the training set. (This technique has recently also been shown to be successful in analysing 30 QSAR models for AMES mutagenicity, submitted for the 2009 QSAR Challenge.[35]) The Bayer Schering group found that all 4 different descriptions of the domain of applicability were adequate to reject compounds for predictions outside the domain of applicability of the relevant model.

The domain of applicability based on the training set descriptors is a useful indicator with which to inform the model user of the likely confidence of

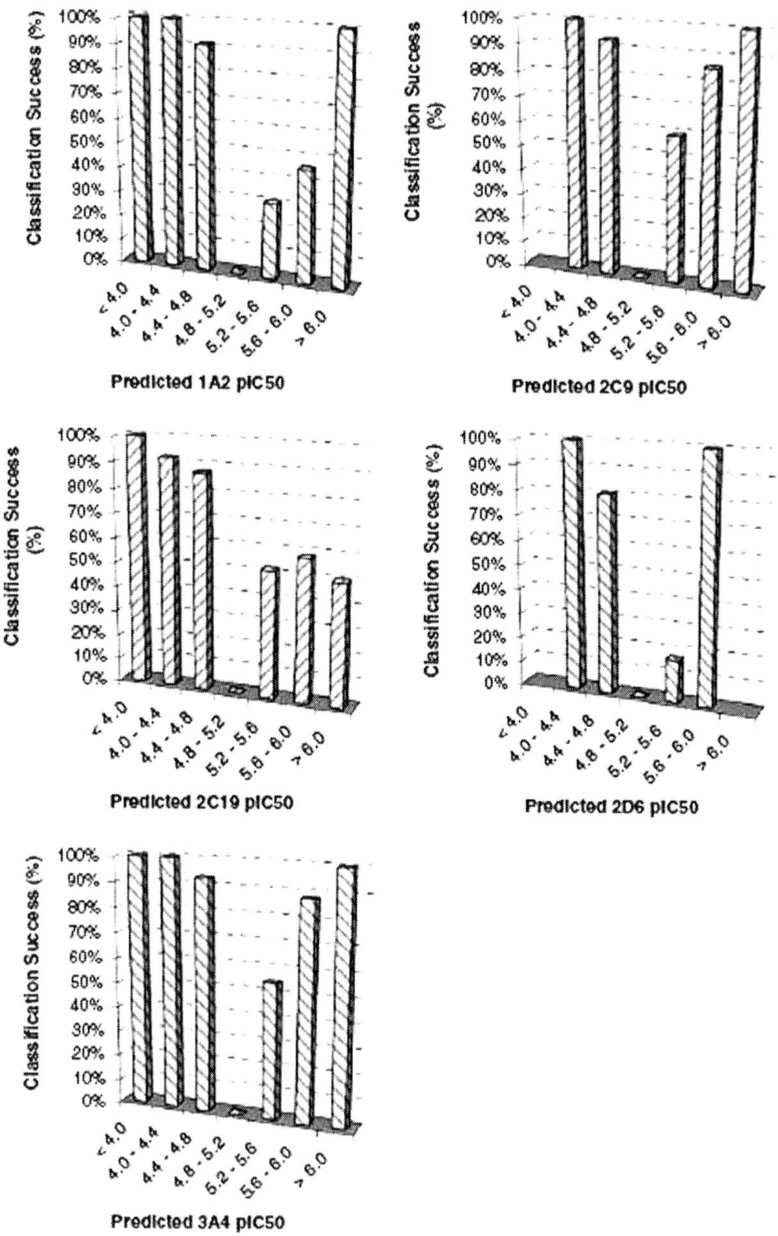

Figure 9.5 The performance of Gleeson's fuzzy classification on CYP P450 models. As the RMSE of CYP models was *ca.* 0.6 log units, with a cut-off of 10 μM between high and low classes, a prediction of pIC50 <4.4 or >5.6, assuming the errors are normally distributed, has only a 15.9% chance of being an incorrect classification. The results from the 5 CYP models support this. The compromise is an increased number of indeterminate classified compounds.

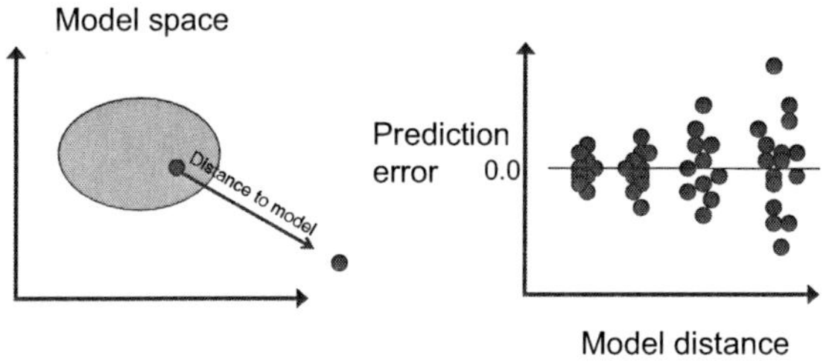

Figure 9.6 The hypothetical relationship between distance to the model space/domain of applicability and prediction error for a test set of compounds.

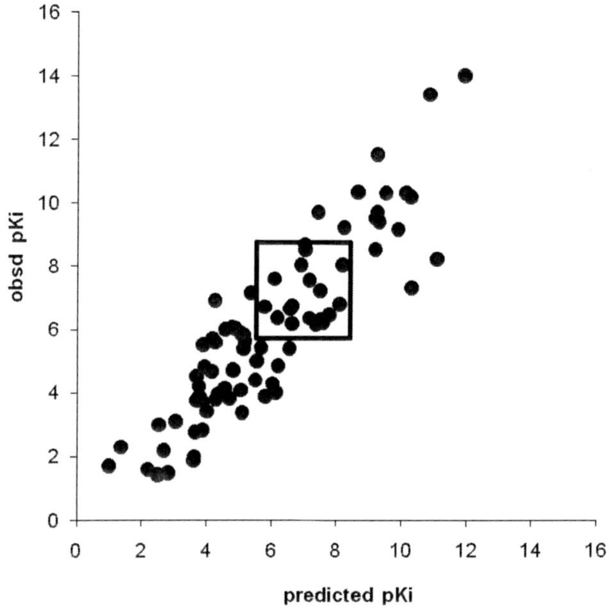

Figure 9.7 Correlation between observed pKi's and those predicted by the LUDI scoring function for its calibration set of 82 protein-ligand complexes. The black box illustrates the range of activity encountered in a drug discovery project from a chemical starting point to a candidate drug.

predicted compounds. This confidence can be calibrated from observation of the errors in prediction of similar compounds from an independent test set and can detect the degradation of models as the test set becomes further from the applicability domain.

But property-based methods may overlook structural features in a test molecule that are not encoded in the learned model, either because the descriptors fail to encode the structural feature in the training set, or the structural feature is unique to the test set compounds. Hence the OECD guidelines recommend consideration of the applicability domain both from a property-based approach and also based on a structural assessment.

9.7 Statistical *vs.* Practical Significance in Prediction

One of the difficulties in assessing the utility of predictive models, from the model builder's point of view, is that different users have different demands on a particular model. An empirical model may have statistical significance and a physics-based model may have true predictivity across many chemotypes, but still may not be practically useful. For example, Figure 9.8 illustrates a typical literature solubility model.[36] The r^2 value for this correlation of predicted *vs.* observed solubility is 0.87 and is highly significant. The residual error on this correlation RMSE = 0.87, which is the region of uncertainty of this model in predicting solubility. Assuming this fitted error is an estimate of the prediction error of the model, and assuming a normal distributing of errors, 68% of the measured solubility predictions of this model would be expected to be within +/− 0.87 log units range (±1 standard deviation). To put this another way, we are only 68% confident we can predict solubility within a 55-fold range. This range of solubility is wide enough to encompass mis-prediction of the solubility of a progressable candidate drug and a problem compound.

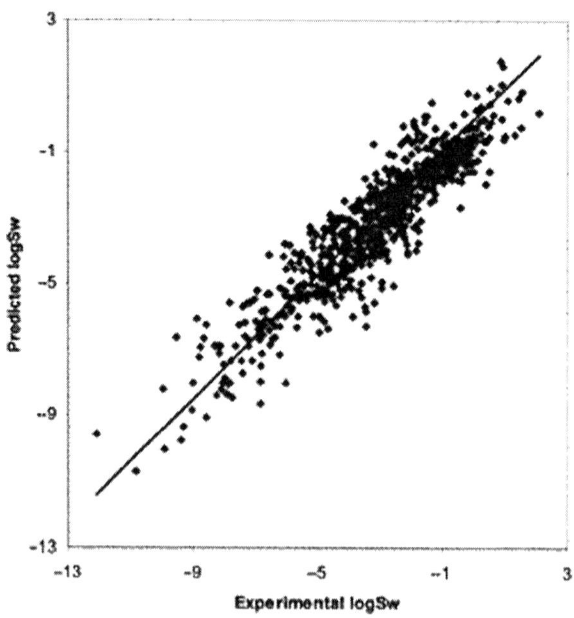

Figure 9.8 A representative solubility QSAR with $r^2 = 0.87$ and RMSE = 0.87.

The well-known LUDI scoring function was developed to predict ligand–protein binding constants from X-ray structures of protein–ligand complexes. It reproduces the pK_i values of 82 protein–ligand complexes from the calibration set, with correlation coefficient = 0.890 ($r^2 = 0.79$), with a standard deviation of 7.3 kJ ml^{-1}, corresponding to 1.4 log units in K_i. Often, drug discovery programs aim to take a micromolar lead to a nanomolar clinical candidate. This range is shown by the box on Figure 9.7. It would appear that LUDI does not have the resolution to discriminate a lead from a potential drug in the discovery program. But it is possible that either of these models may perform better (or worse!) than the average performance of these validation sets for a particular chemical series. This is something that users must assess for themselves, using their own test sets and the known statistics of the model.

Equally, a model with relatively poor predictivity may be useful in enriching a hit-set with compounds of more favourable properties. Scoring functions for the prediction of protein–ligand affinities, even with a prediction error of ±4–8 kcal mol^{-1} have been found to be very useful for identifying actives for potential testing from very large virtual libraries that would be impracticable to screen using experimental techniques.[37]

Recently, Segall and Chadwick have highlighted the often overlooked question in applying a well validated predictive model: "how useful will the model be in making a practical decision?". Another way of expressing this is, how predictive does the model need to be to make a difference in decision-making? Their paper highlights the importance of using prior probabilities in the application of predictive models in decision-making. The impact of priors was illustrated in a conceptual model illustrating different scenarios in applying an *in silico* model and an *in vitro* model for an adverse event in a screening cascade to the costs of compound progression. The model parameterisation was as follows:

- *in vitro* model with a 95% success rate in identifying true positive and true negative rate
- *in silico* model with an 80% success rate in identifying true positive and true negative rate
- cost per compound of *in silico* screen 1 unit
- cost per compound of *in vitro* screen 100 units
- downstream cost to prove safety 5000 units
- net value of safe compound 10 000 units.

Changing the prior probability of an adverse event from 30% to 40% switched the optimal screening cascade from a sentinel approach (using the *in vitro* model to screen negatives predicted from the *in silico* model) to an *in vitro* only strategy. This switch occurred because the inaccuracies in the *in silico* model at the higher 40% prior prevalence of an adverse event meant that the cost of progressing compounds that are likely to yield expensive failures outweighs the savings on filtering out compounds that are likely to fail. In order for users to experiment with this simple model themselves, the authors made available a mathematical model to allow scientists to experiment with the concept of priors to decision making.[38] The authors also illustrated the approach to multi-parameter optimisation, often applied by scientists in prosecuting drug discovery programmes, as implemented in the Stardrop™ platform.[39]

9.8 Future Outlook

The essential principles of model quality assessment are mature and well documented and aim to provide confidence in the model's ability to predict the future. The principles are not always well-applied. The complexity of the machine learning techniques and the easy availability of large numbers of descriptors can seduce or confuse model builders and users into an over-optimistic assessment of a model's quality. Describing the relationship between prediction accuracy and precision, and the distance of future compounds from the model domain of applicability defines the central uncertainty. Many different descriptions of the domain of applicability have been used and a number have shown utility in relating the distance to the error in prediction with some datasets. Further work in this area is warranted and indeed recommended by the OECD guidelines in QSAR model validation.

Model builders are beginning to reduce the distance between the model and the prediction set by updating models, using auto-updating modelling methods. Automation enables the subjectivity of the human model builder to be removed and potentially allows good objective modelling and validation practice to be encoded within the automation algorithm. It also allows models to be tailored to projects or chemical series.

Quantitative model builders can provide diagnostic statistics and indicators of model quality, including Q^2, errors in prediction of temporal test sets, and distance to model. They can also provide plots of residuals for visual analysis to guide use, and update the models to ensure they are as up-to-date as possible. Ultimately, it is model users who need to translate these quality measures into practical significance. Automation may reduce the workload for model builders and enable them to work more closely with model users to ensure that the practical utility of models is maximised within the contexts of the models quality to predict.

Machine learners have been successful in developing predictive empirical QSAR models to guide drug discovery programmes, but the finite size of training sets, deficiencies in descriptors to encode structural variation and the limitations of learners to encode that information in a mathematical model limits the generalisability of many of these predictive models. However, some empirical models are approaching the generalisability of a physics-based model. For instance, if the learner was able to identify, through correlation analysis, the true underlying controlling physical properties of an endpoint, the model may be expected to generalise across all chemistries. Some models of solubility, protein binding, and hERG inhibition appear to be able to generalise across chemotypes to some degree, as these endpoints depend to a large extent upon bulk properties such as charge-type and lipophilicity, which can be reliably estimated across many chemotypes. The ClogP algorithm itself was derived empirically, as described in Chapter 7, but with a good deal of physico-chemical intuition. It appears to have identified some of the underlying physical basis of *n*-octanol–water partitioning. So empirical models can be a good source of understanding for medicinal and computational chemists, of the underlying physics of the system being studied. QSAR models may have come full circle—in that empiricism can lead to true physical understanding of the system, may allow the system to be understood on a deeper level, may in turn allow the chemist to rationally decide *in cerebro* which compound to make next.

References

1. A. Tropsha, P. Gramatica, V. K. Gombar, *QSAR Comb. Sci.* 2003, **22**(1), 69–77.
2. T. R. Stouch, J. R. Kenyon, S. R. Johnson, X.-Q. Chen, A. Doweyko and Y. Li, *J. Comput. Aided Mol. Des.*, 2003, **17**(2-4), 83–92.
3. OECD Environment Health and Safety Publication series on Testing and Assessment No. 69, *Guidance Document on the Validation of (Quantitative) Structure-Acticity Relationships [(Q)SAR] Models*, ENV/JM/MONO (2007)2.
4. http://www.oecd.org/document/23/0,3343,en_2649_34365_33957015_1_1_1_1,00.html, ast accessed January 2011.
5. J. M. Sonnergaard, *Int. J. Pharm.*, 2006, **321**(1-2), 12–17.
6. F. J. Anscombe, *Am. Stat.*, 1973, **27**, 17–21.
7. J. Lee, *Brit. J. Ind. Med.*, 1992, **49**(7), 526–527.
8. J. G. Topliss, R. J. Costello, *J. Med. Chem.* 1972, **15**, 1066–1068.
9. D. W. Salt, S. Ajmani, R. Crichton and D. J. Livingstone, *J. Chem. Inf. Model.*, 2007, **47**(1), 143–149.
10. C. Kramer, C. S. Tautermann, D. J. Livingstone, D. W. Salt, D. C. Whitley, B. Beck, T. Clark, *J. Chem. Inf. Model.*, 2009, **49**(1), 28–34.
11. A. Golbraikh, A. Tropsha, *J. Mol. Graph. Model.*, 2002, **20**(4), 269–276.
12. R. Wehrens, H. Putter, L. M. C. Buydens, *Chemometr. Intell. Lab.*, 2000, **54**(1), 35–52.
13. M. Baroni, G. Costantino, G. Cruciani, D. Riganelli, R. Valigi and S. Clementi, *Quant. Struct. Act. Relat.*, 1993, **12**(1), 9–20.
14. H. Kubinyi, in *Handbook of Chemoinformatics*, ed. J. Gasteiger, Wiley-VCH, Weinheim, 2003, vol. 4, pp. 1532–1554.
15. C. Ruecker, G. Ruecker and M. Meringer, *J. Chem. Inf. Model.*, 2007, **47**(6), 2345–2357.
16. A. C Rencher, F. C. Pun, *Technometrics*, 1980, **22**, 49–53.
17. A. R. Katritzky, M. Kuanar, S. Slavov, C. D. Hall, M. Karelson, I. Kahn and D. A. Dobchev, *Chem. Rev.*, 2010, **110**(10), 5714–5789.
18. D. L. Selwood, D. J. Livingstone, J. C. W. Comley, A. B. O'Dowd, A. T. Hudson, P. Jackson, K. S. Jandu, V. S. Rose and J. N. Stables, *J. Med. Chem.*, 1990, **33**(1), 136–42.

19. D. M. Hawkins, S. C. Basak and S. D. Mills, *J. Chem. Inf. Comput. Sci.*, 2003, **43**(2), 579–586.
20. J. T. Leonard and K. Roy, *QSAR Comb. Sci.*, 2006, **25**, 235–251.
21. A. O. Aptula, N. G. Jeliazkova, T. W. Schultz and M. T. D. Cronin, *QSAR Comb. Sci.*, 2005, **24**(3), 385–396.
22. K. Roy, *Expert Opin. Drug Dis.*, 2007, **2**(12), 1567–1577.
23. G. Schuurmann, R.-U. Ebert J. Chen, B. Wang and R. Kuhne, *J. Chem. Inf. Model.*, 2008, **48**(11), 2140–2145.
24. V. Consonni, D. Ballabio and R. Todeschini, *J. Chemometr.*, 2010, **24**(3-4), 194–201.
25. S. Rodgers, A. M. Davis, H. van de Waterbeemd and N. Tomkinson, *Molecular Informatics*, 2011, **30**(1-3), 256–266.
26. D. E. Clark, *Expert Opin.Drug Dis.*, 2007, **2**(11), 1423–1429.
27. J. A. Cooper, R. Saracci and P. Cole, *Br. J. Cancer*, 1979, **39**, 87–89.
28. J. R. Landi and G. C. Koch, *Biometrics*, 1977, **33**, 159–174.
29. R. Blagus and L. Lusa, *BMC Bioinformatics*, 2010, **11**, 523.
30. M. P. Gleeson, A. M. Davis, K. K. Chohan, S. W. Paine, S. Boyer, C. L. Gavaghan, C. H Arnby, C. Kankkonen and N. Albertson, *J. Comput. Aided Mol. Des.*, 2007, **21**(10-11), 559–573.
31. R. D. Clark and D. J. Webster-Clark, *J. Comput. Aided. Mol. Des.*, 2008, **22**(3-4), 141–146.
32. S. Weaver, M. P. Gleeson, *J. Mol. Graph. Model.*, 2008, **26**(8), 1315–1326.
33. P. Bruneau, *J. Chem. Inf. Comput. Sci.* 2001, **41**, 1605–1616.
34. T. S. Schroeter, A. Schwaighofer, S. Mika, A. T. Laak, D. Suelzle, U. Ganzer, N. Heinrich and K-R. Müller, *J. Comput. Aided. Mol. Des.*, 2007, **21**, 651–664.
35. I. Sushko, S. Novotarskyi, R. Koerner, A. K. Pandey, A. Cherkasov, J. Li, P. Gramatica, K. Hansen, T. Schroeter, K.-R. Mueller, L. Xi, H. Liu, X. Yao, T. Oeberg, F. Hormozdiari, P. Dao, C. Sahinalp, R. Todeschini, P. Polishchuk, A. Artemenko, V. Kuz'min, T. M. Martin, D. M. Young, D. Fourches, E. Muratov, A. Tropsha, I. Baskin, I. D. Horvath, G. Marcou, A. Varnek, V. V. Prokopenko and I. V. Tetko, *J. Chem. Inf. Model.*, 2010, **50**(12), 2094–2111.
36. A. Cheng and K. M. Merz, *J. Med. Chem.*, 2003, **46**(17), 3572–3580.
37. J. Michel and J. W. Essex, J. *Comput. Aided Mol. Des.*, 2010, **24**(8), 639–658.
38. http://www.tessella.com/screening-strategy-explorer.
39. www.optibrium.com.

CHAPTER 10

Application of Modelling Techniques

DAVID E. LEAHY

Molplex Ltd, i6, Charlotte Square, Newcastle Upon Tyne, UK

10.1 Introduction

A quantitative structure–activity relationship (QSAR) is defined by its embrace of quantitative modelling to build relationships between a molecule's structure and its properties. It was brought into being by Corwin Hansch's application of linear regression to biological activity data. These early studies were firmly rooted in physical organic chemistry, in particular the Linear Free Energy assumption, and modelled by straightforward linear regression. However, in the years following, QSAR practitioners have largely abandoned physical organic chemistry and the use of quantitative modelling as a form of hypothesis testing. Instead, they have eagerly adopted computational chemistry-generated descriptors and new waves of machine learning methodology, developed by both statistical and computer science communities, much like "the Borrowers" borrowed from the houses they lived in.

Restricted to substituent constants extracted from experiments and hand-crafted into tables, graphs and simple regression models, early QSAR was very limited in scope, but had the advantage that the practitioner became intimately connected to the data, and the patterns that emerged from an analysis could be translated into a coherent chemical explanation.[1] QSAR modelling was an inductive scientific process, aimed at creating understanding by creating a mental model of the SAR, based on sound chemical concepts, such as electron

RSC Drug Discovery Series No. 13
Drug Design Strategies: Quantitative Approaches
Edited by David J. Livingstone and Andrew M. Davis

Published by the Royal Society of Chemistry, www.rsc.org

distribution, charge and polarisability. Quantification in this case was a means to an end, the end being prediction based on a human expert's interpretation of the data in terms of concepts that could be extrapolated to new and untested structures. Over the intervening years, multiple developments have largely replaced hand-crafted modelling, tending towards idealised software systems being reduced to a large green button saying "press here to get model".

There has been a profound shift in the capabilities available to us as QSAR modellers, particularly in the expansion of readily calculated chemical descriptors and data mining methods, but at its heart remains the question of purpose. Are we seeking models that the human expert can understand and interpret? Or models that are useful, validated and predictive? It is easy to say both, of course, but the simplicity required for a human to understand a model can significantly reduce the chance of finding a model, and the use of powerful but complex learning methods can deliver valid and predictive models which are not interpretable in any chemical sense.

For new practitioners, faced with the overload of methodology, there is the much bigger question of which method is best? Or, which method is best for my compound set? Unfortunately there are no rigorously clear answers, and although many have developed successful strategies through years of experience, the combinatorial space of chemical space, data sets and properties, descriptors and data mining methods is so vast that there are not yet rigorously validated answers to these questions.

The most important question to ask prior to starting work on a QSAR modelling exercise is what decision will be taken based on the model results? What are you trying to achieve and what data do you have?

This chapter does not provide an answer to the question, often asked of experienced QSAR specialists, of what is the "best" modelling method to use. There isn't an answer to this question. The goal is to describe a framework for understanding the options available, and to point to more recent work that seeks to identify more rigorous quantifiable metrics for assessing performance of data mining techniques in QSAR modelling.

10.2 Quantitative Modelling Methods

10.2.1 Overview

There is now an overwhelming number of learning methods for the extraction of implicit, previously unknown, useful information from data. These methods have emerged from different communities and as a result, tend to use different terminology. Many of the statistical regression, data reduction and clustering methods were originally described as "chemometrics", although this term is now less widely used and mostly focussed on diversity and experimental design. With the availability of large numbers of chemical descriptors giving "short, fat" datasets (where the number of columns of independent variables becomes much greater than the number of compounds, each representing one row in the dataset), regression methods are no longer valid. Regression search methods, such as stepwise regression, have been used, but more typically a data reduction method, such as principal component analysis, non-linear mapping or partial least squares, can be used to reduce the number of independent variables to a smaller number of orthogonal variables that can then be used in the regression.

The terms "data mining" and "machine learning" have come from the computer science research community and tend to be conflated alongside "modelling". However, it may be better to view "data mining" as the generic term, with machine learning, clustering and correlation analysis as examples of specialisations of data mining.

Data mining then is the finding of patterns in data, some of which may be obvious or only weak relationships while what we are looking for ideally are strong patterns that are likely to generalise and which are stable to imperfections in the data and small numbers of exceptions. Many of the rule-based and other data mining feature selection methods are particularly sensitive to this.

10.2.2 Knowledge Representations

In modelling, we describe chemical structures in terms of features which are either structural, molecular properties or descriptors calculated using one or more computational chemistry programs. These features are most often numeric, that is they are a continuous variable taking some real value. Since real values take different scales and therefore have different ranges, it is common practice to autoscale or normalise the real values using the mean and standard deviation of the training set. Most often this will be done for the range within the training set, although it is probably theoretically more valid to use a range determined for a wider range of structures— perhaps some nominal drug-like database— otherwise the range, and therefore the significance of the feature, could vary significantly from one set to another. Indeed one of the issues with normalisation is that where there is a small variation in a particular feature of a given internal training set, normalisation can exaggerate the variation and potential significance of that feature. Normalisation shouldn't be avoided however, since different chemical features are genuinely estimated on different scales because their physical units of measurement are different. One of things a good feature selection method[2] should do is to check for these cases of small variation within the training set and eliminate that feature from the study. If a feature selection method isn't used, then some standard set of structures with a diversity large enough to cover the chemical space of interest—not just for the model building, but also the predictions—should be established in order to generate the variance values to be used in the normalisation.

Features can also be symbolic or categorical, such as in the presence or absence of a particular sub-structural feature, often captured as chemical fingerprints. Of course the modeller can choose to express numeric data in terms of intervals. In the early stages of a project, a target of 1 μMolar inhibition might be set as a cut off point for defining activity in a lead-finding screen.

Numeric data can be transformed into intervals such as high, medium or low, and these values are then used, with appropriate modelling methods, in the analysis. Symbolic features can be nominal, where the activity is expressed on some arbitrary scale such as 1-5, and where there is an order. However, the distinction between neighbouring points is not necessarily the same, or nominal, where the activity is described by a code which is a purely qualitative descriptor. These and other ways of expressing the activity are less commonly used in QSAR modelling than numeric and interval data, unlike in more general data mining where they are of greater significance.

The quality of data provided for QSAR modelling varies throughout the development of a project, but so do the questions being asked. It is the modeller's prerogative to select the form and methods that are appropriate to the decisions necessary at each stage of the project, rather than to assume there is one, "best", way. Typically, data is in fact a hybrid of continuous data, pseudocontinuous and nominal. Activity is often reported as inhibitory concentration at 50% effect and vary from the nano to micromolar but are also nominal since compounds with no observed effect are reported as "not active". That is, they show no response in the test system. Activities are sometimes reported as "% inhibition at 10 or 100 μMolar", which is what is meant by pseudocontinuous; continuous in that these data can be simply transformed and put on the same scale as the genuinely continuous, but since this is effectively equivalent to a one point curve-fit (which will horrify biologist colleagues), there is probably little to be gained from this. The question of how to handle this is not a technical QSAR issue, it is a project issue. At an early stage in the project, it may be better to define less than 10 μMolar as active, and greater than 10 μMolar as inactive, and use one of a number of modelling methods suited to this kind of interval data. Later in the project this can change to less than 1 μMolar as active and even later, during optimisation, as continuous data only (where those compounds with activity greater than 1 μMolar are discarded). The question to ask colleagues in the team is what decision they would take based on the model results. If a model predicts (with high validity) that a novel chemical structure is likely to be more active than 10 μMolar, would the chemists synthesise it? If this is no longer critical decision-making information, the model is not adequate for the task. All these concerns argue that QSAR modellers need a flexible toolkit of data representations and modelling methods to best suit the state of the project. The essential point is that when put into practice, QSAR is a process of building a knowledge representation that assists decision making, and this requires flexibility and adaptation in the selection of methodology.

Another issue to consider prior to selecting a modelling method is the form of the chemical structure to be used. With the very large, industrial-scale processes now commonplace in large companies, this is practically impossible, and there is an effective separation between the processes by which structures are registered and used. However, models can be strongly affected by the rules used to assign tautomerism, charge and the selection of the dominant conformation, and it is important to at least check, even if this is only possible for a sub-set of the data. These are not, however, issues of the modelling process; they reflect on difficult chemical questions of how to represent structures for QSAR modelling which is largely unsolved. We should at least aim for consistency of representation across a model–building set, so that, for example, all ionisable groups are kept in the neutral form.

10.2.3 Data Mining Survey

Data mining methods are widely available and can often be highly sophisticated algorithms that use advanced techniques from computer science and artificial intelligence.[3] However, simple and intuitive methods can often work well, without much loss in predictive ability. With small datasets, where the focus is developing interpretable models, these simple methods may be the best first approach, perhaps as part of a conscious effort to explore the data. In any case it is useful to have some benchmark result against which the performance of more complex, computationally expensive and difficult to interpret methods can be compared. The simplest naive model is prediction by the mean of the dataset, which is in effect prediction without using a model, and this can serve as a useful reality check and comparator.

Simple rule induction methods can also perform well compared to more sophisticated examples,[4] particularly if the descriptors or attributes are well chosen. A rule is simple to express and interpretable, for example:

If Log P > 2 then compound is active

Or

If (Amide) then compound is active

A rule is discovered by taking each attribute in turn, counting how often each class appears and making the rule assign the most frequent class to that attribute value, minimising the error rate. For example, by counting the frequency of occurrence and error rate for different possible error rate rules, such as Log P > 0; Log P > 1; Log P > 2, we select the rule that has the minimum error rate to create the rule. The rule set is formed by repeating the process for all the available attributes and assuming that all contribute equally. Rule methods can handle missing values by assigning these as an attribute and numeric data can be partitioned into categories. With many attributes, the algorithm can become inefficient and it is wise to minimise the number of attributes used in the analysis. However, we can use a larger set of attributes if we don't assume that all attributes are equally probable. This is the basis for the statistical analogue of rule induction, which can be highly effective for noisy data, such as that from high throughput screening.[5] In the Naive Bayes method, the attribute is assigned a likelihood, based on occurrence in the data set. The rule is estimated from the conditional probability obtained by combining all the probabilities from the values of the different attributes. It is called "naive" because it assumes the independence of all attributes;

nevertheless, the method can often work well and, given the use of interpretable attributes (*e*.g. the presence and absence of structural features), will give models that might be interpreted by the chemist in the team.[6]

These simple rule induction and probability methods are straightforward to use and understand, and can often outperform more sophisticated approaches. They are sensitive to being skewed by redundant variables however, as well as by dependencies between attributes and the non-normal distributions of attribute values. These problems can be overcome by the careful selection of a sub-set of attributes and the use of other estimation methods for a more appropriate distribution for each attribute, although this is probably not necessary for most analyses.

Decision trees[7] are an elaboration of the simple rule sets, but are also ways of representing rules discovered in the data. In the case of a decision tree, the rules are placed in a tree hierarchy and evaluated down the branches of the tree from the root to the leaf. The algorithm selects one attribute as root and creates each attribute value as branch, splitting the full dataset so that each leaf contains the compounds that have that value. The process is repeated recursively, using only the sub-set of compounds at each node, so that eventually all compounds end up allocated to one leaf and the algorithm stops at a leaf where all the compounds are members of the same class. The algorithm uses an information measure to select which attribute to use next at each branch, in order to get to the shallowest trees, and the largest allocation of compounds per node. The decision tree method is a top-down, divide and conquer approach to model building. Alternative rule building employs a covering approach, where rules are sought for each class in turn. Visually, this can be seen as a way of splitting the feature space in such a way that the rules independently cover all the classes. Although similar, covering algorithms differ from decision trees in that they focus on adding rules that have maximum accuracy in separating out a class, whereas trees add tests to maximise separation between classes.

Rule-based methods, such as decision trees and covering algorithms, work best for nominal attributes. Numeric attributes can be converted into nominal attributes of course, by defining cut off limits, but they can also be analysed using linear regression models. Linear models are a natural and heavily used method for the analysis of numeric data and almost all early QSAR modelling work was done with linear regression. Even today, with the widespread availability of more sophisticated machine learning methods, linear regression remains popular, particularly where the practitioner is seeking an interpretable model. A QSAR derived from a linear regression using only a limited number of chemical features can inform the medicinal chemist's understanding of what drives structure–activity. Linear regression requires a "long, thin" dataset, one where there are significantly more instances of compounds than features. It is also important that the features are not highly correlated as, for example, with simple physical properties such as Log P and MR (molar refractivity). This has led to the use of data reduction methods, such as principal components analysis, that orthogonalise the features and allow the selection of a smaller number, consistent with the size of the training dataset. Linear regression can also be used to classify data by assigning real values to each class and carrying out the regression on each class separately. A variant of this approach is logistic regression, which transforms the class data into a logistic curve prior to regression analysis.

Linear regression is always a sensible modelling method in QSAR modelling since it provides a baseline modelling approach that has the potential to lead to chemical insight. This reduces the chance of finding a validated and predictive model, but is a very useful way of looking at QSAR data. Although data reduction methods such as principal component analysis combined with regression, either separately or combined—as in a partial least squares analysis—can overcome the problem of "short, fat datasets", models are dependent on latent variables, which are unlikely to be chemically interpretable. If models that are not interpretable are acceptable, then partial least squares regression is recommended as a robust and widely used approach.

The use of linear regression to classify compounds can be visualised as selecting a hyperplane. In the case of data for which we are trying to model two separate classes, such as active and inactive, an algorithm which iteratively adjusts the weights against the data set features, known as the perceptron learning rule, can be used to generate a linear separation model, which successfully divides the data into its two classes. If the data is not cleanly separable, as in the case of most modelling exercises in QSAR, the updating algorithm is not able to converge on a single solution. A change in weighting for one feature in one round of iteration may improve the separation for part of the dataset, while making it worse for other compounds. Therefore limits need to be set on the convergence algorithm. The perceptron leaning algorithm is the foundational concept for model building by neural networks. Here, each input feature is represented by a weighted node and there is one output node, connected by a network of "neurons", where each connection is weighted. In this case, the perceptron algorithm is used to iteratively adjust the weights of the network nodes in order to optimise the agreement of output values with the observations.[8]

Kernel methods, which include support vector machines and Gaussian processes, transform the data into a higher dimensional space, where it is possible to construct one or more hyperplanes for separation of classes or regression. These methods are more mathematically rigorous than neural networks and have in recent years been widely used in QSAR modeling.[9–14]

Gaussian process modelling works by a similar functional transformation of the data[15] and has also begun to be deployed by QSAR practitioners.[16–18] It is thought to be particularly good at building models with little expert supervision, because it allows statistically sound alternatives to both model discovery and validation, therefore is well suited to automated QSAR model discovery and updating.

Inductive Logic Programming (ILP) is a distinctly different modelling method from all others described here. Most of these applications rely on attribute-based learning. However, attribute-based learning is limited to non-relational descriptions of objects in the sense that the learned descriptions do not specify relations among the objects' parts. ILP uses relational descriptions that extend the attribute-pair to predicate logic representations, and the modelling algorithms are then deployed to induce logic relations that can be used predictively to predict chemical properties.[19]

Genetic programming uses an Abstract Syntax Tree (AST) representation of a process, such as a program or mathematical equation, acting on symbols. Genetic programming methods evolve novel trees by node mutation and crossover of sub-trees, in order to maximise the predictive quality of the tree.[20,21] The method is stochastic in that it generates many possible combinations of models which are selected by fitness during an iterative process, which it is hoped will discover optimum solutions. A distinct advantage of the method is the potential for generating explicit and interpretable models, although this assumes the selection of a limited set of features in the symbolic representations.[22]

Bayesian networks are models that are represented as directed acyclic graphs, with features placed at nodes in the graph and the edges representing probabilistic dependencies between the nodes.[23] The Bayesian network learning problem includes the discovery of the graph network structure, as well as the conditional probabilities attached to each edge in the graph, so as to maximise the predictive quality of the network. However, given an implementation the method has real potential for generating predictive and interpretable models and although there are a few partly relevant publications,[24] they have not been widely applied in QSAR modelling.

In instance based learning the training data is used directly and predictions are done by taking some consensus value of the nearest training set points. Sometimes called model independent methods, or naive models, they are conceptually simple and explicit examples of the well known similarity principle, which is the hypothesis that chemically similar compounds have similar properties. The diversity of methods available derives from the different choices to be made regarding how chemical space is defined and which distance metric is used.

The K-nearest neighbours method has been widely adopted[25] and uses different property spaces so that predictions are carried out by first selecting the k (3 is often used) nearest compounds in the chemical space and then taking a mean of their values as the estimate. Any distance metric can be used but it is normally, although not always,[26] Euclidean. For very large chemical structure sets, the pre-processing of the data to generate a nearest neighbour list is very time consuming, but efficient algorithms such as kd trees have been developed to overcome this problem.[26] Kd tree algorithms and variants, like the ball tree method,[27] can also be used to speed up clustering methods since at some point, each point needs to be allocated to a nearest cluster mean and this can be done in the same way.

10.2.4 *Meta*-Model Application

We can think of *meta*-modelling as concerning the discovery of models, and there has been rapidly increasing interest in applications over the last few years as practitioners look to exploit the wealth of alternative methods to improve model quality, deal with the need for updating as new data becomes available and tackle the generation of multiple models that can partition data into locally valid subsets. The use of ensembles of models[28] in this way raises additional questions of validation and consensus, or the selection of models for prediction.

In a typical QSAR publication it is now more usual to see some attempt to compare modelling methods against one or more datasets.[9,29,30,31] Some workers have developed automated systems to do this, describing it as "combi" or combinatorial QSAR.[32–34]

The *meta*-modelling approach can automate the surveying of model space and partition local models. An ensemble of models can be more stable and reliably predictive. Since a key element of the QSAR practitioner's expertise is the selection of the appropriate method, *meta*-modelling can lead to more reliable automation of model discovery, as is often required in the context of large and continuously operated screening campaigns.

Bagging[28] is an example of ensemble modelling and is fairly straightforward since it is simply the repeat application of a model building method to multiple subsets of the data, sampled from the original set, and the use of a replacement technique to fill in the removed data point. In bagging, the method used is the same and the applications are independent of each other. For prediction the models can be used to vote on the value or through some other weighted combination of value.

In boosting,[35] which also uses the same modelling method, a weighting is chosen based on how difficult the point is to fit.[36] As the weight is increased in successive rounds of model building for those points that don't fit the model and so boosted into being take account.[29] Random forests use boosting to create select multiple decision trees with a random selection of features at each node as the tree is grown.[37]

In stacking, an ensemble of models is built, using different methods, and a model of the models is created, which weights the predictions from the different models.

Temporal, or time-series, QSAR manages updating of QSAR models explicitly[38,39] and is well suited to the continuous data generation processes built into the large pharmaceutical company infrastructure.

Competitive workflow[40,41] is a software architecture for explicitly engineered exploration of combinatorial model space and continuous updating with new data and new methods. The Discovery Bus implementation of competitive workflow has been used[42] to generate 13 000 stable and valid models for over 3000 QSAR datasets in an entirely automated process. This implementation of competitive workflow captures QSAR modelling in the form of a best practice workflow, where each node in the workflow is implemented as multiple alternative or competing methods, such as those for data transformation, descriptor calculation feature filtering and model building. The end result is that all possible models are generated from all possible combinations of methods at each stage of the workflow. The models are then selected (again, automatically) by standard tests for model stability and validation.[43] The addition of new data or methods automatically triggers the downstream processing and generation of new models that automates continuous updating of the knowledge base. This is a brute force search of combinatorial or *meta*-model space, which can be refined by tree pruning, by filtering out model pathways that are not generating better solutions or through other search optimisation techniques. It is a significant step towards the automation of QSAR modelling expertise, and by explicitly generating all possible model pathways for all available datasets (subject to practical constraints), it holds out the prospect of answering the difficult questions posed at the beginning of the chapter. Those questions are: which methods are best for which data? And for which purpose? By creating a database of models[42] generated by combinations of multiple methods for very large and diverse data series that can be queried for performance metrics, there is the prospect of a rigorous approach to what we might tentatively label "*Meta* QSAR", the analysis of the performance of QSAR modelling methodology. *Meta*-QSAR analysis based on such a database of models, that can be routinely updated as new methods are published, could be the basis of a new direction for the subject.

10.3 Conclusions

QSAR modelling is a mature science where recent research is directed at *meta*-modelling and ensemble approaches to rigorous QSAR model discovery, in an environment where the availability of data and techniques is beyond the capacity of specialists to analyse. Newer systems engineering approaches to automation of QSAR modelling[39,41] hold out the prospect of a rigorous understanding of the relative strengths of different techniques. However, for today's practitioners the selection of technique to suit the data and the purpose remains the critical issue. Advice is easy to give of course, but reflects the specific experience of individuals, as well practical issues such as the availability of software and the level of skill of the practitioner. For small datasets where there is a strong requirement for interpretation, reducing the input features to a relatively small (less than 40) set of features, all of which are fully understandable in chemical terms or as measurable physical properties, is essential. Linear methods such as PLS and PCA regression, or stepwise regression to select important features, are relatively easy to use and models have well determined statistical properties. Rule-based methods such as decision trees, genetic programming and ensembles such as Random Forests are particularly attractive in these cases. The focus on smaller datasets and reduced feature sets will help in the subjective process of understanding the chemical structure–property data and the creation of knowledge in order to advise the project team on chemical ways forward.

The increasing use of automation in QSAR and ensemble methods reflects the way that most practitioners working in industry have adapted to the explosion of new data through automation of laboratory processes. This shift away from traditional low throughput QSAR coupled to interpretation is now the standard approach to QSAR in industry and has been accompanied by a shift in expertise and skills. Sometimes derided as "black box" and "unscientific", the indiscriminate use of thousands of descriptors followed by automated feature selection[44] methods and ensemble *meta*-learning is generally found to be more likely to give stable and valid models. In addition, model ensembles are found to be more predictive, assuming that they are genuinely capturing different features of the problem and/or modelling different regions of the chemical space. They can be promoted on the basis of pragmatism at least, but also raise new and interesting scientific questions, albeit ones from different domains and requiring different skill sets and experience.

References

1. C. D Selassie and R. P. Verma, *Briefings Bioinf.*, 2002, **3**(1), 73.
2. Y. Liu, *J. Chem. Inf. Comput. Sci.*, 2004, **44**, 1823.
3. I. H. Witten and E. Frank, *Data Mining: Practical Machine Learning Tools and Techniques*, Morgan Kaufmann, 2005.
4. Y. Lee, B. G. Buchanan, and J. M. Aronis, *Mach. Learn.*, 1998, **30**, 217.
5. M. Glick, A. E. Klon, P. Acklin and J. W. Davies, *J. Biomol. Screening*, 2004, **9**, 32.
6. J. Heal, B. Hamilton and J. Sheridan, *QSAR World*, 2007, 1–3.
7. C. Y. Ma, F. V. Buontempo and X. Z. Wang, *Comput. Aided Chem. Eng.*, 2008, **25**, 581.
8. I. V. Tetko, V. Y. Tanchuk, T. N. Kasheva and A. E. Villa, *J. Chem. Inf. Comput. Sci.*, 2001, **41**, 1488.
9. X. J. Yao, A. Panaye, J. P. Doucet, R. S. Zhang, H. F. Chen, M. C. Liu, Z. D. Hu and B. T. Fan, *J. Chem. Inf. Comput. Sci.*, 2004, **44**, 1257.
10. B. Niu, W.-C. Lu, S.-S. Yang, Y.-D. Cai and G.-Z. Li, *Acta Pharmacol. Sin.*, 2007, **28**, 1075.
11. R. Darnag, E. L. Mostapha Mazouz, A. Schmitzer, D. Villemin, A. Jarid, and D. Cherqaoui, *Eur. J. Med. Chem.*, 2010, **45**, 1590.
12. H. Tang, X. S. Wang, X.-P. Huang, B. L. Roth, K. V. Butler, A. P. Kozikowski, M. Jung and A. Tropsha, *J. Chem. Inf. Model.*, 2009, **49**, 461.
13. H.-F. Chen, *Anal. Chim. Acta*, 2008, **624**, 203.
14. E. Pourbasheer, S. Riahi, M. R. Ganjali and P. Norouzi, *European Eur. J. Med. Chem.*, 2010, **45**, 1087.
15. C. E. Rasmussen and C. K. I. Williams, *Gaussian Processes for Machine Learning*, The MIT Press, 2006.
16. O. Obrezanova, G. Csanyi, J. M. R. Gola and M. D. Segall, *J. Chem. Inf. Model.*, 2007, **47**, 1847.
17. M. Seeger, *Int. J. Neural. Syst.*, 2004, **14**, 69.
18. O. Obrezanova and M. D. Segall, *J. Chem. Inf. Model*, 2010, **50**, 1053.
19. R. D. King, S. H. Muggleton, A. Srinivasan and M. J. Sternberg, *Proc.-Natl. Acad. Sci. U. S. A.*, 1996, **93**, 438.
20. D. P. Searson, D. E. Leahy and M. J. Willis, in *Proceedings of the International MultiConference of Engineers and Computer Scientists*, 2010, Vol. 1, IMECS 2010 Hong Kong.

21. F. Archetti, S. Lanzeni, E. Messina and L. Vanneschi, *Genetic programming for humanoral bioavailability of drugs*, in, *Proceedings of the Genetic and Evolutionary Computation Conference*, ed. M. Keijzer *et al.*, GECCO 2006, ACM Press, 2006, Vol. 1, pp. 255–262.
22. R. Poli and W. B. Langdon, Springer-Verlag, p. 276, *http://www.lulu.com.*
23. J. Pearl, *Wiley Interdiscip. Rev. Comput.Stat.*, 1995, **1**, 307.
24. B. Ellis and W. H. Wong, *J Am. Stat. Assoc.*, 2008, **103**, 778.
25. E. Karakoc, A. Cherkasov and S. C. Sahinalp, *Bioinformatics*, 2006, **22**, e243.
26. D. K. Agrafiotis and V. S. Lobanov, *J. Chem. Inf. Model.*, 1999, **39**, 51.
27. T. Liu, A. W. Moore and A. Gray, *J. Mach. Learn. Res.*, 2006, **7**, 1135.
28. T. G. Dietterich, *Lecture Notes Comput. Sci.*, 2000, **1857**, 1.
29. C. L. Bruce, J. L. Melville, S. D. Pickett and J. D. Hirst, *J. Chem. Inf. Model*, 2007, **47**, 219.
30. E. Karakoc, S. C. Sahinalp, and A. Cherkasov, *J. Chem. Inf. Model*, 2006, **46**, 2167.
31. M. Seierstad and D. K. Agrafiotis, *Chem. Biol. Drug Des.*, 2006, **67**, 284.
32. L. Zhang, H. Zhu, T. I. Oprea, A. Golbraikh and A. Tropsha, *Pharm. Res.*, 2008, **25**, 1902.
33. M. Olah, C. Bologa, and T. I. Oprea, *J. Comput. Aided Mol. Des.*, 2004, **18**, 437.
34. P. de Cerqueira Lima, A. Golbraikh, S. Oloff, Y. Xiao and A. Tropsha, *J. Chem. Inf. Model.*, 2006, **46**, 1245.
35. H. Drucker, C. Cortes, L. D. Jackel, Y. Lecun and V. Vapnik, *Neural Comput.*, 1994, **6**, 1289.
36. V. Svetnik, T. Wang, C. Tong, A. Liaw, R. P. Sheridan and Q. Song, *J. Chem. Inf. Model.*, 2005, **45**, 786.
37. V. Svetnik, A. Liaw, C. Tong, J. C. Culberson, R. P. Sheridan and B. P. Feuston, *J. Chem. Inf. Comput. Sci.*, 2003, **43**, 1947.
38. S. L. Rodgers, A. M. Davis, N. P. Tomkinson and H. van de Waterbeemd, *J. Chem. Inf. Model.*, 2007, **47**, 2401.
39. S. L. Rodgers, A. M. Davis and H. van de Waterbeemd, *QSAR Comb. Sci.*, 2007, **26**, 511.
40. J. Cartmell, S. Enoch, D. Krstajic and D. E. Leahy, *Design. Comput. Aided Mol. Des.*, 2005, **19**, 821.
41. J. Cartmell, D. Krstajic and D. E. Leahy, *Curr. Opin. Drug Discovery Dev.*, 2007, **10**, 347.
42. D. E. Leahy and D. P. Searson, www.openqsar.org, 2009.
43. A. Golbraikh and A. Tropsha, *J. Mol. Graph. Model.*, 2002, **20**, 269.
44. M. A. Hall, PhD Dissertation, Hamilton NZ Waikato University, Department of Computer Science, 1998.

CHAPTER 11

Expert Systems: The Use of Expert Systems in Drug Design-Toxicity and Metabolism

ANTHONY LONG,* KRISTINA FIELDING, NICOLE MCSWEENEY, MARTIN PAYNE AND ELLA SMORACZEWSKA

Lhasa Limited, 22-23 Blenheim Terrace, Woodhouse Lane, Leeds, West Yorkshire, LS2 9HD, UK

11.1 Introduction

The development and use of knowledge based expert systems in the chemical sciences has a long and rich history, beginning in the 1960s, and has recently been eruditely reviewed in an excellent monograph by Judson.[1] This volume contains a thorough historical perspective of developments in the field, a comprehensive technical review of the current state-of-the-art and an insightful, objective review of possible future directions. It is not our intention in this current volume to duplicate sections of this work but rather to reinforce some of the material therein in respect of the expert system approach to the prediction of toxicity and xenobiotic metabolism. With reference to examples and case studies, we will endeavour to outline the use of such systems in the field of drug design and discovery. We will, out of necessity, as they are the systems best known to us, mainly confine our discussion to two programs only: Derek for Windows and Meteor, used, for the prediction of toxicity and xenobiotic metabolism respectively. Other programs are available and have been

RSC Drug Discovery Series No. 13
Drug Design Strategies: Quantitative Approaches
Edited by David J. Livingstone and Andrew M. Davis

Published by the Royal Society of Chemistry, www.rsc.org

reviewed,[1] but it is not our intention to present here a comparative feature and performances analysis in relation to alternative systems.

Expert systems may be thought of as programs in which a reasoning (or inference) engine solves problems (or makes predictions) by applying rules from a knowledge base in response to single or multiple queries (or hypotheses). Such rules may be a combination of two types: factual and heuristic, and these are, in the main part (but not exclusively), non-numerical. It seems strange to talk about non-numerical approaches in a book dedicated to quantitative methods, but in reality, human decision making is based on a combination of qualitative and quantitative reasoning, therefore it is right that an expert system (which should be thought of as a decision support tool) also uses these two types of input. As an example, imagine a situation in which a patient is describing symptoms to the family doctor. If the patient presents with mild symptoms such as runny nose, sore throat, headache, congestion and so on, it would not be difficult for the expert medical practitioner (or even a non-expert) to diagnose the problem as a common cold. The doctor has applied both factual (his medical knowledge in relation to an usual combination of well-known symptoms, both described and observed) as well as, perhaps, heuristic rules (*"it's early December and there's a lot of it about"*) in arriving at the diagnosis. For a less common presentation involving, for example, an unusual combination of rarer symptoms, diagnosis may be more difficult. In such a case the practitioner may note the verbally described symptoms and any other indications, such as skin colour or pallor, sweating, shaking, and so on, but may also need recourse to taking the patient's temperature, pulse and blood pressure, recording an electrocardiogram, sending blood or urine samples away for analyses of biomarker concentrations, and so on. In other words, the doctor needs to use a combination of qualitative and quantitative indicators in the process of making his diagnosis. Furthermore, the results of one set of tests may indicate which tests need to be performed next, *i.e.* the diagnostic procedure may be multilayered. Later in this chapter we present an example of the use of a numerical method to improve upon the accuracy of a prediction made by the expert system (in this case Meteor) alone. The important thing to emphasise and appreciate here is that the prediction is improved because the qualitative and quantitative approaches, as in the case of the medical diagnosis, are complementary, and there is no aspect of consensus modelling involved. In this respect and for this reason, we believe that some discussion of expert systems in a book focussed on quantitative methods is appropriate. We will also touch on the use of numerical methods in some of the evaluation protocols used in the refinement and improvement of the knowledge bases. We begin with a brief description of Derek for Windows.

11.2 A Description of Derek for Windows

The Derek for Windows toxicity prediction system has recently been reviewed.[2] Predictions are based on the screening of a drawn or imported chemical structure or structures for the presence of two-dimensional toxicophores ("alerts") known by the program to be associated with one or more toxicological endpoints. Such processing is conducted with (or optionally without) automated chemical appreciation of tautomers and can be adjusted by the user to give results, for example, for given species or endpoints. The toxicophores are contained in the program's knowledge base as structural alerts that additionally contain information on the basis of their derivation: references (including a brief review of published qualitative and quantitative structure-activity relationships (QSARs), where appropriate), a definition of its scope based on available data (public and proprietary) and such qualitative structure activity relationships (SARs) derived from it, comments on the current understanding of (or speculations on) toxicological mechanisms, species specificity and, in some instances, advice on the applicability domain of the defined SAR and validation studies conducted on the alert. Alerts may additionally be linked to one or more supporting examples. All of this information is made available through the on-screen results display or printable reports, generated in a range of formats (Figures 11.1 and 11.2).

Based on standard assays, such as an Ames test, the predictions for each endpoint are presented with qualitative confidence levels for positive results. Confidence is expressed as, for example, "plausible" or "equivocal" in association with a particular alert and can be modulated, for example, to "probable", reflecting identification of actual data on the query chemical (a matching

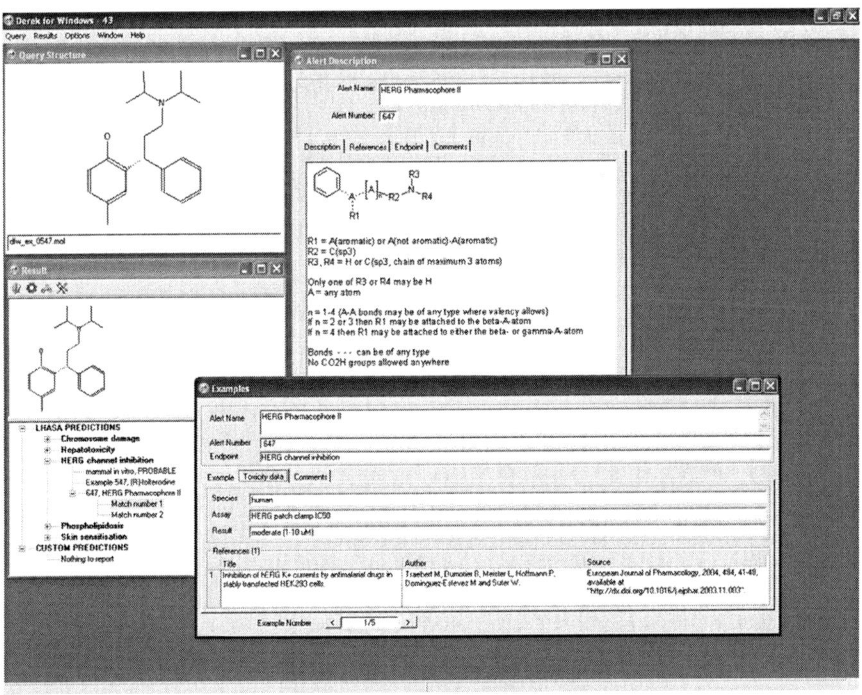

Figure 11.1 Derek for Windows on-screen results displays.

HERG channel inhibition

Rule name: Rule 643
Rule 643: If [known HERG patch clamp IC50 strong (less than 1 uM) in hamster] is [certain] then [HERG channel inhibition in vitro] is [Species dependent variable 25]
Comments: If a chemical is known to possess a strong (less than 1 uM) in vitro HERG patch clamp IC50 in the hamster then it is considered certain that the chemical will cause HERG channel inhibition in vitro in hamsters, probable in mammals other than the hamster and impossible in bacteria.

Rule name: Rule 974
Rule 974: If [alert 647] is [certain] then [HERG channel inhibition in vitro] is [Species dependent variable 22]
Comments: If a chemical contains alert 647 then it is considered plausible that the chemical will cause HERG channel inhibition in vitro in mammals and impossible in bacteria.

Alert overview: 647 HERG Pharmacophore II
Comments:
This alert describes a structure-based pharmacophore developed primarily from compounds that have been reported to be moderate or strong inhibitors of the HERG (human ether-a-go-go-related gene) potassium channel [Pearlstein et al, Cavalli et al, Ekins et al].

A number of pharmaceutical drugs have been shown to block the HERG channel in whole-cell patch clamp electrophysiology assays in a variety of cell types. Examples include norastemizole [Zhou et al], chloroquine [Traebert et al], imipramine [Teschemacher et al], amitriptyline [Tie] and (R)-tolterodine [Kang et al].

HERG encodes for the alpha-subunit of a potassium channel which is thought to carry the rapid component of the delayed rectifier current IKr in the heart [Mitcheson et al, Tristani-Firouzi and Sanguinetti].

Example 5. (R)-tolterodine
CAS Number: 124937-51-5

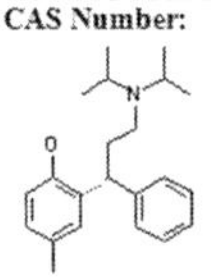

Test Data: (R)-tolterodine
Species: hamster
Assay: HERG patch clamp IC50
Result: strong (less than 1 uM)
References:
Title: Cardiac ion channel effects of tolterodine.
Author: Kang J, Chen XL, Wang H, Ji J, Reynolds W, Lim S, Hendrix J and Rampe D.
Source: Journal of Pharmacology and Experimental Therapeutics, 2004, 308, 935-940, available at "http://dx.doi.org/10.1124/jpet.103.062182".

Figure 11.2 Edited extract from a Derek for Windows rich text format report.

supporting example) in the system, by species extrapolation or, less commonly, by rules relating the values of physicochemical properties to toxicity. The underlying theory of the reasoning has been described and illustrated with examples.[3] The aggregated arguments for and against a given prediction and their resolution into an overall confidence level is summarised in a reasoning display (Figure 11.3).

It is possible to modulate the likelihood levels by relationships with experimental or calculated physicochemical (or other) parameters such as octanol–water partition coefficients. Such calculators may be external to the main program. For example, within the skin sensitisation endpoint, a very low skin permeability coefficient (Log Kp),as calculated from the molecular weight and the octanol–water partition coefficient estimation using the method of Potts and Guy,[4] reduces the default skin sensitisation alert likelihood from "plausible" to "equivocal" (Figure 11.4).

11.2.1 Rapid Prototype Alerts

Derek is a relatively mature product with a structure–activity toxicity knowledge base developed, refined and expanded over the last twenty years to include

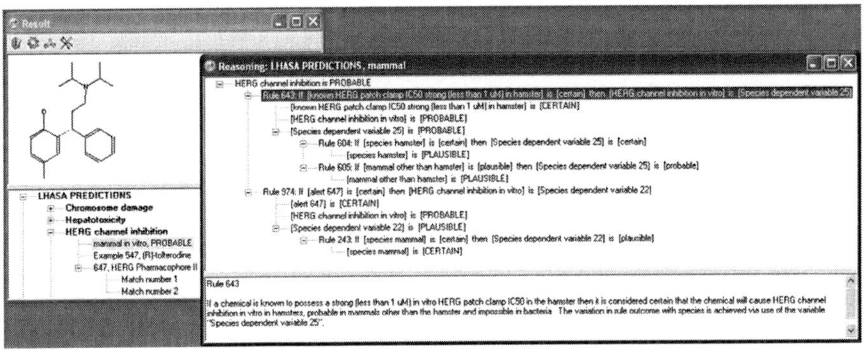

Figure 11.3 Derek for Windows on-screen reasoning display.

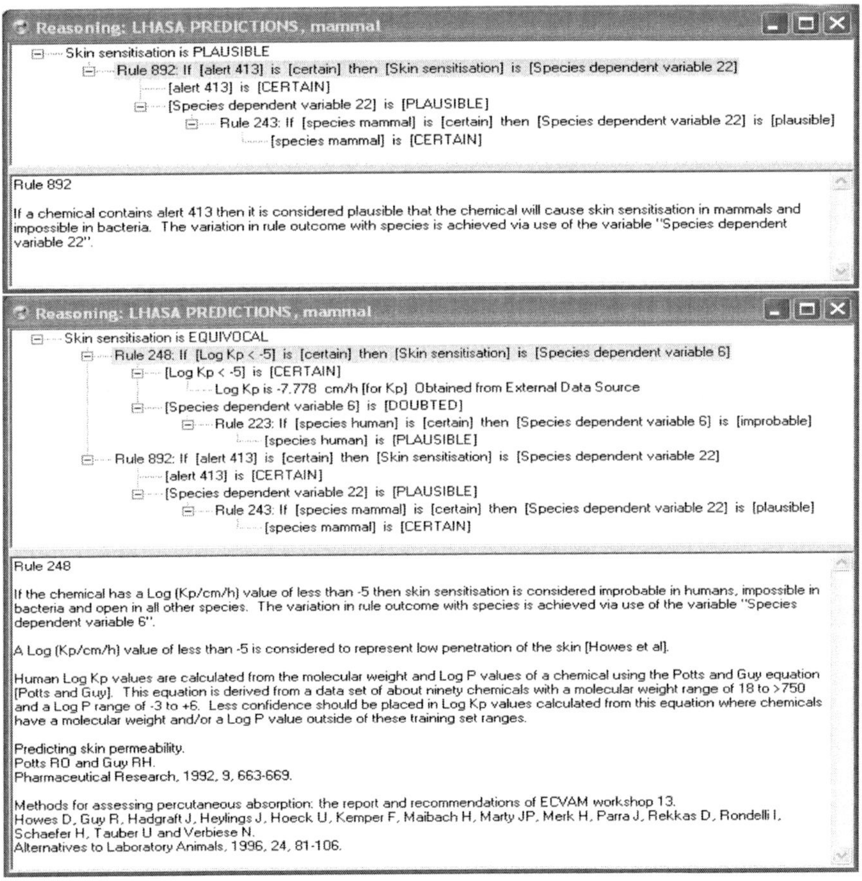

Figure 11.4 Derek for Windows reasoning display showing the reduction in confidence on the basis of a Log KP estimation.

an increasing range of more complex endpoints such as teratogenicity, hERG channel inhibition, chromosome aberration and mitochondrial toxicity. The predicted toxicology has been greatly extended from the original core endpoints of skin sensitisation, bacterial (Ames) mutagenicity and genotoxic carcinogenicity, which have a high dependence on chemical reactivity with protein or DNA. Well researched alerts refined by a mechanistic rationale take considerable time, effort and expertise to implement and develop. They may also require a significant amount of data to define the scope or applicability domain. In many instances, data is limited and the SAR unknown or poorly characterised. However, use can still be made of clusters of related chemicals with similar toxicity through a rapid prototype approach—in which the relevant common structural features, derived by similarity clustering (machine-assisted or by eye) and the application of a basic knowledge of metabolism—to identify metabolic precursors, allowing an initial rapid prototype alert to be defined. Rapid prototype alerts are typically developed within the Derek for Windows knowledge base in order to describe initial classes of structures associated with toxicological effects identified from a large dataset. They have a specific scope and limited comments but do not contain references or examples. They are useful as an indication of a potential toxicophore.[5] For example, eight rapid prototypes for bradycardia were developed on this basis and included in the 2009 knowledge base release (version 11). Based on the features of small clusters of between two and forty-two chemicals, alerts formed around a model chemical or class have been included. For bradycardia, most of the model compounds form the alerts, for example, those described as "esmolol-like", "vecuronium-like", "remifentanil-like", "ropivacaine-like", with some generic treatment or allowed variation of the substituents. Similar rapid protocol alerts have been developed for chromosome damage *in vitro* (four alerts), hepatotoxicity (thirty-two alerts), kidney disorders (four alerts) and nephrotoxicity (forty-two alerts). The thirty-two rapid prototype alerts for hepatotoxicity (and those for nephrotoxicity), unlike those for bradycardia, consist of smaller chemical functionalities such as "diphenyl sulfide, sulfoxide or sulfone" or "thiotetrazole or derivative" or "3-cyanopyridine" and generally have a greater than fifty percent association with hepatotoxicity in the database examined. Presently, the rapid prototype alerts make predictions that are reported as having a likelihood of "equivocal" in mammals (including man), representing the level of confidence the rule-writer has in the supporting data for this alert when compared to the usual level of "plausible" for more fully researched, data-rich or mechanistically more informed alerts.

11.2.2 Applicability Domains

Definition of the applicability domain of a structure–activity relationship is a requirement for most regulatory purposes. This rule came out of a recommendation of a set of principles at a workshop in Setubal, Portugal in 2002,[6] which were adopted by the Organisation for Economic Co-operation and Development (OECD) in 2004. The principles reflect the general "local" nature of many structure–activity relationships and the frequently limited structural scope of chemicals used in their derivation (the "training set"). The true applicability domain encompasses limits due to variable toxicokinetics (absorption, metabolism, distribution and elimination) and the nature of the toxic mechanisms which may include specific non-covalent interactions with, for example, receptors, ion-channels or enzymes, all of whose structural requirements for ligand binding are difficult to characterise (pharmacophore/toxicophore approaches are traditionally used). In practice, most QSAR approaches define applicability domains from the "training" sets. Applicability domains have been described by a "stepwise" (or onion-shell structure) approach consisting of: (a) physicochemico property requirements (*e.g.* determined from the properties of the training set), (b) structural requirements for accurate prediction (based on the structural diversity for available data), (c) structural features determined by known mechanistic requirements (such as reactive functional groups) and (d) the requirements of metabolic activation and biotransformations.[7]

Derek alerts, to varying extents, include an implicit applicability domain in the definition of the scope.[8] Reactivity based alerts are often assumed to have unrestricted applicability across chemical space except when there are additional limitations applied by absorption considerations, as for skin sensitisation prediction (calculated skin permeability coefficients $>10^{-5}$ cm sec^{-1}). Where non-covalent interactions contribute to toxicity (*e.g.* for hERG channel inhibition, teratogenicity, hepatotoxicity and some chromosome aberration alerts) some restrictions that more closely reflect the structures of known examples, where possible, are placed on the scope. It is not generally possible to define the applicability domain scope exactly in this way and in setting the scope of an alert, a judgement is made to cover the maximal chemical space without severely compromising the accuracy. When, for a given end-point, there is no prediction of toxicity, the level of confidence of that prediction will depend on whether the chemical is sufficiently similar to chemicals of known toxicity and, specifically, contains no new chemical functionality not previously considered, that may potentially give rise to activity or influence toxicity due to other parts of the molecule firing the alert. Particularly for such circumstances, we are currently investing in research to develop a means for providing advice on whether a chemical is within an alert's or an end-point's applicability domain. Some initial results have been published.[9]

11.3 A Description of Meteor

Like Derek for Windows, the Meteor xenobiotic metabolism system has also recently been reviewed.[2] It shares many commonalities with Derek for Windows (and indeed was developed from a common computer-code base) in terms of, for example, the processing of query structures with different constraints, the viewing of results and the generation of reports using different filters. The Meteor knowledge base is composed of a biotransformation dictionary, rules and example metabolic reactions. The biotransformation dictionary consists of structure–metabolism relationships (SMRs) expressed as generic reaction descriptions (biotransformations), and these can be thought of as the equivalent of the Derek for Windows alerts but with the incorporation of metabolites and their reaction intermediates, which are displayed in the form of sizable and scalable reaction trees and results tables (Figure 11.5).

Biotransformations are supported by comments, which contain information on the scope of the biotransformation and the SMR as well as an outline of the applicability of the biotransformation (usually a summary of the range of chemical space in which the biotransformation has been observed). Comments will also contain information on: the mechanism of the chemical transformation (which can involve a number of distinct steps and include reaction intermediates); the enzymes involved in the process (and isozymes where this has been established); some commentary on whether the metabolites associated with the biotransformation are expected to be major, minor or detectable only at trace levels; the general quality of structural characterisation of such metabolites and any species selectivity that is known. Biotransformations will usually contain citations of books, reviews and primary literature sources and may additionally have supporting example reactions linked to them.

The rules in the knowledge base are used by the program's reasoning engine to discriminate between all possible metabolic outcomes and the most likely

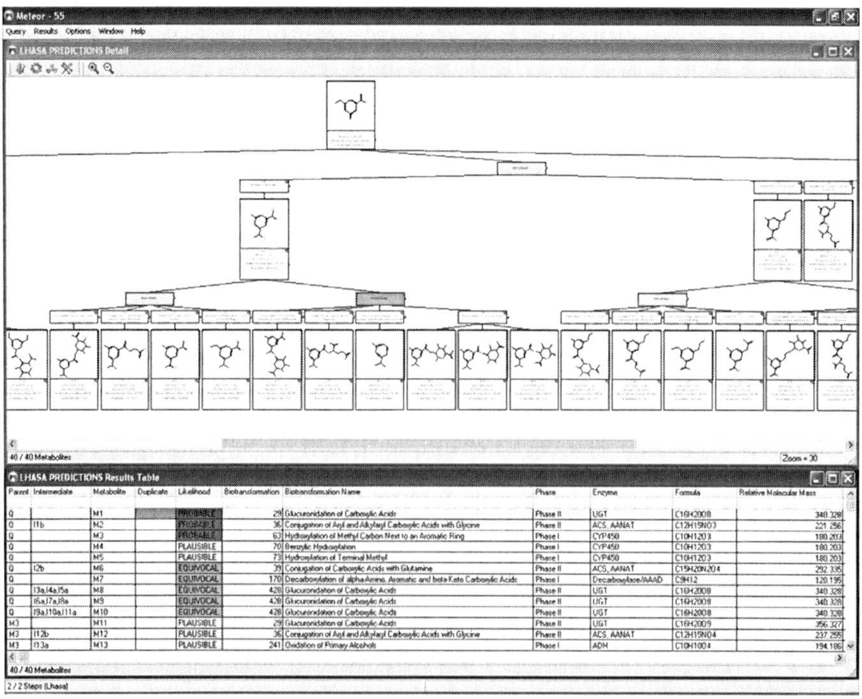

Figure 11.5 Meteor on-screen results display showing the metabolic tree and the summary table.

ones and are of two types: absolute and relative. Absolute reasoning rules determine the likelihood of a particular biotransformation taking place and use the same uncertainty terms as Derek for Windows: "probable," "plausible," "equivocal," "doubted," or "improbable". Relative reasoning rules evaluate the comparative likelihood of two different biotransformations taking place and therefore allow rules about potentially competing reactions to be written, a requirement that Meteor does not share with Derek for Windows. An example of a (somewhat simplistic) absolute reasoning rule would be, "*N*-demethylation is probable," whereas an example of a relative reasoning rule might be, "*N*-demethylation is more likely than *O*-demethylation." Currently, relative reasoning in Meteor is unconditional, therefore only two types of rule can be written. Where a preference exists, a rule of the type, "biotransformation 1 is more likely than biotransformation 2" is indicated. Where no preference exists and the viability of two biotransformations is judged the same, a rule of equality ("biotransformation 1 is as likely as biotransformation 2") can be written. Further details of the relative reasoning methodology used in Meteor have been described elsewhere[10] and will not be discussed further here.

The situation with absolute reasoning rules is more complex since the likelihood of biotransformations can be reasoned depending on factual and heuristic elements such as rates of occurrence in the literature and any known inter-species variations, as well as numerical factors such as substrate lipophilicity and molecular weight. These can be calculated internally or obtained through adaptors to external packages in the same way as for Derek for Windows.

As an example, the effectiveness of glycine conjugation with carboxylic acids to give hippuric acids is known to be dependent on the molecular weight of the substrate. Low molecular weight substrates, such as very simple benzoic acids, often give hippuric acids readily, whereas higher molecular weight acids tend to favour the formation of acyl glucuronides or other metabolic products. The change in outcome for two substrates (both of high lipophilicity) one of high and one of low molecular weight (a reduction in confidence from "probable" to "equivocal") can be traced through the reasoning display (Figure 11.6). For this biotransformation, it can be seen that the outcome may also be moderated by the lipophilicity of the substrate, illustrating the fact that Meteor, like Derek for Windows, is capable of reasoning with multiple numerical input parameters as well as with factual and heuristic data.

11.3.1 Integration of Meteor with Other Programs

The need for reasoning in Meteor is self-evident. In an unconstrained analysis, an unmanageable number of metabolic pathways would be generated and this is not always useful or desirable. Rather than allowing this to happen, reasoning affords a method by which biotransformations can be prioritised and the more likely results, rather than all possible outcomes, displayed. However, users of the program may set the processing constraints such that a query can

Figure 11.6 Meteor on-screen reasoning display showing the dependence of molecular weight and lipophilicity on the likelihood of glycine conjugation with carboxylic acids.

be analysed using different reasoning levels, and this gives the user degrees of flexibility. For example, setting the absolute reasoning threshold to "equivocal" or "doubted", displays putative and increasingly more speculative metabolites as well as the likely ones. Raising the level to "plausible" or "probable", results in the display of the more likely metabolites. Recent studies[11–13] have shown a satisfying trend in performance improvement, with vastly decreasing occurrences of false negative predictions at the level of "equivocal" or higher and an acceptable level of unconfirmed positive predictions at these same levels. Like alerts in Derek for Windows, Meteor biotransformations are constantly reviewed and updated. Many of them have sophisticated definitions of scope which includes, for example, restrictions based on lack of observed reactivity due to high steric encumbrance or poor electronic properties, as well as such logical interpolation and extrapolation as is judged appropriate by the expert developer. Despite this level of sophistication, it is often observed that many biotransformations are predicted at the same levels of combined absolute and relative reasoning, but in reality, and for reasons sometimes difficult to discern by eye, not all such metabolic conversions are equally facile. Enhanced specificity in prediction is obviously desirable, and the use of quantitative estimations of reactivity parameters may be useful in discriminating between such coequally rated processes.

SMARTCyp[14] is an *in silico* method that predicts sites of cytochrome P450 (CYP) -mediated metabolism of drug-like molecules. To estimate these susceptibilities, SMARTCyp uses an algorithm, which involves an atomic reactivity descriptor and an atomic accessibility descriptor. A combination of these two descriptors for each atom gives an overall score for that atom, where a low score represents a high probability that the atom will be a site of metabolism. The program then ranks each potentially susceptible atomic centre from most to least likely. The results of a recent comparative evaluation of Meteor and SMARTCyp[15] showed that SMARTCyp has the potential to discriminate between biotransformations that are ranked equally by Meteor. Such metabolites are not always all confirmed either positive predictions or unconfirmed positives. Meteor predicts three *O*-demethylation reactions for the substrate M445526 (Figure 11.7), all at the same level. In reality, only two of the possible three metabolites are confirmed experimental positives: those arising from biotransformation at methoxy 1 and 2. Biotransformation at methoxy 3 is not observed and interestingly SMARTCyp ranks this site of metabolism lower than the other two. A second example concerns hydroxylation of alicyclic methylene groups in tramadol. Here, there are four potential sites of metabolism in the cyclohexane ring (methylenes 1, 2, 3 and 4). Meteor predicts biotransformation at the same reasoning level for all four sites, however only one metabolite, corresponding to metabolism at methylene 2, is observed. Metabolites arising from biotransformation at methylenes 1, 3 and 4 are

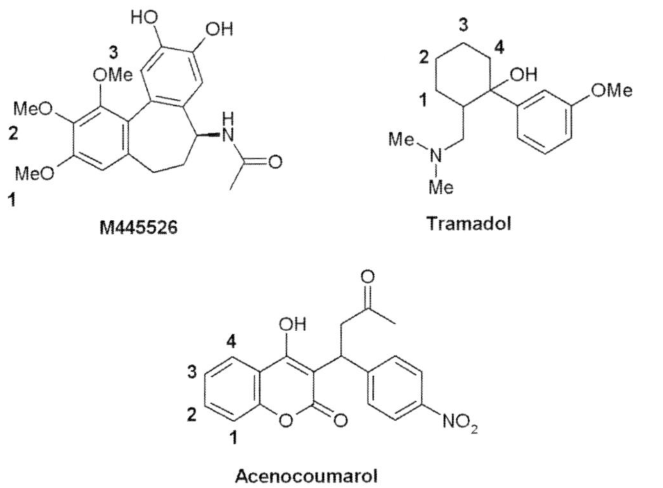

Figure 11.7 Some predicted sites of metabolism for M445526, tramadol and acenocoumarol.

unconfirmed positive. Interestingly, SMARTCyp rates methylene 2 as having a higher potential for biotransformation than the other three, in accordance with the experimental observation. In a final example, acenocoumarol, a similar analysis of the aromatic hydroxylation scenario displays a similar but inverse pattern. All four hydroxylation sites in the fused benzene ring of acenocoumarol (carbons 1, 2, 3 and 4) are predicted at the same absolute reasoning level. Of these four sites, metabolites corresponding to biotransformation at three of them (atoms 1, 2 and 3) are observed. The metabolite corresponding to biotransformation at atom 4 is not observed. Again, of these four sites, SMARTCyp ranks atom 4 lower than the other three. This preliminary study has indicated that a quantitative model can refine the results of the expert system prediction. Further studies are aimed at establishing the model space in which such quantitative support can be beneficial and informative.

11.4 The Expert System for Toxicity Prediction in Drug Design and Discovery

Drug discovery can cost approximately US $1.8 billion and often takes over 10 years. Out of thousands of potential lead compounds, less than 0.1% pass the safety stages.[16] Of the few that are taken as potential leads, there is often only one resulting successful candidate. During drug design, compounds will undergo refinement, testing for good absorption, distribution, metabolism and excretion (ADME) and toxicity testing. Promising candidates will then undergo formulation and testing for shelf-life. After this, it is no wonder that many compounds fail to make the cut. In many cases, lead compounds are discovered by chance or from natural products and previously marketed drugs (the *"me too"* approach), but often a rational approach to drug design is needed in order to reduce risks for pharmaceutical companies later on in development. Perhaps due to increased legislature as well as increased regulation on animal testing, there has been a greater need than before for alternative methods for gaining toxicity information. Toxicity is obviously a major concern throughout drug development as the final product is likely to be used on human patients. As a result, safety in terms of toxicity is high on the agenda, as is ensuring any adverse affects are identified as soon as possible. The last decade has seen a steady rise in the use of *in silico* tools in order to aid the drug discovery process, particularly in the area of toxicity. Derek for Windows, like many expert systems, uses rules and reasoning, as described earlier. This provides the user with transparency (including references, comments, examples and so on) and also highlights, in on-screen displays and, optionally, reports, the toxicophore in the structure that is involved in causing toxicity.

11.4.1 Storage of Toxicological Knowledge

Pharmaceutical companies test thousands of compounds during each drug discovery process, where new data becomes available for many of these compounds. Due to these processes being split across different sites, as well as across research groups working on alternative drugs and areas, a vast amount of this knowledge and learning may get lost or is not always passed on to colleagues. Expert systems such as Derek for Windows can be used in order to improve communication, not only within departments, but across other research groups, geographical locations and different projects. Derek for Windows contains an editor within its knowledge base, where toxicologists can input rules and alerts which then become available for all in-house users. This not only provides good data storage for future use (thus preventing repetitive experiments) but also enables Derek for Windows to predict more accurately for in-house compounds, therefore reducing any potential false positives and perhaps worse, false negatives. As well as this, there is a data sharing initiative within Lhasa that encourages companies to collaborate and provide new SARs that can be implemented into Derek for Windows to take into account rare or unknown toxicities. This further enables a reduction in animal testing, as well as saving pharmaceutical companies time and money in repetitive and unnecessary testing.

11.4.2 Lead Identification

The idea of high throughput screening is to rapidly sort through thousands of compounds into those that bind to the target (*e.g.* receptor, enzyme) with a high potency (strength of effect) and high activity (a particular biological or pharmacological effect), and those that don't. ADME also needs to be observed. This method can show how selective each compound is for the target of interest. If a potential new molecular entity (NME) is proving not to be specific in its target, this could prove to be an area of concern later in the design process as it could lead to toxic interactions. This second process is referred to as cross-screening. Once highly potent compounds have been found, expert systems can be used during this stage to assess for toxicity. This removes potentially toxic compounds right at the start of discovery, reducing the risk of late stage failures that would cost pharmaceutical companies billions as well as a significant delay in their development. This process can be referred to as 'flagging'. During late stage screening, expert systems can also be used to rank and prioritise compounds on the basis of safety and therefore which should be tested first. During this stage, more chemicals are discarded and those that pass are then taken into lead optimisation and modification. In pharmaceutical companies, most emphasis is put on genotoxic alerts. This is due to increased legislation in this area (see also Section 11.4.8).

11.4.3 Lead Optimisation

During the later stages of development and lead optimisation, other factors (such as how the drug metabolises) are taken in to account. Additionally, any impurities and intermediates are considered and structures are modified to increase potency and specificity to the target. These areas can all lead to toxicity and therefore need to be tightly scrutinised for any such evidence. As Derek for Windows highlights the toxicophore involved in causing toxicity, it can also be used in order to aid structure refinement, as well as providing more information on the mechanisms behind the toxicity. During this stage, many companies look into using in-house literature and data as well as references used to support Derek for Windows alerts, in order to fully evaluate each compound. During this process, other expert systems (such as Meteor for metabolism prediction, and toxicity databases such as Vitic[2]) can be used to gain further insight into potential areas of concern, enabling scientists to know what structures to avoid and which areas are more prone to metabolism.

11.4.4 Pre-Clinical and Clinical Stages

The failure rate due to toxicity during early stage pre-clinical testing is thought to be around 20%.[16] It is during these final stages of development that other toxic endpoints, such as skin sensitisation (if it applies) and phototoxicity, will be considered. It is also during these stages that formulation scientists need to start considering the packaging and excipient compatibility concerned with the drug. There are also stability considerations and shelf-life testing. Again, toxicity will be a concern, as the active pharmaceutical ingredient (API) may react with an excipient, causing toxicity. Expert systems can be used here for final checks and to provide evidence that certain areas do not require further testing, or to explain why other alerts have been rejected for unlikely toxicity.

11.4.5 Occupational Health Assessment

Although the main focus has seemingly been on drug design, it is also important to note that many scientists involved in drug discovery, especially at the chemical synthesis stages, can be at risk of exposure to harmful chemicals such as the drug or synthetic intermediates. Derek for Windows can be used to assess such occupational hazards. An example of this is skin sensitisation analysis. Previous to this, tests such as the guinea pig maximization test (GPMT) were used to assess the dangers of skin sensitisation. Derek for Windows, together with use of the local lymph node assay (LLNA), which uses less animals than the GPMT, has reduced the need for this test to the point of it almost being completely discarded. As a result, it has enabled a reduction in animal testing by approximately two thirds for organisations that apply this strategy.[17] Following such a paradigm, computer analyses are used prior to testing, and if companies are still unsure, the LLNA can be used to confirm the safety of the relevant compounds. Many of these predictions can be further supported by alerts which use Ames test results as well as physicochemical properties to instill confidence to its users.

11.4.6 Environmental Safety

Due to stricter legislation, environmental issues are becoming more of a concern to pharmaceutical companies. The future is likely to see expert systems being used to a greater degree for predicting the impact of certain chemicals on the environment.

11.4.7 Support of Regulatory Applications

Although *in silico* tools have been around for several decades and widely used in the pharmaceutical industry since around 1996, the use of expert systems by regulatory authorities only really started around the early 2000s. Expert systems are now widely accepted within regulatory authorities and are even mentioned in several guidelines provided by bodies such as the European Medicines Agency (EMA)[18] and the US Food and Drug Administration (FDA),[19] as well as in International Conference on Harmonisation (ICH) for Registration of Pharmaceuticals for Human Use[20] guidelines. Many regulators choose to use expert systems as a final check for new drug applications (NDAs), in order to ensure the toxicity profile is not missing anything potentially harmful to human health. As the use of *in silico* tools has rapidly increased within pharmaceutical companies, many submit reports from software such as Derek for Windows to regulatory authorities, to further support claims that they have taken into account all possibilities of toxicity, as well as to provide evidence that no further testing is required. As a result of these regulatory submissions, many regulators now also use expert systems in order to help them interpret and clarify the submissions. The use of Derek for Windows is likely to increase as regulatory controls become increasingly tight. We may expect to see this use becoming even more important in the cosmetics and personal products industries and in chemical companies, as Registration, Evaluation, Authorisation and Restriction of Chemicals (REACH) regulations and animal testing restrictions begin to take become implemented.

11.4.8 The Use of Expert Systems to Assess Genotoxic Impurities

At the beginning of the 21st century, it was starting to be recognised that it was not only the investigational new drugs (INDs) and NDAs themselves that needed to be assessed for toxicity during clinical development, but also the impurities created as a result of their development. When conducting risk assessment for INDs and NDAs, it is important to understand the actual overall effects of the substance on the end target; in the case of pharmaceutical products, the end target is usually human beings. These effects include both the required effect, for example inhibiting the replication of human cytomegalovirus (CMV) to treat HIV, and possible side effects (such as carcinogenicity).

The ICH published the first guidelines to address how impurities should be treated, although these contained no specific guidance on acceptable levels for genotoxic impurities other than to recognise the fact that unusually toxic impurities may require tighter limits of control.[20] The threshold of toxicological concern (TTC) is a belief that for individual chemicals with known toxicological profiles, there are levels of exposure which are considered safe. The concept proposes that a low level of exposure with a negligible risk can be identified for many chemicals, including those of unknown toxicity, based on knowledge of their chemical structures.[21] This idea has been developed gradually over the years so that extensive analysis of actual toxicity data is now considered when deciding these levels. However, whilst this concept was evolving, pharmaceutical companies took it into their own hands to categorise impurities using a variety of methods, including SARs. This was outlined by the Pharmaceutical Research and Manufacturers of America (PhRMA) (Figure 11.8). In recent years, the EMA and the FDA have both published guidelines on how to treat genotoxic impurities, and both consider the levels at which the genotoxic effect of these impurities is considered to be safe, or where the risks are outweighed by the benefit. The EMA guidelines establish that a TTC level of 1.5 μg per day intake of a genotoxic impurity is considered to be associated with a low cancer risk.[18] This is corroborated by the FDA draft guidance.[19]

The FDA draft guidance clearly states that if an impurity that is present at levels below the ICH qualification thresholds is identified, the impurity should be evaluated for genotoxicity and carcinogenicity based on SAR assessments (that is, whether there are structural alerts present or not). This evaluation can be conducted by a review of the available literature or through a computational toxicology assessment; commonly used software includes MDL-QSAR, MC4PC and Derek for Windows. The EMA guidance supports this by stating that potential genotoxic impurities should be identified with the guidance of existing genotoxicity data or the presence of structural alerts. It has also been suggested that when a potential impurity contains structural alerts, additional genotoxicity testing of the impurity, typically in a bacterial reverse mutation assay, should be considered.[22,23]

Since the introduction of these guidelines, many pharmaceutical companies are now focussing more on establishing *in silico* results for their impurities before entering their NDAs and INDs into actual *in vitro* tests.

Although individual companies each have their own, slightly different, working procedures, a generalised prioritisation strategy for impurities can be

Class 1 Impurities: Impurities known to be both genotoxic (mutagenic) and carcinogenic.

Class 2 impurities: Impurities known to be genotoxic (mutagenic), but with unknown carcinogenic potential.

Class 3 impurities: Alerting structure, unrelated to parent structure and of unknown genotoxic (mutagenic) potential.

Class 4 Impurities: Alerting structure, related to the parent Active Pharmaceutical Ingredient (API).

Class 5 Impurities: No alerting structure or indication of genotoxic potential.

Figure 11.8 PhRMA classification of impurites.

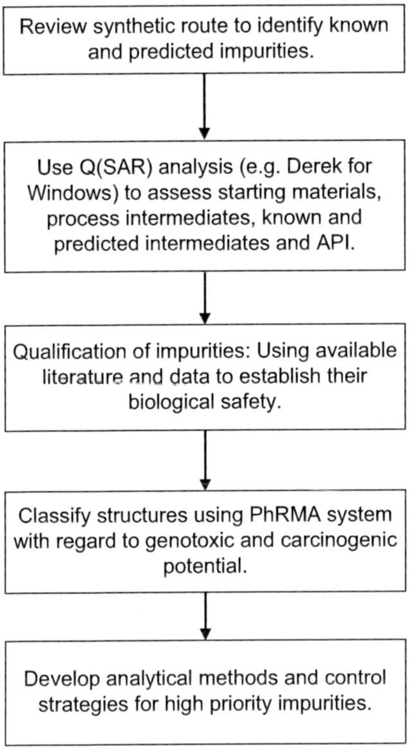

Figure 11.9 Generalised prioiritisation strategy for dealing with potentially genotoxic impurities.

outlined (Figure 11.9). This is used to establish which impurities need analytical methods developed and quality control procedures established to reduce their genotoxic potential. Q(SAR) models are generally used at early stages to help this classification process.

To facilitate further harmonisation, in June 2011, there was a steering committee meeting of two new expert working groups to discuss the ICH Guidelines M7 (Assessment and Control of DNA Reactive (Mutagenic) Impurities in Pharmaceuticals to Limit Potential Carcinogenic Risk) and S10 (Photosafety Evaluation). The ongoing aim is to reach a consensus on the procedure for conducting a (Q)SAR evaluation, including how to judge what constitutes an appropriate (Q)SAR model, how to have confidence in a negative result from a database and how to incorporate public and proprietary sources of data into an integrated assessment. This will mean that all organisations involved in the development of drugs will be assessing their impurities to the same transparent standard.

Expert systems can however be used throughout the development process and in a variety of ways to reduce the risk of cancer from genotoxic impurities. The ICH Q3A/B guidelines indicate that qualification studies can be conducted on the API that contains the impurity, although the FDA recommends that assays are used where the impurity is in isolation. It is not until the impurities have been classified that they are submitted for further, more expensive and time consuming *in vitro* testing. The classifications are shown in Figure 11.8, where class 1 and 2 are given the highest priority for developing analytical methods and controls. Impurities assigned to class 3 by the use of SAR systems alone are treated as genotoxic until they are assessed otherwise using the Ames test, although the FDA does state that if an impurity activates a genotoxic structural alert within Derek for Windows but can be qualified at levels lower than the daily exposure threshold (1.5 μg per day) then no further testing is required.

One example of an instance where Q(SAR) models have been used in industry to support higher specifications involved a company in 2005, before regulatory bodies had approved the use of Q(SAR) systems. They were testing an impurity that was a dimer of the drug substance (DS) and had been qualified by conducting toxicity testing on the DS but was only found at a level of $<1\%$. They assessed the impurity as an isolated entity using both Derek for Windows and MCASE after discovering that it was not stable enough to be tested directly using *in vitro* methods as it could not be isolated. No structural concerns emerged from these *in silico* analyses; Derek for Windows predicted skin sensitisation as equivocal for mammals, as it did for the DS itself. This was supported by fully explained reasoning and a comments section that contained the mechanism of action within the Derek for Windows program. This tool also has the added benefit that it provides information about current published literature relating to the associated structural class. The regulating bodies took this as acceptable evidence.

Although there are no formalised TTCs for genotoxic metabolites as there are for impurities, strategies for the risk assessment of genotoxic metabolites in humans have been proposed.[24]

11.5 The Expert System for Metabolism Prediction in Drug Design and Discovery

Although not every drug-related toxicity is metabolism-based, active metabolites of a parent drug circulating in the body can be effectors of both desirable and adverse pharmacodynamic effects.[25] Predicting and understanding the metabolic profile of a drug intended for human use is of major importance for drug discovery and development, and this is also becoming increasingly important in risk assessment and regulatory review.[26] The chemistry and biochemistry of metabolic reactions and the consequences of such reactions on activation and inactivation (toxification and detoxification), predictions of drug metabolism (pro-drug and soft-drug design), and changes in physicochemical properties (acidity, basicity, lipophilicity, *etc.*) resulting from biotransformation can affect both the pharmacokinetic (distribution and excretion in particular) and pharmacodynamic (efficacious and adverse pharmacology) behaviour of a drug. The ability to predict and anticipate the biotransformation of any given compound, including sites of metabolism and the quantities, distribution and nature of metabolites generated, is highly desirable.

Available methods for the prediction of metabolism can be classified as local or global.[12] Local predictive methods are applicable to simple biological systems such as a single enzyme or a single reaction; they include quantitative structure–metabolism relationships (QSMRs), three-dimensional quantitative structure–activity relationships (3D-QSAR) and comparative molecular field analysis (CoMFA), quantum-mechanical calculations, molecular modelling of enzymes and ligand docking (pharmacophore modelling) and combinations of the these methods. Global methods are applicable to versatile biological systems (multiple enzymes and reactions) and to series of compounds with broad chemical diversity; they include knowledge based expert systems, such as Meteor, which we discuss in this chapter.

11.5.1 Meteor and the 'Metabolites in Safety Testing' Guidelines

In many cases, a drug itself does not reveal any toxicological concerns. Once metabolised however, several toxic concerns can result, often in the form of reactive intermediates. Drug metabolism therefore often plays an essential role in drug safety and clinical efficacy. A major concern within pre-clinical testing is whether animal studies are showing accurate toxicity in relation to human toxicity, and if not, what the next step should be. The metabolite in safety testing (MIST) guidelines, published by the FDA,[27] apply to small molecule drug products and suggest how to test and identify potential toxic metabolites. These metabolites require further non-clinical toxicity testing as they are often found to be present in greater quantities in humans, or are identified uniquely in humans in standard non-clinical tests. Drugs for severe life-threatening conditions such as cancer are exceptions to these guidelines as the risk–benefit is modified and lowered to take into account the severity of the disease. Currently the MIST guidelines are still that, guidelines. They are simply a recommendation to drug manufacturers, where the FDA are expressing their current views on this topic.

Current testing strategies are usually sufficient when the testing is done in at least one species which has a similar metabolic profile to humans. This often consists of assessing drug exposure in mainly parent drug plasma concentrations. Similarities however are not always consistent and often situations can occur where metabolites have not been identified, or if they have, their evaluation has not been thorough. As it is not the norm for potential metabolites to be evaluated across a variation of species, there can often be cases in late drug discovery where these metabolites have gone unnoticed in causing toxicity and adverse drug reactions (ADRs). It is important to note that although tests are becoming more diverse and sensitive for assessing metabolites (through the use of mass spectrometers, for example), lack of testing paradigms is perhaps a major contributor to insufficient knowledge on potential toxic metabolites and the mechanisms involved in their role within the toxicity of drugs.

Meteor allows users to choose between showing phase I and phase II metabolites. This is extremely important within the guidelines as it is phase I metabolites that are most likely to be of concern due to being chemically or pharmacologically reactive. They are often reactive intermediates which could potentially bind to an alternative target, such as a protein or receptor, or even to the target receptor, in both cases resulting in toxicity or undesired side-effects. In rare cases, it is a particular concern when these metabolites are formed in humans and not the test species. More likely however is the occurrence of these metabolites at higher concentrations in humans than in animals. Therefore the MIST guidelines state that if a metabolite forms in at least one species at adequate exposure levels (greater than or equal to human exposure) it can be presumed that the metabolite's contribution to drug toxicity has been established. As most of the attention is directed to phase I, there is also the option in Meteor to stop the analysis at this point.

Many metabolites can be short-lived and therefore hard to identify. During testing, there are options to synthesise these metabolites and administer them directly, in order to identify the effect of intermediates. However, this is not simulating the exact effects of the drug and can therefore be misleading. These intermediates could form a toxic compound and therefore it is important that these are investigated further. Meteor can show intermediates that are potentially adduct-forming (Figure 11.10). Previously uncharacterised but presumed intermediates are included, to give an idea of what could be present in order for a particular biotransformation to make mechanistic sense. Showing that a metabolite is pharmacologically inactive at the target site does not confirm it is safe. Cause for concern is usually considered, according to MIST, when human metabolites are formed at greater than 10% of parent drug systemic exposure at steady state.

It is currently not standard practice to test each metabolite individually across different species. As a result, some metabolites may be pivotal in causing the toxicity of an NME, but this may remain overlooked. Meteor gives individual information on each proposed metabolite from a parent. This can provide the user with important information on what action to take to modify the API, such as adding or modifying substituents with a view to reducing the metabolic liabilities at those sites.

The MIST guidelines state that it is important to take into account the similarity of the metabolite to the parent; the chemical class, solubility and stability in the stomach (if it is to be administered orally); phase I *versus* phase II metabolism and the relative amounts detected in humans compared to animals. The similarity can be compared as Meteor represents the metabolites in structure format within a results tree. These can further be exported to, for example, an SD file.

In order to gain an idea of toxicity of each metabolite and to be more specific about the endpoints of concern, Meteor is able to link to Derek for Windows, which is able to provide toxicity information related to specific species. This can

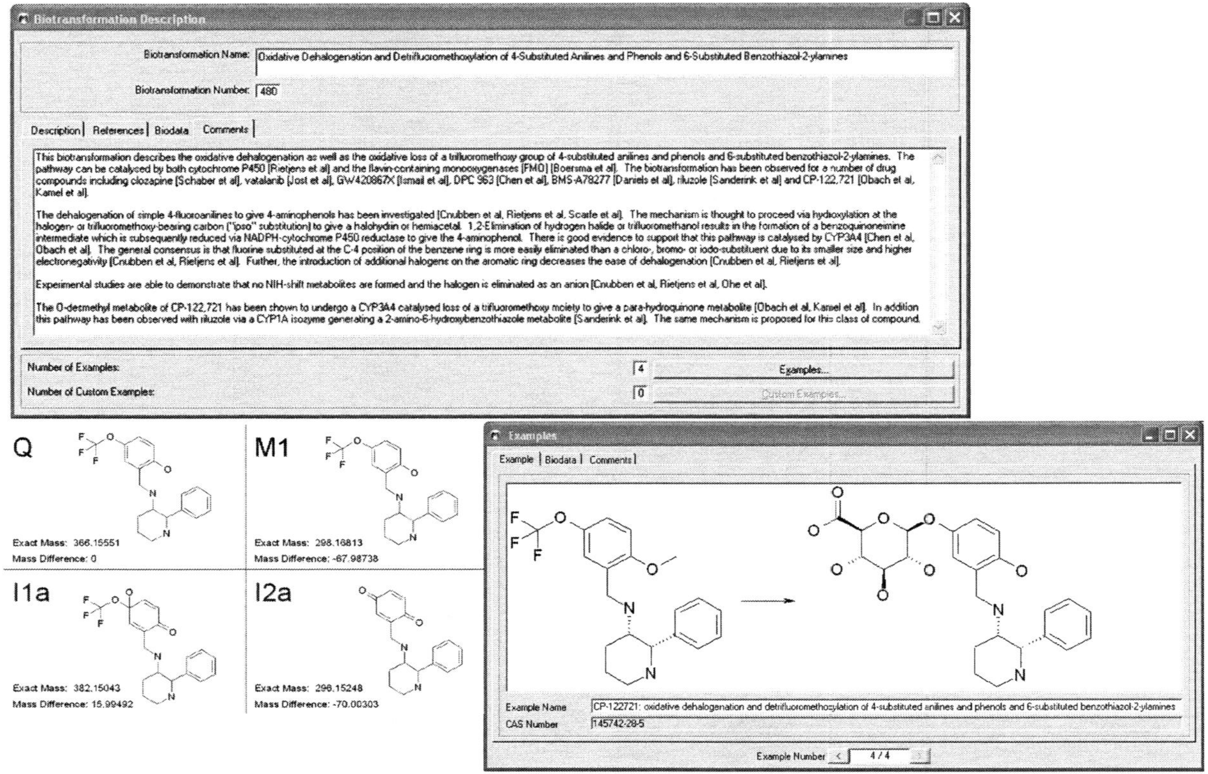

Figure 11.10 Comments and example for a biotransformation generating potentially adduct-forming intermediates (I1a and I2a).

aid in building a profile not only of the drug, but of potential toxic metabolites, which may be safe in dogs and bacteria but not in humans.

An important point to consider is that Meteor predicts general mammalian metabolism. Where there is relevant data, this will be noted (as mentioned previously) within the comments and examples. Species-specific (or selective) predictions are difficult, since most variation is quantitative rather than qualitative and predictive quantification of metabolites remains a significant challenge. Also, species-specific data tends to be available on a substrate-specific rather than a reaction type-selective basis, and this makes generalisation difficult. As a rule, if a metabolite is toxic in animals, the MIST guidelines advise that this will be a concern in humans, as long as the species is similar in nature in terms of metabolism profile. The ability to predict across species in a non-selective way is also useful. In a recent evaluation exercise, we compared the predicted biotransformations of seventeen hepatotoxic drugs with known metabolic profiles.[28] For one of these drugs, (zomeripac, a cyclo-oxygenase 2 inhibitor) Meteor predicted the zomeripac-1-*O*-acyl glucuronide as a probable metabolite. Acyl glucuronide metabolites of carboxylic acids have been implicated as chemical initiating agents in hepatobiliary adverse effects and have been shown to bind irreversibly with plasma albumin. Correlation of the prediction for zomeripac with the known metabolism was good: over 56% of the administered dose was detected as the acyl glucuronide in humans and monkey. But, significantly, in rats, rabbits and hamsters, this metabolite was detected at trace levels at best and often not at all. This indicates the usefulness of *in silico* tools over and above relying solely on *in vitro* rodent liver microsome or hepatocyte experiments.

If a metabolite is only present in low doses in humans, it could still be potentially toxic. *In situ*ations such as this where the opposite is true, further dose testing is required. However, a metabolite that is identified in animals but not in humans is unlikely to be of concern, even if it is toxic in particular species. In this case Meteor is unable to provide information on dose dependencies for reasons discussed.

11.5.2 The Use of Meteor to Accelerate Metabolite Identification

One of the most popular applications of *in silico* metabolism simulation is as an aid in metabolite structure elucidation during analysis of samples collected in the course of metabolic studies. Typically, metabolite characterisation is attempted using tandem liquid chromatography-mass spectrometry (LC-MS) techniques. Technological advances during the past decade have greatly improved the analytical capabilities for the detection, identification and characterisation of metabolites, and this has lead to a better understanding of the role metabolites play in drug safety assessment. As a result, a wealth of data can be obtained during analysis, so any tools that assist in the interpretation or allow targeted data acquisition are very useful to the analyst. Many vendor supplied LC-MS data analysis tools contain an empirical list of mass differences for common phase I and phase II metabolic reactions, however they are not predictive and manual selection of the metabolites is required by the user.

The FDA suggests that the metabolic profile of a drug be identified during the drug development process.[27] *In vitro* studies utilise liver microsomes, liver slices or hepatocytes from animals and humans before initiation of clinical trials. Results from *in vivo* metabolism studies should be available early in the drug development process since they confirm results from *in vitro* studies or reveal quantitative or qualitative differences in metabolism across species, which would cause a safety concern. Human *in vivo* metabolism is carried out at a later stage.

Several attempts have been made to use multiple tools in tandem to help accelerate metabolite identification. In the early 2000s, the analytical chemistry experts at BioRad were using their accumulated expertise in converting several large handbooks of spectral data into electronic format to generate KnowItAll, a one stop solution for absorption, distribution, metabolism, excretion and toxicity (ADMET) using CompuDrug's MetabolExpert. Most of their efforts however concentrated on generating consensus models for toxicity prediction. In the absence of having a single tool to provide all the answers, published experiments typically use metabolite prediction software such as Meteor, combined with fragmentation prediction software, such as Apex (Sierra Analytics) and MS Fragmenter (ACD/Labs) and the metabolite detection software that is supplied with the mass spectrometer being used in the experiment. There are several publications describing such methods, typically focussing on a single well characterised drug to see if this strategy can add value to the metabolite elucidation process. This section explores a few of these examples.

11.5.2.1 Case Study 1: Indinavir

Sierra Analytics and Applied Biosystems[29] investigated the metabolic profile of indinavir, an HIV protease inhibitor with a very short half-life (2 hours), readily metabolised via CYPs to produce phase I and subsequent II products. Aliquots (10 μL) of indinavir incubated with rat liver S9 fractions were separated by online liquid chromatography and analysed with an Applied Biosystems hybrid triple quadrupole/linear ion trap mass spectrometer in information dependent acquisition (IDA) mode. This consisted of an enhanced MS survey scan followed by enhanced product ion scans for the two most intense parent ions as dependent MS/MS experiments.

Six phase I metabolites were identified by analysing the spectral data alongside an expert chemist's review of the public literature. Meteor (version 7) predicted five of the six metabolites, along with products arising from sulfonation, glucuronidation, hydrolysis and other pathways. Meteor did not predict M3 (Figure 11.11), which is thought to arise from the sequential application of two phase I biotransformations on the parent *N*-dealkylation and benzylic hydroxylation. Biotransformation likelihood can be attenuated by consideration of whether the process is the first step in a sequence or a subsequent step. On this occasion, the likelihood of the second phase I process

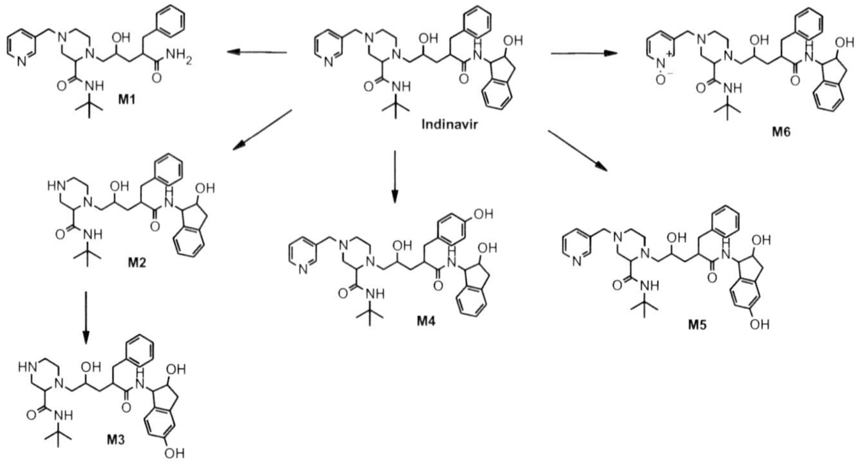

Metabolite	Seen Experimentally Biotransformation Name	Predicted by Meteor (Biotransformation Number, Absolute Reasoning Level)
M1	Oxidative N-Dealkylation	244, Probable
M2	Oxidative N-Dealkylation	243, Probable
M3	Hydroxylation of Fused Benzenes	*Not predicted*
M4	Para Hydroxylation of Monosubstituted Benzene Compounds	78, Plausible
M5	Hydroxylation of Fused Benzenes	209, Plausible
M6	N-Oxidation of Aromatic Nitrogen	102, plausible

Figure 11.11 Indinavir phase I *in vitro* metabolites identified with a Q-TRAP™ instrument matched with the Meteor (version 7) prediction.

fell below the chosen threshold ("plausible") for the analysis and therefore was not generated, displayed or reported.

The APEX software, capable of reading raw data from most industry standard mass spectrometers, automatically processed the experimental LC-MS and MS–MS IDA dataset. It then used its comprehensive scoring and correlation algorithm to generate a set of predicted parent ion isotope patterns and product ion fragmentation spectra by importing an SD file of Meteor-predicted metabolites for indinavir (Figure 11.12). By correlation of the predicted and experimental spectra, APEX determined that the most probable predicted metabolites were the same five as previously identified. Reported retention times and relative concentrations were consistent with the manual analysis. Rejected structures were either inconsistent with observed parent ions or had poor correlation between predicted and observed fragmentation spectra.

Overall, the use of a Q-TRAP Instrument with IDA demonstrated powerful capabilities of high throughput quality data acquisition. APEX metabolite identification examines the isotope patterns of the targeted ions to give more accurate results, increasing the confidence level of the automation. Meteor showed satisfactory prediction abilities that could be improved with more knowledge about more complicated or combined biotransformation pathways. In fact, in Meteor (version 12, the current version), version, M3 is predicted, albeit *via* the reverse mechanism to the one proposed (*i.e.* aromatic ring

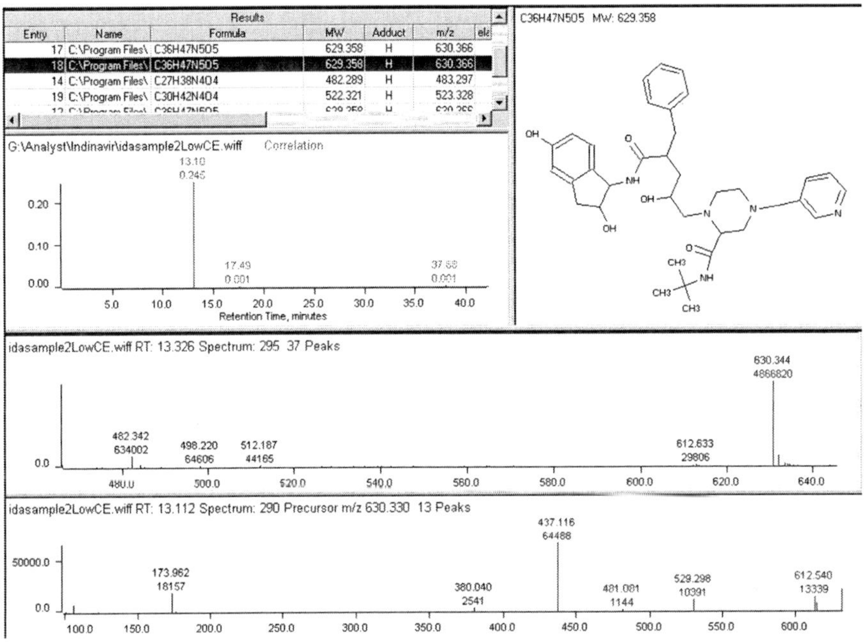

Figure 11.12 APEX Identification of indinavir metabolite M5 (from Figure 11.11). Details of the match of predicted and experimental spectra are given in the table at the upper left of the Figure. The correlation score chromatogram, based on combined MS and MS/MS match is shown below the table. The Q-TRAP EMS spectrum, with match to the predicted isotope cluster is in the center, and corresponding EPI MS/MS spectrum with matches to the predicted fragmentation on the bottom.

hydroxylation followed by *N*-dealkylation, via M5 rather than M2). Combination of the three tools demonstrated great successes with the indinavir project.

11.5.2.2 Case Study 2: Quetiapine

The University of Helsinki is in the process of automating liquid chromatography-time-of-flight mass spectrometry (LC-ToFMS) methods, which yield accurate mass results despite the complexity of the biological matrix, to screen for toxic metabolites in urine samples. They published a method that detected thirteen phase I metabolites (Figure 11.13) from the antipsychotic drug, quetiapine, without needing to rely on the use of reference standards,[30] which are rarely available for new drugs. Without reference standards, it is not possible to differentiate compounds with the same molecular formula, so they have used a combination of Meteor (to predict the metabolites), Bruker MetaboliteDetect (to detect metabolites) and ACD/MS Fragmenter (to predict fragmentation patterns). Quetiapine was chosen as it is often seen in forensic toxicology

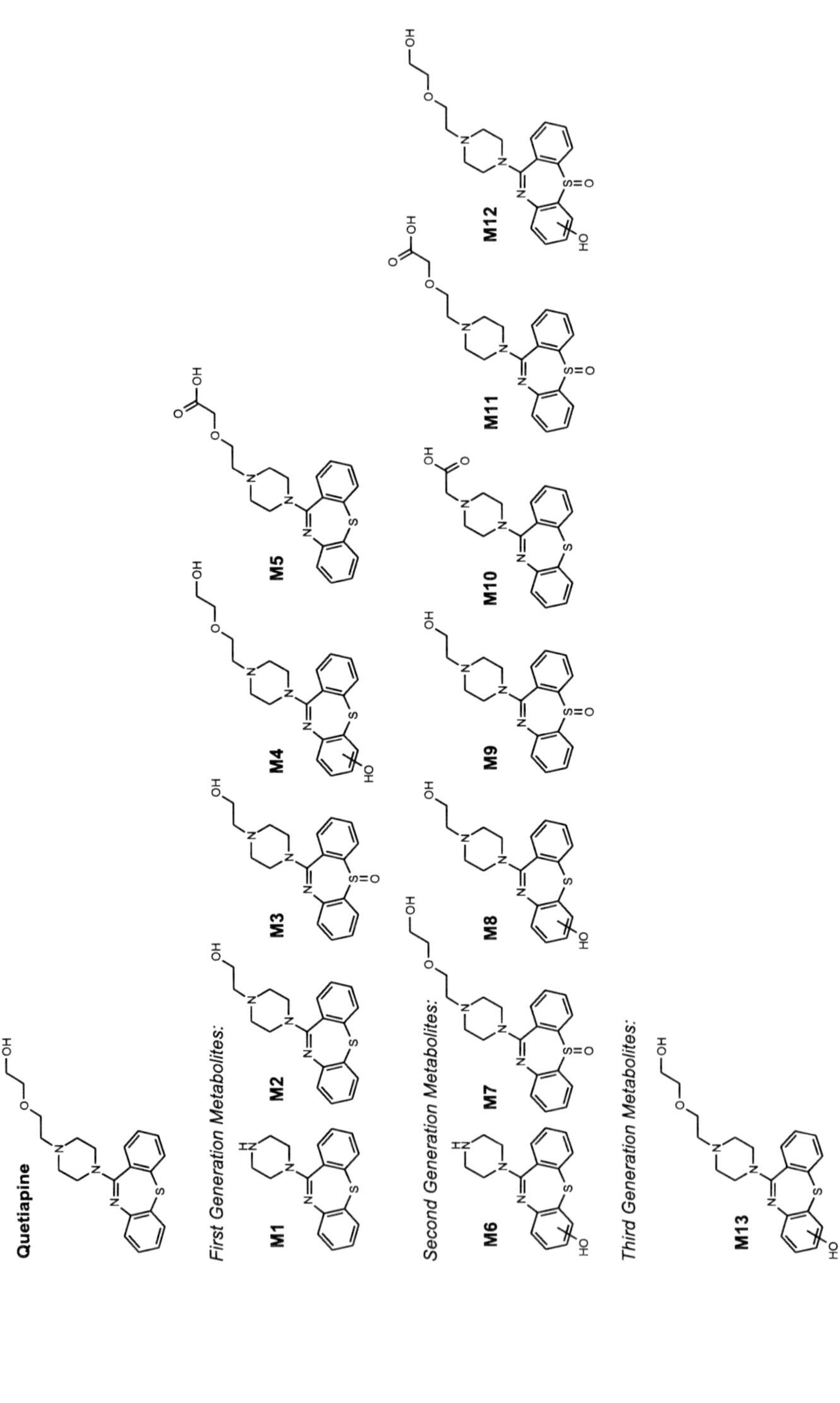

Figure 11.13 Metabolites of quetiapine.

casework. It is an antipsychotic agent with complex pharmacology affecting multiple receptor families (D_3, D_4, 5-HT_{2A}, 5-HT_{2C}, 5-HT_7, α_1 and α_2-adrenergic receptors) with a wide range of IC_{50} values and can be clinically indicated for conditions ranging from mild sleep and anxiety disorders to schizophrenia and acute manic episodes of bipolar disorder. Although it is extensively metabolised, its chromatographic behaviour is not well documented. The experiment used human urine samples collected at autopsy.

Meteor (version 10) predicted fourteen probable metabolites, under default processing constraints, of which eight were detected by LC-ToFMS. However, it did not predict any of the hydroxyl metabolites under default processing constraints (for reasons already mentioned, as they are generated by application of sequential phase I biotransformations) but did suggest further oxidation of sulfoxides to sulfones. The literature reports varying numbers of metabolites for this compound but does not show structures for all of them. The main issue is that the sulfoxyl and hydroxyl metabolites have the same mass, hence the use of ACD/MS Fragmenter to assign peaks to fragments.

Overall, this approach showed that it could be a useful automated method to help the university continue to develop an extensive database containing exact monoisotopic masses, verified retention times of drugs and their urinary metabolites for LC-ToFMS drug screening. This kind of work provides a method of finding the knowledge gaps within *in silico* prediction tools. In fact, as a result of this paper, further knowledge work has been carried out on Meteor over the past two years and the current version predicts eleven of the proposed thirteen metabolites, missing M6 and M8 (see Figure 11.13) under default processing constraints.

11.5.2.3 Application in Drug Discovery

Over the years, many pharmaceutical companies have built a vast collection of metabolism study data from the development of all their compounds, consisting of studies using test systems ranging from rodent microsomal to human *in vivo*. Historically, before the use of metabolite prediction tools, full scan LC-MS datasets would be interrogated in vendor provided MS software by experts using their accumulated knowledge, alongside any information provided from drug discovery, to determine the potential metabolite structures and match them up with chromatographic profiles. This was very time consuming and involved significant experience or educated guess work. Drug metabolism and pharmacokinetics (DMPK) departments are now trying to automate the process to improve efficiency and accuracy. One organisation's process[31] (Figure 11.14) involves analysing all new compounds in Meteor (generally with the absolute reasoning limit set to "plausible" and the relative reasoning threshold set to level two) to generate a prediction tree, which is frequently quite large. The expected accurate mass of metabolites generated by Meteor is used to automatically sort through the LC-MS data and generate accurate mass chromatograms for each potential metabolite detected in the dataset. The same Meteor-generated list of metabolites can be used to direct data acquisition. If

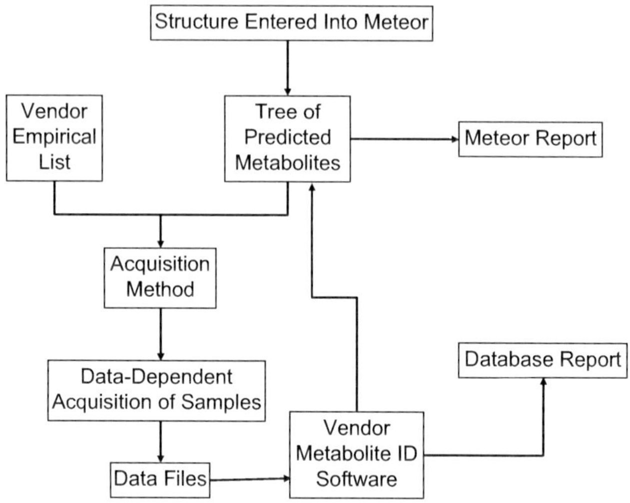

Figure 11.14 A metabolite identification workflow diagram.

an ion of predicted mass and correct intensity is detected, the MS instrument is directed to automatically acquire a product ion scan of that ion to further elucidate the structure. The accepted peaks are formatted to send back to Meteor in order to assign structures to detected metabolites. If the exact position of metabolism is required, the work is scaled up in order to carry out nuclear magnetic resonance (NMR) experiments, which is expensive, time consuming and requires a lot more sample.

To aid both drug discovery and development, some pharmaceutical organisations use exclusively high resolution (exact mass) mass spectrometry for metabolite identification. An *in silico* prediction tool would only be used if there was clear evidence of exotic metabolism, not to investigate obvious mass changes of +16 (hydroxylation, epoxidation), –14 (demethylation), *etc*. Exotic metabolism can be either the situation where a very unusual biotransformation has occurred and there is no precedence for it, or biotransformations occur in an unusual order or combination. If software can help for exotic metabolism, it does not matter if the analysis takes all week and generates tens of thousands of metabolites, as long as they can be searched and retrieved effectively. This is quite a high expectation of a knowledge based system, which would only be able to accommodate such specific information through in-house data sharing projects, potentially at the expense of generating rather large metabolism trees.

There have been several cases of similar studies,[31–34] showing the use of *in silico* metabolism prediction combined with vendor supplied LC-MS data analysis tools, with varying degrees of success. Generally the prediction package will provide several suggestions for metabolites and miss some key pathways. This process is very beneficial to the continuous improvement of the software as missed pathways can be added, and those that are over-predicting can either be set at a lower likelihood or have the chemistry modified to accommodate metabolism mediating factors such as steric hindrance or poor enzyme fit. Some vendors have *in silico* prediction fully integrated into their metabolite identification software in order to enhance their existing lists of expected biotransformations; others have created application notes to describe how to manually link the two systems together to enhance the laboratory workflows. Some have either yet to decide how best to link the tools together, or feel their existing software is already fit-for-purpose.

Work is continuing to improve the predictions of the Meteor prediction tool and to help the pharmaceutical industry streamline their workflows more efficiently by working closely with vendors to facilitate this.

11.6 Conclusion and Future Perspectives

This chapter has given a very brief overview of some of the useful features of knowledge based expert systems for the prediction of toxicity and xenobiotic metabolism and has attempted to highlight the application of some of these features in drug discovery and development. We have further touched on some related areas of interest, such as the use of these systems in other domains like forensic toxicology, occupational health, environmental safety and regulatory affairs. We have mentioned the complementarity of some quantitative and qualitative methods and suggested that their concerted use can increase predictive capabilities to a level higher than may be attainable by individual methods in isolation. We have further touched on the integration of more than one expert system; metabolites (and indeed biotransformation intermediates) generated by Meteor can be automatically batch-processed through Derek for Windows to gain an assessment of their toxicity potential. Some metabolites (and reaction intermediates in particular) can, of course, be very short-lived or present in low concentrations. Conjunctive use of quantitative methods for the estimation of physicochemical properties and chemical reactivity parameters here would aid in judging the feasibility of metabolites and intermediates (quantitative estimation of metabolic biotransformation rates and metabolite yields remains a significant and, as yet, unrealised challenge), their distribution and excretion potentials and their propensity to bind to plasma proteins and cellular macromolecules. Such information, once calculated, has the potential to be shared between expert systems in a higher level reasoning model, but again we do not allude here to consensus models, rather to the marriage of complementary systems designed to perform different functions. The use of the expert system for toxicology prediction in drug discovery is now mainstream and that of the expert system for metabolism prediction is gaining rapid acceptance - this, despite the fact that it is extremely difficult (and, indeed, probably inappropriate) to validate an expert system in the same rigorously statistical way as for a quantitative method; factual and heuristic knowledge cannot be treated in such ways.[1] Expert opinion rarely comes ready-packed in containers where numerical values, ranges and thresholds are known, predefined, agreed upon and validated. As an example, an airline pilot may be armed with such information as: "*. . . . . severe turbulence may be expected at altitudes between 25 000 and 35 000 feet on this bit of your flight-path . . . . .* ". An intelligent and experienced pilot (hopefully an expert) may reason that he should fly at some altitude that is either lower or higher than the stated range. What would, perhaps, be somewhat alarming is a situation where the pilot considered an altitude of 24 999 or 35 001 feet as safe in these circumstances. Although we freely admit to knowing nothing about aviation, clearly margins of uncertainty and the lack of precision in the predictive model need to be taken into account, and we would hope that the pilot would factor in heuristic knowledge, based on his experience, in the process of deciding on a reasonable and safe altitude. It has been stated that: "*an expert system gives the same answers as you would expect an expert to give – including the wrong ones*".[1] That said, the evaluation of an expert system, by which we seek to establish its general usefulness and applicability, should not be avoided as a futile exercise. Indeed, such exercises bring two main benefits. Firstly, dissemination of the findings of such exercises, if favourable, help establish a degree of confidence within the user community, and secondly, they highlight areas of weakness and deficiency within the knowledge base. This informs the expert community in general, and the knowledge base development team in particular, on areas to focus on and prioritise for new research initiatives.

Recent efforts in Derek for Windows knowledge base development have involved the introduction of validation comments into individual alerts, as previously mentioned, which allows validation data for an alert to be displayed when it is activated by a query compound. Such data assist the user in understanding the reliability of an alert and contribute to compliance with the OECD principles for (Q)SAR validation. The activation of an alert across multiple datasets is recorded and the experimentally confirmed positive compounds occurring within those sets reported as a percentage of the number of compounds activating the alert (the positive predictivity). Derek for Windows is not a "yes/no" system and therefore some approximations need to be made and some thresholds set, for example predictions of "equivocal" or above may be considered a "positive" prediction. In this protocol no account is taken of compounds which contain multiple alerting features. The validation is applied to the alerts themselves and not to the endpoint under consideration; the approach is, necessarily, a reductionist rather than a holistic one but has proven useful in shaping knowledge base development strategy.

Like Derek for Windows, Meteor is not a "yes/no" system and because its predictions are multilayered and are not route of administration-, time-, dose-, species- or assay-specific, thresholds are arbitrary and difficult to set and would, in any case, be reasonably meaningless. Our evaluation methods to date have focussed on examining ratios of unconfirmed positive to congruent (experimentally confirmed) predictions and on ratios of congruent to false negative predictions under different processing conditions. Understanding the results of this approach gives an expectation of the sort of sensitivity/specificity balance that can be expected under variable combinations of absolute and relative reasoning levels. Evaluation of Meteor performance against metabolic profiles published in the literature has highlighted subsets of biotransformations, biotransformation types and individual biotransformations, where increases in selectivity in their performance leading to a reduction in the number of predicted unconfirmed positive biotransformations is indicated. There are convincing arguments as to why a reasonable level of unconfirmed positive biotransformations should be tolerated, and we have previously discussed some of these.[28] As long as predictions are not vague, unreasonable or overly optimistic (and do not exponentiate into an overwhelming and uncontrollable morass of information), the presence of experimentally unconfirmed biotransformations within the simulated profile brings a diverse, comprehensive and mechanistic richness to the understanding of the possible behaviour of the compound. This level of understanding can reach over and above the conclusions that can be arrived at from observations determined experimentally, bringing didactic value and potential ideas for further experimentation, and filling gaps in data and knowledge.

For both toxicology prediction and xenobiotic metabolism simulation, expert systems will need to draw increasingly on improving quantitative methods (in well understood model space) to refine and improve the predictions that they make. Qualitative and quantitative methods are complementary and should not be used in consensus approaches. A future in which both sit side by side in comfortable co-existence, gaining wide usage and acceptance, is highly desirable, and indeed perhaps essential.

References

1. P. Judson, *Knowledge-based Expert Systems in Chemistry: Not Counting on Computers*, RSC Publishing, Cambridge, UK, 2009.
2. C. A. Marchant, K. A. Briggs and A. Long, *Toxicol. Mech. Methods*, 2008, **18**, 177.
3. P. N. Judson, C. A. Marchant and J. D. Vessey, *J. Chem. Inf. Comput. Sci.*, 2003, **43**, 1364.
4. R. O. Potts and R. H. Guy, *Pharm. Res.*, 1992, **9**, 663.
5. E. J. Matthews, N. L. Kruhlak, R. D. Benz, D. Aragones Sabate, C. A. Marchant and J. F. Contrera, *Regul. Toxicol. Pharmacol.*, 2009, **54**, 43.
6. A. P. Worth and M. T. D. Cronin, *Altern. Lab. Anim.*, 2004, Supplement **1**, 703.
7. S. Dimitrov, G. Dimitrova, T. Pavlov, N. Dimitrova, G. Patlewicz, J. Niemela and O. Mekenyan, *J. Chem. Inf. Model.*, 2005, **45**, 839.
8. C. A. Marchant, L. Fisk, R. R. Note, M. L. Patel and D. Suarez, *Chem. Biodivers.*, 2009, **6**, 2107.
9. C. M. Ellison, S. J. Enoch, M. T. Cronin, J. C. Madden and P. Judson, *Altern. Lab. Anim.*, 2009, **37**, 533.
10. W. G. Button, P. N. Judson, A. Long and J. D. Vessey, *J. Chem. Inf. Comput. Sci.*, 2003, **43**, 1371.
11. B. Testa, A. L. Balmat and A. Long, *Pure Appl. Chem.*, 2004, **76**, 907.

12. B. Testa, A. L. Balmat, A. Long and P. N. Judson, *Chem. Biodivers.*, 2005, **2**, 872.
13. A. Long, E. Murray and B. Testa, 2011, *Manuscript in Preparation.*
14. P. Rydberg, D. E. Gloriam, J. Zaretzi, C. Brenman and L. Olsen, *ACS Med. Chem. Lett.*, 2010, **1**, 96.
15. A. Long and P. Rydberg, *Unpublished Results.*
16. S. Ekins (ed.), *Computational Toxicology. Risk Assessment for Pharmaceutical and Environmental Chemicals*, John Wiley and Sons, Hoboken, USA, 2007.
17. D. A. Basketter, E. W. Scholes, M. Chamberlain and M. D. Barratt, *Food. Chem. Toxicol.*, 1995, **33**, 1051.
18. EMA guideline document, 2006, available at: http://www.ema.europa.eu/docs/en_GB/document_library/Scientific_guideline/2009/09/WC500002903.pdf.
19. FDA guideline for industry document, 2008, available at: *http://www.fda.gov/downloads/Drugs/GuidanceComplianceRegulatoryInformation/Guidances/ucm079235.pdf.*
20. ICH document Q3A(R2), 2006, available at: *http://private.ich.org/LOB/media/MEDIA422.pdf.*
21. R. Kroes, J. Kleiner and A. Renwick, *Toxicol. Sci.*, 2005, **86**, 226.
22. K. L. Dobo, N. Greene, M. O. Cyr, S. Caron and W. W. Ku, Regul. *Toxicol. Pharmacol.*, 2006, **44**, 282.
23. L. Muller, R. J. Mauthe, C. M. Riley, M. M. Andino, D. De Antonis, C. Beels, J. DeGeorge, A. G. M. De Knaep, D. Ellison, J. A. Fagerland, R. Frank, B. Fritschel, S. Galloway, E. Harpur, C. D. N. Humfrey, A. S. J. Jacks, N. Jagota, J. Mackinnon, G. Mohan, D. K. Ness, M. R. O'Donovan, M. D. Smith, G. Vudathala and L. Yotti, *Regul. Toxicol. Pharmacol.*, 2006, **44**, 198.
24. K. L. Dobo, R. S. Obach, D. Luffer-Atlas and J. P. Bercu, *Chem. Res. Toxicol.*, 2009, **22**, 348.
25. K. Park, D. P. Williams, D. J. Naisbitt, N. R. Kitteringham and M. Pirmohamed, *Toxicol. Appl. Pharmacol.*, 2005, **207,** 425.
26. S.Coecke, H. Ahr, B. J. Blaauboer, S. Bremer, S. Casati, J. Castell, R. Combes, R. Corvi, C. L. Crespi, M. L. Cunningham, G. Elaut, B. Eletti, A. Freidig, A. Gennari, J. F. Ghersi-Egea, A. Guillouzo, T. Hartung, P. Hoet, M. Ingelman-Sundberg, S. Munn, W. Janssens, B. Ladstetter, D. Leahy, A. Long, A. Meneguz, M. Monshouwer, S. Morath, F. Nagelkerke, O. Pelkonen, J. Ponti, P. Prieto, L. Richert, E. Sabbioni, B. Schaack, W. Steiling, E. Testai, J. A. Vericat and A. Worth., *Altern. Lab. Anim.,* 2006, **34**, 1.
27. FDA guideline for industry document, 2008, available at: *http://www.fda.gov/downloads/Drugs/GuidanceComplianceRegulatoryInformation/Guidances/ucm079266.pdf.*
28. L. G. Valerio Jr and A. Long, Curr. Drug. Discov. Technol., 2010, **7**, 170.
29. C. Hao, S. Campbell, D. Stranz D and N. McSweeney, presented at the 52nd ASMS Conference on Mass Spectrometry, Nashville, 2004.

30. A. Pelander, E. Tyrkko and I. Ojanpera, *Rapid Commun. Mass Spectrom.*, 2009, **23**, 506.
31. J. W. Firth, L. M. Boyling, G. L. Shackleton and S. E. Ives, *Drug Metab. Rev.*, 2009, **41** (Supplement 3), 117.
32. S. E. Ives, K. Gjervig-Jensen and N. McSweeney, presented at the 55th ASMS Conference on Mass Spectrometry, Indianapolis, 2007.
33. R. J. Mortishire-Smith, D. O' Connor, J. M. Castro-Perez and J. Kirby, *Rapid Commun. Mass Spectrom.*, 2005, **19**, 2659.
34. S. Christensen, S. E. Ives, A. M. Jacobsen and D. Spencer Jones, presented at the 58th ASMS Conference on Mass Spectrometry, Salt Lake City, 2010.

CHAPTER 12

Ligand-Based Modeling of Toxicity

SEAN EKINS,[a,b,c] SANDHYA KORTAGERE,[d] MATTHEW D. KRASOWSKI,[e] ANTONY J. WILLIAMS,[f] JINGHAI JAMES XU[g] AND MICHAEL ZIENTEK[h]

[a] Collaborations in Chemistry, 5616 Hilltop Needmore Road, Fuquay-Varina, NC 27526, USA; [b] Department of Pharmaceutical Sciences University of Maryland, MD 21201, USA; [c] Department of Pharmacology University of Medicine & Dentistry of New Jersey (UMDNJ)-Robert Wood Johnson Medical School, 675 Hoes Lane, Piscataway, NJ 08854, USA; [d] Department of Microbiology and Immunology, Drexel University College of Medicine, Philadelphia, PA 19129, USA; [e] Department of Pathology, University of Iowa Hospitals and Clinics, Iowa City, IA, 52242, USA; [f] Royal Society of Chemistry, 904 Tamaras Circle, Wake Forest, NC-27587, USA. (AJW); [g] Knowledge Discovery and Knowledge Management (KDKM), Merck & Co., Inc., 126 E. Lincoln Ave, Rahway, NJ 07065, USA. (JJX); [h] Pfizer Global Research & Development Pharmacokinetics, Dynamics & Drug Metabolism, San Diego CA, USA

12.1 Introduction

To describe a topic to which a whole book could be devoted in just one chapter is a difficult task. One has to address which topics to include and which to omit. For example, there have been several books that address ligand-based approaches for modeling various toxicity endpoints in detail over many chapters, and the reader is recommended to read these to see the broad diversity of computational approaches applied to date.[1–3] A simple explanation of the

RSC Drug Discovery Series No. 13
Drug Design Strategies: Quantitative Approaches
Edited by David J. Livingstone and Andrew M. Davis

Published by the Royal Society of Chemistry, www.rsc.org

goal of these methods is to take the structures of molecules tested against some toxicity endpoint and try to relate this to molecular descriptors as either a continuous or discrete value. This chapter will offer just a brief overview of the areas on which we have published collaboratively in recent years, thereby providing a sample of some models and methods. We shall describe ligand-based models for cardiotoxicity, *e.g.* the human ether-a-go-go-related gene (hERG) potassium channel, and models for hepatotoxicity, *e.g.* the pregnane X receptor (PXR), as well as drug metabolizing enzymes (P450 3A4), and other more complex endpoints, for drug-induced liver injury (DILI). We and others have been interested in computational modeling of toxicity-related proteins such as these in order to identify potential liabilities in molecules of clinical interest as early as possible.

Due to space restrictions this chapter will not cover in any detail the structure-based or docking methods which we have compared extensively with machine learning methods in the case of PXR.[4–6] During the chapter we will highlight how our understanding of molecular causes of toxicity has allowed quantitative modeling, how various pharmaceutical companies actively use such computational models and their impact on drug discovery and toxicity-related attrition.

12.2 hERG

In cardiac tissue, a reduction in flow through any of several potassium channels is associated with QT interval prolongation.[7] The most common channel linked to drug-induced QT interval prolongation in humans is the rapid component of the delayed rectifier potassium current (I_{Kr}), one of the components driving ventricular repolarization and influencing the QT interval in human myocardium.[8,9] The identification of the hERG gene in 1995[8] and the fact that the I_{Kr} is a co-assembly of hERG α–subunits encoded by the *KCNH2* gene and MiRP1 β-subunits encoded by the *KCNE* gene, provide important evidence that changes in potassium conductance cause prolongation of the QT interval. Mutations in hERG have also been found to lead to one form of hereditary long QT syndrome, LTQ2.[10,11] Many drugs associated with QT interval prolongation have also been found to block hERG,[12] and when unintended this has prompted withdrawal of drugs from the market.

12.2.1 hERG Pharmacophores

Ligand-based modeling has been applied extensively to understanding structure–activity relationships (SARs) of hERG channel blockers. One of the first structure–activity studies performed on compounds that cause QT prolongation was by Morgan and Sullivan.[13] This work (and several other studies) was carried out prior to the discovery of hERG and it is now clear that the mechanism of action for most, if not all of the compounds, involves hERG blockade. Another example of early hERG pharmacophore modeling appears to be the work by Matyus *et al.* [14] on elucidation of the pharmacophore for blockers of I_{Kr} current as Class III antiarrhythmic agents. Eleven ligands were divided into two sets. The first set included the six most active agents, spanning a 1.5 log activity range, and the second set contained the less active compounds. Starting structures were obtained by energy minimization in the Tripos[15] force field. The sets of conformations were generated by the Multisearch option in the distance comparison (DISCO)[15] interface. Fifty-three pharmacophores from DISCO were analyzed and the best model was selected. Liu *et al.*[16] investigated activity data in the isolated guinea pig atrium assay for a series of 17 dofetilide analogs using comparative molecular field analysis (CoMFA) and comparative molecular similarity analysis (CoMSiA). With dofetilide as a reference, its low energy conformations were used as templates for molecule alignment. The optimized CoMFA model produced a correlation with $q^2 = 0.695$ (a later study[17] involved three different techniques: CoMFA, CoMSiA, and hologram quantitative structure–activity relationships (HQSARs)[18] for predicting the activity of 11 newly synthesized compounds).

The first hERG pharmacophore published since the renewed interest in predicting I_{Kr} blockade was described by Ekins *et al.*[19] using 15 molecules from the literature. It contained four hydrophobes that surround the central positive ionizable feature and produced an r^2 value of 0.90. The model was further applied to predict IC_{50} values for a test set of 22 predominantly antipsychotic compounds known to inhibit hERG ($r^2 = 0.83$). A pharmacophore for antipsychotic drugs, with three hydrophobic features and a ring aromatic feature, was also referred to and later published.[20] 'Local' pharmacophore models have also been generated around sertindole analogs.[21] For example, publications containing [^{3}H]-dofetilide binding data for the $5HT_{2A}$ class of molecules,[22,23] 3-aminopyrrolidinone farnesyltransferase inhibitors,[24] have also been used to produce individual pharmacophores that were combined, suggesting common areas of positive ionizable features and hydrophobicity from the aromatic rings.[25,26] A recent homology modeling and docking study has suggested that molecules may bind orthogonally to the hERG channel pore, and this may rationalize these various early hERG pharmacophores.[27]

The initial training set from the original pharmacophore study was a foundation for later hERG models. It was first expanded to include 66 molecules and used with a recursive partitioning method (ChemTree) with path-length descriptors, resulting in an observed *versus* predicted correlation of $r^2 = 0.86$. This model produced a correlation of $r^2 = 0.67$ when tested with a set of 25 additional molecules from the literature.[28] An updated version of this model was also built using 99 literature molecules and has been used more recently to rank the 23 sertindole analogs generated by Pearlstein *et al.*[21] (Spearman rho = 0.74, $P < 0.0001$, $r^2 = 0.53$).[26] Further testing of this tree model containing 99 molecules with 35 diverse molecules[29] showed a relatively low, though statistically significant, correlation ($r^2 = 0.33$, Spearman rho 0.55, $p = 0.0006$). However, using the Tanimoto similarity analysis of the test set molecules with Accelrys Accord descriptors (Tanimoto similarity >0.77) resulted in 18 remaining molecules with improved correlation statistics ($r^2 = 0.83$, Spearman rho = 0.75, $p = 0.0003$).[29] There have been many other examples of hERG QSAR models and some are summarized in Table 12.1.

Table 12.1 hERG QSAR models.

Method	*Training set*	*Test set*	*Notes*	*Reference*
CoMFA	$N = 31$ training set produced a correlation with $r^2 = 0.95$ ($q^2 = 0.77$).	$N = 6$ additional compounds ($r^2_{pred} = 0.74$).	Pharmacophore contains three aromatic moieties connected through a nitrogen function that is a tertiary amine. The nitrogen and the aromatic moieties are separated by distances of 5.2–9.1 Å, 5.7–7.3 Å, and 4.6–7.6 Å.	172
CoMSiA	$N = 18$ sertindole analogs ($q^2 = 0.57$).	$N = 4$ sertindole analogs.	Decreasing the positive charge on the central nitrogen and increasing the steric bulk on the hydrophobic end of the molecule reduce hERG blocking activity.	21
HQSAR model	Literature data on $N = 55$ compounds.	$N = 13$ compounds ($r^2 = 0.81$, $SSE = 0.67$).	HQSAR model created that made use of 2D fragment fingerprints.	173
Catalyst	$N = 34$ compounds from the literature spanning a range of IC_{50} values.	$N = 21$ compounds, split into three groups based on the hERG activity level (<1 μM, 1–100 μM, and >100) μM). The model test set $r^2 = 0.71$, with all highly active (<1 μM) compounds predicted correctly.	The best pharmacophore hypothesis contained a positive ionizable feature, two aromatic rings, and a hydrophobic group.	174
QSAR	150 descriptors were calculated for $N = 19$ structurally diverse hERG blockers from the literature.		Multiple linear regression using two descriptors—log*D* and D_{max} (the maximum diameter of molecules)—were: $pIC_{50}(hERG) = 0.58 \log D + 0.30\ D_{max} - 0.36$ ($r^2 = 0.87$, $q^2 = 0.81$).	175
Support vector regression model	261 counts of structurally diverse 2D fragments. The best model was built for $N = 71$ known hERG blockers ($q^2 = 0.64$).	$N = 19$ additional compounds produced $r^2 = 0.85$ and $rmse = 0.60$ Predictive power for 20 in-house ligands ($r^2 = 0.29$, $rmse = 1.26$).	The best predictions were made for compounds with higher similarity to the training set.	176

Table 12.1 *(Continued)*

Method	*Training set*	*Test set*	*Notes*	*Reference*
2D QSAR multiple linear regression	N = 104 hERG ligands from different cell lines in the literature and interpretable descriptors such as ClogP, TPSA, diameter, summed surface area of atoms, and partial charges.	Testing was performed using a leave out group N = 18 molecules repeated five fold (average $r^2 = 0.66$, SD 0.85).	An indicator variable was also included to represent the different experimental conditions ($q^2 = 0.67$).	177
Least Median Squares regression	N = 1075 patch clamping IC_{50} data.	N = 1679 IC_{50} and single point data. Test set R2 = 0.54 RMSE = 0.63	Volsurf and other descriptors.	178
Catalyst pharmacophore ensemble/SVM	N = 26 literature data training $r^2 = 0.94$, $q^2 = 0.89$ for 10 fold cross validation	N = 13 $r^2 = 0.94$	3 pharmacophores consisted of hydrophobic, ring aromatic and a positive ionizable feature.	179
ALMOND	N = 31 (data from [180]) ALMOND $q^2 = 0.69$.	N = 6 (data from).[180]	ALMOND predictions less accurate than CoMFA.	181
GFA PLS regression QSAR with 4D fingerprints and MOE descriptors	N = 250 from the literature, $r^2 = 0.58$, $q^2 = 0.54$.	PubChem bioassay database N = 250 active, N = 1703 inactive.	QSAR used to make classifications with 10 μM IC_{50} cut off. 65% accuracy. With a smaller test set (N = 876) the overall accuracy is 82%.	182
Support Vector regression, Gaussian process, random forest	N = 110 literature compounds and N = 550 in house compounds.	3 fold cross validation RMSE ~0.6 in cross validation.	ChemAxon, MOE, VolSurf descriptors. Ensemble of models.	44
Hierachical PLS	N = 1312 molecules; data generated in house with IonWorks HT patch clamping assay.	N = 7520 additional molecules tested over time in the same assay.	Used SELMA, DRONE, VOLSURF, fragment descriptors training set mean RMSE = 0.40, test set RMSE 0.49, within experimental error of assay (0.5 log units).	166

12.2.2 hERG Classification Methods

QSAR methods aim to predict an absolute compound activity. Classification methods on the other hand, attempt to bin compounds by their predicted hERG inhibition. The first example of a hERG-based classification was reported by Roche *et al.*[30] A total of 244 compounds representing the extremes of the data set (<1 μM and >10 μM for actives and inactives respectively) were modeled with a variety of techniques such as substructure analysis, self-organizing maps, partial least squares, and supervised neural networks. The descriptors chosen included pK_a, Ghose-Crippen,[31] TSAR,[32] CATS,[33] Volsurf,[34] and Dragon[35] descriptors. The most accurate classification was based on an artificial neural network. In the validation set containing 95 compounds (57 in-house and 38 literature IC_{50} values), 93% of inactives and 71% of actives were predicted correctly. Other groups have also generated classification models using different methods and molecular descriptors (Table 12.2). More recently, we have expanded the literature hERG data and used shape signature descriptors with support vector machine (SVM) methods and various mapping methods,[36] to generate classification models. The overall accuracy obtained in a set of 30 leave-20-out statistical tests is about 69–74%, comparable to the results of alternative computational models published to date and described in Table 12.2. In an unpublished study, we have also used the same set of 134 molecules with previously described quantitative activity data[29] to generate a Bayesian model with molecular function class fingerprints of maximum diameter 6 (FCFP_6) and eight additional simple descriptors (ALogP, molecular weight, number of aromatic rings, number of H-bond acceptors, number of H-bond donors, number of rings, number of rotatable bonds, and molecular fractional polar surface area). This model had a cross-validated receiver operator characteristic area under the curve (XV ROC AUC) = 0.82. When leave out 20% or 50% × 100 fold internal validation was performed, similar AUC statistics were generated (Table 12.3). Concordance was ~70%, specificity ~75%, and sensitivity ~60%. The descriptors that were important for activity are shown in Figure 12.1a and those features not present in hERG actives are also shown (Figure 12.1b). External testing of this model was also performed on a data set of 1908 molecules (from PubChem) which has been used by other groups previously (Table 12.3). Using the Bayesian model to rank these molecules indicated good enrichments based on the ROC plot (Figure 12.2). For example, approximately 4% of the hits are ranked in the top 1% (N = 19). The same test set data has concordance = 59.3%, specificity = 74.4%, and sensitivity = 34.6% for a test set with mean closest maximal Tanimoto similarity = 0.30 to members of the training set. The advantages of the Bayesian modeling approach are the cost effective fast model generation, and use of interpretable descriptors which aid in molecule modification and prediction of which molecules will have the best chance of success in the clinic.[7] This method has been used for other toxicity endpoints described in the following sections.

The quality of models for hERG may be closely related to the consistency of the data used for modeling. There have been efforts to analyze the hERG

Table 12.2 hERG classification models.

Method	*Training set*	*Test set*	*Notes*	*Reference*
Naïve Bayes classifier	$N = 1979$ compounds with measured hERG activity from the Roche corporate collection. Receiver operator curve (ROC) accuracy of 0.87 was achieved for the training set.	The model was validated on an external set of $N = 66$ drugs, of which 58 were classified correctly (88% accuracy).	218 in-house atom-type descriptors were used to develop the model. $pIC_{50} = 4.52$ was set as a threshold between hERG actives and inactives.	183
Decision tree-based approach	Used three descriptors: ClogP, calculated molar refractivity (CMR), and the pK_a of the most basic nitrogen, to identify hERG blockers within an in-house data set.		With $IC_{50} = 130$ nM as a cutoff, factors suggestive of hERG activity were determined to be ClogP ≥ 3.7, $110 \leq \text{CMR} < 176$, and pK_a max ≥ 7.3.	184
Combined 2-D/3-D procedure	Upon 50-fold cross-validation of the model on a literature data set containing $N = 85$ actives (threshold HERG $IC_{50} = 40$ μM) and 329 inactives, 71% of hERG actives and 85% of hERG. inactives were correctly identified.	Five of eight (62.5%) hERG blockers were identified correctly in a $N = 15$ compound in-house validation set. Most of the statistically significant pharmacophores from the ensemble procedure were three-feature [aromatic]–[positive charge]–[hydrophobe] combinations similar to those reported by Cavalli *et al.*[172] However, a novel three point pharmacophore containing a hydrogen bond acceptor was also proposed.	A 2D topological similarity screen utilizing atom pair[185] descriptors and an amalgamated similarity metric termed TOPO was combined with a 3D pharmacophore ensemble procedure in a 'veto' format to provide a single binary hERG classification model. A molecule flagged by either component of the method was considered hERG active.	186

3D structures of the antipsychotics were generated starting from known small molecule X-ray structures, followed by solvation and minimization.	N = 17 antipsychotic drugs, all associated with reports of torsadogenic cardiotoxicity.	The variable moiety present in 14 of the 17 ligands is a hydrogen bond acceptor, either a carbonyl oxygen or a heteroaromatic nitrogen.	The distance between a basic nitrogen and a second variable moiety, converged in the range between 4.32 Å and 5.50 Å (average = 4.87 Å).	187
1D profile	N = 10 known hERG blockers.	Enrichment factors for a set of N = 92 hERG blockers ranged from 6 to 8 in the top 1–5%.	Tested for the ability to discriminate between hERG blockers and MDDR-derived decoys.	188
SVM	N = 73 drugs with known hERG IC_{50} values.	Two different separation boundaries were tried ($pIC_{50} = 4.4$ and $pIC_{50} = 6$), with 95% and 90% accuracy for the classification, respectively. The model also predicted known cardiovascular side effects with an accuracy of 70% when tested using an external set.	Seven 2D MOE[189] descriptors and 51 fragment count descriptors (subset of the 166-bit MACCS keys) were calculated.	190
Genetic programming (GP)	N = 124 compounds from the literature. A total of 618 descriptors were used.	The best models achieved 85–90% accuracy in predicting the validation set when $IC_{50} < 1$ μM was used as a threshold for hERG-active ligands.	Model performance was judged using Akaike fitness criterion and the ROC.	191
MOE pharmacophores	N = 194 uncharged hERG blockers planar patch clamping data $IC_{50} < 10$ μM	N = 519 in house ligands with $IC_{50} < 30$ μM 80–85% match the 2 5 point pharmacophores	5 feature and 6 feature pharmacophores derived with hydrophobic and hydrogen bond acceptor groups	192

Table 12.2 *(Continued)*

Method	*Training set*	*Test set*	*Notes*	*Reference*
Sammon maps and Kohonen maps	Training set N = 93 molecules to compare the classification of high (log10 $IC_{50} < 0$) and low affinity (log10 $IC_{50} < 2$) compounds	Test set N = 35. The average classification quality is high for both training and test selections: up to 86% and 95% of compounds were classified correctly in the corresponding data sets.	The Sammon mapping technique outperformed the Kohonen maps for the external test set.	29
SVM and GRIND descriptors	N = 495 compounds from the literature. Thresholds from 1–40 μM.	N = 66 compounds from WOMBAT and PubChem Data on hERG for N = 1948 compounds.	Non linear SVM with 40 μM cutoff shows best accuracy (72%) with WOMBAT data.	193
SOM	N = 37 compounds from the literature.	PDSP hERG data for N = 243 actives and N = 1652 inactives.	188 MOE descriptors. SOM used as a classifier map.	194
Bayesian and counter propagation neural networks	N = 285 from the literature. 3 classes ≥ 10μM, < 1μM and 1–10μM.	80/20 5-fold cross-validation, N = 58 newly selected compounds used as a test set as well as the PubChem data N = 193 actives, N = 1703 inactives.	Used SIBAR Descriptors. 85–88% classification of hERG blockers.	195

Table 12.3 Internal cross-validation statistics for hERG Bayesian classification model.

Validation	*External ROC score*	*Internal ROC score*	*Concordance*	*Specificity*	*Sensitivity*
Leave out 20% × 100	0.77 (0.09)	0.80 (0.03)	71.44 (8.81)	74.84 (13.84)	62.85 (19.63)
Leave out 50% × 100	0.75 (0.04)	0.77 (0.07)	70.43 (5.89)	76.53 (12.48)	57.88 (15.37)

(a)

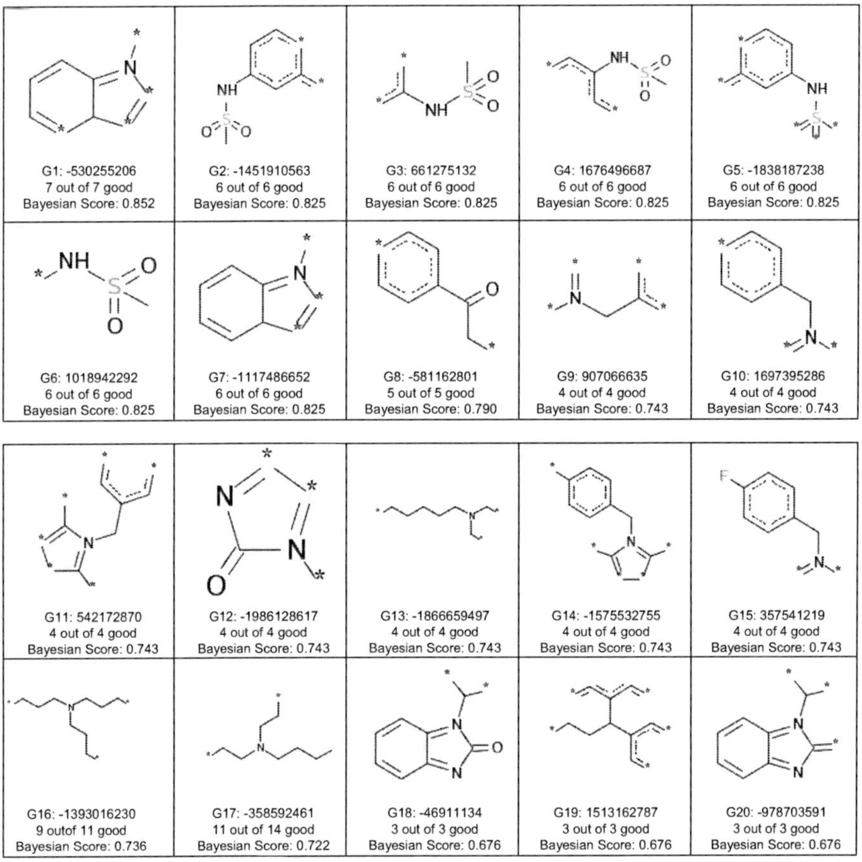

Figure 12.1 (a) Features important for hERG actives with activity < 1 μM. Bayesian model built using 134 molecules with data previously described.[29] (b) Features important for hERG inactives with activity > 1 μM. Bayesian model built using 134 molecules with data previously described.[29]

inhibition data in the literature, which is predominantly from human embryonic kidney 293 (HEK) cells.[38] There are well known variations in IC_{50} values for the same molecule in HEK, Chinese hamster ovary (CHO), and the

(b)

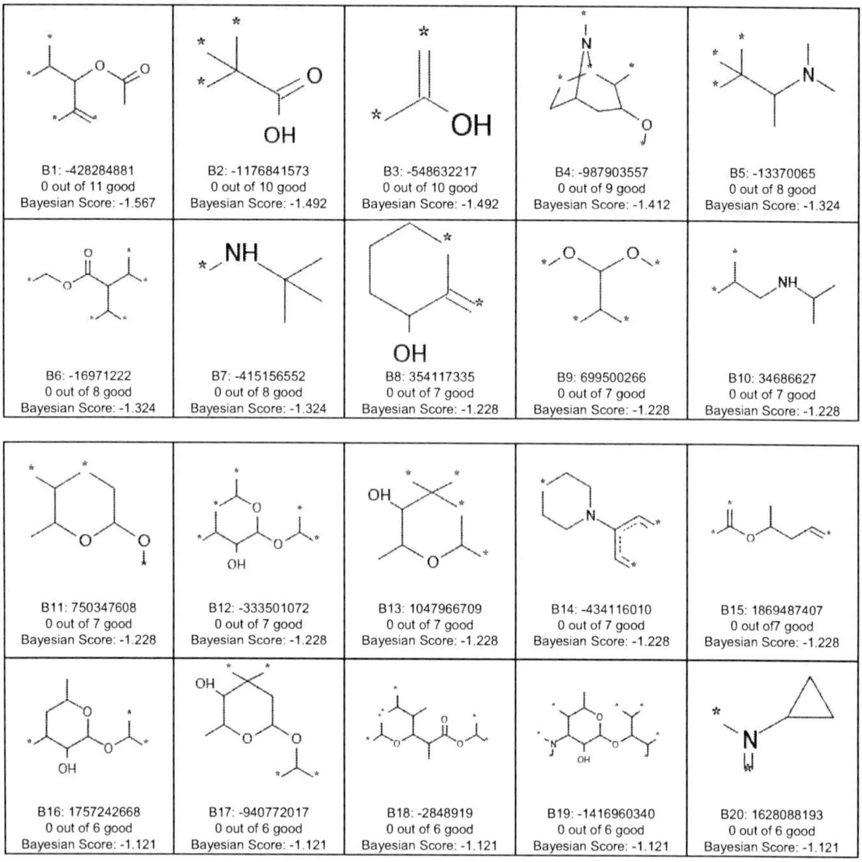

Figure 12.1 Continued

non-mammalian Xenopus oocytes. We have seen that most modeling efforts tend to mix up the various sources of data, and this has been a limitation of literature-based hERG models. There have also been studies that look for relatively simple rules of thumb and one example from GlaxoSmithKline analyzed 35 200 compounds screened in a hERG assay and studied changes in mean pIC_{50}. hERG activity increased with molecular weight, ClogP, and basic or zwitterionic ionization state.[39] Another study by GSK analyzed 11 105 compounds and found that there was a bell-shaped relationship with the number of aromatic rings (less than 2 may be ideal for hERG and P450 3A4 inhibition).[40] AstraZeneca has used a combination of 2D fragment-based QSAR, pharmacophores, homology modeling, and docking for chemokine receptor(CCR) antagonists to improve selectivity and avoid hERG.[41] They found that some hERG binding molecular fragments were pointed toward the selectivity filters and these needed to be avoided. A study by the same group with CCR8 antagonists used site-directed mutagenesis and fragment-based

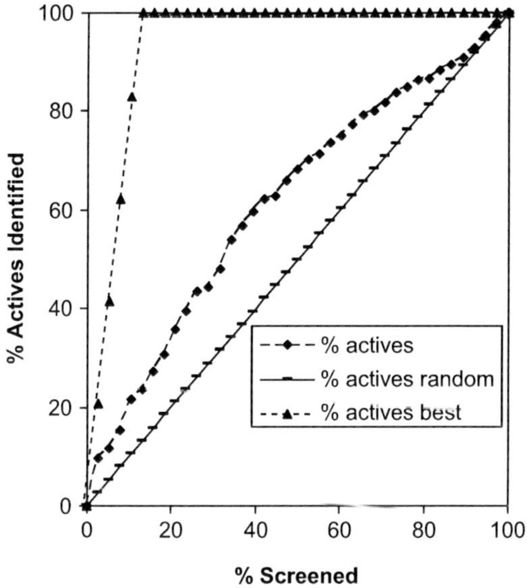

Figure 12.2 ROC plot for hERG model external test set (1908 molecules, 241 active) from PubChem.

QSAR to identify that a bulky lipophilic addition to a molecule clashed with residue F656.[42] A group at Pfizer targeting $5HT_{2A}$ and the serotonin transporter with a series of 66 phenoxyphenyl-methanamine analogs used Hammett values, the number of hydrogen bond acceptors, and the molecular weights for A and B ring substituents to build a SAR for hERG ($r^2 = 0.58$).[43] They found that hERG potency increased with molecular weight and with the number of hydrogen bond acceptors on the A ring in the *para*-position.[43] Researchers at Boehringer Ingelheim have described how they have combined literature data with their own data to build machine learning models for hERG[44] (Table 12.2). To our knowledge, virtually every major pharmaceutical company has attempted to build hERG models, and many have published examples of their use in drug discovery.

12.3 QSAR for Drug Metabolizing Enzymes

Initially, in the then absence of X-ray crystal structures, computational models provided considerable insights for many P450s.[45] Quantitative structure metabolism relationships (QSMRs) were pioneered by Hansch and co-workers[46–49] using very small sets of similar molecules and a few molecular descriptors. Later, Lewis and co-workers provided many QSAR studies for the human P450s, which resulted in a decision tree for classifying human P450 substrates.[50] Lipophilicity expressed as logP or molecular refractivity were the first important molecular properties related to enzyme–substrate binding. These were followed by steric, electronic, and molecular shape properties, which were also found to be important for enzyme binding and transformation. To date, QSAR models have been constructed for virtually all major human P450 enzymes. Software that has been used for 3D-QSAR includes Catalyst (Accelrys, San Diego, CA), DISCO, CoMFA, ALMOND, and GOLPE, which have been described in detail.[51] CoMFA was used to describe key molecular features of ligands for human P450 P1A2,[52] P450 2C9,[53] and, more recently, for P450 2C9 ligands.[54] However, the data sets used were small and likely capture only a very small portion of chemical space. Many of the early models rarely used test sets of molecules to evaluate their predictive capability.

Other QSAR methods have been used to generate predictions. Kohonen maps have also been useful for differentiating high and low affinity P450 3A4 substrates,[55] while neural networks have also been used to predict *N*-dealkylation rates for P450 3A4 and P450 2D6 substrates.[56] Other examples of recent QSAR for P450 inhibition and regioselectivity prediction are shown in Table 12.4.

12.3.1 Pharmacophores for P450s

Computational pharmacophore models have been widely applied to predicting metabolism and interactions with P450s. Pharmacophores have been generated for P450s,[57] including 1A2, 2A6, 2B6, 2C9, 2D6, 3A4, 3A5, and 3A7.[58] The P450 3A enzymes are perhaps the most important human drug metabolizing enzymes,[59] with a very broad substrate specificity metabolizing a very large proportion of marketed drugs. Computational pharmacophores for P450 3A4 have been developed for substrates[60] and inhibitors[60–62] using an array of kinetic constants (K_m, $K_{i\ (apparent)}$, and IC_{50}).[57] The development of benzbromarone analogs that are P450 2C9[63,64] and 2C19[65] inhibitors, with K_i values in the nM range, has enabled the further extension of the pharmacophore /3D-QSAR models which this group has been developing and refining over many years.[54] These analogs pointed to a role for hydrophobic interactions, which had also been addressed in an earlier pharmacophore study.[66]

The computational pharmacophore approach has been used to develop a model for the features of molecules that increase their own metabolism (autoactivators) *via* P450 3A4.[60] This may be important, however the *in vivo* relevance is unclear. The pharmacophore approach has also been used with heteroactivators (molecules that increases the metabolism of another molecule that is metabolized by the same enzyme) of P450 3A4 and P450 2C9 metabolism.[67,68]

Pharmacophores assume a similar binding mode and interaction with the protein, which is unlikely to be the situation in enzymes like P450s as they generally do not indicate reactivity (although some pharmacophores have included features for the site of metabolism).[69]

12.3.2 Metabolic Stability Models

More recently, other types of QSAR methods have been used to generate predictions for metabolic stability as a general property rather than at the individual enzyme level. For example, recursive partitioning is a powerful

Table 12.4 Recent P450 models for inhibition and site of metabolism.

Method	*Training set*	*Test set*	*Notes*	*Reference*
Recursive partitioning, Bayesian classifier, logisitic regression, k-nearest neighbor, support vector machine	N = 4000 CYP 3A4 inhibitors.	N = 470 compounds. Accuracy 82%, precision 76%, recall 75%, kappa = 0.61. An additional test set of 2195 compounds used to show change in predictivity with decreasing Tanimoto similarity. Threshold = 0.7.	Used BCI, MACCS, MolconnZ and typed graph triangle descriptors. Best models with BCI descriptors and SVM.	196
Gaussian kernel weighted k-nearest neighbor	N = 865 CYP2D6 inhibitors, N = 1037 CYP3A4 inhibitors.	N = 288 compounds for CYP2D6 82% correctly predicted, N = 345 compounds for CYP3A4 88% correctly predicted.	MACCS keys were used as descriptors.	197
PLS and regression tree	N = 301 (CYP1A2), N = 457 (CYP2C9), N = 369 (CYP2C19), N = 170 (CYP2D6), N = 463 (CYP3A4).	N = 126 (CYP1A2), = 202 (CYP2C9), N = 168 (CYP2C19), N = 89 (CYP2D6), N = 187 (CYP3A4). CYP1A2, CYP2C9, CYP3A4 PLS models most predictive, while RT models for CYP2C19 and CYP2D6 were most predictive.	Bulk property and fragment descriptors were used, consensus models better than individual models.	165
Multiple stepwise regression	Curcumin analogs N = 20 (CYP1A2), N = 7 (CYP3A4), N = 12 (CYP2C9), N = 6 (CYP2D6).	None described.	137 descriptors using QuaSAR in MOE. Electrostatic and hydrophobic descriptors important for CYP1A2, CYP2C9 and CYP2D6.	198
Mixed-integer programming hyperboxes classification, Bayes Network, Naïve Bayes, Liblinear, LibSVM, RBF network, SMO, Logistic, IBk, Bagging, Ensemble selection, Logit Boost, LMT, NBTree, Random Forest, DTNB	Literature IC_{50} data for CYP2D6 (N = 209), CYP1A2 (N = 160), CYP3A4 (N = 106), CYP2A6 (N = 79), CYP2C9 (N = 69), CYP2C8 (N = 58), CYP2C19 (N = 62).	10-fold cross-validation % accuracy data reported MILP-hyperbox general performs best with % accuracy CYP2D6 (97.05), CY1A2 (91.94), CYP3A4 (89.27), CYP2A6 (88.50), CYP2C9 (83), CYP2C8 (81.67), CYP2C19 (83.17).	1600 descriptors from E-DRAGON along with unsupervised forward selection to select independent descriptors. Ultimately 6 descriptors used in each model.	199

Table 12.4 (*Continued*)

Method	*Training set*	*Test set*	*Notes*	*Reference*
Multiobjective recursive partitioning	N = 161 with classification for metabolism by CYP1A2, CYP2C9, CYP2C19, CYP2E1 and CYP3A4.	Leave out 10%, misclassification rate 12.6%.	Descriptors generated with ADMET Predictor software.	200
k-nearest neighbor, decision tree induction, artificial neural networks, support vector machines	N = 1436 FDA drugs literature data on CYP1A2, CYP2D6 and CYP3A4 classification (substrate, inhibitor or inducer).	10-fold cross-validation correct classification rates 81.7 to 92.9% Trees perform best.	ChemAxon and CDK descriptors.	201
MLR, FA-MLR, PLS, GFA, G/PLS, ANN	N = 28 literature IC_{50}data for CYP3A4.	N = 7.	Descriptors calculated in Cerius 2, G/PLS performs best r^2 0.58. LogP and topological, electronic descriptors were important.	202
Ensemble Catalyticphore based docking and activation energy estimation model	N = 45 substrates for CYP3A4.	N = 25 substrates for CYP3A4.	77.8–81.9% of metabolic sites correctly predicted using a combination of structure based method and activation energy estimation.	203
Probabilistic QSAR models, Binomial PLS and local similarity assessment	N = 560 CYP3A4, N = 526 CYP2D6 substrates.	Data split 70/30 training/test set.	14.2% false positive (CYP3A4) 6.6% false positive (CYP2D6).	204
Activation energy of metabolism reactions with effective atomic descriptors	Training molecules not described (394 metabolic sites for aliphatic hydroxylation, 27 for *N*-dealkylation, 9 for *O*-dealkylation, 85 for aromatic hydroxylation).	N = 46 CYP3A4 substrates.	Regioselectivity predictions 60% correct.	205
Probabilistic QSAR models, logistic PLS and local similarity assessment, SVM and random forest	N = 907 compounds from the literature CYP3A4 IC50. < 40 μM = active IC50 > 60 μM = inactive.	PubChem data on N = 11,060 molecules.	Fragment descriptors were used. As reliability index increases, prediction accuracy increases but fewer compounds belong to the applicability domain. Prediction accuracy may be 93% but only 41% of the PubChem dataset is used.	206

statistical method that can uncover relationships in large complex data to classify objects into categories based on similar activities.[70] A recursive partitioning model containing 875 molecules with human liver microsomal metabolic stability was used to predict and rank the clearance of 41 drugs.[71] Another approach, a k-nearest neighbour statistical model, finds a subspace of the original descriptor space where activity of each compound in the data set is most accurately predicted as the averaged activity of its k nearest neighbors in this subspace. This method has also been used with metabolic stability data from human S9 homogenate for 631 diverse molecules and was able to adequately classify metabolism of a further set of over 100 molecules.[72] A set of 130 calcitriol analogs with *in vitro* metabolic stability was used to develop QSAR models with molecular structure descriptors from QikProp and DiverseSolutions.[73] Variable selection was carried out and partial least squares (PLS) regression models were generated. The models were used for the prediction of the metabolic stability of 244 virtual calcitriol analogs. Twenty molecules were selected for *in vitro* testing and correctly predicted the *in vitro* metabolic stability for 17 of the 20 selected analogs (85 % success rate).[73] The Kohonen map is a multivariate statistical technique which approximates local geometric relationships of a multidimensional property space on a 2D plot.[74] Kohonen maps have also been useful for differentiating high and low affinity P450 3A4 substrates.[55] Neural networks have also been used to predict *N*-dealkylation rates for P450 3A4 and P450 2D6 substrates.[56] The latter work represents a foundation for a software system to predict metabolites and the enzymes involved from an input molecular structure and has also been applied to the differentiation of P450 substrates from non-substrates.[75,76] We have previously used a Kernel partial least squares (K-PLS) algorithm to generate classification models for human phase I and II reactions [77] using literature data.

Pfizer has recently evaluated open source descriptors and model building algorithms using a training set of approximately 50 000 molecules and a test set of approximately 25 000 molecules with human liver microsomal metabolic stability data.[78] A C5.0 decision tree model demonstrated that the Chemistry Development Kit descriptors together with a set of SMARTS keys had good statistics (Kappa = 0.43, sensitivity = 0.57, specificity 0.91, positive predicted value (PPV) = 0.64) equivalent to models built with commercial MOE2D and the same set of SMARTS keys (Kappa = 0.43, sensitivity = 0.58, specificity 0.91, PPV = 0.63). This observation was also confirmed upon extension of the data set to ~193 000 molecules and generation of a continuous model using Cubist. When the continuous predictions and actual values were binned to get a categorical score, an almost identical Kappa statistic (0.42) was observed.[78]

All of these computational techniques represent efforts to predict metabolism focused either on diverse or structurally similar training sets of molecules for metabolic stability, or a particular P450 reaction, and at the same time attempt to address the specific tissues with modelling of these complex enzymes. One of the major limitations of many of the P450 inhibition models described in this section is that they cannot differentiate the other forms of inhibition that are described in the next section.

12.3.3 P450 3A4 Metabolic Intermediate Complex Formation and Time Dependent Inhibition

Molecules can interact with the binding site of P450s, resulting in reversible inhibition.[79,80] Mechanism-based inhibition results either in metabolic products that form heme or protein adducts, or a metabolic inhibitory complex (MIC) that renders the enzyme catalytically inactive.[81] These result in more impactful forms of drug–drug interactions. Mechanism-based enzyme inhibition (MBI) can be considered a subset of time dependent inhibitors (TDI).[82] Mechanism-based types of inhibition have been the focus of several recent reviews, *e.g.* Fontana *et al.* described 59 molecules which were specifically metabolized by P450 3A4 alone to form an intermediate that was able to irreversibly bind to the enzyme.[83] These included acetylenes, furans, thiophenes, conjugated structures, dichloro- and trichloro-ethylenes, methylenedioxyphenol compounds, secondary and tertiary amines, and thiono compounds.[83] Ghanbari *et al.* also collated experimental conditions for many MBI studies, along with physicochemical properties such as pK_a, logP and logD.[84] Hollenberg *et al.* described MBI forming compounds including acetylenes, thiol-containing compounds, arylamines, quinones, furanocoumarins, and tertiary amines.[85] A recent study has suggested a TDI-reactive metabolite screening paradigm.[82]

The P450 3A family of enzymes is widely recognized as the most important for human metabolism of a large proportion of commercially available drugs.[59] It is also frequently assessed in early pharmaceutical drug discovery screens to identify TDI. In a recent study, we have used 54 molecules to assess metabolic intermediate complex (MIC).[86] The study used recombinant P450 3A4 ($+b_5$) and inhibition data were analyzed using multiple computational approaches that included the generation of simple molecular descriptors to identify the relationship between MIC and molecular structure. Comparing MIC and non-MIC forming compounds based on the mean molecular weight showed a significant difference between the two groups of compounds ($p < 0.05$), indicating that larger molecules (likely to be P450 3A4 substrates) are more likely to form MIC *in vitro*. Recursive partitioning, tree-based methods, and logistic regression were used and tested with an external test set of molecules. This study reiterated that P450 3A4 ($+b_5$) mediated MIC formation requires other molecular properties besides primary, secondary, and tertiary amines or methylenedioxyphenyl features, and was dependent on the number of hydrophobic features and hydrogen bond acceptors in the molecule.[86] Recently, we have also used an *in silico–in vitro* approach to predict compounds likely to cause TDI of P450 3A4 in human liver microsomes. We used a Bayesian classification approach[87,88] with simple, interpretable molecular descriptors, as well as FCFP_6,[86] to classify P450 3A4 TDI. The models used between 1853 and 2071 molecules and were tested with molecules excluded from the models. All of the receiver operator characteristic curves show better than random ability to identify the TDI positive molecules.[89] We also found that the indazole substructure is dominant in the features needed for TDI activity, alongside other nitrogen-containing aromatic ring systems (pyrazole and methoxy-aminopyridine). These structures are different to the known structural alerts (such as primary, secondary, and tertiary amines, or methylenedioxyphenyl features) and thus provide information on new structural alerts that could be added to other filtering software. All Bayesian models generated were also evaluated by leaving out either 30% or 50% of the data and rebuilding the model 100 times. In each case, the leave out 30% or 50% testing AUC value was comparable to the leave-one-out approach and these values were very favorable, indicating good model robustness.[89]

Computational methodologies are becoming an increasingly integrated part of the drug discovery process.[3,90–93] However, the advantage of *in silico* approaches, when highly predictive of experimental results, is their ability to virtually assess compounds prior to their synthesis, thus reducing the potential for the development of non-productive compounds.[3,90–92] This *in silico* TDI method also allows for filtering of compounds that absolutely must be screened *in vitro* due to the potential for TDI, and limits the numbers of compounds screened that are low risk.

12.4 Pregnane X Receptor Models

The human pregnane X receptor (PXR)[94–96] is widely known to transcriptionally regulate genes involved in xenobiotic metabolism and excretion, as well as other cellular processes such as apoptosis.[97–101] Human PXR has a very broad specificity for ligands, including endogenous (bile acids, steroid hormones, and fat-soluble vitamins) and exogenous (prescription and herbal drugs, and environmental chemicals) compounds. Activation of human PXR can cause drug–drug interactions or result in an array of physiological effects.[100] Currently there are five high-resolution crystal structures of human PXR[102–106] available (and another to be deposited[107]) in the Protein Data Bank (PDB). The structures have provided atomic level details that have led to a greater understanding of the ligand binding domain (LBD) and the structural features involved in ligand–receptor interactions.[102–108] The cavernous ligand binding pocket (LBP) with a volume >1350 Å^3 accepts molecules of widely varying dimensions and chemical properties and is likely capable of binding small molecules in multiple orientations.[108]

Human PXR agonist pharmacophore models have been shown to possess hydrophobic, hydrogen bond acceptor, and hydrogen bond donor features, consistent with the crystallographic structures of human PXR ligand–receptor complexes.[109–112] These pharmacophore models have predominantly used structurally diverse ligands in the training set and have the limitation in most cases of compiling data from multiple laboratories using different experimental protocols, ultimately forcing the binary classification of ligands for the training sets (*i.e.*, activating *versus* non-activating).

One study has used human PXR activation data for 30 steroidal compounds (including 9 bile acids) to create a pharmacophore with four hydrophobic features and one hydrogen bond acceptor.[110] This pharmacophore contained

5α-androstan-3β-ol (EC_{50} 0.8 μM) which contains one hydrogen bond acceptor, indicating [in contrast to the crystal structure of 17β-estradiol (published EC_{50} 20 μM) bound to human PXR with two hydrogen bonding interactions][107] that hydrophobic interactions may be more important for increased affinity. [110] This, and other pharmacophores, have been used to predict PXR interactions for antibiotics[113] which were verified *in vitro*, suggesting one use for computational approaches in combination with experimental methods.

To date, there have been few attempts to build ligand-based models around a large, structurally narrow set of PXR activators. The absence of large sets of quantitative data for PXR agonists has restricted QSAR models to a relatively small universe of molecules compared to the known drugs, drug-like molecules, endobiotics and xenobiotics in general.[114] Various machine learning methods (*e.g.* SVMs, recursive partitioning *etc.*) that can be used when the biological data is binary in nature (*e.g.* activating or binding *versus* non-activating/non-binding) have been used with PXR. We have previously generated computational models for PXR using recursive partitioning (RP), random forest (RF), and SVM algorithms with VolSurf descriptors. Following 10-fold randomization, the models correctly predicted 82.6–98.9% of activators and 62.0–88.6% of non-activators. All models were tested with a test set ($N = 145$), and the prediction accuracy ranged from 63 to 67% overall.[115] These test set molecules were found to cover the same area in a principal component analysis plot as the training set, suggesting that the predictions were within the applicability domain. A second study used the same training and test sets with molecular descriptors derived from two sources, Shape Signatures and the Molecular Operating Environment (MOE) application software to generate SVM models. The overall test set prediction accuracy for PXR activators with SVM was 72% to 81%.[116]

A large cadre of experimental hPXR data was recently generated for classes of steroidal compounds (namely androstanes, estratrienes, pregnanes, and bile acids/salts)[117] and has been used with an array of ligand-based computational methods, including Bayesian modeling with 2D fingerprints methods.[118] All 115 compounds were used to generate a Bayesian classification model[119] using a definition of active as a compound having an EC_{50} for PXR activation of less than 10 μM. Using FCFP_6 and 8 interpretable descriptors (AlogP, molecular weight, rotatable bonds, number of rings, number of aromatic rings, hydrogen bond acceptor, hydrogen bond donor, and polar surface area) a model was developed with a receiver operator characteristic for leave-one-out cross validation of 0.84. In addition to the leave-one-out cross validation, further validation methods were undertaken. After leaving 20% of the compounds out 100 times, the ROC is 0.84, concordance 73.2%, specificity 69.1%, and sensitivity 84.1%. In comparison to molecular docking methods, ligand-based models performed better in classifying the compounds. The Bayesian method appeared to have good model statistics for internal cross validation of steroids. We have additionally used this model to classify a previously used diverse molecule test set. The Bayesian PXR model was used to rank 123 molecules (65 activators and 58 non activators). Out of the top 30 molecules scored and ranked with this model, 20 (75%) were classified as activators (EC_{50} <100 μM).[118] All PXR positive contributing substructures were essentially hydrophobic, while PXR negative contributing features possessed hydroxyl or other substitutions which are likely not optimally placed to facilitate interactions with hydrogen bonding features in PXR. Therefore possession of these hydrogen bond acceptor and donor features indicated in the steroidal substructures appears to be related to loss of PXR activation.

The Bayesian approach using fingerprints and 117 structural descriptors was also used recently with the large diverse training set comprising 177 compounds. The classifier was used to screen a subset of FDA approved drugs, followed by testing of a few compounds (17 compounds from the top 25) with a cell-based luciferase reporter assay for evaluation of chemical-mediated hPXR activation in HepG2 cells. The reporter assay confirmed 9 drugs as novel PXR activators: fluticasone, nimodipine, nisoldipine, beclomethasone, finasteride, flunisolide, megestrol, secobarbital, and aminoglutethimide.[120] Such ligand-based Bayesian approaches with a diverse training set may be more useful than a narrow structural series of steroidal compounds, which was previously used for database searching a set of pesticides and other industrial chemicals.[4] These global models for PXR could also be of value in selecting compounds for *in vitro* screening. We are not aware of anyone exhaustively screening FDA approved drugs (computationally) for potential PXR agonists and then testing compounds. Such an approach may be useful to understand PXR mediated drug–drug interactions more comprehensively.

12.5 Drug Induced Liver Injury Models

There have been many examples where computational QSARs or machine learning methods have been used for predicting hepatotoxicity[121,122] or drug–drug interactions.[89,114,123,124] Drug metabolism in the liver can convert some drugs into highly reactive intermediates[125–128] and consequently drug-induced liver injury (DILI). DILI is the number one reason why drugs are not approved or are withdrawn from the market after approval.[129] Idiosyncratic liver injury is much harder to predict from the preclinical *in vitro* or *in vivo* situation, so we frequently become aware of such problems once a drug reaches large patient populations in the clinic and this is generally too late for the drug developer to identify an alternative and safer drug molecule. One study assembled a list of approximately 300 drugs and chemicals, with a classification scheme based on human clinical data for hepatotoxicity, for the purpose of evaluating an *in vitro* testing methodology based on cellular imaging of primary human hepatocyte cultures.[130] It was found that the 100-fold C_{max} scaling factor represented a reasonable threshold to differentiate safe *versus* toxic drugs for an orally dosed drug and with regard to hepatotoxicity.[130] The concordance of the *in vitro* human hepatocyte imaging assay technology (HIAT) applied to about 300 drugs and chemicals, is about 75% with regard to clinical hepatotoxicity, and has very few false positives.[130] An alternative is to use the clinical DILI data to create a computational model and then validate it with enough compounds to provide confidence in its predictive ability so that it can be used as a prescreen before *in vitro* testing.

One recent computational study used 74 compounds to create classification models based on linear discriminant analysis (LDA), artificial neural networks (ANNs), and machine learning algorithms (OneRs).[131] These modeling techniques were found to produce models with reasonable internal cross validation statistics (accuracy, sensitivity, and specificity of over 84%, 78%, and 90%, respectively). These models were then tested (6 and 13 compounds respectively) with over 80% accuracy. A second study compiled a data set of compounds with liver effects in different species and then used binary QSAR models to predict effects in humans. The resultant SVM models had good predictive power, as assessed by external 5-fold cross validation procedures, and 78% accuracy for a set of 18 compounds.[132] A third study created a knowledge base with structural alerts from 1266 chemicals. The alerts created were used to predict 626 Pfizer compounds (sensitivity 46%, specificity 73%, and concordance 56%).[133] In a recent study, we have used a training set of 295 compounds and a test set of 237 molecules with a Bayesian classification approach[87,88] using interpretable molecular descriptors and extended connectivity functional class fingerprints of maximum diameter 6 (ECFC_6).[86] The cross validated receiver operator characteristic area under the curve (XV ROC AUC) for the model with 295 molecules was 0.86. The Bayesian model generated was also evaluated by leaving out either 10%, 30%, or 50% of the data and rebuilding the model 100 times in order to generate the XV ROC AUC. In each case, the leave out 10%, 30%, or 50% testing AUC value was comparable to the leave-one-out approach and these values were very favorable, indicating good model robustness. The mean concordance >57%, specificity >61%, and sensitivity >52% did not seem to differ depending on the degree of data left out.[134] Molecular features such as long aliphatic chains, phenols, ketones, diols, α-methyl styrene (represents a polymer monomer), conjugated structures, cyclohexenones, and amides predominated in DILI active compounds.[134]

The Bayesian model was tested with 237 new compounds. The concordance ~60%, specificity 67%, and sensitivity 56% were comparable with the internal validation statistics. A subset of 37 compounds of most interest clinically showed similar testing values with a concordance greater than 63%.[134] Compounds of most clinical interest are defined as well-known hepatotoxic drugs, plus their less hepatotoxic comparators. These less hepatotoxic comparators are approved drugs that typically share a portion of the chemical core structure with the hepatotoxic drugs. The purpose of this test set was to explore whether the Bayesian *in silico* method could differentiate differences in DILI potential between or among closely related compounds, as this is likely the most useful case in the real-world drug discovery setting.

A principal component analysis (PCA) using simple molecular descriptors showed that the training and test sets overlapped. Focusing on compounds with a Tanimoto similarity greater than 0.7 resulted in a test set of 28 compounds, which had Matthews correlation coefficient and concordance statistics that were similar to those obtained for the complete test set. The specificity increased to 80% and sensitivity decreased to 50% in this case.[134]

We have also compared the total data set of 532 compounds with a set of 77 recently launched small-molecule drugs from the period 2006–2010, extracted from the Prous Integrity database.[134] Again, we find these molecules are distributed throughout the combined training and test set, representative of overlap which is also suggested from the mean physicochemical property values. These combined analyses would suggest that the test and training set used for the DILI model is biased towards some "dated chemistry", and is in fact representative of medicinal chemistry space, which led to newly launched small-molecule drugs.

This example represents the first large scale testing of a machine learning model for DILI that uses a similarly sized training and test set. The overall concordance of the model is lower (~60–64% depending on test set size) than that observed previously for *in vitro* HIAT (75%[130]). However, the test set statistics are similar to those reported elsewhere using structural alerts.[133] This work suggests that current data on compounds can be used to predict with reasonable accuracy future compounds and their potential for DILI. Such models could be used before the *in vitro* HIAT assay, which takes more resources to conduct than *in silico* technologies.

12.6 From Ligands to Pharmacophores to Networks to Understand Toxicity

There have been many studies establishing relationships between ligand molecular structure and broad biological activity, both on and off-target toxicity effects.[135–138] One could even consider the many pharmacophore-based studies for toxicity related proteins and 3D database searching efforts to find new inhibitors as early attempts at understanding off-target effects and adverse drug reactions. For example, we have used pharmacophores for various transporters, such as the human peptide transporter 1 (HPEPT1),[139] P-glycoprotein (P-gp),[140] human organic cation/carnitine transporter (hOCTN2),[141,142] and human apical sodium-dependent bile acid transporter (ASBT)[143] to search a subset of FDA approved drugs and identify previously unknown inhibitors upon *in vitro* testing. In particular, inhibition of hPEPT1 or P-gp are involved in drug–drug interactions,[140] hOCTN2 inhibition may cause rhabdomyolysis,[141,142] and drugs inhibiting ASBT have potential to promote several adverse drug reactions (ADRs) including colon cancer.[143]

In the last five years 2D ligand-based approaches have increasingly used sophisticated algorithms and networks. Fliri *et al.* at Pfizer used biological spectra for a cross section of the proteome.[144] They used hierarchical clustering of the spectra similarity and created a relationship between structure and bioactivity, before extending this to identify receptor agonist and antagonist profiles.[145] The same group applied a probabilistic approach to link adverse effects for drugs with biological spectra (similar molecules had overlapping profiles, in the same way that they had similar biological spectra), linking preclinical with clinical effects.[146] A different group at Pfizer created a global mapping of pharmacological space, focusing on a polypharmacology network of 200 000 molecules with activity against 698 proteins.[147] A further published study created a drug–target network of approved drug data for 890 molecules from DrugBank and Online Mendelian Inheritance in Man (OMIM), with over half of these molecules forming the largest connected network with multiple target proteins (illustrating polypharmacology or promiscuity).[148] Such networks might help understand likely nodes involved in toxicity and add to the similarity maps for enzymes, receptors[149] and human polypharmacology networks[147] that have also been developed to date. A recent study from Abbott even introduced a sequence-independent kinome inhibitor interaction map[150] while another established links between over 44 000 small molecules and 160 cardiovascular targets, with kinases having on average 7 connections to cardiovascular targets.[151] An example from Berg *et al.*, has merged chemical target and pathway toxicity mechanisms that can be defined from profiling in primary human cell systems, covering many readouts and enabling known reference compounds to be mapped by functional similarity.[152]

A complementary approach taken by a group at Novartis uses chemical substructures relevant to toxicology related adverse effects[153] for approximately 4000 molecules with side effect information from the World Drug Index. The same group related promiscuity of compounds to their safety,[154] as compounds hitting several targets cause more undesired effects. More recently, the group has related over 4000 medical dictionary for regulatory activities (MedDRA) terms for adverse drug reactions for over 1800 drugs using the ECFP_4 descriptors and Bayesian models.[155,156] This resulted in a map of ADRs in chemical space and an approach that could be used to predict *in silico* the ADR likelihood for new molecules, based on substructures. Interestingly the similarity ensemble analysis described by Keiser *et al.* also used the ECFP_4 descriptors and Bayesian models to predict off-target effects of 3665 FDA approved drugs and investigational compounds.[157] Decision tree induction has also been used to predict ADR for 507 drugs from the Swiss drugs registry, models that at least internally looked predictive.[158]

These various methods may be useful not only for understanding and identifying toxicity mechanisms but also, potentially, for repurposing molecules for new uses that could focus and accelerate *in vitro* screening efforts,[159–163] as we have previously described with transporters.[139–143]

12.7 Discussion

We have provided an overview of several ligand-based models for human toxicity that we have been involved in generating over the last few years, with a particular focus on Bayesian models. This represents a snapshot of the efforts to build computational models for toxicity. The value of such models may be greater in early drug discovery to triage virtual compounds, to determine which scaffolds to buy or make, and select which compounds to screen *in vitro*. Such *in silico* models are also fast and enable real-time decision making when used as part of a molecular properties "dashboard" or in a multiple optimization approach.[164] While not all computational toxicology models may tell you what molecular features are undesirable, they at least represent a way to deal with far more compounds than one could process manually in a realistic timeframe.

We have described cases where pharmaceutical companies have published data sets or models of interest for computational toxicology. In recent years we have seen GSK, BMS, and AstraZeneca publish widely (hERG, ADME rules, *etc.*),[39,165,166] and Pfizer has used very large training sets for ADME properties,[78] hERG, CYPs,[167] *etc.* Pfizer researchers have also stated that integrated *in silico–in vitro* approaches have reduced early ADME *in vitro* screening by 30%, resulting in substantial cost savings.[89] A review from Roche has summarized available methods for computational toxicology predictions and emphasized the importance of hERG models due to the cost of screening. Such models could serve to highlight adverse drug issues for decision making without necessarily killing compounds immediately.[168] Even relatively simple approaches may be insightful to companies. For example, Pfizer scientists looked at the relationship between physicochemical properties and animal *in vivo* toleration for 245 preclinical compounds. ClogP <3 and total polar surface area >75Å^{2}were seen as preferable, with fewer toxicity findings.[169] This toxicity risk may be due to increased promiscuity at different targets.[169,170] Researchers at Novartis have shown, using Bayesian models, that marketed drugs have a lower promiscuity than those that have failed during clinical development.[171]

As the pharmaceutical companies massively restructure, there will be increasing focus on efficiency in drug discovery. We may also see a strategy that emphasizes computational toxicology more heavily than in the past, as a younger generation of scientists (more aware of knowledge management and data mining and generally more proficient in using computers and technologies), comes to the fore. Computational toxicology is not some magic art, it is data driven and should be thought of as a decision making aid (in the same way that we use preclinical *in vitro* and *in vivo* or even clinical trial data), rather than panacea. Computational toxicology data will have truly come of age when it is accorded the same level of respect as any experimental biology or pharmacology data in a decision making context. That day may be approaching very rapidly.

Acknowledgements

S.E. gratefully acknowledges the many collaborators involved in the computational toxicology projects described and Accelrys Inc. for providing Discovery Studio used for the various Bayesian models described.

References

1. *Predictive toxicology*, ed. C. Helma, Taylor and Francis, Boca Raton, 2005.
2. *Predicting Chemical Toxicity and Fate*, ed. M. T. D. Cronin and D. J. Livingstone, CRC Press, Boca Raton, 2004.

3. *Computational Toxicology: Risk Assessment for Pharmaceutical and Environmental Chemicals*, ed. S. Ekins, John Wiley and Sons, Hoboken, 2007.
4. S. Kortagere, M. D. Krasowski, E. J. Reschly, M. Venkatesh, S. Mani and S. Ekins, *Environ. Health Perspect.*, 2010, **118**, 1412.
5. S. Kortagere, D. Chekmarev, W. J. Welsh and S. Ekins, *Pharm. Res.*, 2009, **26**, 1001.
6. A. Khandelwal, M. D. Krasowski, E. J. Reschly, M. W. Sinz, P. W. Swaan and S. Ekins, *Chem. Res. Toxicol.*, 2008, **21**, 1457.
7. D. M. Barry, H. Xu, R. B. Schuessler and J. M. Nerbonne, *Circ. Res.*, 1998, **83**, 560.
8. M. C. Trudeau, J. W. Warmke, B. Ganetzky and G. A. Robertson, *Science*, 1995, **269**, 92.
9. J. W. Warmke and B. Ganetzky, *Proc Natl. Acad. Sci. U. S. A.*, 1994, **91**, 3438.
10. M. C. Sanguinetti, C. Jiang, M. E. Curran and M. T. Keating, *Cell*, 1995, **81**, 299.
11. M. E. Curran, I. Splawski, K.W. Timothy, G. M. Vincent, E. D. Green and M. T. Keating, *Cell*, 1995, **80**, 795.
12. H. Suessbrich, R. Schonherr, S. H. Heinemann, B. Attali, F. Lang and A. E. Busch, *Br. J. Pharmacol.*, 1997, **120**, 968.
13. T. K. Morgan Jr and M. E. Sullivan, *Prog. Med. Chem.*, 1992, **29**, 65.
14. P. Matyus, A. P. Borosy, A. Varro, J. G. Papp, D. Barlocco and G. Cignarella, *Int. J. Quantum Chem.*, 1998, **69**, 21.
15. Tripos Inc., St. Louis, MO, 1997.
16. H. Liu, M. Ji, H. Jiang, L. Liu, W. Hua, K. Chen and R. Ji, *Bioorg. Med. Chem. Lett.*, 2000, **10**, 2153.
17. H. Liu, M. Ji, X. Luo, J. Shen, X. Huang, W. Hua, H. Jiang and K. Chen, *J. Med. Chem.*, 2002, **45**, 2953.
18. W. Tong, D. R. Lowis, R. Perkins, Y. Chen, W. J. Welsh, D. W. Goddette, T. W. Heritage and D. M. Sheehan, *J. Chem. Inf. Comput. Sci.*, 1998, **38**, 669.
19. S. Ekins, W. J. Crumb, R. D. Sarazan, J. H. Wikel and S. A. Wrighton, *J. Pharmacol. Exp. Ther.*, 2002, **301**, 427.
20. W. J. Crumb Jr, S. Ekins, D. Sarazan, J. H. Wikel, S. A. Wrighton, C. Carlson and C. M. Beasley, *Pharm. Res.*, 2006, **23**, 1133.
21. R. A. Pearlstein, R. J. Vaz, J. Kang, X. L. Chen, M. Preobrazhenskaya, A. E. Shchekotikhin, A. M. Korolev, L. N. Lysenkova, O. V. Miroshnikova, J. Hendrix and D. Rampe, *Bioorg. Med. Chem. Lett.*, 2003, **13**, 1829.
22. S. R. Fletcher, F. Burkamp, P. Blurton, S .K. Cheng, R. Clarkson, D. O'Connor, D. Spinks, M. Tudge, M. B. van Niel, S. Patel, K. Chapman, R. Marwood, S. Shepheard, G. Bentley, G. P. Cook, L. J. Bristow, J. L. Castro, P .H. Hutson and A. M. MacLeod, *J. Med. Chem.*, 2002, **45**, 492.
23. M. Rowley, D. J. Hallett, S. Goodacre, C. Moyes, J. Crawforth, T. J. Sparey, S. Patel, R. Marwood, S. Thomas, L. Hitzel, D. O' Connor, N. Szeto, J. L. Castro, P. H. Hutson and A. M. MacLeod, *J. Med. Chem.*, 2001, **44**, 1603.

24. I. M. Bell, S. N. Gallicchio, M. Abrams, L. S. Beese, D. C. Beshore, H. Bhimnathwala, M. J. Bogusky, C. A. Buser, J. C. Culberson, J. Davide, M. Ellis-Hutchings, C. Fernandes, J. B. Gibbs, S. L. Graham, K. A. Hamilton, G. D. Hartman, D. C. Heimbrook, C. F. Homnick, H. E. Huber, J. R. Huff, K. Kassahun, K. S. Koblan, N. E. Kohl, R. B. Lobell, J. J. Lynch Jr, R. Robinson, A. D. Rodrigues, J. S. Taylor, E. S. Walsh, T. M. Williams and C. B. Zartman, *J. Med. Chem.*, 2002, **45**, 2388.
25. S. Ekins and P. W. Swaan, *Rev. Comput. Chem.*, 2004, **20**, 333.
26. S. Ekins, *Drug Discovery Today*, 2004, **9**, 276.
27. U. Zachariae, F. Giordanetto and A. G. Leach, *J. Med. Chem.*, 2009, **52**, 4266.
28. S. Ekins, *Biochem. Soc. Trans.*, 2003, **31**, 611.
29. S. Ekins, K. V. Balakin, N. Savchuk and Y. Ivanenkov, *J. Med. Chem.*, 2006, **49**, 5059.
30. O. Roche, G. Trube, J. Zuegge, P. Pflimlin, A. Alanine and G. Schneider, *ChemBioChem*, 2002, **3**, 455.
31. A. K. Ghose and G. M. Crippen, *J. Comput. Chem.*, 1986, **7**, 565.
32. G. Chen, E. Banoglu and M. W. Duffel, *Chem. Res. Toxicol.*, 1996, **9**, 67.
33. G. Schneider, W. Neidhart, T. Giller and G. Schmid, *Angew. Chem. Int. Ed.*, 1999, **38**, 2894.
34. H. Chen, A. G. Fantel and M. R. Juchau, *Drug Metab. Dispos.*, 2000, **28**, 1051.
35. H. Chen, W. Howald and M. Juchau, *Drug Metab. Dispos.*, 2000, **28**, 315.
36. D. S. Chekmarev, V. Kholodovych, K. V. Balakin, Y. Ivanenkov, S. Ekins and W. J. Welsh, *Chem. Res. Toxicol.*, 2008, **21**, 1304.
37. P. W. Swaan and S. Ekins, *Drug Discovery Today*, 2005, **10**, 1191.
38. S. Polak, B. Wisniowska and J. Brandys, *J. Appl. Toxicol.*, 2009, **29**, 183.
39. M. P. Gleeson, *J. Med. Chem.*, 2008, **51**, 817.
40. T. J. Ritchie and S. J. Macdonald, *Drug Discovery Today*, 2009, **14**, 1011.
41. I. Shamovsky, S. Connolly, L. David, S. Ivanova, B. Norden, B. Springthorpe and K. Urbahns, *J. Med. Chem.*, 2008, **51**, 1162.
42. I. Shamovsky, C. de Graaf, L. Alderin, M. Bengtsson, H. Bladh, L. Borjesson, S. Connolly, H. J. Dyke, M. van den Heuvel, H. Johansson, B. G. Josefsson, A. Kristoffersson, T. Linnanen, A. Lisius, R. Mannikko, B. Norden, S. Price, L. Ripa, D. Rognan, A. Rosendahl, M. Skrinjar and K. Urbahns, *J. Med. Chem.*, 2009, **52**, 7706.
43. S. Mente, R. Gallaschun, A. Schmidt, L. Lebel, M. Vanase-Frawley and A. Fliri, *Bioorg. Med. Chem. Lett.*, 2008, **18**, 6088.
44. K. Hansen, F. Rathke, T. Schroeter, G. Rast, T. Fox, J. M. Kriegl and S. Mika, *J. Chem. Inf. Model.*, 2009, **49**, 1486.
45. C. de Graaf, N. P. Vermeulen and K. A. Feenstra, *J. Med. Chem.*, 2005, **48**, 2725.
46. C. Hansch, *Drug. Metabol. Rev.*, 1972, **1**, 1.
47. C. Hansch, *Drug. Metabol. Rev.*, 1984, **15**, 1279.
48. C. Hansch, E. J. Lien and F. Helmer, *Arch. Biochem. Biophys.*, 1968, **128**, 319.

49. C. Hansch and L. Zhang, *Drug Metabol. Rev.*, 1993, **25**, 1.
50. D. F. V. Lewis, *Biochem. Pharmacol.*, 2000, **60**, 293.
51. S. Ekins and P. W. Swaan, *Rev. Comput. Chem.*, 2004, **20**, 333.
52. U. Fuhr, G. Strobl, F. Manaut, E.-M. Anders, F. Sorgel, E. Lopez-de-brinas, D. T. W. Chu, A. G. Pernet, G. Mahr, F. Sanz and A. H. Staib, *Mol. Pharmacol.*, 1993, **43**, 191.
53. J. P. Jones and K. R. Korzekwa, *Predicting the Rates and Regioselectivity of Reactions Mediated by the P450 Superfamily*, Academic Press Inc., New York, 1996.
54. C. W. Locuson and J. L. Wahlstrom, *Drug. Metab. Dispos.*, 2005, **33**, 873.
55. K. V. Balakin, S. Ekins, A. Bugrim, Y. A. Ivanenkov, D. Korolev, Y. Nikolsky, S. A. Skorenko, A. A. Ivashchenko, N. P. Savchuk and T. Nikolskaya, *Drug Metab. Dispos.*, 2004, **32**, 1183.
56. K. V. Balakin, S. Ekins, A. Bugrim, Y. A. Ivanenkov, D. Korolev, Y. Nikolsky, A. A. Ivashchenko, N. P. Savchuk and T. Nikolskaya, *Drug Metab. Dispos.*, 2004, **32**, 1111.
57. S. Ekins, M. de Groot and J. P. Jones, *Drug Metab. Dispos.*, 2001, **29**, 936.
58. L. J. Jolivette and S. Ekins, *Adv. Clin. Chem.*, 2007, **43**, 131.
59. S. A. Wrighton, E. G. Schuetz, K. E. Thummel, D. D. Shen, K. R. Korzekwa and P. B. Watkins, *Drug Metabol. Rev.*, 2000, **32**, 339.
60. S. Ekins and S. A. Wrighton, *Drug Metabol. Rev.*, 1999, **31**, 719.
61. S. Ekins, J. Berbaum and R. K. Harrison, *Drug Metab. Dispos.*, 2003, **31**, 1077.
62. S. Ekins, D. M. Stresser and J. A. Williams, *Trends Pharmacol. Sci.*, 2003, **24**, 191.
63. C. W. Locuson 2nd, J. L. Wahlstrom, D. A. Rock, D. A. Rock and J. P. Jones, *Drug Metab. Dispos*,. 2003, **31**, 967.
64. C. W. Locuson 2nd, D. A. Rock and J. P. Jones, *Biochemistry*, 2004, **43**, 6948.
65. C. W. Locuson 2nd, H. Suzuki, A. E. Rettie and J. P. Jones, *J. Med. Chem.*, 2004, **47**, 6768.
66. S. Ekins, G. Bravi, S. Binkley, J. S. Gillespie, B. J. Ring, J. H. Wikel and S. A. Wrighton, *Drug Metab. Dispos.*, 2000, **28**, 994.
67. A. C. Egnell, C. Eriksson, N. Albertson, B. Houston and S. Boyer, *J. Pharmacol. Exp. Ther.*, 2003, **307**, 878.
68. A. C. Egnell, J. B. Houston and C. S. Boyer, *J. Pharmacol. Exp. Ther.*, 2005, **312**, 916.
69. Q. Wang and J. R. Halpert, *Drug Metab. Dispos.*, 2002, **30**, 86.
70. A. Rusinko 3rd, M. W. Farmen, C. G. Lambert, P. L. Brown and S. S. Young, *J. Chem. Inf. Comput. Sci.*, 1999, **39**, 1017.
71. S. Ekins, *Biochem. Soc. Trans.*, 2003, **31**, 611.
72. M. Shen, Y. Xiao, A. Golbraikh, V. K. Gombar and A. Tropsha, *J. Med. Chem.*, 2003, **46**, 3013.
73. B. F. Jensen, M. D. Sorensen, A. M. Kissmeyer, F. Bjorkling, K. Sonne, S. B. Engelsen and L. Norgaard, *J. Comput. Aided Mol. Des.*, 2003, **17**, 849.

74. T. Kohonen, *Self-Organisation and Associative Memory*, Springer-Verlag, Berlin, 1989.
75. D. Korolev, K. V. Balakin, Y. Nikolsky, E. Kirillov, Y. A. Ivanenkov, N. P. Savchuk, A. A. Ivashchenko and T. Nikolskaya, *J. Med. Chem.*, 2003, **46**, 3631.
76. A. Bugrim, T. Nikolskaya and Y. Nikolsky, *Drug Discovery Today*, 2004, **9**, 127.
77. S. Ekins, in *Computer Applications in Pharmaceutical Research and Development*, ed. S. Ekins, John Wiley and Sons, Hoboken, 2006, ch. 18, pp. 445–468.
78. R. R. Gupta, E. M. Gifford, T. Liston, C. L. Waller, B. Bunin and S. Ekins, *Drug Metab. Dispos.*, 2010, **38**, 2083.
79. D. Newton, R. Wang and A. Lu, *Drug Metab. Dispos.*, 1995, **23**, 154.
80. H. Vanden Bossche, L. Koymans and H. Moereels, *Pharmacol. Ther.*, 1995, **67**, 79.
81. R. Silverman, *Mechanism-Based Enzyme Inactivation: Chemistry and Enzymology*, CRC Press, Boca Raton, 1988, 3.
82. K. H. Grimes, J. Bird, D. Ferguson and R. J. Riley, *Eur. J. Pharm. Sci.*, 2009, **36**, 175.
83. E. Fontana, P. M. Dansette and S. M. Poli, *Curr. Drug. Metab.*, 2005, **6**, 413.
84. F. Ghanbari, K. Rowland-Yeo, J. C. Bloomer, S. E. Clarke, M. S. Lennard, G. T. Tucker and A. Rostami-Hodjegan, *Curr. Drug. Metab.*, 2006, **7**, 315.
85. P. F. Hollenberg, U. M. Kent and N. N. Bumpus, *Chem. Res. Toxicol..*, 2008, **21**, 189.
86. D. R. Jones, S. Ekins, L. Li and S. D. Hall, *Drug Metab. Dispos.*, 2007, **35**, 1466.
87. X. Y. Xia, E. G. Maliski, P. Gallant and D. Rogers, *J. Med. Chem.*, 2004, **47**, 4463.
88. A. Bender, Ph.D. Thesis, University of Cambridge, Cambridge, 2005.
89. M. Zientek, C. Stoner, R. Ayscue, J. Klug-McLeod, Y. Jiang, M. West, C. Collins and S. Ekins, *Chem. Res. Toxicol.*, 2010, **23**, 664.
90. S. Ekins, in *Computer Applications in Pharmaceutical Research and Development*, ed. B. Wang, John Wiley and Sons, Hoboken, 2006.
91. S. Ekins, J. Mestres and B. Testa, *Br. J. Pharmacol.*, 2007, **152**, 21.
92. S. Ekins, J. Mestres and B. Testa, *Br. J. Pharmacol.*, 2007, **152**, 9.
93. A. Nath, M. A. Zientek, B. J. Burke, Y. Jiang and W. M. Atkins, *Drug Metab. Dispos.*, 2010, **38**, 2195.
94. G. Bertilsson, J. Heidrich, K. Svensson, M. Asman, L. Jendeberg, M. Sydow-Backman, R. Ohlsson, H. Postlind, P. Blomquist and A. Berkenstam, *Proc. Natl. Acad. Sci. U. S. A.*, 1998, **95**, 12208.
95. B. Blumberg, W. Sabbagh Jr., H. Juguilon, J. Bolado Jr., C. M. van Meter, E. S. Ong and R. M. Evans, *Genes. Dev.*, 1998, **12**, 3195.
96. S. A. Kliewer, J. T. Moore, L. Wade, J. L. Staudinger, M.A. Watson, S. A. Jones, D. D. McKee, B. B. Oliver, T. M. Willson, R. H. Zetterstrom, T. Perlmann and J. M. Lehmann, *Cell*, 1998, **92**, 73.

97. S. Verma, M. M. Tabb and B. Blumberg, *BMC Cancer*, 2009, **9**, 3.
98. D. Gupta, M. Venkatesh, H. Wang, S. Kim, M. Sinz, G. L. Goldberg, K. Whitney, C. Longley and S. Mani, *Clin. Cancer Res.*, 2008, **14**, 5332.
99. J. Zhou, M. Liu, Y. Zhai and W. Xie, *Mol. Endocrinol.*, 2008, **22**, 868.
100. A. Biswas, S. Mani, M. R. Redinbo, M. D. Krasowski, H. Li and S. Ekins, *Pharm. Res.*, 2009, **26**, 1807.
101. B. L. Urquhart, R. G. Tirona and R. B. Kim, *J. Clin. Pharmacol.*, 2007, **47**, 566.
102. R. E. Watkins, P. R. Davis-Searles, M. H. Lambert and M. R. Redinbo, *J. Mol. Biol.*, 2003, **331**, 815.
103. R. E. Watkins, J. M. Maglich, L. B. Moore, G. B. Wisely, S. M. Noble, P. R. Davis-Searles, M. H. Lambert, S. A. Kliewer and M. R. Redinbo, *Biochemistry*, 2003, **42**, 1430.
104. Y. Xue, E. Chao, W. J. Zuercher, T. M. Willson, J. L. Collins and M. R. Redinbo, *Bioorg. Med. Chem.*, 2007, **15**, 2156.
105. J. E. Chrencik, J. Orans, L. B. Moore, Y. Xue, L. Peng, J. L. Collins, G. B. Wisely, M. H. Lambert, S. A. Kliewer and M. R. Redinbo, *Mol. Endocrinol.*, 2005, **19**, 1125.
106. D. G. Teotico, J. J. Bischof, L. Peng, S. A. Kliewer and M. R. Redinbo, *Mol. Pharmacol.*, 2008, **74**, 1512.
107. Y. Xue, L. B. Moore, J. Orans, L. Peng, S. Bencharit, S. A. Kliewer and M. R. Redinbo, *Mol. Endocrinol.*, 2007, **21**, 1028.
108. R. E. Watkins, G. B. Wisely, L. B. Moore, J. L. Collins, M. H. Lambert, S. P. Williams, T. M. Willson, S. A. Kliewer and M. R. Redinbo, *Science*, 2001, **292**, 2329.
109. K. Bachmann, H. Patel, Z. Batayneh, J. Slama, D. White, J. Posey, S. Ekins, D. Gold and L. Sambucetti, *Pharmacol. Res.*, 2004, **50**, 237.
110. S. Ekins, C. Chang, S. Mani, M. D. Krasowski, E. J. Reschly, M. Iyer, V. Kholodovych, N. Ai, W. J. Welsh, M. Sinz, P. W. Swaan, R. Patel and K. Bachmann, *Mol. Pharmacol.*, 2007, **72**, 592.
111. S. Ekins and J. A. Erickson, *Drug Metab. Dispos.*, 2002, **30**, 96.
112. D. Schuster and T. Langer, *J. Chem. Inf. Model*, 2005, **45**, 431.
113. K. Yasuda, A. Ranade, R. Venkataramanan, S. Strom, J. Chupka, S. Ekins, E. Schuetz and K. Bachmann, *Drug Metab. Dispos.*, 2008, **36**, 1689.
114. C. Y. Ung, H. Li, C. W. Yap and Y. Z. Chen, *Mol. Pharmacol.*, 2007, **71**, 158.
115. A. Khandelwal, M. D. Krasowski, E. J. Reschly, M. W. Sinz, P. W. Swaan and S. Ekins, *Chem. Res. Toxicol.*, 2008, **21**, 1457.
116. S. Kortagere, D. Chekmarev, W. J. Welsh and S. Ekins, *Pharm. Res.*, 2009, **26**, 1001.
117. S. Ekins, E. J. Reschly, L. R. Hagey and M. D. Krasowski, *BMC Evol. Biol.*, 2008, **8**, 103.
118. S. Ekins, S. Kortagere, M. Iyer, E. J. Reschly, M. A. Lill, M. Redinbo and M. D. Krasowski, *PLoS Comput. Biol.*, 2009, **5**, e1000594.
119. M. Hassan, R. D. Brown, S. Varma-O' Brien and D. Rogers, *Mol. Diversity*, 2006, **10**, 283.

120. Y. Pan, L. Li, G. Kim, S. Ekins, H. Wang and P.W. Swaan, *Drug Metab. Dispos.*, 2011, **39**, 337.
121. R. D. Clark, P. R. Wolohan, E. E. Hodgkin, J. H. Kelly and N. L. Sussman, *J. Mol. Graph. Model.*, 2004, **22**, 487.
122. A. Cheng and S. L. Dixon, *J. Comput. Aided Mol. Des.*, 2003, **17**, 811.
123. J. D. Marechal, J. Yu, S. Brown, I. Kapelioukh, E. M. Rankin, C. R. Wolf, G. C. Roberts, M. J. Paine and M. J. Sutcliffe, *Drug. Metab Dispos.*, 2006, **34**, 534.
124. S. Ekins, C. L. Waller, P. W. Swaan, G. Cruciani, S. A. Wrighton and J. H. Wikel, *J. Pharmacol. Toxicol.*, 2000, **44**, 251.
125. U. A. Boelsterli, H. K. Ho, S. Zhou and K. Y. Leow, *Curr. Drug Metab.*, 2006, **7**, 715.
126. K. Kassahun, P. G. Pearson, W. Tang, I. McIntosh, K. Leung, C. Elmore, D. Dean, R. Wang, G. Doss and T. A. Baillie, *Chem. Res. Toxicol.*, 2001, **14**, 62.
127. J. L. Walgren, M. D. Mitchell and D. C. Thompson, *Crit. Rev. Toxicol.*, 2005, **35**, 325.
128. B. K. Park, N. R. Kitteringham, J. L. Maggs, M. Pirmohamed and D. P. Williams, *Annu. Rev. Pharmacol.*, 2005, **45**, 177.
129. D. Schuster, C. Laggner and T. Langer, *Curr. Pharm. Des.*, 2005, **11**, 3545.
130. J. J. Xu, P. V. Henstock, M. C. Dunn, A. R. Smith, J. R. Chabot and D. de Graaf, *Toxicol. Sci.*, 2008, **105**, 97.
131. M. Cruz-Monteagudo, M. N. Cordeiro and F. Borges, *J. Comput. Chem.*, 2008, **29**, 533.
132. D. Fourches, J. C. Barnes, N. C. Day, P. Bradley, J. Z. Reed and A. Tropsha, *Chem. Res. Toxicol.*, 2010, **23**, 171–183.
133. N. Greene, L. Fisk, R. T. Naven, R. R. Note, M. L. Patel and D. J. Pelletier, *Chem. Res. Toxicol.*, 2010, **23**, 1215.
134. S. Ekins, A. J. Williams and J. J. Xu, *Drug. Metab. Dispos.*, 2010, **38**, 2302.
135. L. M. Kauvar, D. L. Higgins, H. O. Villar, J. R. Sportsman, A. Engqvist-Goldstein, R. Bukar, K. E. Bauer, H. Dilley and D. M. Rocke, *Chem. Biol.*, 1995, **2**, 107.
136. L. M. Kauvar and E. Laborde, *Curr. Opin. Drug Discovery Dev.*, 1998, **1**, 66.
137. L. M. Kauvar, H. O. Villar, J. R. Sportsman, D. L. Higgins and D. E. J. Schmidt, *J Chromatogr. B*, 1998, **715**, 93.
138. A. Ma'ayan, S. L. Jenkins, J. Goldfarb and R. Lyengar, *Mt Sinai J. Med.*, 2007, **74**, 27.
139. S. Ekins, J. S. Johnston, P. Bahadduri, V. M. D'Souzza, A. Ray, C. Chang and P. W. Swaan, *Pharm. Res.*, 2005, **22**, 512.
140. C. Chang, P. M. Bahadduri, J. E. Polli, P. W. Swaan and S. Ekins, *Drug Metab. Dispos.*, 2006, **34**, 1976.
141. L. Diao, S. Ekins and J. E. Polli, *Mol. Pharmacol.*, 2010, **7**, 2120.
142. L. Diao, S. Ekins and J. E. Polli, *Pharm. Res.*, 2009, **26**, 1890.

143. X. Zheng, S. Ekins, J.-P. Rauffman and J. E. Polli, *Mol. Pharmacol.*, 2009, **6**, 1591.
144. A. F. Fliri, W. T. Loging, P. F. Thadeio and R. A. Volkmann, *Proc. Natl. Acad. Sci. U. S. A*, 2005, **102**, 261.
145. A. F. Fliri, W. T. Loging, P. F. Thadeio and R. A. Volkmann, *J. Med. Chem.*, 2005, **48**, 6918.
146. A. F. Fliri, W. T. Loging, P. F. Thadeio and R. A. Volkmann, *Nat. Chem Biol.*, 2005, **1**, 389.
147. G. V. Paolini, R. H. Shapland, W. P. van Hoorn, J. S. Mason and A. L. Hopkins, *Nat. Biotechnol.*, 2006, **24**, 805.
148. M. A. Yildirim, K. I. Goh, M. E. Cusick, A. L. Barabasi and M. Vidal, *Nat. Biotechnol.*, 2007, **25**, 1119.
149. M. J. Keiser, B. L. Roth, B. N. Armbruster, P. Ernsberger, J. J. Irwin and B. K. Shoichet, *Nat. Biotechnol.*, 2007, **25**, 197.
150. J. T. Metz and P. J. Hajduk, *Curr. Opin. Chem. Biol.*, 2010, **14**, 498.
151. M. Cases and J. Mestres, *Drug Discovery Today*, 2009, **14**, 479.
152. E. L. Berg, J. Yang, J. Melrose, D. Nguyen, S. Privat, E. Rosler, E. J. Kunkel and S. Ekins, *J. Pharmacol. Toxicol.*, 2010, **61**, 3.
153. A. Bender, J. Scheiber, M. Glick, J.W. Davies, K. Azzaoui, J. Hamon, L. Urban, S. Whitebread and J. L. Jenkins, *ChemMedChem*, 2007, **2**, 861.
154. K. Azzaoui, J. Hamon, B. Faller, S. Whitebread, E. Jacoby, A. Bender, J. L. Jenkins and L. Urban, *ChemMedChem*, 2007, **2**, 874.
155. J. Scheiber, B. Chen, M. Milik, S. C. Sukuru, A. Bender, D. Mikhailov, S. Whitebread, J. Hamon, K. Azzaoui, L. Urban, M. Glick, J. W. Davies and J. L. Jenkins, *J. Chem. Inf. Model.*, 2009, **49**, 308.
156. J. Scheiber, J. L. Jenkins, S. C. Sukuru, A. Bender, D. Mikhailov, M. Milik, K. Azzaoui, S. Whitebread, J. Hamon, L. Urban, M. Glick and J.W. Davies, *J. Med. Chem.*, 2009, **52**, 3103.
157. M. J. Keiser, V. Setola, J. J. Irwin, C. Laggner, A. I. Abbas, S. J. Hufeisen, N. H. Jensen, M. B. Kuijer, R. C. Matos, T. B. Tran, R. Whaley, R. A. Glennon, J. Hert, K. L. Thomas, D. D. Edwards, B. K. Shoichet and B. L. Roth, *Nature*, 2009, **462**, 175.
158. F. Hammann, H. Gutmann, N. Vogt, C. Helma and J. Drewe, *Clin. Pharmacol. Ther.*, 2010, **88**, 52.
159. C. R. Chong and D. J. Sullivan Jr., *Nature*, 2007, **448**, 645.
160. N. H. Jensen and B. L. Roth, *Comb. Chem. High Throughput Screening*, 2008, **11**, 420.
161. R. T. Strachan, G. Ferrara and B. L. Roth, *Drug Discovery Today*, 2006, **11**, 708.
162. K. A. O'Connor and B. L. Roth, *Nat. Rev. Drug Discovery*, 2005, **4**, 1005.
163. B. L. Roth, E. Lopez, S. Beischel, R. B. Westkaemper and J. M. Evans, *Pharmacol. Ther.*, 2004, **102**, 99.
164. S. Ekins, J. D. Honeycutt and J. T. Metz, *Drug Discovery Today*, 2010, **15**, 451.

165. M. P. Gleeson, A. M. Davis, K. K. Chohan, S. W. Paine, S. Boyer, C. L. Gavaghan, C. H. Arnby, C. Kankkonen and N. Albertson, *J. Comput. Aided Mol. Des.*, 2007, **21**, 559.
166. C. L. Gavaghan, C. H. Arnby, N. Blomberg, G. Strandlund and S. Boyer, *J Comput. Aided Mol. Des.*, 2007, **21**, 189.
167. S. E. O' Brien and M. J. de Groot, *J. Med. Chem.*, 2005, **48**, 1287.
168. W. Muster, A. Breidenbach, H. Fischer, S. Kirchner, L. Muller and A. Pahler, *Drug Discovery Today*, 2008, **13**, 303.
169. J. D. Hughes, J. Blagg, D. A. Price, S. Bailey, G. A. Decrescenzo, R. V. Devraj, E. Ellsworth, Y. M. Fobian, M. E. Gibbs, R. W. Gilles, N. Greene, E. Huang, T. Krieger-Burke, J. Loesel, T. Wager, L. Whiteley and Y. Zhang, *Bioorg. Med. Chem. Lett.*, 2008, **18**, 4872.
170. D. A. Price, J. Blagg, L. Jones, N. Greene and T. Wager, *Expert. Opin. Drug Met.*, 2009, **5**, 92.
171. K. Azzaoui, J. Hamon, B. Faller, S. Whitebread, E. Jacoby, A. Bender, J. L. Jenkins and L. Urban, *ChemMedChem*, 2007, **2**, 874.
172. A. Cavalli, E. Poluzzi, F. De Ponti and M. Recanatini, *J. Med. Chem.*, 2002, **45**, 3844.
173. G. M. Keseru, *Bioorg. Med. Chem. Lett.*, 2003, **13**, 2773.
174. L. P. Du, K. C. Tsai, M. Y. Li, Q. D. You and L. Xia, *Bioorg. Med. Chem. Lett.*, 2004, **14**, 4771.
175. A. O. Aptula and M. T. Cronin, *SAR QSAR Environ. Res.*, 2004, **15**, 399.
176. M. Song and M. Clark, *J. Chem. Inf. Model.*, 2006, **46**, 392.
177. D. Fernandez, A. Ghanta, G. W. Kauffman and M. C. Sanguinetti, *J. Biol. Chem.*, 2004, **279**, 10120.
178. S. R. Johnson, H. Yue, M. L. Conder, H. Shi, A. M. Doweyko, J. Lloyd and P. Levesque, *Bioorg. Med. Chem.*, 2007, **15**, 6182.
179. M. K. Leong, *Chem. Res. Toxicol.*, 2007, **20**, 217.
180. A. Cavalli, E. Poluzzi, F. De Ponti and M. Recanatini, *J. Med. Chem.*, 2002, **45**, 3844.
181. G. Ermondi, S. Visentin and G. Caron, *Eur. J. Med. Chem.*, 2009, **44**, 1926.
182. B. H. Su, M. Y. Shen, E. X. Esposito, A. J. Hopfinger and Y. J. Tseng, *J. Chem. Inf. Model.*, 2010, **50**, 1304.
183. H. Sun, *ChemMedChem*, 2006, **1**, 315.
184. C. Buyck, in *EuroQSAR 2002. Designing Drugs and Crop Protectants: Processes, Problems, and Solutions*, ed. M. Ford, D. Livingstone, J. Dearden and H. Van de Waterbeemd, Blackwell Publishing, Oxford, UK, 2003, 86.
185. R. E. Carhart, D. H. Smith and R. Venkataraghavan, *J. Chem. Inf. Comput. Sci.*, 1985, **25**, 64.
186. A. M. Aronov and B. B. Goldman, *Bioorg. Med. Chem.*, 2004, **12**, 2307.
187. L. Testai, A. M. Bianucci, I. Massarelli, M. C. Breschi, E. Martinotti and V. Calderone, *Curr. Med. Chem.*, 2004, **11**, 2691.
188. N. Wang, R. K. DeLisle and D. J. Diller, *J. Med. Chem.*, 2005, **48**, 6980.
189. Chemical Computing Group Inc., Montreal, Canada, 2002.

190. M. Tobita, T. Nishikawa and R. Nagashima, *Bioorg. Med. Chem. Lett.*, 2005, **15**, 2886.
191. W. Bains, A. Basman and C. White, *Prog. Biophys. Mol. Bio.*, 2004, **86**, 205.
192. A. M. Aronov, *J. Med. Chem.*, 2006, **49**, 6917.
193. Q. Li, F. S. Jorgensen, T. Oprea, S. Brunak and O. Taboureau, *Mol. Pharmacol.*, 2008, **5**, 117.
194. S. Hidaka, H. Yamasaki, Y. Ohmayu, A. Matsuura, K. Okamoto, N. Kawashita and T. Takagi, *J. Toxicol. Sci.*, 2010, **35**, 393.
195. K. M. Thai and G. F. Ecker, *Mol. Diversity*, 2009, **13**, 321.
196. R. Arimoto, M. A. Prasad and E. M. Gifford, *J. Biomol. Screen.*, 2005, **10**, 197.
197. B. F. Jensen, C. Vind, S. B. Padkjaer, P. B. Brockhoff and H. H. Refsgaard, *J. Med. Chem.*, 2007, **50**, 501.
198. R. Appiah-Opong, I. de Esch, J. N. Commandeur, M. Andarini and N. P. Vermeulen, *Eur. J. Med. Chem.*, 2008, **43**, 1621.
199. O. Dagliyan, I. H. Kavakli and M. Turkay, *J. Chem. Inf. Model.*, 2009, **49**, 2403.
200. F. Yamashita, H. Hara, T. Ito and M. Hashida, *J. Chem. Inf. Model.*, 2008, **48**, 364.
201. F. Hammann, H. Gutmann, U. Baumann, C. Helma and J. Drewe, *Mol. Pharmacol.*, 2009, **6**, 1920.
202. K. Roy and P. Pratim Roy, *Eur. J. Med. Chem.*, 2009, **44**, 2913.
203. W. S. Oh, D. N. Kim, J. Jung, K. H. Cho and K. T. No, *J. Chem. Inf. Model.*, 2008, **48**, 591.
204. J. Dapkunas, A. Sazonovas and P. Japertas, *Chem. Biodivers.*, 2009, **6**, 2101.
205. D. N. Kim, K. H. Cho, W. S. Oh, C. J. Lee, S. K. Lee, J. Jung and K. T. No, *J. Chem. Inf. Model.*, 2009, **49**, 1643.
206. R. Didziapetris, J. Dapkunas, A. Sazonovas and P. Japertas, *J. Comput. Aided Mol. Des.*, 2010, **24**, 891.

CHAPTER 13

ADME(T) Predictions in Drug Discovery

ANDREW M. DAVIS[a] AND ROBERT J. RILEY[b]

[a] AstraZeneca Respiratory and Inflammation Innovative Medicines, Pepparedsleden 1, 43 150 Mölndal, Sweden; [b] AstraZeneca New Opportunities Innovative Medicines, Alderley Park, Macclesfield, Cheshire, UK, SK10 4TG

13.1 Introduction

While the drug industry is an apparently mature industry, we are still discovering the fundamental principles underlying the identification of a successful marketable drug. In the ages before modern science, at best "medicines" were identified through trial and error and at worst were based upon superstition and quackery. For example, the ancient Greeks used opium for pain relief, and morphine is still a major drug today. In 1887, William Withering identified the active ingredient in an old wives' remedy for dropsy as foxglove, and digoxin is still widely used for heart failure today. In some cases, happy coincidence meant that trial and error and quackery combined to yield useful remedies. For example, in a letter to the Royal Society in 1763, the Rev Edward Stone described how the "doctrine of Signatures" led to a very interesting discovery. "As this tree (Willow) delights in a moist and wet soil, where agues chiefly abound, the general maxim, that many natural maladies carry their cure along with them, or that their cures lie not far from their causes was so very apposite to this particular case, that I could not help applying it". In the 1830s, salicylic acid was identified as the active ingredient in Willow. The first synthesis of

RSC Drug Discovery Series No. 13
Drug Design Strategies: Quantitative Approaches
Edited by David J. Livingstone and Andrew M. Davis

Published by the Royal Society of Chemistry, www.rsc.org

salicylic acid was achieved in 1859 by Kolbe, from which in 1897 Hoffman synthesised acetylsalicylic acid (Bayer). It was not until many years later in 1971, that Vane discovered the mode of action of Hoffman's discovery, aspirin.[1]

Trial and error approaches were important in the early stages of the evolution of modern drug industry. For example, sulfonamide antibiotics were discovered through *in vivo* screening of coal tar dyes, based on the hypothesis that some dyes, which preferentially bind to bacteria and other parasites, might also interfere with their function. The discovery of penicillin by Alexander Fleming also occurred through random events, of a discarded agar plate, an open window and a chance observation.

Interestingly, one might argue that to a large extent, trial and error defines the modern pharmaceutical industry today. We use high throughput screening to find chemical starting points. We use screening cascades to empirically filter good from bad compounds based on current drug-like dogma. We use a cascade of development tests, both preclinically and clinically, to find drugs that can be dosed at suitable doses and dose frequency, and with acceptable margins against side-effects, to potentially provide therapeutic benefit to patients.

Clearly, we can no longer survive as a business if our major source of innovation is deemed to be trial and error, or a "game of chance" in a high volume, high attrition model. Our success in this process, from identifying a compound worthy of entering preclinical development to a successful drug registration, is currently as low as 6%, and may be falling further. With commercial, regulatory and other pressures growing, trial and error approaches are becoming commercially non-viable. However, we are beginning to learn some of the secrets of successful drug design. One of these secrets, uncovered in the past 20 years, is that absorption, distribution, metabolism, elimination (and toxicology) [ADME(T)] research, or as it is otherwise known, drug metabolism and pharmacokinetic (DMPK)—research, can be a quantitative, predictive science.

In 1988, an eye-opening article appeared in the literature.[2] It was to change the course of drug discovery and development. A study of the fate of development compounds between 1964 and 1985 from the then 7 UK-owned pharmaceutical companies, showed that a poor pharmacokinetic profile (including poor bioavailability) was the single biggest factor in development attrition. This contributed almost 40% of all development failures (Figure 13.1).

This observation catalyzed a change in how drug metabolism and pharmacokinetics was applied in the mid 1990s. Scientists started to examine how the descriptive nature of development drug metabolism and pharmacokinetics could be applied in a predictive way into drugs at the design stage.

The discipline of DMPK has always been a drug development activity. In this realm, it still plays a pivotal role in describing the kinetic profile of the drug in both preclinical species and humans, where this information guides the selection of safe therapeutic doses for man for the first clinical studies, as well guiding the choice of dose frequency in patients. It is also important to describe accumulation, distribution and elimination profiles of the drug, its metabolic fate and the profiles of those key metabolites. The use of predictive tools to indicate metabolic fate is described in Chapter 11. The study of competition

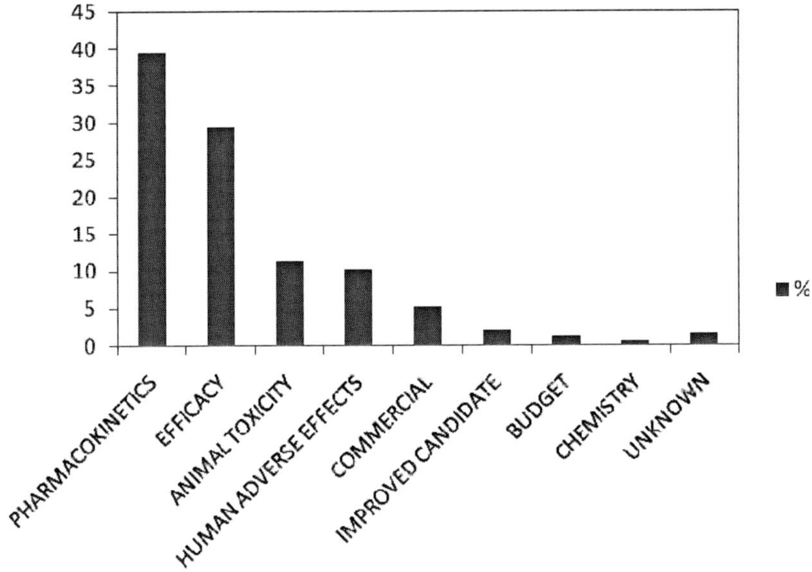

Figure 13.1 Reasons for withdrawal of drugs from development from 1964–1985. Data taken from ref. 3.

between drugs for the binding sites of metabolising enzymes or drug transporters provides understanding of the potential and consequences of drug–drug interactions. An understanding of variability in exposure is also a pivotal contribution of DMPK in development, where the drug will expose an increasingly large population.

By its nature, DMPK is a mathematical discipline that describes the concentration–time profile of a drug and links this to its efficacy response and safety profile. From its mathematical basis, DMPK has developed as a truly predictive quantitative discipline. A number of fundamental DMPK principles and mathematical equations have underpinned this transformation.

These principles include the role of physico-chemical properties; the free drug hypothesis; and the concept of clearance and scaling. Pharmacokinetic–pharmacodynamic simply stated these principles are:

- Physico-chemical properties underpin many of the structure–activity relationships of absorption, distribution, metabolism and elimination. The control of physico-chemical properties is a first and major step in solving many DMPK problems.
- The free drug hypothesis states that only the free unbound drug is available to drive efficacy and free drug is in equilibrium throughout the body. Hence the free plasma concentration can be used as a surrogate for the free concentration of the drug at its site of metabolism or activity.
- A study of metabolism at the enzyme, sub-cellular or cellular level can be the basis, through scaling, to a prediction of whole body clearance

through empirical correlation, simple liver mathematical models such as the well-stirred model, through to more complex physiologically based pharmacokinetic (PBPK) mathematical models.

In the following sections, we will illustrate how these principles have been applied. We will highlight the fundamental mathematical models, that with increasing confidence have allowed us to identify compounds projected to exhibit good human pharmacokinetic profiles from simple preclinical data. These mathematical relationships may also provide the basis for a purely *in silico* prediction of human pharmacokinetics when coupled with robust *in silico* models of DMPK endpoints.

The successful prediction of a human pharmacokinetic profile would require the prediction of the extent and rate of absorption, the rate of clearance of the drug from plasma, the extent of distribution of drug around the body and, from the clearance and distribution parameters, the half-life of the drug in the body. If the pharmacokinetic profile can be related to the degree of biological target engagement required to drive efficacy, then the dose and dose frequency to drive efficacy can be predicted. The pharmacokinetic–pharmacodynamic relationship can then be compared to exposures of drug known to give toxicology in preclinical species, and hence a potential therapeutic margin can be estimated and a safe dosing regimen defined. So prediction of absorption, clearance and drug distribution together provide a foundation for a quantitative DMPK strategy.

A recent shift in emphasis has been from simple DMPK parameters to a more physiological interrogation of the data. Such physiological based pharmacokinetic modelling may provide mechanistic links to understand the influence of drug delivery formulations, or even the relationship between efficacy and toxicity at the tissue level. However literature examples of the benefit of such detailed analysis are sparse, or even lacking.

13.2 Predicting Human Absorption

Before a drug can exert its action, it must be released from its formulation, absorbed from its site of administration and distributed to its site of action. Drugs showing poor absorption lead to high dose and often lead to variability in exposure across a population. This has been a cause of attrition for clinical development candidates. *In vivo* preclinical species can provide information on the extent and rates of absorption, but as physiology and gastrointestinal transit times differ, they provide a level of confidence only. For example, it is a common occurrence for drugs to show better absorption in dogs than they do in rat or man.[3] Also estimates of drug absorption from *in vivo* measurements in rats has been shown in a number of studies to be a poor model of human drug absorption.

A number of *in vitro* systems to model absorption have been developed and have gained widespread use through their successful prediction of human absorption. These include the Caco-2 cell monolayer model,[4] Ussing chambers to measure drug transit across gastrointestinal tissue, and even pure physicochemical membrane permeability assays such as the PAMPA assay.[5]

The introduction of the Caco-2 cell monolayer permeability assay has had a significant influence in identifying absorption problems early in the drug discovery process.[4] This colonic carcinoma cell line has provided a good model for at least the rates of the passive component of drug absorption, and validation studies have shown good relationships between permeabilities determined in the Caco-2 cell monolayer assay and human drug absorption. As a rule of thumb, and barring efflux (which is observed as a difference in permeability from apical-basal relative to basal-apical side of the monolayer), an apparent permeability Papp $> 1 \times 10^{-6}$ cm s^{-1} in a typical Caco-2 assay, would correlate with good human absorption. The growing database of Caco-2 permeabilities in many companies have also provided datasets for in silico QSAR models to allow drug absorption to be predicted directly from chemical structure.[6,7]

13.2.1 The Biopharmaceutics Classification System

Although introduced in 1991, Caco-2 cell monolayer permeability is still a widely used model for predicting human absorption. Indeed, it has been used as a surrogate for human absorption in the FDA Biopharmaceutics Classification System (BCS) guidelines for bioequivalence study waiver, which are based on high solubility and high human absorption.[8] "When the *in vivo* dissolution of an immediate release solid oral dosage form is rapid in relation to gastric emptying, and the drug has high permeability, the rate and extent of drug absorption is unlikely to be dependent on drug dissolution and/or gastrointestinal transit time. Under such circumstances, demonstration of *in vivo* bioavailability or bioeqivalence may not be necessary for drug products containing class 1 drug substances, as long as the inactive ingredients used in the dosage form do not significantly affect absorption of the active ingredients."[9] A drug is classified as having high solubility if the highest dose can dissolve in 250 cm^3 of aqueous media of pH 1–7, and is classified as high permeability if the human fraction absorbed is predicted to be $>90\%$ based on *in vitro* epithelial cell biolayer permeability measurements or mass transfer rate across a human intestinal membrane.

The 4 classes of the BCS represent:[9]

1. high solubility/high permeability
2. low solubility/high permeability
3. high solubility/low permeability
4. low solubility/low permeability

A class 1 drug can obtain a waiver for bioequivalence studies for immediate release solid dose oral formulations if: (1) the drug is stable in the gastrointestinal tract; (2) excipients used in the immediate release (IR) solid oral dosage forms have no significant effect on the rate and extent of oral drug absorption; (3) the drug does not have a narrow therapeutic index; and (4) the product is designed not to be absorbed in the oral cavity.

13.2.2 The ACAT Model

The advanced compartmental absorption and transit (ACAT) model is a physiology based mathematical model of drug absorption that includes factors such as drug dissolution rate, solubility, membrane permeability, transit time and compartmental pH used to model the absorption profile of drugs. It was originally developed by Amidon and colleagues.[10–11] It has been extended and found widespread use in packages such as GastroPlus™ and Idea™, which can be used to predict human absorption/absorption rates, and the effects of food, and formulations on gastric absorption. It has even been shown that *in silico* estimates of permeability based on Caco-2 data, coupled with Gastroplus™, enables the correct classification of BCS absorption class 70% of the time, with experimental estimates of permeability providing little improvement on this success.[12]

The results from the Gastroplus™ package are increasingly being used as justification for applications for biowaiver with the FDA BCS.[13] For example, Gastroplus™ simulations were used to support the case of etoricoxib solid dose oral forms being classified as BCS class 1 rather than class 2, due to the *modeled* solubility behaviour in the pH gradient of the gastrointestinal tract.[14] Similar justifications have been proposed for formulations of carbamazepine.[15]

13.2.3 The Biopharmaceutics Drug Disposition Classification System

Benet and co-workers have questioned how the BCS system is applied by the FDA and other agencies.[16] In the original publication by Amidon,[17] the relationship between high jejunal permeability for 30 drugs and 4 non-drugs, and the extent of absorption, was demonstrated. High permeability is a kinetic measure, while extent of absorption in man is a thermodynamic measure. Agencies appear to use permeability and extent of absorption interchangeably in the BCS classification. However, Benet highlights that compounds accepted as high absorption by the FDA can have lower permeability than the standard compound used to define the high permeability boundary in the BCS, metoprolol. In a survey of 130 drugs classified into the 4 quadrants of the BCS, they found that the major route of elimination of class 1 and class 2 drugs was metabolism, while the main route of elimination for class 3 and class 4 drugs was renal or biliary elimination. At least for drugs on the market, metabolism could be used as a surrogate for the extent of absorption in the BCS. As extent of metabolism is better defined than extent of absorption, Benet and Wu suggested the Biopharmaceutics Drug Disposition Classification System (BDDCS) be based on solubility-metabolism rather than solublility-permeability. The cut-off between class 1 and class 2 was based on an extent of metabolism of

70%. This removed ambiguities in classification based on different estimates of permeability as metabolism is often so much better defined. Benet and Wu found that the metabolism criteria successfully predicted high *versus* low permeability of 27 out of 29 drugs whose jejunal permeability had been (laboriously) determined. They suggested that the FDA add extent of metabolism as an alternative method to extent of absorption in determining class 1 compounds. In 2008, the EMEA revised draft of "Guideline on the Investigation of Bioequivalence" includes metabolism as a criterion for permeability.[18] Although potentially useful in clarifying class 1 biowaivers, the BDDCS was originally developed to make early predictions of likely transporter and drug–drug interactions. Class 1 high solubility, high permeability/extensively metabolised compounds will be unlikely to suffer significant transporter and drug–drug interactions in the GI tract or liver, while for class 2, low soluble high permeability/extensively metabolised compounds, efflux transporters will have a predominant effect in the gut. For class 3 and class 4 drugs, uptake transporter, efflux transporter and uptake–efflux transporter interplay will be of major importance. As metabolism can be predicted from *in vitro* systems (discussed later), the BDDCS can be used prospectively early in a drug discovery program.

13.3 Predicting Human Drug Distribution

Volume of distribution is a pharmacokinetic term, derived from analysing the plasma-time curve of a drug, and is the hypothetical volume in which a drug would have to distribute to give the observed plasma concentration. It modulates the half-life of the drug with a given clearance.

$$t_{1/2} = 0.693 \times (\mathrm{V}/\mathrm{Cl}) \tag{13.1}$$

Mathematically, it can be described by the Gilette equation where volume of distribution (at steady state, V_{ss}) is given by eqn (13.2).

$$V_{ss,man} = V_{p,man} + \frac{f_{up,man}}{f_{ut,man}} V_{t,man} \tag{13.2}$$

Where V_p = plasma volume
V_t = tissue volume
f_{up} = fraction unbound in plasma
f_{ut} = fraction unbound in tissue.

A more physiological approach was taken by Oie and Tozer, who expressed V_{ss} according to eqn (13.3).[19]

$$V_{SS} = V_p(1 + R_{E/I}) + f_{up}V_p(V_E/V_p - R_{E/I}) + \frac{V_R f_{up}}{f_{ut}} \tag{13.3}$$

In eqn (3), f_{up} is the fraction unbound in plasma, f_{ut} is the fraction unbound in tissues and $R_{E/I}$ is the extravascular/intravascular ratio of binding proteins (usually 1.4 for albumin). V_P, V_E, and V_R refer to the volumes of plasma, extracellular fluid and remainder fluid, with values of 0.0436, 0.151, and 0.38 L kg^{-1} respectively in humans.

Most predictive methods of volume of distribution are based on *in vivo* measurements from preclinical species, scaled to the prediction in man. It is observed that volume of distribution is fairly constant across species,[20] with major deviations only occurring when changes in plasma protein binding are observed across species. Our laboratory has shown that for many drugs, whose volumes had been measured in mouse, rat or dog, one can describe the changes in volume of distribution across all these species with a single correlation to human volume of distribution, when these measurements have been corrected for changes in protein binding across species (Figure 13.2).

The average fold error (AFE) across this correlation was 1.36.[21] External validation of this approach predicted the human volume of distribution for an additional 9 compounds with an AFE of 1.83. It was recommend that volume of distribution from rat and dog is sufficient to provide a useful prediction of human volume, and that measurement in further species is only warranted if deviations unrelated to plasma protein binding differences are observed in the first two species. Deviations may be caused by active processes which may not scale/translate directly across species.

A number of direct *in silico* models to predict volume of distribution have been published, which allow volume of distribution across species to be predicted directly from chemical structure.[22,23]

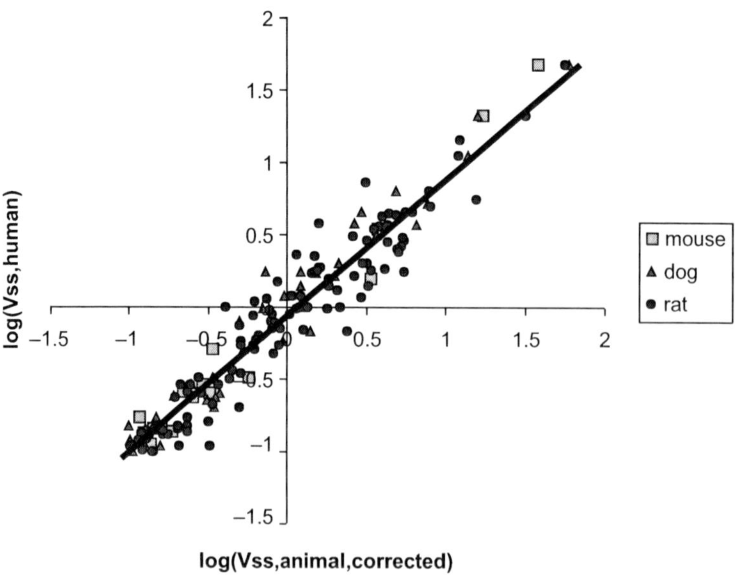

Figure 13.2 Plot of log(Vss, human) *versus* log(Vss, animal corrected) corrected for protein binding differences across species, for acids, neutrals and bases in mouse, rat, dog and human. Data from our own laboratory and reference 47.

13.4 Predicting Human Hepatic

Many mechanisms of clearance exist to remove a drug from the body. Total clearance is the sum of these different clearance processes [eqn (13.4)], which can be described through *in vivo* measurements

$$Cl_{\text{total}} = Cl_{\text{hepatic}} + Cl_{\text{biliary}} + Cl_{\text{renal}} + Cl_{\text{other}} \tag{13.4}$$

Thus total clearance can be estimated by independently predicting each of the major clearance processes indicated. The liver is by far the main route of clearance for most drugs, therefore it has been the focus for much of the development of pharmacokinetic modeling methods.

13.4.1 The Well-Stirred Model

The well-stirred model [eqn (13.5)] is a simplified model of hepatic drug clearance and relates *in vivo* intrinsic clearance, which is the clearance of the drug experiences at the enzyme level, to blood flow and liver distribution.[24] The *in vivo* Cl_{int} can be estimated from an *in vitro* Cl_{int} measurement by "scaling" the *in vitro* measurement to the *in vivo* situation, using estimates of microsomal protein yield per gram of liver and liver weights (for microsomal estimates of Cl_{int}) or hepatocyte numbers per gram of liver for *in vitro* Cl_{int} values derived from hepatocytes.[25] See eqn (13.5).

$$Cl = \frac{f_{unbound} Cl_{\text{int}} Q}{f_{unbound} Cl_{\text{int}} + Q} \tag{13.5}$$

Where,

$f_{unbound}$ = fraction unbound in plasma
Cl_{int} = *in vivo* intrinsic clearance
Q = hepatic blood flow

A number of variants of the well-stirred model have been published, differing in their assumptions concerning the distribution of drugs within the liver. These differing assumptions largely affect compounds with high clearance, but as oral drugs are designed to have low liver extraction ratios, the models converge in this area of most interest.

The well-stirred model therefore allows prediction of *in vivo* clearance from intrinsic clearance estimates, generated from simple enzyme, sub-cellular or cellular assays.

This model for predicting hepatic clearance has allowed simple enzyme (*e.g.* cytochrome P450 enzymes for phase 1 metabolism), sub-cellular (microsomes for phase 1 metabolism) or cellular assays (hepatocytes for phase 1 and 2 metabolism) to be established in drug discovery cascades to enable the identification of low clearance compounds, with the ability to predict the consequent expected *in vivo* clearance. Results from *in vitro* assays based on rat or dog microsomes or hepatocytes for a particular compound can be validated with subsequent *in vivo* measurements, thereby providing confidence that we may also predict from human microsomes or hepatocytes to human *in vivo* clearance, for compounds known to be predominantly cleared by hepatic drug metabolism. Such confidence is paramount before entering expensive clinical development, where these principles are ultimately validated.

These *in vitro–in vivo* methods have become more sophisticated, for example to allow for non-specific binding effects in the *in vitro* assay which otherwise may cause an underestimate of clearance.[26]

The well-stirred model, including corrections for non-specific binding is shown in eqn (13.6).

$$Cl_h = \frac{(Q_h \times f_{ub} \times (Cl_{\text{int}^*}/fu_{inc}))}{(Q_h + f_{ub} \times (Cl_{\text{int}^*}/fu_{inc}))} \tag{13.6}$$

Where,

Cl_h = hepatic clearance
fu_b = fraction unbound in blood
fu_{inc} = fraction unbound in incubation
Cl_{int^*} = $Cl_{int} \times$ scaling factor (standard value 2652 millions of hepatocytes per kg body weight)

This is particularly useful for certain drug classes, such as acids, as shown in Figure 13.3, where high protein binding may drastically affect the free drug

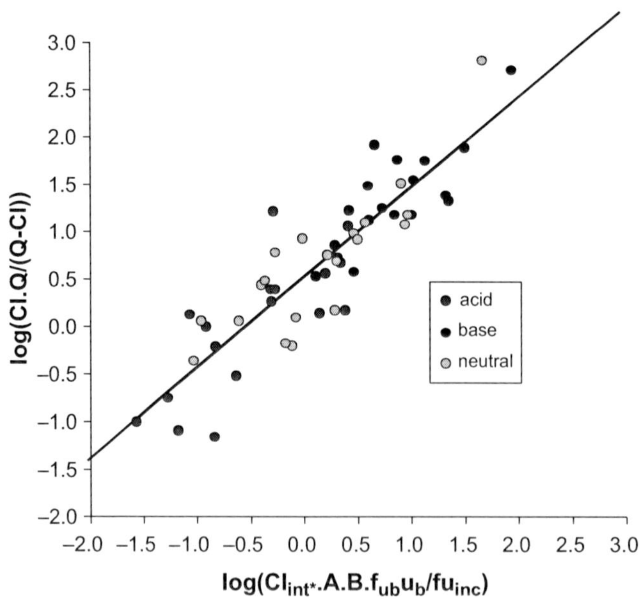

Figure 13.3 Prediction of *in vivo* human metabolic clearance from *in vitro* human hepatocyte metabolism data, correcting for non-specific binding in the incubation. Data taken from ref. 45.

concentration in the assay, where fetal calf serum, for example, may be added to stabilize the hepatocyte preparation.

QSAR models for fu_{inc} have now been reported and implemented early in drug discovery to good effect, with measurements being proposed as a compound progresses towards development[27,28] Some inter-laboratory comparisons have been reported and suggested that some modifications may be required for more lipophilic compounds.[29–31] However, considerations of drug-likeness, good physical properties and concerns around promiscuity should mean that ventures into such regions are relatively few for oral drugs.

Our laboratory has shown that accounting for non-specific binding in the *in vitro* incubation allowed the prediction of clearance for 21 diverse drugs from *in vitro* data, with an average fold error in prediction of 3-fold across many drug classes, over a wide range of lipophilicities and across all charge types (Figure 13.3). Recent research has focused on including transporter mediated uptake in the scaling prediction from hepatocytes.[32] Transporter-mediated hepatocyte uptake, if the drug's membrane permeability is poor, may result in higher intracellular concentrations relative to extracellular concentrations, leading to increased metabolism. This effect is underestimated in standard hepatocyte assays, risking a significant under-prediction of *in-vivo* intrinsic clearance.

13.4.2 Physiologically Based Pharmacokinetic Models

A further development of the *in vitro–in vivo* correlation approach is provided by physiologically based pharmacokinetic models. These attempt to expand the 1 compartment liver model to a multi-compartment system to predict the complete pharmacokinetic profile in the whole organism. Often only one compartment (the liver) is responsible for clearance; the others represent different physiologically distinct structures that can act as sites for drug distribution. The compartments are joined by mathematical "plumbing" with physiogically relevant blood flows linking them together into a system. Key unknown parameters are the partition coefficients (K_ps) for the drug into each compartment. These K_ps are often estimated from one species to allow the extrapolation of the PK profile to another species. By the nature of their demand for parameterization, PBPK models are not as widely applied as simple compartmental approaches but have value in analyzing the often multi-phase concentration time profiles exhibited by some drugs, and in the detailed studies often undertaken as drugs near a selection for development phase. Despite the impact of mechanistic tissue composition calculations to predict K_p and V_{ss} terms,[33] literature reports of detailed profile analyses using PBPK remain relatively scarce and recent analyses still suggest that knowledge from a least a single preclinical species performs better than true *ab initio* predictions.[34] Roche has shown that PBPK modeling was able to predict the AUC of 19 in-house compounds within 2-fold of the observed value, compared to only 46% using the Dedrick method.[35] As an illustration of the maturity in application, it is still of interest, and publishable, when the pharmacokinetic profile an old drug such as midazolam can be predicted across 4 species (albeit unusual species) for DMPK studies of chicken, turkey, pheasant and quail.[36] Several commercial PBPK packages are now available, including GastroPlus™, PKSim™, Cloe PK™ and SimCYP™.

13.4.3 Allometry

Allometry is an empirical approach to predicting pharmacokinetic parameters across species. It has developed from a much wider discipline that correlates shape to a power-function of some measure of size or mass. For example, "Kleibers law" states that the basal metabolic rate across animal species scales to the power 0.75 to body weight.[37]

Allometry is a widely used technique in drug development. Numerous physical and physiological parameters, and certain pharmacokinetic parameters (Y) have been found to be related to body weight according to eqn (13.7) and (13.8).

$$Y = a \cdot W^b \tag{13.7}$$

Or, in its logarithmically transformed form:

$$\ln Y = \ln(a) + b\ln(W) \tag{13.8}$$

where Y is the parameter of interest (*e.g.* clearance, volume of distribution *etc.*), W is the body weight, a is the allometric coefficient and b is the allometric exponent. When plotted on a log–log scale, the relationship is a straight line with intercept ln(a) and slope b. The physical nature of parameters a and b are different in that a is drug dependent and b is dependent on the nature of the variable being analyzed, *e.g.*, Cl, V or $t_{1/2}$. The exponents for flow rate dependent parameters (*e.g.* clearance or glomerular filtration rate) are normally around 0.7 and for organ size (*e.g.* volume of distribution) around unity.

A number of adjustments have been proposed for allometric relationships when the slopes are extreme. When slopes of flow dependent allometric relationships exceed 0.7 (between 0.7 and 1.0) a further correction is often applied. The correction, suggested by Mahmood,[38] uses the "maximum lifespan potential" (MLP) and has the mathematical form described by eqn (13.9) and (13.10).

$$\text{Log}(\text{Cl} \times \text{MLP}) = \text{a.log}(\text{body wt}) + \text{constant} \tag{13.9}$$

$$\text{Where MLP} = 185.4 \times \text{BRW}^{0.636} \times \text{BW}^{-0.225} \tag{13.10}$$

Where BRW = animal species brain weight, and BW = animal body weight in kg.

The overall effect of MLP correction serves to reduce the slope of the allometric relationship by 0.3, as the MLP correction has a greater effect on the higher body weight species. Why and when the application of this correction is scientifically valid is not clear, and can best be viewed as an empirical adjustment.

There is continuing debate over the utility of allometry in predicting human pharmacokinetics from preclinical animal studies. On a fundamental scientific level, it might be reasonable to expect an allometric relationship to hold for high clearance compounds, where organ blood flow may be the limiting factor in clearance. However, for low clearance compounds, it seems more reasonable not to expect an allometric relationship. This is because clearance is likely more dependent on molecular recognition and turnover rates at the sites of metabolising enzymes, rather than than blood flow to the liver. Enzymology differences across species are likely to be significant. Indeed it is not uncommon for errors associated with allometry to be huge for individual compounds.[39] Additionally and importantly, allometry has been criticised on a purely statistical basis.[40,41]

13.4.4 Predicting Other Human Clearance Mechanisms

Renal clearance prediction has been recently reviewed.[42] Significant renal clearance is largely the realm of polar compounds, which are not significantly passively reabsorbed after filtration through the renal glomerulus. Most reported methods for renal clearance rely on allometric scaling across species which appear relatively successful, as its a flow dependent property. It can also be predicted from renal clearance measured in rat corrected for differences in glomerular filtration rate and plasma protein binding. Where active secretion occurs, again poor permeability preventing reabsorption is a dominating factor. Hence both passive and active renal clearance have been successfully predicted using *in silico* models largely based on bulk physical properties.[43] Recent advances in our knowledge of species' differences in renal clearance support the dog as more predictive of human renal clearance across many chemical classes.[44]

Other clearance mechanisms, such as biliary clearance, are less well understood.

13.5 Predicting Human Bioavailability

Bioavailability is a measure of the extent of a drug reaching the systemic circulation from its point of administration. From an oral dose therefore, it depends both on the degree of absorption and first pass clearance experienced by the dose. Assuming good estimates of absorption and predicted clearance are available, the bioavailability (F) can be predicted using eqn (13.11).

$$F_{\text{human}} = \left(\frac{Q_{\text{human}} - \text{blood } Cl_{\text{human}}}{Q_{\text{human}}}\right) Abs_{\text{human}} \qquad (13.11)$$

13.6 Predicting Human Dose and Dose Frequency

Pharmacokinetic-pharmacodynamic (PKPD) modeling makes a quantitative link between the pharmacokinetic profile of the drug and the concentration required to significantly occupy the target receptor and drive efficacy. The free drug hypothesis would lead to the assumption that the free concentration in the plasma controls efficacy, or in other words, the free concentration in the vicinity of the target is equal to the free concentration in the plasma. A sensible dosing paradigm would be to keep the receptor significantly occupied with drug throughout the dosing interval. The minimal effective concentration (MEC) required to significantly occupy the receptor at $C_{\min}$ would be given by eqn (13.12).

$$MEC = \frac{CF10^{-\text{potency}}}{fu_{\text{human}}} \ (\text{mol l}^{-1}) \tag{13.12}$$

CF is the fold over the K_i you wish to achieve. For simple ligand–target interactions, a 3-fold concentration factor would equate to a 75% receptor occupancy, while a 10-fold factor would equate to approximately 90% receptor occupancy.

If one assumes simple, one compartment models for drug elimination, and instantaneous absorption, the dose required to maintain the minimum concentration before re-dose above the MEC is described by eqn (13.13).

$$Dose(mg/kg/day) = \frac{(24/\tau)MEC \cdot V_{ss}(\exp(k_{el}\tau) - 1)}{F} \tag{13.13}$$

If the absorption rate constant is incorporated, the expression becomes eqn (13.14).

$$Dose\,(\text{mg/kg/day}) = \frac{24MEC\ Vss_{\text{human}}(k_a - k_{el})}{\tau F_{\text{human}} k_a \left[\dfrac{1}{1-\exp(-k_{el}\tau)} - \dfrac{1}{1-\exp(-k_a\tau)}\right]} \tag{13.14}$$

Vss is volume of distribution at steady state, F is oral bioavailability, k_a is the absorption rate constant, k_{el} is the elimination rate constant and τ is the dosing interval. If the units of MEC are mg/mL, V_{ss} expressed as mL/kg, bioavailability F expressed as a fraction, k_a and k_{el} in hours^{-1}, and dose frequency in hours, then the dose will be in units mg/kg/day.

Applying eqn (13.14), we were able to show that the clinically used dose of 28 drugs could be predicted to within an error of 3-fold of the reported value.[45] We also showed that out of 28 drugs in the dataset, most are dosed to provide a minimum effective plasma concentration of 1–3 over the *in vitro* potency of the target corrected for protein binding (45% drugs ≤1-fold, 75% of drugs ≤3-fold) (Figure 13.4).

Furthermore, compounds which were dosed to give MEC >3-fold had good reasons to do so. For example, ritonavir requires to be maintained at IC95 in order to sufficiently reduce viral load.

Therefore if no other information exists, it is reasonable to set the MEC to between 1 and 3 times the unbound $p\text{IC}_{50}/p\text{A}_2$ as measured from plasma. This

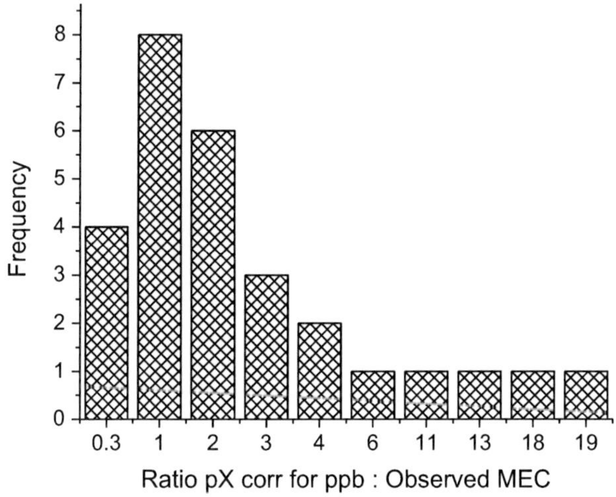

Figure 13.4 Relationship between pIC_{50}/pA_2, corrected for protein binding and minimum effective concentration for 28 marketed drugs. Data taken from ref. 47.

important observation provides further confidence that a human effective dose could be predicted from only preclinical measurements and knowledge of the intrinsic potency at the molecular target. It also supports one of the fundamental principles mentioned earlier, that efficacy is driven by the free drug concentration and that for permeable compounds, the free drug concentration in plasma at equilibrium is similar to the free drug concentration at the location of the target receptor. This may not be the case for poorly permeable compounds and/or where the compound is a substrate for transporter proteins, which may either accumulate the drug at the target protein site, or exclude it from that compartment. Other information within a project, such as PKPD relationships (assuming species crossover) and data from preclinical or clinical competitor activity, may help further predict MEC. Prior knowledge of likely MEC places drug discovery projects in a strong position, with respect to biomarkers in patients, better and faster design of proof of principle studies, better interpretation of PKPD and improved confidence in predicted safety margins.

The $C_{\max}$ is an important parameter in predicting margins *versus* some key off target potencies, such as hERG inhibition, and assuming instantaneous absorption is given by eqn (13.15). We found $C_{\max}$ could also be predicted within 2-fold of the clinically observed value, as shown in Figure 13.5.[47]

$$C_{max,ss} = \frac{F \cdot Dose}{V_{ss}(1 - e^{-k_{el}t})} \tag{13.15}$$

Because of the assumption of instantaneous absorption in the model, $C_{\max}$ tended to be slightly overpredicted, which may be the most reassuring prediction to make from a toxicology point of view.

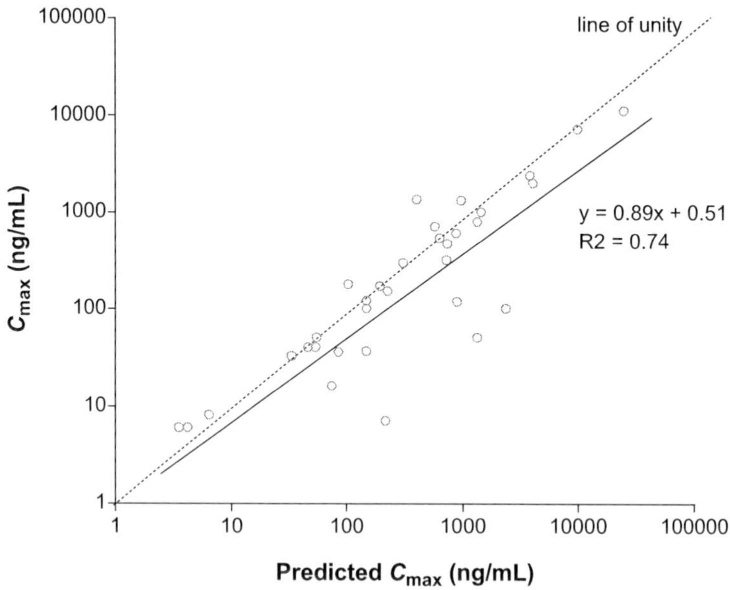

Figure 13.5 Relationship between predicted and observed C_{max} for 28 drugs – data taken from ref. 47.

13.7 Predicting Drug–Drug Interactions

As the enzymology of drug metabolism is unraveling, it is now possible to predict clinically likely drug–drug interactions direct from simple *in vitro* assays. Many drugs are metabolized, at least in the first step, through oxidation by cytochrome P450s. As drug therapy for important diseases frequently involves a drug combination, the potential for competition for metabolism at the individual P450 level is a distinct possibility. Drugs that are substrates with high binding affinities, or indeed inhibitors, can grossly affect the concentration–time profiles of co-administered drugs. As these enzymes are now available cloned and expressed to support both screening and X-ray crystallographic analysis, the enzymology of drug metabolism can be studied from both a kinetic and a structural point of view.

Quantitative calculation of drug–drug interactions would require knowledge of the binding constants of co-administered drugs, together with knowledge of their concentrations at the site of the enzme. Applying the free drug hypothesis, the target concentration can be approximated by the free plasma concentration, and eqn (13.16) can be used to predict the likelihood of a drug–drug interaction, which is often known as the static $[I]/K_i$ approach.[46]

$$AUC_{ratio} = 1 + [I]/K_i \tag{13.16}$$

The $[I]/K_i$ approach has been extended to consider the free drug at the site of the enzyme $[I]_{In,ub}$, and incorporates adjustment for the fraction metabolized by a particular CYP.

$$AUC\,ratio = \frac{1}{\dfrac{fm_{CYP}}{1+[I]_{in,ub}/K_{i,ub}} + (1 - fm_{cyp})} \tag{13.17}$$

The difficulty in choosing the appropriate free concentration for the term $[I]_{in,ub}$ is one reason for the popularity of the simplified I/K_i model. Ito has shown that the estimate of the maximum unbound concentration at the inlet to the liver resulted in fewer false positive drug–drug interaction predictions.[47]

$$[I]_{in,max} = [I]_{max} + \frac{K_a.Dose.F_a}{Q_h} \tag{13.18}$$

Riley *et al.* showed that, for 15 known competitive inhibitor drugs, combining information from automated recombinant cytochrome P450 assays against 8 Cyps, achieved a true positive prediction drug–drug interaction rate of 77%, where the predicted impact is >2-fold increase in area under the plasma time curve of the "victim" drug.[48] Riley also examined SimCYP™, which applies a PBPK approach to include the time-variant change in concentration of the inhibitor. They found, for the same dataset, SimCYP™ was able to achieve a true positive rate of 94%.

Another validation study compared the SimCYP™ approach to the $[I]/K_i$ approach, an extended $[I]/K_i$ approach incorporating fraction metabolised by the cytochrome P450, fm^{CYP}, intestinal availability for CYP3A4 substrates and mechanism based inactivation, and the mechanism-dynamic model of SimCYP™, Overall, the SimCYP™ approach showed an improvement over the other two. Moreover, where an interaction >2-fold existed, the SimCYP™ approach was able to predict the change in area under the curve for the victim drug in 64% of trials, compared to only 53% and 17% for the $[I]/K_i$ and extended $[i]/K_i$ approach, respectively.[49]

Additionally, knowledge of the genetic prevalence of polymorphisms in these key enzymes has allowed the enzymological approach to be extrapolated to consider the population pharmacokinetics of drug treatment. The program SimCYP™ applies this information to take the predictivity a further step towards modeling real patient population pharmacokinetics[50,51] SimCYP™ is a product of an academic industry collaboration that originated from the University Sheffield, United Kingdom (www.simcyp.com). Independent SimCYP™ validation studies have shown it to be 88–90% successful in predicting drug-drug interactions for 30 drugs on midazolam metabolism with mean fold eror of 1.59 and 1.47 fold error using steady-state or time based approaches. The validation set contained drugs that were not only competitive reversible inhibitors of CYP3A4 but also inactivators and inducers[52]

An increasing number of *in silico* models are available to predict CYP-P450 inhibition. CYP 3A4 appears the most non-specific, where both inhibition and turnover appear largely controlled by logP as shown in Figure 13.6. Unhindered, unprotonated nitrogen acceptors (*e.g.* pyridines and imidazoles) are more potent CYP 3A4 inhibitors for a particular logP.[53] Other CYP-P450

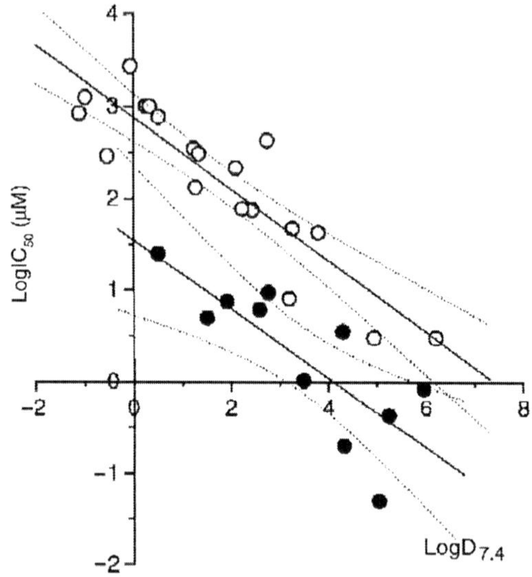

Figure 13.6 Relationship between lipophilicity (logD7.4) and CYP3A4 inhibition for 2 subpopulations of chemistry within the dataset identified by the presence (filled circles) or absence (open circles) of a sterically unhindered pyridine, imidazole, or triazole function. The broken lines are 95% confidence intervals. Reproduced with permission ref. 53.

structure–activity relationships are more structure dependent and less bulk-property dependent, which results in more difficulty in generating precise property based QSAR models.[54]

These models open up the opportunity to predict compounds likely to cause drug-drug interactions directly from structure.

13.8 Physico-Chemical Control

The key DMPK parameters of absorption, distribution, metabolism and elimination have been described mathematically in the previous sections, with indications of both the accuracy and precision of predictions that can be made from these models using *in vitro* data, or cross-species scaling using *in vivo* data. As highlighted in each section, a number of endpoints can be modeled *in silico*, directly from chemical structure, and many of the relationships are underlined by control of logP. Chapter 2 reinforces the fundamental importance of controlling logP from the point of view of development attrition, which includes of course attrition due to poor DMPK properties. In the context of DPMK, Gleeson described simple physical property rules of thumb—based on molecular weight, lipophilicity and charge type—that underpin the key DMPK endpoints of absorption, distribution, CYP-P450 mediated metabolism and inhibition. They found that almost universally, all DMPK parameters degrade on increasing molecular weight and logP.[55] A fundamental rule of drug design therefore appears to be control of logP at all costs, and many of the common reasons for attrition can be mitigated.

13.9 Conclusions and Future Outlook

Drug metabolism and pharmacokinetics continues to evolve as a discipline. As was mentioned in the introduction, in 1988, 40% of all drug development failures were attributed to poor pharmacokinetics and poor bioavailability. By 2000, these DMPK factors were reported to contribute less than 10%. So the quantitative approaches outlined in this chapter appear to have been successfully applied, with significant impact on reducing attrition. Increasingly, DMPK screening has moved from *in vivo* to *in vitro*, to *in silico*, allowing true DMPK control directly from chemical structure. Recent registrations also show sources of new drugs as pro-drugs and metabolites of existing, marketed drugs, so DMPK as a discipline is also a direct source of innovation, intellectual property and revenue.

Despite the considerable progress made, DMPK remains a key focus directly and indirectly for compound quality and its link to development attrition. The detailed understanding we now have of phase I metabolism *via* cytochrome P450s needs to extend to an understanding of molecular mechanisms underpinning P450 induction, phase 2 metabolism *via* the UDP-glucuronosyltransferases and other conjugases, and a molecular understanding of drug transporters that are being found to play a pivotal role in controlling the pharmacokinetic profile of an increasingly large number of drugs. Further work in DMPK also now focuses on non-metabolic, transporter-driven clearance, tissue uptake and the link between uptake, tissue exposure and toxicology.

Unfortunately the reduction in development attrition due to poor pharmacokinetics has failed to propagate through into an overall increase in drug development success rates. In 2000, the single biggest cause for development failure was reported as lack of efficacy, which had shown no improvement since the 1988 review. It seems the success in reducing pharmacokinetic attrition, which normally occurs in phase 1 development, has enabled candidate drugs to progress further into more expensive phase 2 proof of principle studies, where we uncover the weaknesses of our biological hypotheses linking molecular drug targets to their relevance in disease.[56] Further effort is now being focused to exploit translational science to strengthen our target rationale. Drug metabolism and pharmacokinetics may also have an important role to play in understanding and controlling attrition at this stage, where a link between exposure and the pharmacodynamic effect needs to be demonstrated. The degree of biological target engagement by the drug, and the length of time the biological target is engaged by the drug, are critical success factors in maximizing our chances of demonstrating efficacy in the disease. The tools of quantitative DMPK, through modeling and simulation, allow predictions to be made, describing the pharmacokinetic–pharmacodynamic relationship, facilitating the design of both preclinical and clinical studies. PKPD modeling and simulation is proving increasingly successful in linking all the way though to the relationship between efficacy and therapeutic safety margins.

Toxicology remains another significant reason for drug development failure and therefore investment in a more mechanistic, molecular approach to toxicology, similar to the one experienced by DMPK in the 1990s, is now warranted. Progress is being made, and some toxicology has been unveiled from empiricism to a scientific molecular understanding. For instance the potentially fatal consequence of inhibition of the hERG cardiac ion channel has now been realized, and investigation of hERG inhibition is now a successful part of drug discovery screening cascades, and good quality *in silico* models are available to predict this liability.[57,58] A further emerging paradigm is the role of phospholipidosis in (hepato)toxicity.

The controlling influence of physical properties on attrition is another emerging paradigm, as described in Chapter 2, as is secondary screening to identify more selective compounds which can be progressed.[59] There are further secrets yet to uncover in successful drug design. We hope we can uncover them before we lose viability as an industry.

References

1. J. R. Vane, *Nature (London) New Biol.*, 1971, **231**, 232.
2. R. A. Prentis, Y. Lis and S. R. Walker, *Br. J. Clin. Pharmacol.*, 1988, **25**, 387.
3. A. M. Davis and R. J. Riley, *Curr. Opin. Chem. Biol.*, 2004, **8**, 378.
4. P. Artursson and J. Karlsson, *Biochem. Biophys. Res. Comm.*, 1991, **175**, 880.
5. M. Kansy, F. Senner and K. Gubernator, *J. Med. Chem.*, 1998, **4**, 1007.
6. P. Paixao, L. F. Gouveia and J. A. G. Morais, *Eur. J. Pharm. Sci.*, 2010, **41**, 107.
7. R. P. Verma, C. Hansch and C. D. Selassie, *J. Comput.-Aided Mol. Des.*, 2007, **21**, 3.
8. G. L. Amidon, H. Lennernäs, V. P. Shah and J. R. Crison, *Pharm. Res.*, 1995, **12**, 413.
9. Food and Drug Administration, *Guidance for Industry: Waiver of In vivo Bioavailability and Bioequivalence Studies for Immediate Release Solid Oral Dosage Forms Based on a Biopharmaceutics Classification System*. Food and Drug Administation, Rockville, 2000.
10. D. M. Oh, R. L. Curl and G. L. Amidon, *Pharm. Res.*, 1993, **10**, 264.
11. L. X. Yu, J. R. Crison and G. L. Amidon, *Int. J. Pharm*, 1996, **140**, 111.
12. N. Parrott and T. Lave, *Eur. J. Pharm. Sci.*, 2002, **17**, 51.
13. M. Tubic-Grozdanis, M. B. Bolger and P. Langguth, *AAPS J.*, 2008, **10**, 213.
14. A. Okumu, M. DiMaso and R. Loebenberg, *Eur. J. Pharm. Biopharm.*, 2009, **72**, 91.
15. I. Kovacevic, J. Parojcic, I. Homsek, M. Tubic-Grozdanis and P. Langguth, *Mol. Pharmaceutics*, 2009, **6**, 40.
16. C.-Y. Wu and L. Z. Benet, *Pharm. Res.*, 2005, **22**, 11.

17. G. L. Amidon, H. Lennernas, V. P. Shah and J. R. Crison, *Pharm. Res.*, 1995, **12**, 413.
18. European Medicines Agency, Guideline on the investigation of bioequivalence. CPMP/WEP/QWP/1401/98. European Medicines Agency Web site (July 2008). 2008: Retrieved from http:// www.emea.europa.eu/pdfs/human/qwp/140198enrev1.pdf.
19. S. Oie and T. Tozer, *J. Pharm. Sci.*, 1979, **68**, 1203.
20. B. Fichtl, A. Von Nieciecki and K. Walter, *Adv. Drug Res.*, 1991, **20**, 117.
21. D. F. McGinnity, J. Collington, R. P. Austin and R. J. Riley, *Curr. Drug Metabol.*, 2007, **8**, 463.
22. M. P. Gleeson, N. J. Waters, S .W. Paine and A. M. Davis, *J. Med. Chem.*, 2006, **49**, 1953.
23. F. Lombardo, R. S. Obach, M. Y. Shalaeva and F. Gao, *J. Med. Chem.*, 2004, **47,** 1242.
24. M. Rowland, L. Z. Benet and C. G. Graham, *J. Pharmacokinet. Biopharm.*, 1973, **1**, 123.
25. J. B. Houston, *Biochem. Pharmacol.*, 1994, **47**, 1469.
26. R. J. Riley, D. F. McGinnity and R. P. Austin, *Drug Metab. Dispos.*, 2005, **33**, 1304.
27. R. P. Austin, P. Barton, S. L. Cockroft, M. C. Wenlock and R. J. Riley, *Drug Metab. Dispos.*, 2002, **30**, 1497.
28. R. P. Austin, P. Barton, S. Mohmed and R. J. Riley, *Drug Metab. Dispos.*, 2005, **33**, 419.
29. P. J. Kilford, M. Gertz, J. Houston and A. Galetin, *Drug Metab. Dispos.*, 2008, **36**, 1194.
30. M. Gertz, P. J. Kilford, J. B. Houston and A. Galetin, *Drug Metab. Dispos.*, 2008, **36**, 535.
31. H. Gao, S. J. Steyn, G. Chang and J. Lin, *Expert Opin. Drug Metab. Toxicol.*, 2010, **6**, 533.
32. M. G. Soars, P. J. H. Webborn and R. J. Riley, *Mol. Pharmaceutics*, 2009, **6**, 1662.
33. T. Rodgers and M. Rowland, *Pharm. Res.*, 2007, **24**, 918.
34. T. Heimbach, S. B. Lakshminarayana, W. Hu and H. Handan, *AAPS J.* 2009, **11**(3), 602.
35. H. M. Jones, N. Parrott, K. Jorga and T. Lave, *Clin. Pharmacokinet.*, 2006, **45**(5), 511.
36. K. A. Cortright, S. E. Wetzlich and A. L. Craigmill, *J. Vet. Pharmacol. Therapeut.*, 2009, **32**(6), 552.
37. M. Kleiber, *Hilgardia*, 1932, **6**, 315.
38. I. Mahmood and J. D. Balian, *J. Pharm. Sci.*, 1996, **85**, 411.
39. U. Fagerholm, *J. Pharm. Pharmacol.*, 2007, **59**, 803.
40. H. Tang, A. Hussain, M. Leal, E. Fluhler and M. Mayersohn, *J. Pharm. Sci.*, 2011, **100**, 402.
41. H. Tang, A. Hussain A, M. Leal, M. Mayersohn and E. Fluhler, *Drug Metab. Dispos.*, 2007, **35**, 1886.

42. T. Lave, K. Chapman, P. Goldsmith and M. Rowland, *Expert Opin. Drug Metab.Toxicol.*, 2009, **5**, 1039.
43. S. W. Paine, P. Barton, J. Bird, R. Denton, K. Menochet, A. Smith, N. P. Tomkinson and K. K. Chohan, *J. Mol. Graphics Modell.*, 2010, **29**, 529.
44. S. W. Paine, K. Menochet , R. Denton, D. F. McGinnity and R. J. Riley, *Drug Metab. Dispos.*, 2011, **39**, 1008.
45. D. F. McGinnity, J. Collington, R. P. Austin, R. J. Riley, *Current Drug Metab.*, 2007, **8**, 463.
46. K. Ito, H. S. Brown and J. B. Houston, *Br. J. Clin. Pharmacol.*, 2004, **57**, 473.
47. K. Ito, K. Chiba, M. Horikawa, M. Ishigama, N. Mizuno, J. Aoki, Y. Gotoh, T. Iwatsubo, S. Kanamitsu and M. Kato, *AAPS Pharm. Sci.*, 2002, **4**, 1.
48. D. F. McGinnity, N. J. Waters, J. Tucker and R. J. Riley, *Drug Metab. Dispos.*, 2008, **36**, 1126.
49. H. J. Einolf, *Xenobiotica*, 2007, **37**, 1257.
50. E. M. Howgate, Y. K. Rowland, N. J. Proctor, G. T. Tucker and A. Rostami-Hodjegan, *Xenobiotica*, 2006, **36**, 473.
51. S. Inoue, E. M. Howgate, K. Rowland-Yeo, T. Shimada, H. Yamazaki, G. T. Tucker and A. Rostami-Hodjegan, *Xenobiotica*, 2006, **36**, 499.
52. O. A. Fahmi, S. Hurst, D. Plowchalk, J. Cook, F. Guo, K. Youdim, M. Dickins, A. Phipps, A. Darekar, R. Hyland and R. S. Obach, *Drug Metab. Dispos.*, 2009, **37**, 1658.
53. R. J. Riley, A. J. Parker, S. Trigg and C. N. Manners, *Pharm. Res.*, 2001, **18**, 652.
54. M. P. Gleeson, A .M. Davis, K. K. Chohan, S. W. Paine, S. Boyer, C. L. Gavaghan, C. Hasselgren Arnby, C. Kankkonen, N. Albertson, *J. Comput. Aided Mol. Des.*, 2007, **21**, 559.
55. M. P. Gleeson, *J. Med. Chem.* 2008, **51**, 817.
56. S. M. Paul, D. S. Mytelka, C. T. Dunwiddie, C. C. Persinger, B. H. Munos, S. R. Lindborg, A. L. Schacht, *Nat. Rev. Drug Discovery*, 2010, **9**, 203.
57. D. J. Diller, *Curr. Comput.-Aided Drug Des.*, 2009, **5**,106.
58. C. L. Gavaghan, C. Hasselgren Arnby, N. Blomberg, G. Strandlund, S. Boyer, *J. Comput.-Aided Mol. Des.*, 2007, **21**,189.
59. P. D. Leeson, B. Springthorpe, *Nat. Rev. Drug Discovery*, 2007, **6**, 881.

CHAPTER 14

Design of Chemical Libraries

DARREN V. S. GREEN

GlaxoSmithKline R&D, Department of Computational and Structural Chemistry, Gunnels Wood Road, Stevenage SG1 2NY, UK

14.1 Introduction

A compound collection which may be screened to find starting points for medicinal chemistry optimisation has become a vital component of most companies' drug discovery strategies.[1] Such a collection of molecules can be used in its entirety with high throughput screening (HTS), or maybe provide a subset of molecules, selected through knowledge based methods. The scientific questions surrounding how such collections should be designed, and how one may measure the quality of the collection, have been primary drivers for the emergence of the discipline of chemoinformatics over the last two decades.[2,3] It is fair to say that the development of miniaturisation and automation in assay technology and compound handling in many ways created problems for the chemistry community to solve, but some of the underlying philosophies and techniques had begun development precisely because of insufficient throughput in screening. For example, the dissimilarity selection algorithms described by Lajiness *et al.*[4] and the similarity/clustering work of Willett *et al.*[5] were driven by the need to select a small, diverse or representative (the two are not the same, as will be explained) subset which could be screened against novel protein targets using manual or semi-automated assays. The influence of these seminal publications can still be seen in publications more than 20 years later, for example the widespread use of the Tanimoto coefficient to quantify molecular similarity.

RSC Drug Discovery Series No. 13
Drug Design Strategies: Quantitative Approaches
Edited by David J. Livingstone and Andrew M. Davis

Published by the Royal Society of Chemistry, www.rsc.org

To feed the emerging capacity for screening, combinatorial chemistry came to the fore. Driven primarily by the ability to synthesise large numbers of molecules,[6] fuelled by an assumption that medicinal chemists were able to design diverse and drug like libraries without recourse to computational methods, it was some time before chemoinformatics methods made a significant impact. Key to gaining traction was the ability to quantify molecular diversity and similarity, and then to use these to demonstrate, for example, the superiority of design using product structures as opposed to picking reagents.[7] By the mid-1990s, computational approaches had matured sufficiently that several commercial products were available, such as the Diverse Solutions[8] and ChemX[9] suites. With reliable algorithms available for everyday work[10] and screening data to give feedback, chemoinformaticians started to look into more advanced questions, such as the biological relevance of their chemical descriptors, enshrined in the "similar property principle"[11] (chemically similar structures have similar biological properties) whilst integrating the emerging views of "drug likeness" using physiochemical properties, exemplified by the Lipinski rule of 5.[12] This widespread acceptance of multiple ways to describe molecules and chemical libraries led to the adoption of multi-objective methods, particularly the Pareto based algorithms,[13] introducing the concepts of ranking and trade-offs alongside the quantitative measures of similarity and diversity.

The computational community has always had an ambivalent relationship to diversity and HTS, having one foot firmly in the camp of molecular similarity, structure based design and knowledge based methods. Such approaches were vigorously pursued when the early results from combinatorial chemistry ('combichem') and HTS were not particularly good. The development of similarity methods to include pharmacophoric concepts (both 2D and 3D), shape and molecular field[14] has enabled efficient "ligand based" selections to be made from corporate collections, and the design of focussed libraries.[15] Protein structure based methods have continued to be developed,[16] although there is much debate on the relative effectiveness of these techniques compared to ligand based approaches.[17,18] Broader approaches, such as the use of protein families to guide design, have been especially effective in target classes such as kinases,[19] and the concept continues to be refined and extended under the banner of chemogenomics.[20]

The sheer diversity of approaches to the design of a screening collection is enough to confuse a newcomer to the area. Which approaches actually work or, more pertinently, which will work for my needs? This last question has been used by Roth[21] to illustrate a universal truth: diversity measures are irrelevant outside the context of application. State of the art approaches to the design of a screening collection use information about the intended use—for example a portfolio of intended screening targets—to provide this context.[22]

This chapter will cover all of the methodologies mentioned in this introduction, and will illustrate the degree to which it is possible to quantify diversity and similarity and how this data can be used to successfully design compound libraries and collections which are fit for their intended purpose.

14.2 Similarity Methods

To understand the methods used to quantify molecular diversity, it is best to start with those applied to molecular similarity as the two are two sides of the same coin.[23] Modern approaches to molecular similarity can be traced back to the work of Adamson & Bush,[24] who described the use of bit strings—applied as screens to improve the efficiency of substructure searching in chemical databases—to compute molecular similarity.[25] The majority of methods use bit strings as they are able to hold large amounts of information very efficiently and work extremely well alongside the various similarity indices. Early practitioners adopted what are now called "2D fingerprints" as these were readily accessible from chemical database systems. The group at Lederle are credited with the first production similarity system,[26] based on atom pair fingerprints, but many groups adopted fingerprints used by the Daylight[27] and MACCS[28] systems. The Tanimoto similarity index, identified as particularly useful for chemical fingerprints,[29] was and is still a standard, returning a value of between 0 (no similarity) and 1 (identity) when comparing two molecules, A and B [eqn (14.1)].

$$Tanimoto = \frac{A\ and\ B}{A\ or\ B} \tag{14.1}$$

A is the number of bits set in the fingerprint for molecule A, and B is the number set for molecule B. Often misunderstood, the Tanimoto index is not a percent similarity, nor a correlation coefficient. Its behaviour is rather subtle, as there is a dependency of the value on the size of molecule under consideration. An important feature of the Tanimoto is that it relates to the presence of functionality, and pays little attention to the absence of features. For example, consider two molecules, otherwise identical, where one has a carboxylic acid substituent at R1 and the other has hydrogen. A chemist may not consider these two molecules similar due to the absence of the acid in one of them, but if the two molecules are sufficiently large to share many common bits in the fingerprint, the Tanimoto similarity would be high. There are many other similarity indices, such as the Dice, which have slightly different behaviours,[23] but none have displaced the Tanimoto from its position as the method of choice for chemistry systems. As such, the behaviour of the Tanimoto coefficient is very well understood, to the extent that the biological relevance has been quantified by comparison to screening data. For molecules with a similarity of >0.85 Tanimoto (daylight fingerprints) the chance of them sharing a biological activity is $\sim 30\%$.[30] Readers may be surprised that this is so low, however given the simplicity of the description (no conformational, chirality or shape information) it is not unexpected and in fact no other chemical similarity measure has been shown to be consistently superior. More importantly, this measure is extremely useful for understanding how to build a compound collection when using methods based on 2D fingerprints, as will be shown in Section 14.3.

There are many other ways of describing chemical similarity.[31] Perhaps the most impactful methods have invoked the pharmacophore concept.[32] At the minimum level, one can look for molecules which share, say, an acid and a base at a particular distance. This similarity may be enough for the purpose (for example to select compounds which may inhibit an RGD peptide interaction with an Integrin). However, to better describe whole molecule similarity, pharmacophore fingerprints were developed [33]. These have the advantage of encoding all pharmacophore combinations, but the disadvantage (shared by all 3D methods) of sampling across all conformations of the molecule. Without reference of a biologically relevant conformation (for example a binding mode from protein crystallography) these unproductive conformations add noise to the calculation and reduce the effectiveness. However, for molecular similarity applications where a reference conformation is available, such as scaffold hopping, 3D methods can be very powerful and there has been significant progress made in the use of shape,[34] molecular fields[35] and even so-called 4D methods,[36] which incorporate multiple conformations.

A compromise between the 2D structural and 3D pharmacophore descriptors are techniques which implement reduced graph methods, whereby pharmacophoric elements are encoded in the 2D structure, thus gaining the benefit of a description which is more biologically relevant, whilst not adding the complication and noise from 3D conformations. Examples are the CATS,[3] Reduced Graph,[38] and Feature Trees[39] approaches.

An alternative to fingerprint based similarities are those based on BCUTs (**B**urden, **C**AS, **U**niversity of **T**exas).[8] This method uses a modified connectivity matrix (the Burden matrix) onto which are mapped atomic descriptors (such as atomic mass and polarizability) and connectivity information. The eigenvectors of this matrix represent a compressed summary of the information in the matrix and are used to describe a molecule. Typically 5-6 BCUT descriptors suffice to describe the chemical space of a set of molecules, and the space is usually partitioned into distinct "bins", with each molecule assigned to the appropriate partition. In this format, similarity calculations become very simple; molecules which are mapped into the same partition are similar. As an alternative, one could use larger numbers of molecular properties and a correlation vector approach.[40]

There are also methods which are best applied to scaffolds or ring systems. The CAVEAT method[41] was developed to enable researches to design synthetically accessible mimics of complex natural product or cyclic peptide motifs. It focuses on the vector projections of key side chains from the core structure and may be used to predict the similarity between molecular scaffolds.

14.3 Diversity Methods

The concept of molecular diversity became important as high throughput science gained traction in screening and synthesis. The fundamental question to be answered is: how many molecules (and specifically which molecules) are needed to cover "chemical space" such that an HTS campaign will find a lead molecule (or for some of the early evangelists, a drug candidate) for every and any protein target presented to it? For those looking to answer this question, the first uncomfortable truth is the size of accessible chemical space. Starting with estimates of $>10^{60}$ [42] the estimates have come down over the years, but even the most recent and systematic analyses suggest numbers of $>10^{30}$.[43] For those looking at screening collections of no larger than 10^6, this sampling rate of less than 1 in 10^{25} presents a formidable problem which from a purely theoretical sense renders HTS a very risky strategy indeed. That HTS groups regularly find leads for >50% of targets[44] would suggest something other than diversity sampling of chemical space is at work, and this will be discussed further in Section 14.4.

For many groups, the work in hand need not consider the larger philosophical question. There are practical needs, for example which compounds should be first to be weighed into the corporate screening set, which reagents to use in combinatorial libraries and such like. The work of Lajiness *et al.* and Willett *et al.* provided a firm basis from which to proceed. The simplest way to view diversity is *via* the concept of dissimilarity, which in turn could be defined as (1-similarity) if using a similarity index such as the Tanimoto. A dissimilarity selection algorithm aims to select a set of compounds which are maximally dissimilar to each other. The MaxMin algorithm is applied widely and has been the subject of extensive evaluation.[45] For a set of compounds in a database, a lead molecule is selected and placed in the subset to be selected. The distance (in our case dissimilarity) between the selected subset and those remaining in the database is calculated, in this case by the pairwise dissimilarity of every molecule in the subset (K) and every molecule remaining in the database (J), according to eqn (14.2).

$$DIS\,(J, Subset) = \sum DIS(J, K) \qquad (14.2)$$

In this manner, the minimum distance of each molecule J to any member of the existing subset is found. The molecule in the database with the maximum value of the minimum distance to any member of the existing subset is then selected for addition to the subset. In this way, a diverse set of molecules is built up one by one.

Clustering[46] is an alternative to this, where a set of compounds is organised into chemically similar clusters and then representative compounds picked from each cluster to form a set which is both diverse with respect to each other and representative of the starting set of compounds as a whole. These methods can be applied to any of the fingerprint representations discussed.

The methods described provide a means to describe relative diversity between subsets of the same database. However, other methods were needed in order to quantify absolute diversity, which would allow comparisons between two sets of different molecules. Building upon the work of Martin *et al.*,[47] Turner and co-workers[48] described a method which used all the pairwise similarities between all the molecules J, K in a database A of size N [see eqn (14.3)].

$$Diversity(A) = 1 - \frac{\sum_{j=1}^{N(A)} \sum_{K=1}^{N(A)} SIM(J,K)}{N(A)^2} \quad (14.3)$$

However, such algorithms scale at order N^2 with the number of compounds under evaluation and thus the methods began to falter as ever larger libraries of compounds were presented. More efficient methods were sought, such as the use of cosine similarities, which reduces eqn (14.3) to eqn (14.4).

$$Diversity = 1 - \frac{dotproduct(Ac, Ac)}{N^2} \quad (14.4)$$

Ac is centroid fingerprint, computed from all the molecules in database A. Because each molecule is processed only once to produce the centroid fingerprint, the method is fast and applicable to very large databases. But although such methods scaled well (order N), the diversity measures were not well behaved. For example, the addition of a "redundant" molecule in chemical space (a molecule with the same similarity relationship to other molecules in the set as an existing molecule) could cause the diversity measure to either increase or even become zero. Waldman *et al.*[49] introduced 5 requirements for a well behaved diversity function:

1. Adding redundant molecules to a system does not change its diversity
2. Add non-redundant molecules to a system always increases its diversity
3. Space-filling behaviour of diversity space should be preferred
4. Perfect (*i.e.* infinite) filling of a finite descriptor space should result in a finite value for the diversity function
5. If the dissimilarity or distance of one molecule to all others is increased, the diversity of the system should increase. However, as this distance increases to infinity, the diversity should asymptotically approach a constant value.

These behaviours can be seen in most modern diversity functions.[50]

Clustering methods have by and large replaced dissimilarity based algorithms. Although hierarchical methods such as Wards were identified as preferable, it is common to see faster methods quoted. The widely used sphere exclusion clustering algorithm[51] deserves a detailed description. First, a "near neighbours" table (a list of the molecules and the number of analogues they have) is assembled, and given a defined similarity threshold (the ubiquitous Tanimoto coefficient is most often applied, although there is some evidence that alternative coefficients may be superior). The table is ordered and, starting with the molecule with the most near neighbours, it and all the molecules within the defined similarity threshold to it are marked as belonging to the same cluster and removed from the set. Taking the remaining compounds, the molecule with the most near neighbours is selected, a new cluster created from it and molecules similar to it, and the whole process repeated until there are no molecules remaining. The algorithm is fast, intuitive and reliable.

However, the fingerprint driven diversity methods suffer from an inability to describe a bounded chemical space; novel molecules can be added with a concurrent increase in diversity with little indication of how evenly sampled parts of the space is. In this context, partition based methods promise much. The BCUT descriptors are particularly suitable for the definition of a bounded chemical space and have the added bonus that they are quickly calculated. Partition based methods also scale very well in that it is only necessary to calculate which bin the molecule falls into, not to compute all pairwise similarities with the other molecules in the set. Absolute and relative diversity may be computed from bin occupancy, for example the number of bins covered by a compound set.

A combination of partition based selection and 3D pharmacophore fingerprint description was used to implement the pharmacophore-derived queries (PDQ) methodology of Pickett and Mason at Rhone Poulenc Rorer.[33] This influential study was the first to describe a bounded chemical space which could be understood in terms of how small molecules interact with proteins. This concept was taken and refined by the group at NeoGenesis, who produced the Quantised Surface Complementary Diversity (QSCD) methodology.[52] From this they were able to estimate that full coverage of pharmacophorically described space could be achieved by 24 million compounds. This search for a universal compound space still continues with methods such as the ChemGPS system[53] and more recently Molecular Quantum Numbers.[54] It is certainly advantageous for chemists to be able to visualise (on a computer, or in their minds) where coverage is sufficient and where it is sparse, but this often conflicts with the biological relevance of the definitions used, which are often compromised in order to facilitate visualisation. For example, biological activity can be extremely sensitive to the pK_a of a base, or the degree to which an aromatic ring is electron rich or poor, concepts that are not present in the QSCD method and which would add further dimensions to the proposed space, with concurrent increases in the number of molecules that are required to cover it. Equally, the ChemGPS and MQN systems utilise many atom-based and topological properties which are neither intuitive nor give rise to biologically homogenous clusters of molecules.

The aforementioned methods for diversity selection have quite different behaviours, summarised in Figure 14.1. Any will give a diverse set of compounds with reasonable coverage of the chemical space, given that a sufficient number of compounds are selected.

The biological relevance of chemical descriptors was at the heart of the next stage of evolution in diversity evaluation. The "similar property principle" of Johnson and Maggiora[11] states that structurally similar molecules are expected to show similar physical, chemical and biological properties. Patterson *et al.*[55] extended this principle to neighbourhood behaviour; molecules close to each other in a defined property space should show similar physical, chemical and biological properties. Martin and co-workers described a validation study of a

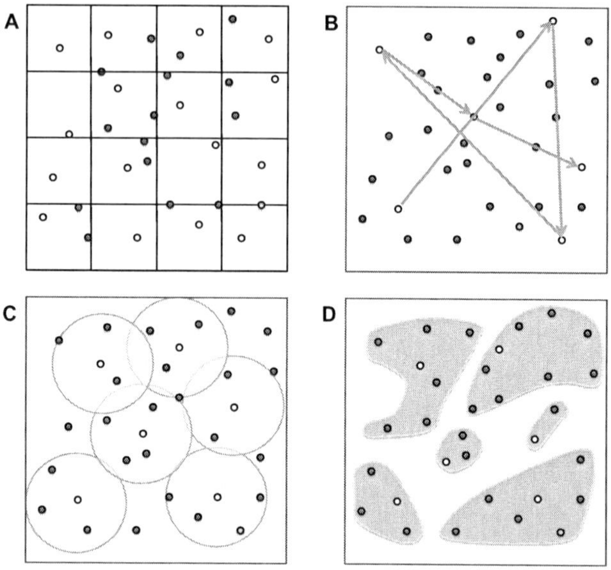

Figure 14.1 Four common selection methods compared. The molecules (represented by dots, those selected are black with a white centre) are distributed in an arbitrary two dimensional property space. A illustrates a cell based selection of one molecule per cell, B a MaxMin dissimilarity selection, C uses sphere exclusion clustering, whilst D invokes a more sophisticated clustering method. This figure is adapted from ref. 56.

sphere exclusion clustering methods with Daylight 2D fingerprints across 115 HTS campaigns.[30] They found that for a relatively conservative similarity figure (Tanimoto 0.85) on average only 30% of molecules in a cluster shared the biological activity uncovered in the HTS (henceforth this will be scaled between 0–1 and called α, so in this case $\alpha = 0.3$). At face value, many chemists may feel that this validates their scepticism for computational diversity measures (indeed there are studies which highlight deficiencies in the methods and conclude that they are only of value in guarding against a poor random selection[56]). However, on a considered evaluation this number is both reasonable and useful. It is well known that simple similarity searches do not return lists of molecules which all share the same biological activity, for reasons that are well documented. What is interesting is that this figure of 30% is significantly higher than is found with other methods, such as 3D methods, where even with a reference conformation to use in virtual screening, an α of 0.3 would be exceptional. This same issue of biological relevance applies to all published methods that describe a universal chemical space. If the biological similarity within a partition is very low, it is difficult to select a subset of compounds using the space with any confidence that the resulting set will exhibit any more biological diversity than would be contained in a random set of compounds. However, this relatively high figure for α of around 0.3 for the 2D clustering method is useful, as it is high enough to be used in wider strategic questions, for example how many

Table 14.1 Effect of selecting multiple members from a single cluster on the probability of finding a hit.

N compounds selected from active cluster	*Pr.(miss active cluster)*
1	0.7
3	0.34
5	0.17
10	0.03
15	0.005

analogues do I have to screen in order to be sure that I would find a particular chemical series if there is some biological activity to be found? For a given value of α, the probability of not finding the series, Pr(miss a lead in a cluster of $N) = (1 - \alpha)^N$. Table 1 expounds the relative probabilities of finding the chemical series as more molecules in the cluster are screened, and shows that 10 molecules per cluster are needed to be 99.9% sure of finding the activity.

Recently, scaffold or framework based descriptions have become popular. This description plays very well to a chemist view, and fits very well with array based chemistry designs. A similar validation experiment to the Martin study has been performed using ring-scaffolds (all the rings in a molecule and their interconnections, but without acyclic appendages),[57] with the conclusion that between 50 and 100 compounds per scaffold are needed to reliably find the activity in a scaffold, 5–10 times more than is required for a cluster in the 2D fingerprint method (*i.e.* α is approximately 0.04). In an elegant paper, Lipkin *et al.*[58] extended the concept to ask what would be needed to see SAR from the screen, in addition to simply finding an active from the cluster. The conclusions are that to reliably see 2 actives from a scaffold, a minimum of 200 compounds per scaffold are needed. To have SAR from 5 active compounds, 650 compounds per scaffold are needed. This result is entirely predictable; a more broad description of chemical similarity leads to a less cohesive biological similarity. This does not preclude use of the method in designing or analysing diversity of a compound collection, as will be described later, but it does impose a reality check on those that would wish to see SAR delivered straight from an HTS.

As scaffold based diversity is particularly sensitive to the number of molecules in it, a novel diversity function has been proposed.[59] PC50C is equal to the percentage of scaffolds in a collection representing 50% of the compounds in a collection. The higher the PC50C, the more diverse the collection is.

For compound sets with a well defined limit, it is possible to enumerate all the scaffolds in the set and to visualise the chemical relationships. In particular, the Bioactivity guided Scaffold tree method (BIOS)[60] provides a visually compelling means of exploring complex structural relationships. As an example, the Structural Classification of Natural Products (SCONP) arranges the scaffolds of the natural products in this tree-like fashion.[61] This provides a concise and chemistry friendly description of natural product diversity,

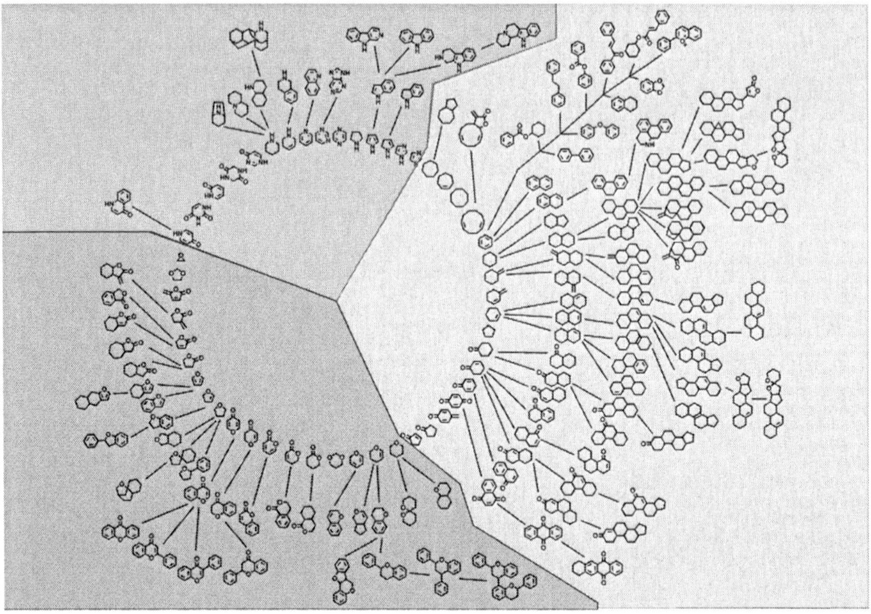

Figure 14.2 The Scaffold tree visualisation of the Structural Classification of Natural Products (SCONP). Reproduced from ref. 61, Waldman *et al.*, *Proc. Natl. Acad. Sci.* 2005, **102**, 17272. Copyright 2005 National Academy of Sciences, U.S.A.

although the method quickly becomes unwieldy for more diverse or larger compound sets. Not only can the diversity and coverage of a collection be measured against the reference set, but it can be visualised on the scaffold tree through judicious use of colour (Figure 14.2).

14.4 Protein Family Approaches

Alongside similarity and diversity approaches to lead identification, the observation that targets can be viewed in these terms has given rise to protein family level design strategies, leading to the current state of the art methods of Chemogenomics.[20,62] This approach attempts to use the similarities between proteins to exploit so-called privileged structures, which are frequently observed to bind such proteins. This knowledge can be used to design sets of molecules for focussed screening, and proponents claim this strategy can deliver leads using fewer resources than diversity based HTS.

Many early literature examples do not employ quantitative design methods, relying on the use of a privileged core and pragmatic choice of reagent with which to decorate the core.[63]

The work of Biofocus has extended the concepts of single privileged scaffold into collections of scaffolds, or "themes", that attempt to cover entire families of proteins such as G-protein coupled receptors (GCPRs)[64] and kinases.[65] In this approach, targets may be classified according to which themes they are known to recognise. The resulting matrix may be used as a fingerprint with which to cluster targets, or as a design resource to suggest how many, and which, themes should be used to build a screening library for a family of targets.

Lowrie *et al.*[66] provide a comparison of how design approaches may vary between target families. For kinases, the availability of 100s of protein crystal structures has enabled a very detailed, atomistic, understanding of the molecular recognition. Cores may be designed or selected with the aid of docking algorithms, with R-groups selected to probe known binding "hot spots", and amino acids that are conserved or varied across the family, in search of selectivity. For 7-transmembrane receptors, which have much less crystal structure information and are inherently more flexible proteins, chemists have tended to use ligand based methods. In this case, a neural net classifier was used to predict "GPCR-likeness" in much the same way as drug-likeness models have been produced (see Section 14.5). Both approaches predict binding to their respective target families, which need to be combined with other considerations such as ADMET, similarity to existing collections of molecules, toxicity predictions and synthetic feasibility, and for this the authors advocate the use of the Pareto design methods discussed in Section 14.6.

There are ever more sophisticated computational approaches emerging, such as *in silico* predictions of polypharmacology across the protein family of interest.[67,68] Gregori-Puigjane and Mestres[68] use a battery of QSAR models to profile *in silico* each member of a virtual library, which creates a predicted ligand–target interaction matrix. A whole library scoring function is then applied, which takes into account both chemical and target diversity [eqn (14.5)].

$$Score = \frac{(T_{\mathrm{cov}} + C_{\mathrm{cov}} + (1 - T_{bias}) + (1 - C_{bias}) + Prom10)}{5} \tag{14.5}$$

T_{cov} is the proportion of targets predicted to bind at least one ligand in the library, C_{cov} is the corresponding proportion of ligands in the library predicted to bind at least one target in the family, the T_{bias} and C_{bias} terms reflect deviation from uniformity of the distributions of target to ligand and ligand to target annotations respectively, whilst Prom10 is the proportion of ligands predicted to bind more than 10 targets in the family.

Unfortunately there are few data to systematically compare protein family strategies to diversity HTS. In fact, in contrast to the thorough analyses published to validate diversity algorithms, there are surprisingly few studies to properly validate target family designs or statistically prove the existence of privileged structures.[69] Indeed an intriguing argument that explains both the theoretically surprising success of HTS and the widespread belief in privileged fragments can be made[70]: chemists mostly make and design molecules based around knowledge of which small molecules are known to bind proteins.

Our screening collections therefore have a "biogenic bias" because we have selected them from ideas based on knowledge of biochemistry. We therefore rediscover binding events through screening, and we ascribe this rediscovery to privileged structure motifs!

14.5 Beyond Simple Diversity

Whatever design/selection strategy is taken, there are other considerations that are commonly applied. The first concerns undesirable substructures. There are many reasons to filter by substructure, for example to remove reactive functionality in compounds being selected for testing; in reagent selection, one would remove functional groups that might interfere with synthetic steps or in HTS analysis, where aggregators or assay technology-specific filters may be available. Diversity selection algorithms can be particularly susceptible to unusual substructure. For example, a set of derivatives of the explosive Pentaerythritol tetranitrate (PETN) will add considerable diversity to a set of marketed drugs, but little benefit to a drug discovery project. Therefore, sets of substructural filters were quickly developed to remove such compounds from consideration for compound purchasing and screening strategies.[71–74] Even with groups of filters available, it became clear that many compounds that could be made by combinatorial chemists, or purchased from early compound vendors who had collated materials from universities, did not "look" right to many chemists, they were not "drug like". "Drug like" is a vague notion and very difficult to articulate in terms of chemical structure (indeed chemists often disagree about what should or should not be filtered from screening collections).[75] After all, drugs are approved based on studies of efficacy, safety and cost effectiveness, not by looking at the chemical structure graph! Therefore, drug structures can range from single atoms (lithium) through what many would consider typical drugs (small heterocyclic, orally administered compounds) through complex natural products to entire proteins. Nonetheless, attempts have been made to quantify a measure of drug likeness by comparing sets of marketed drugs to large sets of presumed non-drugs. These studies typically employ a machine learning method, for example a neural network, to select sets of chemical descriptors and a scoring function which separates the two sets of compounds.[76,77] For example, Ajay *et al.* [78] trained a Baysian neural network to distinguish between compounds in Comprehensive Medicinal Chemistry ("drugs") and the Available Chemicals Directory ("non-drugs") databases. The descriptors employed were a combination of "1D" (logP, molecular weight, numbers of donors and acceptors and number of rotatable bonds) and 2D substructure keys. Testing of the model yielded a correct prediction of ~80% of compounds in the MACCS-II Drug Data Report (MDDR) database of drugs, clinical and preclinical drug candidates (Figure 14.3).

Research in this area continues, with recent approaches using a fragmentation based method that looks at the co-occurrence of structural fragments present in known drugs to yield a drug like score for a novel molecule.[79]

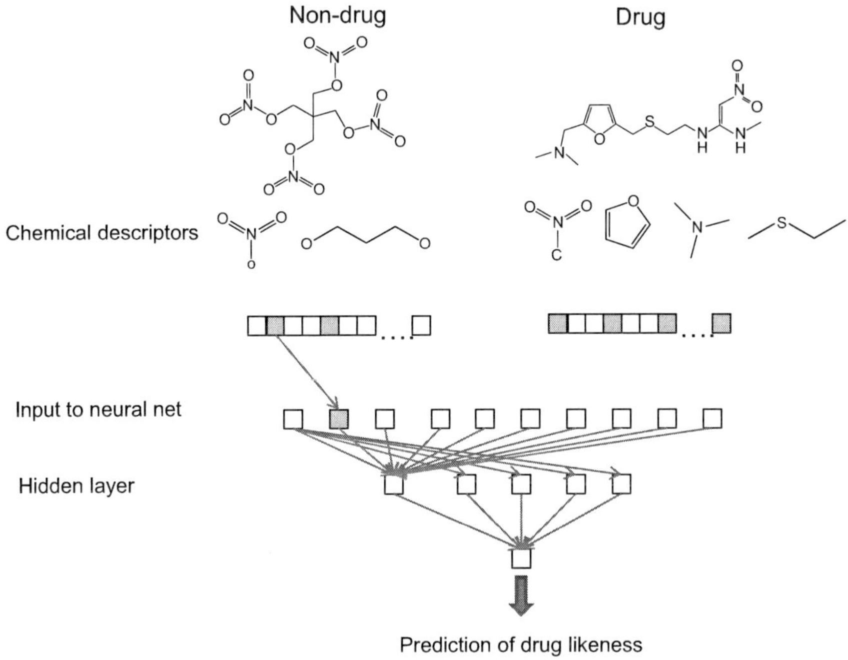

Figure 14.3 An illustrative diagram of how a neural net is trained to distinguish drugs from non-drugs. Descriptors are calculated from the molecules (in this case, presence or absence of a structural fragment to create a binary fingerprint). The descriptors are used as input to the neural net, and the hidden layer learns how to weight the chemical inputs to optimally separate the drugs and non-drugs. Once trained, the network can predict the drug likeness of a novel molecule presented to it.

Physicochemical properties are of great importance to drug likeness and filters in general. Although the importance of parameters such a logP were well known to the industry due to the immense contributions of Hansch and Leo,[80] it took the publication of the Lipinski Rule of 5[12] to remind chemists that orally absorbed drugs tended to have rather modest values of logP, molecular weight and hydrogen bond donors and acceptors. Such parameters had been identified independently by Gillet *et al.* in their drug likeness studies.[76] Although the Rule of 5 became and remains a very commonly applied filter, several researchers noted that the study was performed on drugs, and most applications were in hit or lead discovery. The seminal work by Hann *et al.*[81] proposed a mathematical model of molecular complexity which, when coupled with the observed increase of parameters such as logP and molecular size as leads are optimised into drugs, lead to the proposal that screening collections should be populated with smaller, less complex molecules. This was formalised into rules for "lead likeness"[82] which were similar to the Lipinski parameters but more restrictive. In turn, these studies influenced the early researchers in fragment based drug discovery, leading to even more restrictive frameworks such as the Rule of 3.[83]

Many physicochemical properties can be used as a filter to discard unwanted molecules. However, strict filters can be unsatisfactory. For example, a molecule of molecular weight 399 may be accepted, whilst that of 401 rejected. If the latter molecule contained a novel 3D core, and the former was yet another benzodiazepine or "aryl-amide-aryl-amine" type, one might question the decision. Therefore, researchers adopted more subtle penalty functions.[84] which increase the further away a molecule is from the ideal. Figure 14.4 illustrates the most common penalty functions.

Recently, chemists have looked back to natural products as a source of inspiration,[85,86] and the concept of "natural product like" or "biologically inspired" [87,88] molecules has emerged . Similar approaches to drug likeness studies have been applied,[89] with the findings that natural products tend to be more rigid, have more fused, bridged and spiro ring systems, a lower percentage

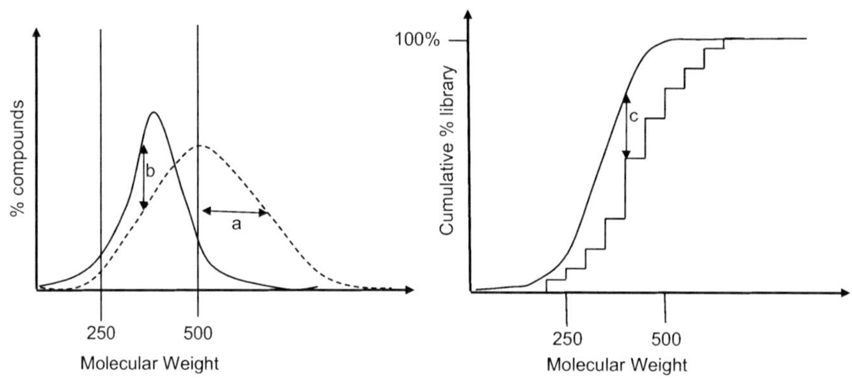

Figure 14.4 Common methods for constraining the property profile of a library to that of a reference set. The graph shows a reference "drug like" set of compounds (bold line) which generally have molecular weight between 200 and 500. The dashed line represents a library with a significant shift towards large molecules (sadly, a frequent occurrence). The first method (left) is to assign a penalty to compounds which fall outside arbitrary limits, in this case set at below 250 and above 500 Daltons, the penalty being proportional to the distance, a, of the compound from the desired limit. The penalty score would be the sum of penalties assigned over all the compounds in the library. The second method is to compute a measure of how close the library profile approximates the reference profile, by computing each distance, b, for all values of molecular weight. This can be accomplished by a root mean square calculation. The third method (right) is to use a cumulative distribution, which enables the use of the χ^2 test, for which the property has to be binned (the figure has molecular weight binned into 50s). The alternative Kolmogorov-Smirnov statistic[76] allows the use of binned or continuous distributions and computes the difference between the distributions by summation of the distance, c, over all points, or bins, in the graph.

of aromatic rings, higher carbon–oxygen bond count and a lower carbon–nitrogen bond count than collections of synthetic heterocyclic molecules (some of these trends had been uncovered by the early drug likeness studies). Intriguingly, the lower percentage of aromatic rings- more accurately, the fraction of sp3 hybridised atoms in a molecule- has been identified as a contributor to attrition in drug development.[90]

Up until this point we have described how molecules may be selected to be similar or diverse and how they may be filtered and scored as to how "like" categories of molecule (drug, lead, fragment, natural product) they are. The following sections will describe how these methods are applied to the design of individual arrays and entire screening collections.

14.6 Combinatorial Design

Combinatorial chemistry and its associated spin off technologies for parallel synthesis, or simply higher throughput synthesis, imparts some practical restrictions on a design. In a classic core template and R-group representation, the most common design is to select a number of reagents at each position that will fit to the synthetic equipment in use, be it 96 well "blocks" or more elaborate automation. The key restriction in design terms is that each reagent at R1 is coupled with every reagent at R2 and every reagent at R3. Therefore, an array with 10 reagents at each position yields $10 \times 10 \times 10 = 1000$ final molecules.

Early array design focussed on selection of diverse reagents. Martin *et al.*[47] used a D-optimal design method to select sets of diverse reagents, with the commonly held assumption that this would lead to diverse products. This hypothesis was disproved by Gillet *et al.*[7], who showed that a selection of diverse reagents did not give the optimal set of diverse products, and subsequent methodologies have focussed on product driven evaluation and design strategies.[84] Nonetheless, the advantages of reagent based selection retained important practical advantages in that for product based selections the combinatorial products need to be enumerated, sometimes leading to very large virtual libraries, which then lead to very slow diversity selection steps. Another issue with product based selection is the combinatorics of the selection process; there are 10^{26} ways of selecting 10×10 reagents from a 100×100 combinatorial virtual library! A basic solution to this was provided by monomer frequency based methods MFA[91] and PLUMS.[92] Products are selected on the basis of diversity or some other score, and reagents are then ranked on the number of times they appear in the selected products. This methodology works well for simple designs, however it soon became apparent that library design was a multiple objective problem. Two early examples of this were from Harpick[93] and Galoped.[94] Harpick used 3-point 3D pharmacophore fingerprints to measure diversity, restrained the design to drug like physicochemical properties profiles and included a reagent cost term. Galoped used a 2D diversity term, physicochemical properties and a requirement that each member of the library had a unique mass, to enable deconvolution of pooled screening. Both methods, and others of the time, applied a weighted sum algorithm to combine the various objectives The Galoped scoring function is shown by eqn (14.6).

$$\begin{aligned}\text{Score} = {} & \text{X.product\_MF\_redundancy} + \text{Y.substituent\_MWt\_redundancy} \\ & + \text{Z.diversity\_redundancy}\end{aligned} \tag{14.6}$$

Galoped uses the concept of redundancy to identify libraries that are sub-optimal. For example, having products with identical molecular formula or molecular weight hinders the deconvolution step (which has a mass spectrometry read out) in use at the time. Equally, having molecules from the same structural cluster introduces redundancy. Therefore the smaller the score the better the library. The weights x, y and z are set by the user.

The Harpick scoring function is shown by eqn (14.7).

$$\begin{aligned}\text{Energy} = {} & (\text{Unique}^{w} \times \text{Conscore} \times \text{Partscore}_{\text{pp}}{}^{x} \times \text{Partscore}_{\text{pa}}{}^{x} \\ & \text{Partscore}_{\text{ha}}{}^{x} \times S)/(\text{Totpharm}^{y} \times \text{Flex}^{z} \times n)\end{aligned} \tag{14.7}$$

Unique is the number of unique 3-point pharmacophores displayed by the whole library, Conscore is a term that can force a design to fill specific 3-point pharmacophores and avoid others (*i.e.* to complement an existing compound collection). Partscore is used to force a distribution of shape related descriptors: heavy atoms (ha), largest triangle perimeter for a pharmacophore in the library (pp) and largest triangle are for a pharmacophore in the library (pa). S is the total number of molecules that pass certain filters, such as a limit of molecular flexibility. Totpharm is the total of pharmacophores displayed by the entire library and Flex is the normalised number of conformations accessed by the library of n molecules. The terms a, b, Y and Z are user defined weights.

With such a scoring function in hand, the combinatorial selection can be optimised by use of stochastic algorithms, particularly simulated annealing and genetic algorithms (stochastic algorithms are needed because the large number of potential combinatorial solutions precludes an exhaustive search across all solutions). A basic genetic algorithm flow, applicable to the Galoped algorithm, is shown in Figure 14.5.

A number of hybrid solutions are proposed,[95] whereby there is a great deal of emphasis on selecting reagents whilst using the associated product properties to score the proposed library. These methods make it quite easy to select a replacement reagent, a common practical problem in the lab, where a reagent may not be delivered or may not react.

Although a great step forward, these methods had a common problem in the restricted formulation of the scoring function. The monomer frequency algorithms are difficult to extend to more than two objectives as they are order dependent and therefore are unable to compare different solutions without great effort. For the stochastic, product based algorithms, a decision has to be made regarding how important each objective is; for example does diversity

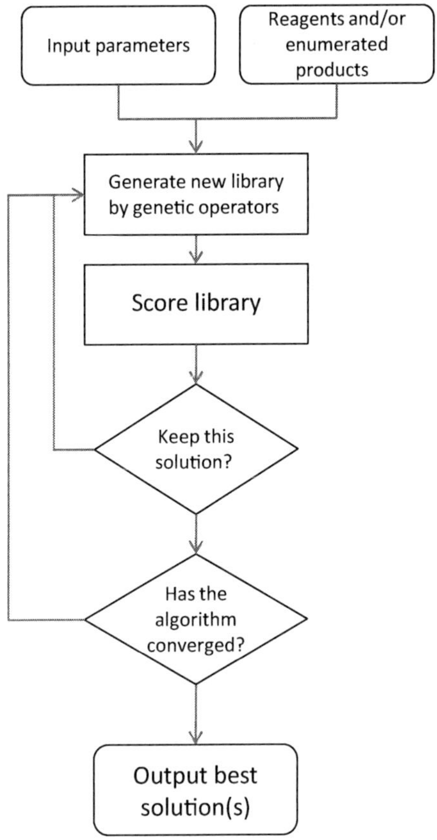

Figure 14.5 The schematic flow for a basic genetic algorithm, as used by the Galoped method for library design.[94]

count twice that of logP? And how do you numerically compare, say, drug likeness to cost? There is no way of determining the weights *a priori*, and as the number of objectives grows, the task of balancing the weights becomes more complex. To add to this problem, the weights that are set govern the solutions that are searched by the algorithm and can actually prevent good solutions being discovered. These are known issues, which have been solved in other industries by the adoption of true multi-objective algorithms, for example multi-objective genetic algorithms (MOGAs).[96] In collaboration with the inventors of MOGA, Gillet *et al.* produced MoSelect,[97] which allows the pursuit of multiple design objectives in a combinatorial library without having to decide weights. Such algorithms are able to do this because a design is not scored in an absolute sense, rather the concept of Pareto optimality is used to rank the design against others in all of the desired objectives. A series of "non-dominated" solutions is found, all of which have an advantage over the other solutions in at least one of the objectives (Figure 14.6). The solutions can then be inspected by the chemist and a favourite design selected.

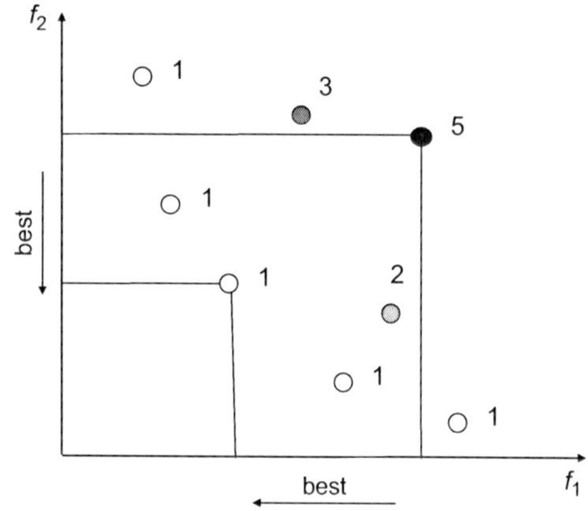

Figure 14.6 An example of Pareto ranking in two objectives, f_1 and f_2 (which could be, for example, diversity and drug likeness). Pareto optimisation seeks to find the set of non-dominated solutions. A non-dominated solution is one where an improvement in one objective results in deterioration in one or more of the other objectives when compared with the other solutions in the population. In Pareto ranking, an individual's rank corresponds to the number of individuals in the current population by which it is dominated. The black coloured library has four libraries that are better than it in one or both of the design criteria, so is given a Pareto rank of 5. The white dots with black outline are non-dominated, so are given a Pareto rank of 1, and represent the set of best solutions found by the algorithm.

A variety of enhancements of and variants on the basic algorithm have since been published.[98]

14.7 Compound Collection Enhancement Workflows

A common application of design methods is in the need to purchase or synthesise molecules which "enhance" a corporate screening collection. The term enhance is ill-defined, but the qualitative meaning is that the additions to the screening collection should make it more capable of yielding a starting point should a target or, more likely, a portfolio of targets, be screened against it.

Novartis have published a detailed account of how commercially available compounds are selected for purchase.[99] First, a set of 250 substructure filters are used to remove undesirable functional groups. Secondly, calculated logP is used to filter (AlogP >8) or penalise (AlogP >6 and ≤ 8) molecules. Molecules are then placed into one of 3 bins according to priority. Priority one are termed "privileged structures"; these are compounds which have a similarity of >0.65 (Tanimoto coefficient using the Unity 2D fingerprints) to a set of biologically relevant reference compounds taken from the MDDR database. Added to these compounds are compounds containing a scaffold which is commonly seen in marketed drugs, or believed to be a privileged structure for specific target classes, or marked to be of particular interest due to its rarity in the Novartis collection. The second priority compounds are those marked neither as privileged nor penalised. The third priority are those compounds that have been penalised due to poor physicochemical properties. Diversity selections proceed in order of priority, with different similarity thresholds chosen (0.95, 0.88 and 0.8 for priority one, two and three respectively), which allows dense sampling of know biologically relevant chemical clusters, with sparser coverage of speculative chemical space. Each molecule is compared to the existing Novartis collection (plus any molecules selected from a higher priority category) and if there are no molecules with a similarity greater than the threshold, the molecule is selected, up to a limit of 10 analogues (the algorithm is essentially a sphere exclusion clustering method).

It is also common to design libraries for addition to a corporate collection. A recent example from Astra Zeneca[100] is illustrative of the approach. A set of spirocyclic diamines were identified as privileged templates, and a bis-azetidine core (Scheme 14.1), which is under represented in the AZ collection and patent literature, was selected.

Suitable reagents (available, not having reactive functionalities or those that would conflict with the proposed synthetic scheme) were identified and a virtual library of 100 299 products enumerated. A series of parallel filters were then applied, with the number of compounds failing each filter shown in brackets: clogP <1 or >4 (27 654); molecular weight >400 (30 094); "Consensus risk score" >6 (the degree to which a molecule is outside preferred ranges for hydrogen bond donors, acceptors and polar surface area. For example, >5 hydrogen donors yields a score of 3) (1470); substructure filters (2736); exact matches (0); molecules >0.8 Tanimoto similarity to a compound existing in the AZ collection (25 845); predictive QSAR models for solubility (5654), hERG (21 254), genotox (938) and reactive metabolites (438). In all, 12 837 compounds passed all filters. The remaining compounds were clustered using both 2D structural and pharmacophoric descriptors. A set of molecules that covered a high proportion of clusters in both descriptor spaces whilst at the same time required no more than 50 reagents for each R-group was selected. At this point, a common technique to apply would be a stochastic selection algorithm, such as MOGA or other Pareto based optimiser. As in this case the problem was sufficiently simple, a stepwise procedure was followed, whereby an initial selection of compounds maximised cluster coverage, whilst a subsequent step added compounds selected, in part, by the reagents needed to synthesise

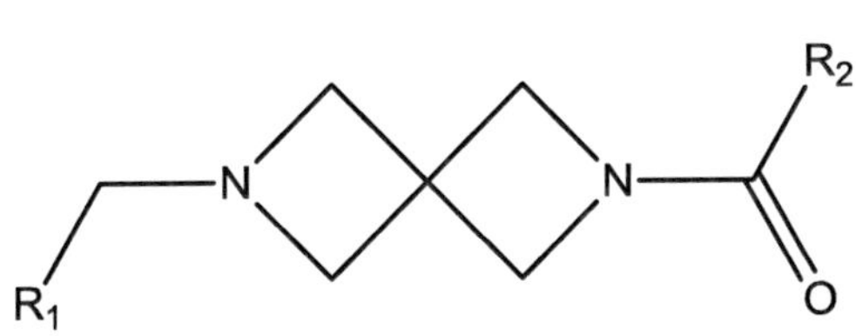

Scheme 14.1 The bis-azetidine core of the library designed by Stocks *et al.*[100]

them. The final design consisted of 50 × R1 and 47 × R2 (2350 compounds), which covers 62% of the structural clusters and 81% of the pharmacophoric clusters. Although pragmatic, this process yields a single solution, whereas a Pareto would yield multiple solutions. All have an advantage in at least one of the design criteria. As a simple example, Figure 14.7 illustrates 3 designs of a

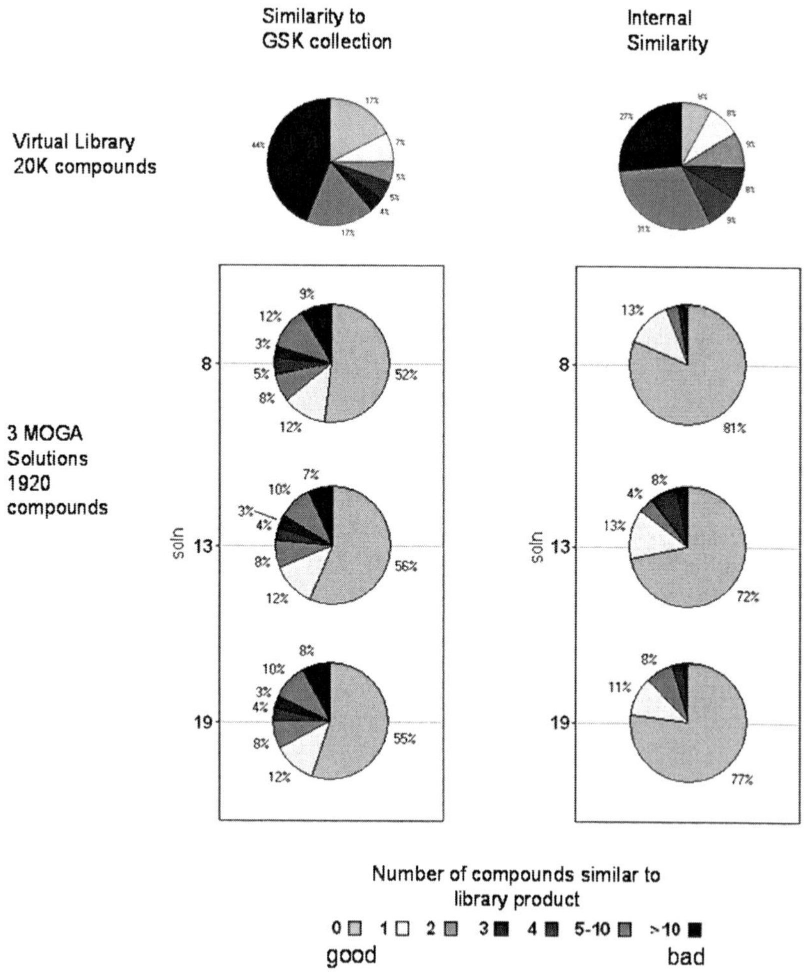

Figure 14.7 An illustration of library optimisation and selection using a multi-objective Pareto algorithm. The virtual library of 20 000 compounds has a large number of compounds similar to existing molecules in the GSK collection, and the library is not structurally diverse, containing many close analogues. Selections of a 1920 combinatorial subset yield multiple solutions that have better complementarity to the GSK collection and are not made up of large numbers of close analogues. However, each library has certain advantages, as they are from the non-dominated set of solutions.

1920 member library selected from a 20 000 virtual library. The two parameters displayed are internal diversity (expressed as the number of close analogues for each molecule in the library) and diversity with respect to the existing GSK screening collection. Each solution has pros and a con, for example solution 8 is the most internally diverse library but is the most similar to the existing GSK collection. Addition of other parameters brings further tradeoffs and choices for the chemist.

14.8 Screening Library Design

The techniques described enable the design of diverse or focussed sets of compounds which can fulfil multiple additional design goals, such as lead likeness and complementarily to existing compounds. These are what could be called the tactical requirements of a design strategy. The strategic requirements are more abstract and more difficult to answer. How many compounds does a screening collection need to contain? What should be the balance between diversity driven design and focussed design? Is the collection diverse enough? As eloquently stated by Roth,[21] such questions are meaningless without context. That context always has to answer the question, "what is the screening collection meant to deliver?" How many starting points, of what type and against what portfolio of targets? In addition, there will be other questions: what screening and compound management capacity is there? Which assay technologies will be used? Will samples be pooled? Do you want SAR to come from initial screening, or will there be iterative cycles of screening?

In the early days of HTS and combichem, such strategic questions were not considered, the focus being on getting the technology to work. It is a sign that the field has matured that there have now been publications addressing these strategic questions.[22,57,58] Mathematical models can be constructed, building on the quantitation of diversity and evaluation of biological relevance described in this chapter. In the case of Harper *et al.*,[22] this is combined with empirical parameterisation based on actual screening results. The model is relatively simple, and makes some straightforward assumptions:

1. Compounds can be clustered into groups of similar structures.
2. Each cluster has a probability Π_i of containing a lead molecule should it be screened against a particular protein, i.
3. If there is a lead molecule to be found in the cluster, each of the other compounds in the cluster has a probability α_i of also exhibiting the biological activity.

In this model, should we understand completely the relationship between compounds and targets, there would be a unique value for Π_i for every cluster and target pair. In reality, we know very little, and the model can use default value of Π_i taken from analysis of HTS campaigns. It is possible (as discussed later) to assign this on the basis of target families where there seems to be a

(a)

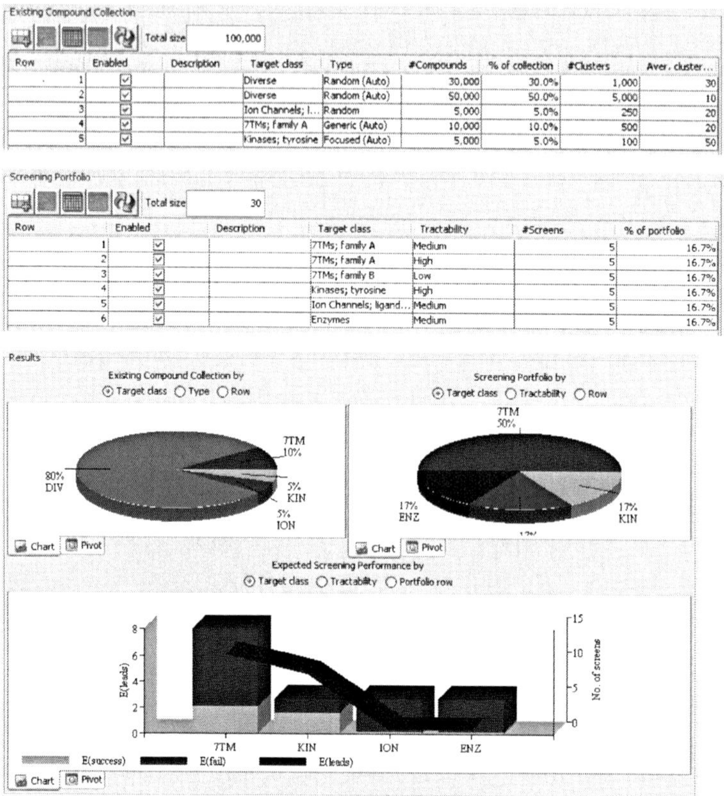
Existing Compound Collection
Total size 100,000
Screening Portfolio
Total size 30
Results
Existing Compound Collection by
Target class Type Row
Screening Portfolio by
Target class Tractability Row
80% DIV
7TM 10%
5% KIN
5% ION
7TM 50%
17% ENZ
17% KIN
Expected Screening Performance by
Target class Tractability Portfolio row
E(leads)
No. of screens
7TM KIN ION ENZ
E(success) E(fail) E(leads)
Chart Pivot

(b)

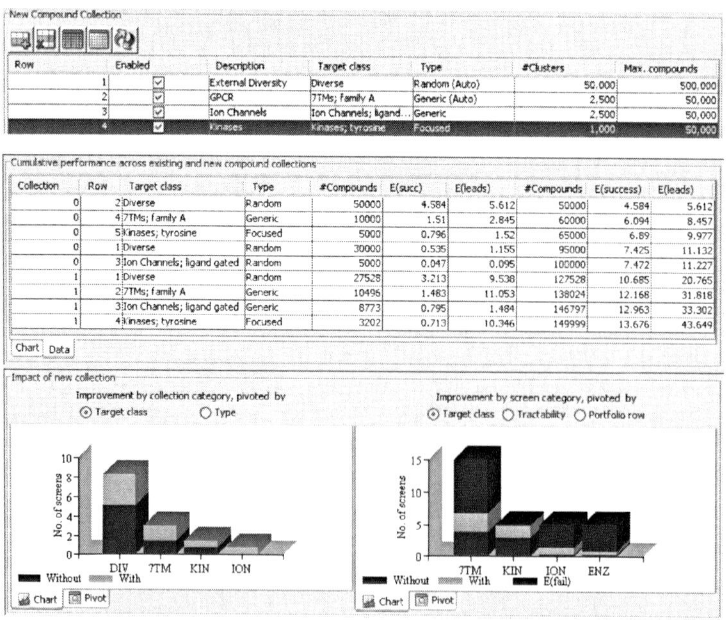
New Compound Collection
Cumulative performance across existing and new compound collections
Chart Data
Impact of new collection
Improvement by collection category, pivoted by
Target class Type
Improvement by screen category, pivoted by
Target class Tractability Portfolio row
No. of screens
DIV 7TM KIN ION
Without With
7TM KIN ION ENZ
Without With E(fail)
Chart Pivot

strong relationship for certain chemotypes across the class, for example kinases. The value of α can be used where there is information. As previously described, for the sphere exclusion/2D fingerprint method of Martin *et al.*,[30] the value is 0.3 and will be used in the example shown in Figure 14.8.

With such a model, it is possible to address strategic questions, for example it follows that a breadth first sampling strategy is always more efficient in that it is more valuable to sample a new cluster than to revisit partially occupied ones. In practice at GSK, this theoretical result is tempered by a preference to not miss a chemotype, given the effort taken to acquire it, and the imperfect match of chemical clustering to biological response. Therefore the α value is used to determine the probability of missing the cluster (see Table 1), and this leads us to select 10 compounds per cluster. With a different clustering technique, with a corresponding difference in α, the number per cluster would be different (for example, 50–100 in the case of Nilakantan *et al.*[57]).

The model lends itself to discussion of "what if" scenarios, and to this end a simulation tool[101] has been produced which enables managers to probe the relationship between current or projected target portfolios and chemistry design streams (Figure 14.8).

First, the current compound collection is annotated by diversity or focussed designs (if focussed, the protein family is indicated), the number of clusters and average cluster size. For this example, it is assumed the company has 30 000 legacy compounds from their projects, 50 000 purchased compounds with only 10 analogues per cluster, and some focussed libraries for their main target families of interest. A target portfolio is created, in this case 30 targets from well known target families (7TM receptors, kinases and ion channels) plus some unclassified enzymes. For each target, the target family and estimated tractability (high, medium or low) is indicated. The model is able to use this information to predict the likely outcome of screening the current collection

Figure 14.8 (a) The simulation tool is populated with an existing compound collection, which is characterised by the number of compounds, the number of clusters represented and the degree of target family design that has been used. A target portfolio is added, annotated by target family and perceived tractability. The simulator can predict the likelihood of success across the portfolio as a whole. This particular compound collection is not predicted to provide very many leads- just 7, and only three targets (2 7TMs, 1 kinase) are predicted to find a lead. (b) The simulator is used to aid decision making as to which type of molecules should be included in a planned addition of 50 000 extra compounds to the screening collection. The potential number of compounds for each type is entered, in this case the possible approaches are targeted libraries at Ion Channels, Kinases and GPCRs, and there is also the possibility of acquiring diverse compounds from external companies. These potential molecules and the existing compound collection are matched to the portfolio of targets, and a subset of the new molecules is chosen to optimise the effect they are predicted to have on additional leads for the target portfolio. This information is displayed in tabular and graphical form, with E(leads) being the expected number of lead series discovered, and E(succ) being the number of targets (or screens) expected to find at least one lead.

against the target portfolio. It predicts only about 7–8 of the targets will find a lead, mostly from the more tractable kinase and 7TM targets, with most targets finding only 1 lead series. The company decides to invest in the screening collection in an attempt to generate more leads. The possible new compound acquisitions are entered, with a choice between purchased diversity compounds or focussed libraries. Alongside this information is the number of molecules that could be added (limited by budget or resources); for this example the number has been set to 50 000. The simulator is then able to determine the optimal combination of chemistry investments to optimise either the number of new leads resulting, or the number of targets predicted to find a hit. The result indicates that the balance of investment should be approx 50–50 diversity–focussed libraries, with significantly more investment in 7TM and ion channel libraries at the expense of new kinase libraries. With this enhanced collection, the model predicts 13–14 targets from the portfolio to find leads, with many of them finding multiple lead series. Clearly, this simulation operates with a fair degree of error, however it does provide qualitative guidance and allows interesting thought experiments. For example, what is the effect of finding a more biologically relevant clustering *i.e.* having a high α value? Crushingly, for those of us who have spent years looking for better chemical descriptors, the model shows very little change in strategy with an increase in α from 0.3, *i.e.* it does not appear to be that important. Having a low alpha (*e.g.* 0.001) is not good in that the expected increase in success when adding new compounds is moderate as the chemical diversity selection has little relation to biological activity. Interestingly, for focussed strategies, if we increase the Π_i value (*i.e.* the design strategies get better) then the model indicates that less focussed molecules are needed, and investment in diversity should increase!

Finally, the model is able to estimate how many clusters are needed to achieve success across a portfolio of targets. It is interesting to see that while HTS success rates are commonly quoted at 50–60% from screening decks of 1–2 million,[44] the model predicts that to increase this to 80% and beyond would need well designed collections of 4–6 million, well outside the capacity of many companies but also well below the theoretical value of 24 million extrapolated from the QSCD methodology.

14.9 Other Considerations

A significant portion of this chapter has been dedicated to addressing design and selection of molecules. There are other considerations that are often overlaid with these criteria. How compounds are sourced can have an effect. For example, there are many compounds available for purchase at reasonable prices but on a non-exclusive basis. The advantages are accessibility, cost, ease of resupply and that often there will be analogues available to probe SAR of any hits. On the downside, an IP strategy will be required early on to mitigate possible competitor activity. There are novel synthetic approaches reaching maturity, such as DNA-encoded libraries[102] and diversity oriented synthesis

(DOS)[103] which offer convenient routes to much larger numbers of compounds. If there is one thing that should be learnt from the early days of combinatorial chemistry, it is that chemists should always judge the quality of these libraries on the chemical structures and biological data produced, not on the elegance of the synthetic chemistry, nor on the unsubstantiated claims of their proponents.

14.10 Discussion

For those seeking to build a screening collection, there exists a vast body of literature to support diversity, protein family and highly focussed approaches. All have strengths and weaknesses.

14.10.1 Diversity

There exist techniques for diversity selection that have been validated by screening data and where the chemical diversity or similarity metric has been shown to have some relationship to biological response. Progress has been made on the construction of substructure filters to remove undesirable functional groups and assay interferers, and concepts such as lead likeness have led to an improvement in the physicochemical profile of compound libraries. However, the filter sets are still imperfect; it is far more difficult to define what would make a high quality lead, or excite a medicinal chemist, than to describe what a poor lead is. For chemical descriptors, for all the efforts of researchers in the past two decades, the standard continues to be a 2D fingerprint, which is proving a very tough benchmark to beat.

14.10.2 Target Families

There are approaches that are able to use sequence, protein structure and privileged structures to generate leads for related targets. The science and thought process can be rather elegant. However, in contrast to HTS, there has not been conclusive data published to show that these methods are able to work consistently. In particular, not enough has been published to show that this model is able to be extended beyond kinases and some classes of 7TMRs, to a significant proportion of druggable targets. There is also the question of cost effectiveness of these more bespoke techniques *versus* the highly automated and efficient diversity practices. Therefore, controlled experiments and wider studies are needed alongside the usual desire for more predictive computational models.

14.10.3 Focussed Screening

Knowledge based selection of compounds for screening is widely validated and is a routine activity in computational groups. The methods are robust and fit for purpose. However, given the knowledge requirements, this approach is of most use in building a compound collection for companies focussed solely on sets of highly related targets. For many companies, these techniques will be an add-on to other lead identification processes, an efficient way to lead hop from literature data, or a technique to employ where it is difficult to develop a high throughput assay.

Overall, compound collections built from a combination of diversity and knowledge based strategies are capable of delivering lead molecules for the majority of targets screened, a great improvement from the early 1990s. However, there remain particular classes of target that are problematic, for example protein–protein interactions. This class of target should not be discussed as a homogenous group as some are clearly tractable. However, there will be some targets for which it is thermodynamically impossible to find a small molecule inhibitor or activator. As we are unable, with the current technologies, to screen a large enough, diverse enough collection of molecules against these targets, we cannot say for certain which targets are truly intractable and which targets have simply not yet met the right molecule.

14.11 Conclusion

The topic of chemical library design, in the singular or collective, has made much progress over the past two decades. Methods have been developed, validated and many incorporated into public domain or commercial software. The chemistry world has looked to other disciplines and adapted similarity coefficients, clustering, machine learning and optimisation methods. With the collective armoury of fragment based discovery, HTS, chemogenomics and now DNA-based libraries, we are well set to discover good quality starting points for the unprecedented targets revealed to us by genetics, epigenetics and systems biology. There is still much of how chemical diversity relates to biological activity that we do not understand, and will never fully understand. It is important that we can accept the imperfections in our theories and adapt our views accordingly. As we move into an era when we screen not with hope but with expectation that we find a small molecule ligand for a target, we should perhaps consider that we understand diversity well enough for our purpose. What we seem to know little about is what makes a safe, efficacious drug candidate, and that is the challenge for the next decade.

References

1. S. W. Djuric, I. Akritopoulou-Zante, P. B. Cox and S. Galasinski, *Annu. Rep. Med. Chem.*, 2010, **45**, 409.
2. M. M. Hann and R. H. Green, *Curr. Opin. Chem. Biol.*, 1999, **3**, 379.
3. P. Willett, *Curr. Opin. Biotechnol.*, 2000, **11**, 85.
4. M. S. Lajiness, *Perspect. Drug Discovery Des.*, 1997, **7**, 65.
5. P. Willett, V. Winterman and D. Bawden, *J. Chem. Inf. Comput. Sci.*, 1986, **26**, 109.

6. E. M. Gordon, M. A. Gallop and D. V. Patel, *Acc. Chem. Res.*, 1996, **29**, 144.
7. V. J. Gillet, P. Willet and J. Bradshaw, *J. Chem. Inf. Comput. Sci.*, 1997, **37**, 731.
8. R. S. Pearlman and K.M. Smith, *Perspect. Drug Discovery Des.*, 1998, **9**, 339.
9. E. K. Davis, in *Molecular Diversity and Combinatorial Chemistry: Libraries and Drug Discovery*, ed. I. M. Chaiken and K. D. Janda, American Chemical Society, Washington DC, 1996, p. 309.
10. Y. C. Martin, *J. Comb. Chem.*, 2001, **3**, 231.
11. M. Randic, in *Concepts and Applications of Molecular Similarity*, ed. M. A. Johnson and G. M. Maggiora, Wiley, New York, 1990, p 77.
12. C. A. Lipinski and F. Lombardo, *Adv. Drug Delivery Rev.*, 1997, **23**, 3.
13. C. A. Nicolaou, N. Brown and C. S. Pattichis, *Curr. Opin. Drug Discovery Dev.*, 2007, **10**, 316.
14. N. Nikolova and J. Jaworska, *QSAR Comb. Sci.*, 2003, **22**, 1006.
15. D. V. S. Green, *Expert. Opin. Drug Discovery*, 2008, **3**, 1011.
16. C. McInnes, *Curr. Opin. Chem. Biol.*, 2007, **11**, 494.
17. G. B. McGaughey, R. P. Sheridan, C. I. Bayly, J. C. Culberson, C. Kreatsoulas, S. Lindsley, V. Maiorov, J.-F Truchon and W. D. Cornell, *J. Chem. Inf. Model.*, 2007, **47**, 1504.
18. A. Nicholls, *J. Comput. Aided. Mol. Des.*, 2008, **22**, 239.
19. M. J. Stoermer, *Med. Chem.*, 2006, **2**, 89–112.
20. D. Rognan, *B. J. Pharmacol.*, 2007, **152**, 38–52.
21. H.-J. Roth, *Curr. Opin. Chem. Biol.*, 2005, **9**, 293.
22. G. Harper, S. D. Pickett and D. V. S. Green, *Comb. Chem. High Throughput Screening*, 2004, **7**, 63.
23. P. Willett, J. M. Barnard and G. Downs, *J. Chem. Inf. Comput. Sci.*, 1998, **38**, 938–996.
24. G. W. Adamson and J. A. Bush., *Inform. Stor. Retr.*, 1973, **9**, 561.
25. G. W. Adamson and J. A. Bush., *J. Chem. Inf. Comput. Sci.*, 1975, **15**, 1.
26. R. E. Carhart, D. H. Smith and R. Venkataraghavan, *J. Chem. Inf. Comput. Sci.*, 1985, **25**, 65.
27. Daylight Information Systems, Santa Fe, NM, USA.
28. MACCS, originally developed by MDL information systems, now part of Accelrys, San Deigo, California.
29. P. Willett and V. Winterman, *Quant. Struct.-Act. Relat.*, 1986, **5**, 18.
30. Y. C. Martin, J. L. Kofron and L. M. Traphagen, *J. Med. Chem.*, 2002, **45**, 4350.
31. R. P. Sheridan and S.K. Kearsley, *Drug Discovery Today*, 2002, **7**, 903.
32. P. Gund, *Prog. Mol. Subcell. Biol.* ,1977, **5**, 117.
33. S. D. Pickett, J. S. Mason and I. M. McLay, *J. Chem. Inf. Comput. Sci.*,1996, **36**, 1214.
34. A. Nicholls, G. B. McGaughey, R. P. Sheridan, A. C. Good, G. Warren, M. Mathieu, S. W. Muchmore, S. P. Brown, J. A. Grant, J. A. Haigh, N. Nevins, A. N. Jain and B. Kelley, *J. Med. Chem.*, 2010, **53**, 3862.
35. T. Cheeseright, M. Mackey, S. Rose and A. Vinter, *Expert Opin. Drug Discovery*, 2007, **2**, 131.

36. J. S. Duca and A. J. Hopfinger, *J. Chem. Inf. Comput. Sci.*, 2001, **41**, 1367.
37. G. Scheider, W. Neidhart, T. Giller and G. Schmid, *Angew. Chem. Int. Ed.*, 1999, **38**, 2894.
38. G. Harper, G. S. Bravi, S. D. Pickett, J. Hussain and D. V. S.Green, *J. Chem. Inf. Comput. Sci.*, 2004, **44**, 2145.
39. M. Rarey and J. S. Dixon, *J. Comput. Aided Mol. Des.*, 1998, **12**, 471.
40. U. Fechner, L. Franke, S. Renner, P. Schneider and G. Schneider, *J. Comput. Aided Mol. Des.*, 2003, **17**, 687.
41. G. Lauri and P. Bartlett, *J. Comput. Aided Mol. Des.*, 1994, **8**, 51.
42. R. S. Bohacek, C. McMartin and W. C. Guida, *Med. Res. Rev.*, 1996, **16**, 3.
43. T. Fink and J.-L. Reymond, *J. Chem. Inf. Model.*, 2007, **47**, 342.
44. R. Macarron, M. N. Banks, D. Bojanic, D. J. Burns, D. A. Cirovic, T. Garyantes, D. V. S. Green, R. P. Hertzberg, W. P. Janzen, J. W. Paslay, U. Schopfer and G. Sitta Sittampalam, *Nat. Rev. Drug Discovery*, 2011, **10**, 188.
45. M. Snarey, N. K. Terrett, P. Willett and D. J. Wilton, *J. Mol. Graphics Modell.*, 1997, **15**, 372.
46. G. M. Downs and J. M. Barnard, in *Reviews in Computational Chemistry*, ed. K. B. Lipkowitz and D. B. Boyd, Wiley-VCH, John Wiley and Sons Inc., Hoboken, New Jersey, USA, 2002, vol. 18, p 1.
47. E. J. Martin, J. M. Blaney, M. A. Siani, D. C. Spellmayer, A. K. Wong and W. H. Moos, *J. Med. Chem.*, 1995, **38**, 1431.
48. D. B. Turner, S. M. Tyrell and P. Willett, *J. Chem. Inf. Comput. Sci.*, 1997, **37**, 18.
49. M. Waldman, H. Li and M. Hassan, *J. Mol. Graphics Modell.*, 2000, **18**, 412.
50. R. Taylor, *J. Chem. Inf. Comput. Sci.*, 1995, **35**, 59.
51. A.-D. Gorse, *Curr. Top. Med. Chem.*, 2006, **6**, 3.
52. E. A. Winter and C. C. Moallemi, *J. Med. Chem.*, 2000, **43**, 1993.
53. T. L. Oprea and J. Gottfries, *J. Comb. Chem.*, 2001, **3**, 157.
54. K. T. Nguyen, L. C. Blum, R. van Deursen and J.-L. Reymond, *ChemMedChem.*, 2009, **4**, 1803.
55. D. E. Patterson, R. D. Cramer, A. M. Ferguson, R. D. Clark and L. E. Weinberger, *J. Med. Chem.*, 1996, **39**, 3049.
56. A. Schuffenhauer and N. Brown, *Drug Discovery Today*, 2006, **3**, 387.
57. R. Nilakantan, F. Immermann and K. Haraki, *Comb. Chem. High Throughput Screening*, 2002, **5**, 105.
58. M. J. Lipkin, A. P. Stevens, D. J. Livingstone and C. J. Harris, *Comb. Chem. High Throughput Screening*, 2008, **11**, 482.
59. M. Krier, G. Bret and D. Rognan, *J. Chem. Inf. Model.*, 2006, **46**, 512.
60. S. Renner, W. A. L. van Otterlo, M. Dominguez Seoane, S. Möcklinghoff, B. Hofmann, S. Wetzel, A. Schuffenhauer, P. Ertl, T. I. Oprea, D. Steinhilber, L. Brunsveld, D. Rauh and H. Waldmann, *Nat. Chem. Biol.*, 2009, **5**, 585.
61. M. A. Koch, A. Schuffenhauer, M. Scheck, S. Wetzel, M. Casaulta, A. Odermatt, P. Ertl and H. Waldmann, *Proc. Natl. Acad. Sci. U.S.A.*, 2005, **102**, 17272.

62. E. Jacoby and A. Mozzarelli, *Curr. Med. Chem.*, 2009, **16**, 4374.
63. S. J. Shuttleworth, R. V. Connors, J. Fu, J. Liu, M. E. Lizarzaburu, W. Qiu, R. Sharma, M. Wañska and A. J. Zhang, *Curr. Med. Chem.*, 2005, **12**, 1239.
64. R. Crossley, *Curr. Top. Med. Chem.*, 2004, **4**, 581.
65. V. Birault, C. J. Harris, J. Le, M. Lipkin, R. Nerella and A. Stevens, *Curr. Med. Chem.*, 2006, **13**, 1735.
66. J. F. Lowrie, R. K. Delisle, D. W. Hobbs and D. J. Diller, *Comb. Chem. High Throughput Screening*, 2004, **7**, 495.
67. A. M. Aronov, B. McClain, C. Stuver Moody and M. A. Murcko, *J. Med. Chem.*, 2008, **51**, 1214.
68. E. Gregori-Puigjane and J. Mestres, *Curr. Opin. Chem. Biol.*, 2008, **12**, 359.
69. D. M. Schnur, M. A. Hermsmeier and A. J. Tebben, *J. Med. Chem.*, 2006, **49**, 2000.
70. J. Hert, J. J. Irwin, C. Laggner, M. J. Keiser and B. K. Shoichet, *Nat. Chem. Biol.*, 2009, **5**, 479.
71. P. S. Charifson and W. P. Walters, *Mol. Diversity*, 2000, **5**, 185.
72. J. R. Huth, R. Mendoza, E. T. Olejniczak, R. W. Johnson, D. A. Cothron, Y. Liu, C. G. Lerner, J. Chen and P. J. Hajduk, *J. Am. Chem. Soc.*, 2005, **127**, 217.
73. M. Hann, B. Hudson, X. Lewell, R. Lifely, L. Miller and N. Ramsden, *J. Chem. Inf. Comput. Sci.*, 1999, **39**, 897.
74. J. B. Baell and G. A. Holloway, *J. Med. Chem.*, 2010, **53**, 2719.
75. M. Lajiness, G. Maggiora, V. Shanmugasundaram, *J. Med. Chem.*, 2004, **47**, 4891.
76. V. J. Gillet, P. Willett and J. Bradshaw, *J. Chem. Inf. Comput. Sci.*, 1998, **38**, 165.
77. J. Sadowski and H. Kubinyi, *J. Med. Chem.*, 1998, **41**, 3325.
78. A. Ajay, W. P. Walters and M. A. Murcko, *J. Med. Chem.* 1998, **41**, 3314.
79. O. Ursu and T. I. Oprea, *J. Chem. Inf. Model.*, 2010, **50**, 1387.
80. C. Hansch, A. Leo and D. Hoekman, in *Exploring QSAR: Hydrophobic, Electronic, and Steric Constants*, American Chemical Society, Washington, DC, 1995.
81. M. M. Hann, A. R. Leach and G. Harper, *J. Chem. Inf. Comput. Sci.*, 2001, **41**, 856.
82. T. I. Oprea, *J. Comp.-Aided. Mol. Des.*, 2002, **16**, 325.
83. M. Congreve, R. Carr, C. Murray and H. Johti, *Drug Discovery Today*, 2003, **8**, 876.
84. D. V. S. Green and S. D. Pickett, *Mini Rev. Med. Chem.*, 2004, **4**, 1067.
85. D. J. Newman, *J. Med. Chem.*, 2008, **51**, 2589.
86. K. Grabowski, K.-H. Baringhaus and G. Schneider, *Nat. Prod. Rep.*, 2008, **25**, 892.
87. M. Kaiser, S. Wetzel, K. Kumar and H. Waldmann, *Cell. Mol. Life Sci.*, 2008, **65**, 1186.

88. K. Kumar and H. Waldmann, *Angew. Chem. Int. Ed.*, 2009, **48**, 3224.
89. P. Ertl, S. Roggo and A. Schffenhauer, *J. Chem. Inf. Model.*, 2008, **48**, 68.
90. F. Lovering, J. Bikker and C. Humblet, *J. Med. Chem.*, 2009, **52**, 6752.
91. W. Zheng, S. J. Cho and A. Tropsha, *J. Chem. Inf. Comput. Sci.*, 1998, **38**, 251.
92. G. Bravi, D. V. S. Green, M. M. Hann and A. R. Leach, *J. Chem. Inf. Comput. Sci.*, 2000, **40**, 1441.
93. A. C. Good and R. A. Lewis, *J. Med. Chem.*, 1997, **40**, 3926.
94. R. D. Brown and Y. C. Martin, *J. Med. Chem.*, 1997, **40**, 2304.
95. J.-F. Truchon and C. I. Bayly, *J. Chem. Inf. Model.*, 2006, **46**, 1536.
96. C. M. Fonseca and P. Fleming, in *Genetic Algorithms: Proceedings of the Fifth International Conference*, ed. S. Forrest, Morgan Kaufmann, San Mateo, 1993, p. 416.
97. V. J. Gillet, W. Khatib, P. Willett, P. J. Fleming and D. V. S. Green, *J. Chem. Inf. Comput. Sci.*, 2002, **42**, 375.
98. T. Wright, V. J. Gillet, D. V. S. Green and S. D. Pickett, *J. Chem. Inf. Comput. Sci.*, 2003, **43**, 381.
99. A. Schuffenhauer, M. Popov, U. Schopfer, P. Acklin, J. Stanek and E. Jacoby, *Comb. Chem. High Throughput Screening*, 2004, **7**, 771.
100. M. J. Stocks, G. R. H. Wilden, G. Pairaudeau, M. W. D. Perry, J. Steele and J. P. Stonehouse, *ChemMedChem.*, 2009, **4**, 800.
101. D. Morley, Enspiral Discovery, 2005.
102. M. A. Clark, R. A. Acharya, C. C. Arico-Muendel, S. L. Belyanskaya, D. R. Benjamin, N. R. Carlson, P. A. Centrella, C. H. Chiu, S. P. Creaser, J. W. Cuozzo, C. P. Davie, Y. Ding, G. J. Franklin, K. D. Franzen, M. L. Gefter, S. P. Hale, N. J. V. Hansen, D. I. Israel, J. Jiang, M. J. Kavarana, M. S. Kelley, C. S. Kollmann, F. Li, K. Lind, S. Mataruse, P. F. Medeiros, J. A. Messer, P. Myers, H. O'Keefe, M. C. Oliff, C. E. Rise, A. L. Satz, S. R. Skinner, J. L. Svendsen, L. Tang, K. van Vloten, R. W. Wagner, G. Yao, B. Zhao and B. A. Morgan, *Nat. Chem. Biol.*, 2009, **9**, 647.
103. S. Dandapani and L. A. Marcaurelle, *Curr. Opin. Chem. Biol.*, 2010, **14**, 362.

CHAPTER 15

The Impact of Genomics, Systems Biology, and Bioinformatics on Drug and Target Discovery: Challenge and Opportunity

DARREN R. FLOWER

Life and Health Sciences, Aston University, Aston Triangle, Birmingham, United Kingdom, B5 7ET

15.1 Introduction

Since its inception, the nature of drug discovery has changed, and changed dramatically; and it continues to change today; and it is set to change, and change again and again, well into the future. Such dynamic and transformational behaviour is characteristic of so-called *high-tech* industries. In this particular context, and over the past few decades, genomics, and those technologies that exploit it or are built upon it, have had a dramatic, nay an unprecedented, impact on the process by which drug discovery is undertaken. That much is clear; that much we know. To what extent this has made the process or drug discovery more productive and more successful is very much open to question. This chapter endeavours to explore the impact of genomics upon drug discovery. It is very much a personal reflection, partial and incomplete, yet hopefully useful also.

RSC Drug Discovery Series No. 13
Drug Design Strategies: Quantitative Approaches
Edited by David J. Livingstone and Andrew M. Davis

Published by the Royal Society of Chemistry, www.rsc.org

Drug discovery is both a scientific discipline in its own right and a vital, even pre-eminent, component of the economic milieu. As an industry, the business of pharmaceutical development, manufacture, and marketing is prone to rapid and traumatic restructuring, typically resulting in significant job losses promulgated as economies of scale. Much passes unnoticed by the public as the media is generally hostile to the perceived duplicity of the pharmaceutical industry and consequently oblivious to the fate of those with which it chooses to dispense. When a car factory or a coal mine closes, it is headline news. The closure of a pharmaceutical research and development site will be fortunate indeed even to gain the barest of mentions. As a science, drug design and development is again not seen in the best of lights, despite the millions, perhaps billions, whose lives are saved or immeasurably improved by the products of the pharmaceutical endeavour: drugs, vaccines, and sundry biologics.

Much as the world recently found itself in financial crisis, so has the pharmaceutical industry, but for rather different reasons. These include: spiralling costs and pricing pressures; frequent patent expiries, with over 30 small molecule drugs to come off-patent in the next five years; changes in regulation and regulatory structure; and the threat from generic medicines, revenues for which are expected to exceed $US84 billion by the end of 2010. Couple this to dwindling corporate drug pipelines, resulting from the failure to exploit new technology and new targets emerging from post-genomic science. All have worked together to thwart the global pharmaceutical endeavour.

There have been many emergent solutions to this dilemma. One seeks to blame the business model under which the pharmaceutical industry once thrived. The so-called "soup to nuts" model, predicated upon the notion that we can, within one single, monolithic organisation, concentrate and sequestrate the whole panoply of drug discovery activities. This has resulted in the creation of many academic centres focussing on drug discovery, with its need for significant inward investment from philanthropists, business, and public funding. Yet governments are happy to spend trillions of dollars on funding standing armies, creating weapons of mass destruction and the like. Meanwhile, comparatively, they spend very little funding the discovery and development of the drugs and vaccines which could save countless lives, and thus stabilise whole fragmenting societies.

Another solution blames the imposition of inappropriate business imperatives onto the scientific process of drug identification and optimisation. This has led some to seek to replace "action" for "thought", substituting the blunt instrument of indiscriminate high-throughput approaches for thorough medicinal chemistry, supported, but never supplanted, by careful computational informatics and modelling. The pharmaceutical industry spends untold millions of person-hours and well-audited billions of dollars in the quest to systematise drug discovery. For example, during the year 2003, the top 10 pharmaceutical companies spent an unprecedented $36 billion on research & development (R&D). Thus while one would be unwise to completely disregard the importance of serendipity in drug design and discovery, the current goal of drug discovery is, through the application of advanced technology and the adoption of good practice, the elimination of luck and its replacement with guaranteed success. So, receptor-orientated, mechanism-driven research has replaced targetless pathologies as the primary focus of the discovery process. Sophisticated medium and high-throughput synthesis and *in vitro* screening technologies have largely displaced individual "hand-crafted" one-off assays. Part of this process has centred on exploiting genomics and the post-genomic sciences developing and devolving from it.

15.1.1 Genomics and its Inheritors

Genomics, as a word, derives from the German *Genom,* which combines *Gen* from gene with *-om* from *Chromosom*, the German form of chromosome. Originally, the term "genome" meant the complete haploid set of chromosomes of an organism, but later took on other meanings; it has come to mean the the genetic material of an organism; that is, complete set of protein-encoding genes, functionless pseudo-genes, regulatory sequences, and all noncoding DNA segments, of an organism or organelle. The word genome now symbolises the DNA sequence, or rather sequences, encoding an organism; as such it is the essence of self and is synonymous with DNA, at least in the sense that the term is used in the vernacular. Dictionaries trace the use of the word genome to at least 1926, but it may have been in use for some time prior to that.

In the 1970s, the group led by Walter Fiers at the University of Ghent determined the first gene sequence: Bacteriophage MS2 coat protein.[1–2] The first DNA-based genome to be sequenced in its entirety was that of bacteriophage Φ-X174 (5,368 bp), sequenced by Frederick Sanger's group in 1977.[3] The first fully-sequenced independent organism was *Haemophilus influenzae* (1.8 Mb) in 1995.[4] As of today, 1000s of viruses, hundreds of bacterial species, and dozens of eukaryotes have been sequenced. Meta-genomics may yet raise these numbers for microorganisms by some orders of magnitude. Just trying to track such increases seems ever more futile, so quickly do new genomes appear. In just a few years, a genome sequence has gone from a true achievement capable of stopping the scientific world in its tracks to becoming mundane and workmanlike. Eventually, genomic sequencing may take its place as a workaday laboratory exercise akin to running a western blot. Soon it may become the subject of a single postgraduate student's thesis. A little later, and an undergraduate might need to sequence ten or twenty genomes just to finish his or her final year project.

Despite the gargantuan investment of time and resources in genomic science across the world over the last thirty years, we are yet to give definitive and unequivocal answers to some very simple questions. For example, how many human genes are there? This should be the most straightforward of questions, but, like so many aspects of biology, it certainly is not.[5] The putative, predicted, and promulgated size of the human genome has been deflating for years, decreasing first from early and inaccurate figures in excess of 100 000 to about 40 000 genes, and then to figures close to 20 000. A more trustworthy estimate from 2006 put human protein coding genes in the region of 25 043,[6] while a

2007 estimate places the value at about 20 488, with hundreds of genes yet to be properly pushed, filed, stamped, indexed, briefed, debriefed, and numbered.[7]

In what follows, the part shall sometimes stand for the whole, for our use of the term genomics as a catch-all will subsume many associated disciplines and terms; to do otherwise will render a periphrastic and sesquipedalian disquisition even more circumlocutory than it might otherwise be. From genomics devolves a whole slew of high-throughput characterisations of the functioning organism, of which the best known and most familiar are the transcriptome and the proteome. The principal underlying idea is that of undertaking work in parallel, addressing questions not individually through single, yet complex and convoluted experiments, but rather *en masse* through sophisticated procedures examining not isolated biological phenomena but rather thousands, or even tens of thousands, of such phenomena.

15.1.2 An Explosion of -OMEs

In the last decade or two, riding on the back of genomics' success, there has been an explosion in the elaboration of -omes, and the concomitant development of so-called -omic technologies focussing upon them. This mirrors and echoes the use and re-use of the suffix -ology in past decades: psychology, sociology, and the like bred dozens of similar uses, both serious and facetious. Within this mass of harmonious and conflicting definitions of different -omes and -omic sciences, the website http://www.genomicglossaries.com/content/omes.asp lists innumerable examples.

Perhaps the most pertinent examples are these: the transcriptome and transcriptomics (high-throughput techniques to measure mRNA), the proteome and proteomics (high-throughput techniques to measure actual express protein), the metabolome and metabolomics (similar techniques for primary and secondary metabolites), the peptidome and peptidomics (high-throughput analysis of the dynamic complement of peptides present in the cell generated by germline encoding and proteolysis), glycome and glycomics (similar techniques for the dynamic complement of carbohydrate and glyocproteins), and the interactome and interactomics (the analysis of protein–protein interactions).

The use of -ome and -omics long ago spread beyond the strictly biological. Such an example is petroleomics, the study of all the components of petroleum and how they affect the properties of crude oil. Consideration of this -ome is particularly instructive. Proteins, like components of petroleum, were originally classified using solubility: albumins (soluble in water and dilute ionic solution), globulins (insoluble or sparingly soluble in distilled water), prolamins (insoluble in water but soluble in aqueous ethanol), glutelins (soluble only in dilute acid or base solution), and scleroproteins (insoluble in most solvents), amongst others. Today, proteins are classified using their sequences and structures. Petroleomics seeks to emulate our burgeoning understanding of the thousands of components of a biological cell by attempting to catalogue the thousands of compounds contributing to the chemical and physical properties of crude oil. Such analogies should not be pushed too far, however. As an amorphous solid, glass is often compared to biological tissue, but the random and disordered nature of glass in no way corresponds to the exquisite self-organisation seen in the highly compartmentalised eukaryotic cell.

If we were so minded, we could add another, all-embracing "-ome" to this increasingly heady mix: the "infome"; this idea seeks to escape the narrow confines of sequence or structure data, being the sum of all biological and chemical knowledge. It is a goal which challenges the growth of knowledge management as it seeks to treat this huge, heterogeneous volume of data. The pharmaceutical company is one of the few organisations that can, within molecular bioscience, anticipate, through its immanent scale, to pursue such an objective.

15.2 Drug Discovery Technology and Genomics

Modern bioscience, especially pharmaceutical drug discovery, is as much technology-driven as it is hypothesis-led. Biologists have traditionally viewed hypothesis-driven research as the only mediator of scientific veracity. Inductive or data-driven approaches have long been deprecated as a consequence, viewed as marginal or irrelevant. The development of technology is again viewed as irrelevant in this context. The two are increasingly seen as complementary by those with the profundity of insight to grasp these self-evident truths. New tools and new technologies are constantly emerging from the fecund minds of scientists and engineers, and they are clearly essential in advancing the biology, chemistry, and associated science, all of which are to a greater or lesser extent now increasingly data-rich and multi-disciplinary. Since many fields are hypothesis-poor but data-rich, computational data analysis may generate new, viable, testable hypotheses.

There has been a synergistic coincidence operating in pharmaceutical research since the inception of large-scale, integrated genome sequencing in the late 1980s. This has been the emergence of high-throughput techniques that simultaneously mirror and respond to the needs of genomic and post-genomic science. We all know about high-throughput screening, which uses robots and computer control to evaluate potentially enormous numbers of small molecules as potential drugs through the indiscriminate screening of extant compound collections and deliberately synthesised compound libraries.

Combinatorial libraries were meant to deliver not just hits but leads, even, as a senior manager once told me, drugs ready for clinical trials. Unfortunately, as time progressed, it became increasingly clear that across company after company, screen after screen, little or nothing was actually coming out of these exercises. Enthusiasm ebbed, as it usually does when initial expectations are proved to be nothing more than hyperbole. Hit rates were low, much lower than screening historical in-house collections. Those hits that were found were not particularly good, nor particularly clean, particularly when tested as mixtures. High-throughput screening is not perfect. For many assays and screening systems, signal-to-noise and reproducibility issues remain pertinent even today, and they are compounded by problems of compound storage and stability. Yet these techniques remain entrenched. Computational techniques, such as virtual screening, have also been developed as analogous approaches, and most informed observers view the experimental and computational as convergent and complementary techniques, able to partially compensate for the deficiencies of the other. In what follows, we elaborate on how the potential of genomics has acted as a spur to the development of other technologies relevant to drug discovery.

15.2.1 Target Discovery in the Genomic Era

In sum, the genome of an organism is the sequence of the DNA that encodes it. Part of a genome will code for genes; these genes will in turn make mRNA, which will make proteins. These proteins will undertake most, but not all, duties within the cell. Part of the genome regulates the expression of DNA as mRNA and proteins, rather than coding for them directly. Life, and the biochemistry of which it is ultimately comprised, is built from the interactions of proteins, and the study of protein–protein interactions is fast becoming a central feature of molecular bioscience. Life—at least life as we understand it[8]—is an emergent property, arising from its underlying biochemistry, and this emergence occurs at many levels; life is thus a multi-scale phenomena. Here, the term scale takes on connotations of both time and physical extension; the processes that ultimately amount to life exist upon many time scales (femtosecond bond str*etching* to nanosecond helix to coil transitions to microsecond protein folding events) and many length scales (molecular, nanonscale, mesoscale, and upwards).

Life exhibits a hierarchical cascade of levels, rising from the world of atoms and molecules through that of cells, tissues, and organs to that of whole organisms. Beyond discrete organisms there are distributed co-operative structures such as communities and ecosystems, which again arise at many levels, from the quorum sensing of bacteria up through the rain forest to encompass the whole biosphere, which itself extends from the edge of space to the deep sea and downward still to those subterranean bacterial colonies miles beneath the surface of the earth. At each such scale, we observe biology displaying emergent complexity, so that we cannot with any ease or surety infer the behaviour of one level from the level or levels from which it is built.

Attempting to address this complexity is the emergent meta-discipline of systems biology. Rather than deal with biochemistry *reductio ad absurdum*, comprising simple mixtures of albeit large numbers of components, systems biology instead views biochemical, and thus by extension biological, systems as that: complex, interacting, and integrated systems that can only be properly understood by drawing together understanding at many levels of analysis. We shall have more to say on this in the following pages.

Looking backwards rather than forwards, genomics has had what many would see as the diametrically opposite effect: reducing the discovery of drugs to the discovery of compounds active at a particular receptor. Driven by the genomic discovery of receptors, many perceive such a focus as inappropriate.

Clearly, ignoring the ability—or rather the prevalent inability—of many candidate drugs to reach their site of action has indeed been an observation worth making. Yet we need to balance the various and important demands placed upon a drug molecule and not place one in the ascendant for personal or subjective reasons. By deprecating the receptor in the light of absorption, distribution, metabolism, and excretion (ADME) and rule-of-five arguments, is very much to jettison the baby with the bath water.

And so the hunt for novel and validated molecular targets—be they receptors or enzymes—within key species (humans, livestock, or microbial pathogens) remains a key aspect of drug discovery at its broadest. At the heart of target discovery lies the ability to accurately identify genes; this remains a problem unsolved. When dealing with entire genome sequences, the need for software tools, able to automate the laborious process of scanning millions upon millions of base pairs, is essential. Databases of nucleic acid and protein sequences and structures have now become available on an unparalleled, multi-genomic scale. To capitalise on this, attention has focused on the ability of such databases to accurately compare active sites across a range of related proteins, allowing us to select and validate biological targets, to control drug selectivity and verify biological hypotheses more precisely. When we move from the genome to the proteome, gene-finding becomes protein-finding; with the prevalence of post-translational processing of all kinds the task becomes ten to one hundred times harder.

However, and with reference to the need to balance off-target with on-target activities (*i.e.* ADME versus potency and selectivity), there has been burgeoning interest in the identification of intrinsically "druggable" targets. So what exactly is a druggable receptor? This is dependent on the drug of interest; the properties required of a short-acting drug are very different to those of long-acting, orally-bioavailable medicine. The G-Protein coupled receptor (GPCR) is an archetypal druggable target, with its small, hydrophobic, internal binding site and crucial physiological roles. By druggable, we mean proteins exhibiting a hydrophobic binding site of defined proportions, leading to the development of drugs of the right size and appropriate physico-chemical properties.

The term druggable relates to both the receptor structure and also to the provenence of a protein family as a source of successful drug targets. Estimates put the number of druggable receptors somewhere in the region of 2000 to 4000.[9–10] Of these, about 10% have been extensively examined to date, leaving many, many receptors left to explore. Beyond the human genome, there are other druggable receptors now receiving the attention of pharmaceutical companies. Bacteria, fungi, viruses, and parasites are all viable targets for drug intervention. As the number of antibiotic resistant pathogens increases, the hunt for new anti-microbial compounds, and thus the number of druggable microbial receptors, will also expand.

Set the task of discovering new, previously unknown druggable receptors, how would we go about it? In particular, how would we find a GPCR? The first step towards functional annotation of a new GPCR sequence usually involves searching a primary sequence database with pair-wise similarity tools. Such searches can reveal clear similarities between the query sequence and a set of other sequences in the database. An ideal result will show unequivocal similarity to a well-characterised protein over its whole length. However, an output will regularly reveal no true hits. The usual scenario falls somewhere between these extremes, producing a list of partial matches which will be to either uncharacterised proteins or have dubious annotations. The difficulty lies in the reliable inference of descent from a shared ancestor and thus extrapolation to a common biological function.

The increasing size of sequence databases increases the probability that relatively high-scoring, yet random, matches will be made. Even if a verifiable match is made, it is difficult for pair-wise similarity methods to distinguish paralogues from orthologues. Moreover, low-complexity matches can dominate search outputs. The multi-domain nature of proteins is also a problem; when matching to multi-domain proteins, it is not always clear which domain corresponds to the search query. Thus achieving trustworthy functional assignments remains a daunting problem. It has become established practice to combine single sequence search strategies to include searches of motif- or domain-databases in order to identify protein family databases. Since family discriminators can often detect weaker similarity, and can simultaneously usefully exploit the differences between sequences as well as their similarities, the integration of searching family databases into wider search protocols can lead to a more sensitive and more selective strategy than simply using global sequence searching alone.[11] Once found, the full power of informatics and high-throughput testing can be brought to bear on the validation of the target.

The search for new targets is neither stalled nor prorogued but continues apace. The hunt is on for new targets within newly indentified pathogenic microorganisms, as well as in already well-characterised organisms such as humans and clinically or geographically interesting human sub-populations, other eukaryotic organisms of economic interest, such as companion animals and farm livestock, and well-studied microbial pathogens for which drug and vaccine treatments are sub-optimal, such as the causative agents of TB, AIDS, or Malaria. Likewise, the combination of genomics and structural biology has provided, and continues to provide, great insight into not only primary drug targets but also the structural basis for the important specific interactions between drugs and genes involved directly in mediating Absorption Distribution Metabolism Excretion and Toxicology (ADMEt) properties. There are many such examples of proteins which bind or metabolise drugs in a highly specific, rather than a bulk or non-specific, manner, amendable to investigation by structural biology and manipulation by structure-based drug design. Spatially indiscriminate lipophilicity, and like characteristics, are not the sole properties determining ADMEt. Pertinent and well-understood examples of specific protein interactions include, but are in no way limited to, the following: human serum albumin or α-1-acid glycoprotein, both of which bind a wide variety of drugs in the serum; a tranche of metabolic enzymes, such as a host of cytochrome P450s; and a plethora of transmembrane transport proteins.

Focussing on the last of these classes of ADMEt proteins provides us with a prime example of the identification of new targets in the genomic era. Breast cancer resistance protein (BCRP), a 75 kDa polytopic plasma membrane protein that forms part of the G-subfamiliy of the ATP-binding cassette (or ABC) family of membrane transporters, is thus also known as ABCG2. It functions primarily as a xenobiotic transporter and appears to undertake many pivotal tasks in multi-drug resistance to chemotherapeutic drugs including topotecan, mitoxantrone, camptothecin, and methotrexate. BCRP also plays a key role in the enhanced excretion of xenobiotic compounds of many kinds, transporting compounds such as nitrofurantoin, prazosin, glyburide, and 2-amino-1-methyl-6-phenylimidazo-4,5–b-pyridine. In a landmark study, BCRP was identified within the multi-drug resistant human breast cancer cell line MCF-7/AdrVp—known not to express other multi-drug transporters, such as P-glycoprotein or multi-drug resistance protein 1—by using PCR-based differential-display RNA fingerprint to identify mRNA transcribes differentially over-expressed within MCF-7/AdrVp cells when compared against MCF-7 cells.[12]

Finally, and in sum, it would seem appropriate to also mention the increasingly widely-recognised phenomenon that many proteins have multiple functions[13–16] and that these can vary profoundly between cellular compartments, indicating that we can simultaneously seek out new functions for established targets as well as seeking to identify new therapeutic roles for established drugs, partly on that basis.

15.2.2 Transcriptomics and Microarrays

The transcriptome is the complement of messenger RNAs (mRNAs) transcribed from a genome. This is a dynamic set of molecules which are constantly changing with time in response to the conditions experienced by the cell. Transcriptomics is the scientific study of the transcriptome; it examines and explores mRNA expression levels. It was once thought that life could be understood by identifying each and every protein and then determining its function. It has since emerged that this was an over-simplified view of how cells behave. Genes have many promoters and their expression is tightly regulated by a complex and integrated system.

The microarray is the principal tool of transcriptomics. It allows for the analysis of gene expression and consists of a glass slide containing samples of 100s or 1000s of genes, organized in a regular and reticulated pattern of spots. Spots are either DNA, cDNA, or an oligonucleotide or oligo, which is a short sequence of single-stranded DNA, usually 5 to 50 nucleotides in length. The location of spots is used to identify each specific sequence. A microarray works by utilising an mRNA molecule's capacity to hybridise, or bind specifically, to its progenitor DNA template. By using a microarray coupled to sophisticated instrumental and computer analysis, we can determine the gene expression profile of 100s or 1000s of genes by quantifying the degree of bound mRNA. Microarrays can assess gene expression levels in a single sample or they can *inter alia* compare gene expression in different tissue or cell types; between cells treated in different ways, as mediated by drugs or microbial components or signalling molecules; or between healthy and diseased tissues.

Methods for analysing a single mRNA sample are known as single-channel, or one-colour, methods, while so-called two-channel, or two-colour, methods analyse two mRNA samples. In single-channel experiments, an mRNA sample is labelled with a single fluorescent dye and the hybridised array scanned for each dye. Single-channel methods are normally used for the more homogenous commercially available microarrays. In two-colour methods, the mRNA is labelled with two fluorescent dyes of different colours, usually green Cy3 and red Cy5. The two labelled samples bind competitively to an array, allowing the ratio between mRNA levels to be determined by scanning at two wavelengths corresponding to the two colours. Two-channel methods were developed for spotted DNA microarrays as bound dye varies significantly from spot to spot. They possess a particular issue with signal-to-noise. Comparing two samples in this way can give high precision yet comparison of two arrays significantly increases the error, since extra measured values are needed for calculation. To help avoid this and the need for additional microarray experiments, experimental design is used.

Processing, analysing, and thus establishing the veracity of such changes, is a science in itself. Any approach based on the analysis of microarrays will require a significant investment in their bioinformatic analysis; computational transcriptomics is a still-evolving sub-discipline of bioinformatics that has as much to do with rigorous statistics as it does with informatics analysis of biological data. At a high conceptual level, we can readily discern 3 principal means of tracking mRNA profiles to discover or elaborate candidate gene function.

First, mRNA expression levels measured under different experimental conditions are compared in order to establish those genes which are regulated similarly; these will likely be involved in mediating similar biological pathways, be that regulatory, signalling, or metabolic. This approach is known as the "guilt-by-association" method. Second, the identification of genes regulated differently under distinct conditions; such as in different cell types or when treated with different chemical or biological agents. Even genes not causally involved may still find uses as disease biomarkers.

The third approach is rather different. It uses tens of thousands of mRNA expression profiles as a collective descriptor of the behaviour of a biochemical system. For example, the profile generated by a cell treated with a drug is compared to that from an untreated cell, allowing us to describe the effect of the compound. Comparison of profiles generated in this way also allows for the identification or elucidation of compounds with similar mechanisms, or can act as a drug discovery tool in itself. When we are seeking drugs that act at the system level, interacting strongly with many receptors and more weakly with many other proteins, this procedure is a potentially powerful means of lead hopping or hit discovery.

The insatiable demand for minimal-cost sequencing technology, able to support the voracious and unquenchable needs of genomics, has lead to the development of so-called next generation sequencing: high-throughput approaches that parallelise sequencing, generating thousands or even millions of sequences simultaneously. Many believe that this fast, efficient, low-cost, direct method will ultimately put an end to microarray-based experiments. On the other hand, there are many microarray applications which extend the boundaries of microarray experiments beyond transcriptomics, for example: profiling DNA level rather than mRNA, including resequencing, detection of single nucleotide polymorphisms or single nucleotide polymorphisms (SNPs), identification of DNA methylation, and detection of specific chromosomal binding sites for proteins. Several of these have developed into large, important, and highly active areas in their own right. Moreover, the development of peptide, protein, and antibody arrays indicates how the underlying methodological approach can be adapted to other areas of scientific endeavour.

Most observed sequence variability seems to result from so-called single nucleotide polymorphisms (or SNPs), which are single base-pair mutations occurring at frequencies of about 1 per 1000 base pairs. Some SNPs will have dramatic effects on gene function, others none at all. Approximately 1 in 100 SNPs will alter protein coding regions within a gene, changing the encoded amino acid within the expressed protein. These are usually called coding SNPs. Again, such changes may have no apparent effect, or enhanced activity, or may completely inactivate protein function. Most changes will have an indirect and cumulative effect on protein function by modulating protein degradation. Some SNPs may generate stop codons, leading to the premature truncation at the protein translation stage. SNPs may also be present in the promoter and 5′-regulatory region of a gene and thus influence gene transcription by altering transcription factor binding. Other SNPs may alter or create splice variants, or alter untranslated regions within transcribed *mRNA*, thus affecting protein formation by modulating mRNA degradation rates.

The key thing of course is to link the occurrence, or rather widespread recurrence, that some would refer to as conservation, of SNPs and other variable genetic structures to altered phenotypes, principally drug–protein interactions. There are several ways of doing this, but the principal means has rapidly become the genome-wide association studies, or GWAS, and are supplanting genetic linkage disequilibrium analysis, which has been used traditionally to identify disease-causing genetic variation. More generally, disease-association studies, both hypothesis-driven candidate gene-based and hypothesis-generating GWAS approaches, have identified genetic variations associated with both Mendelian and more complex diseases, such as asthma or cancer. GWAS is proving of particular use in elucidating common yet discombobulating multi-factorial diseases, including asthma, diabetes, heart disease, cancer, and mental illnesses.

Chronic disease often has a significant genetic component and can result from both Mendelian and multi-factorial inheritance. Multi-factorial disease arises from variation in many different genes and is typically coupled to significant environmental contribution. Mendelian or monogenic disease arises from mutations in a single dominant copy of a gene or in both recessive copies. The search for genes underlying Mendelian disorders has been spectacularly successful, yet, until quite recently, the equivalent search for susceptibility genes linked to multi-factorial disease has yielded few examples. Today, the number of such linkages is beginning to escalate with considerable celerity, throwing light on prevalent diseases such as stroke,[17] as well as less common conditions. For example, Caronia *et al.* identified several genes implicated in functional hypothalamic amenorrhea, a reversible type of gonadotropin-releasing hormone deficiency, typically induced by nutritional deficiencies, psychological distress, or excessive exercise.[18] These included mutations identified within the following: the fibroblast growth factor receptor 1 gene FGFR1, the prokineticin receptor 2 gene PROKR2, the GnRH receptor gene GNRHR, and the Kallmann syndrome 1 sequence gene KAL1. Unfortunately, inheritance in multi-factorial disease is sufficiently complex that the subtle interplay of genes, modifier genes, and causative SNPs, required for the emergence of a pathological phenotype, does not always facilitate straightforward deduction.

Generally speaking, genome-wide association studies, also known as GWAS, involves comparison of two groups: a so-called patient group, suffering with a disease or syndrome or, more generally, possessing an identifiable phenotype, and a control group, comprised of otherwise similar subjects free of the disease-state or phenotype. DNA, typically *via* a blood sample or mouth swab, is taken from each individual and analysed using several techniques including automated PCR and microarray methods. SNPs or other genetic markers can be pre-selected or sampled semi-randomly. The results of this process are analysed statistically, using a variety of sophisticated methods able to deal with the vastness of data collated from thousands of sites within the genome across hundreds of patients. If a SNP or allele or indel or whatever, is significantly more frequent in the patient group compared to the control, the variation is said to be "associated" with the phenotype or particular pathological state. An allele is the name given to one of a set of distinct inherited sequences for a gene. Human Major Histocompatability Complex (MHC) proteins, for example, have thousands of conserved alleles with distinct but related biological properties. An indel is a shorthand term for insertion or deletion (the loss or gain of a sequence position rather than a change in sequence); these positions are identified from large, reliable, multiple sequence alignments. Not all genetics variants will be directly causative, simply coincident with the causal variation or variations. Additional steps, such as specific re-sequencing of a genomic region to clarify the precise genetic change involved, may thus be required.

Often, such approaches identify genes not previously known to be associated with a particular disease state or adverse drug reaction. Emergent refinements to GWAS methodologies include a more efficient iterative approach. Initially, a well-characterised set of markers, such as SNPs distributed across a genome, are assayed and correlated with drug or disease phenotypes. Regions showing the highest correlation are then scanned at higher resolution until a gene is identified. As large-capacity whole-genome SNP arrays have become available, whole genome genotyping can be accomplished in a single operation. The information generated can be both enormous in size and enormous in analytical difficulty, since any links present are unlikely to be in any way clear cut. Association studies suffer from a very high false-positive rate: 500 000 SNPs could generate 25 000 false-positives with 0.05 α error. This can be offset by statistical adjustment using a Dunn-Bonferroni correction, which addresses the multiple comparisons problem. Such corrections assume that testing *n* hypotheses progresses by testing each hypothesis at a significance level *1/n* that of a single hypothesis. This statistical stratagem can eliminate many weaker associations, but the phenomenon remains a confounding complication.

A recent study exemplifies some of these ideas in a most illuminating manner. Wang *et al.* investigated the link between gene expression with known pathways mediating the so-called antioxidant response and genetic variation, making use of microarray-based experimental and computational transcriptomic analysis.[19] In a relatively small scale study, the authors were able to show that the pathway regulated by transcription factor NRF2 behaved differently in patient and control groups, demonstrating that the expression of various genes in the NRF2-regulated pathway was correlated to that of a specific binding partner of NRF2. They also discovered potential regulatory SNPs in NRF2 binding sites. This study both suggests disease-related SNPs that could be used as biomarkers for lung cancer and identifies several potential targets for possible therapeutic intervention by drugs: enzymes responsible for the metabolism of carcinogens, such as members of the aldo-keto reductase (AKR) superfamily, AKR1C1, AKR1C2, and AKR1C3.

Beyond the genome, sit the epigenome, epigenetics, and epigenomics. Epigenetics relates to the interplay between the environment, the organism, and the genome. Several molecular mechanisms mediate epigenetics,[20] including DNA methylation, so-called histone remodelling, and genomic imprinting. This occurs in mammals and higher plants, where there is a significant maternal investment in each offspring. In imprinting, which happens for a few genes, an allele from one parent is silenced. This mechanism probably evolved through competition over the allocation of resources to descendents. Yet as only a single copy of each imprinted gene is inherited, it is sensitive to any epigenetic modification induced by environmental change.

Epigenetic changes to gene expression are transmitted by so-called non-Mendelian mechanisms of inheritance. The evolutionary rationale for epigenetic inheritance suggests that it affords a rapid way of adapting to transient environmental changes, without needing underlying genes to undergo Darwinian selection. Epigenetic mechanisms arise from genome evolution and create additional or alternative mechanisms, capable of accelerating the capacity of an organism to persist and survive, to alter and adapt, and to reproduce. Epigenetic changes in gene expression may create novel phenotypes, acting through many generations, exerting selective pressure on some genes and engendering long-term changes in the genome. Epigenetic inheritance can be viewed not as an evolutionary alternative complementing natural selection by mutation, but rather as a driving force behind sturdy and enduring genetic change.

15.2.3 Proteomics and Mass Spectrometry

The dictionary defines the neologism proteome as the entire complement of proteins that is (or might be) expressed by a cell, tissue, or organism, and traces the use of this word to 1995 (the point at which it appears in the primary literature)[21] but was extant long before that. Proteomics is the science of the proteome. The proteome constantly changes; it is ever in flux. The proteome is, conceptually, biology; proteins make nature happen. Genes are the quintessence of inheritance, yet it is only through the medium of the proteome that they are able to propagate themselves. In a key sense, the proteome is very much more dynamic than the genome; it varies according to the cell type and the functional state of the cell. In addition, the proteome shows characteristic perturbations in response to disease and external stimuli. The proteome is also much larger than the genome from which it is derived, at least in terms of the number of distinct gene products *versus* the number of genes. This fact is what makes the human genome as small as it presently seems to be. The proteome is generated and regulated by the genome both directly and indirectly, by epigenetic mechanisms and, crucially, by the proteome acting upon itself, as well as simultaneously acting upon the genome and transcriptome. Because the proteome is larger, it comes closer and then exceeds the original estimates for the size of genome.

This difference in size between the proteome and the genome is partly mediated by the phenomenon of splice variants,[22] as well as the existence of inteins (or protein-splicing elements) which are able to catalyse their excision from surrounding sequences (known as exteins), creating new proteins where exteins are directly linked by peptide bonds.[23–26] Such splicing mostly occurs in coding regions. The most commonly seen such event is exon skipping, whereby an exon, or set of continuous exons, is permuted in distinct mRNAs. Less frequent are the use of donor and acceptor sites and intron retention. Thus alternative splicing and RNA editing has come to replace the presupposed one-to-one correspondence of gene to protein, and in its stead we see that one gene can be manifest as a plethora of alternate proteins. Alternative splicing is a major factor driving proteomic diversity and provides a partial rationale for the gulf between the small size of the human genome and its implicit and increasingly explicit complexity. Many of these mechanisms seem to modulate and modify each other, compounding their ability to increase the number of potential proteins. The estimated size of the proteome is perhaps 100 to 1000 times the number of genes. The proteome varies according to cell type and the cell's functional state.

Coupled to that the ability of one protein to undertake several functions,[13–16] as well as other proteins having different functions in different cellular compartments, means that the functional diversity implicit within the proteome is large enough for us to hold faith with the notion that the function, action, and behaviour of organisms, notably ourselves, arises solely from its physical structure as encoded by the genome, without the need to invoke vitalism or other ideas stemming from intelligent design.

Proteomics, as a scientific discipline, is relatively new but is based upon rather older techniques, combining sophisticated analytical methods, such as 2D electrophoresis and mass spectrometry (MS), with bioinformatics. Thus proteomics is the study of gene expression at a functional level.

A comprehensive description of the proteome seeks to catalogue all proteins but provide insight into protein expression under defined conditions, the occurrence of post-translational modifications, as well as the distribution of proteins within the cell and its various compartments.[27]

A forerunner to the current proteome paradigm was the concept adumbrated by Anderson and Anderson:[28] the "human protein index". They wished to characterise all the proteins expressed by a cell, by using high-resolution, 2D electrophoresis (2DE). They thought that the human protein index would prove useful in clinical chemistry, pathology, and toxicology. In its proteomic form, this conceit has proved all too true. Hitherto, proteomics has been seen as a primarily analytical science, which combines multi-dimensional poly-acrylamide gel electrophoresis techniques with sensitive biological MS, supported by rapidly growing protein and DNA databases, to effect the high-throughput identification of protein populations from different cell types or cells experiencing different environmental conditions. As we have said, the unambiguous identification of a protein is a prerequisite to their full functional investigation. For example, by using matrix-assisted laser desorption ionisation (MALDI)-MS-based proteomic analysis coupled to photo-affinity labelling, Murray *et al.* were able to identify the target of a series of compounds with profound immunoregulatory behaviour activity in the inhibition of phorbol 12-myristate 13-acetate (PMA)/ionomycin-induced human peripheral blood mononuclear cell (PBMC) proliferation[29] as MCT1, a member of the monocarboxylate transporter family, responsible for transport of lactate and other small monocarboxylates. MCT1 is up-regulated on T lymphocyte activation to facilitate cellular lactate egress due to exacerbated glycolysis. Thus MCT1 inhibition represents a novel approach to non-cytotoxic immunosuppression, since inhibition of MCT1-mediated lactate efflux induces lactate accumulation and thus inhibits glycolysis, which in turn suppresses T lymphocyte proliferation during the early immune response.

This identification is typically effected through MALDI-MS. This can link sequences in databases to protein spots on gels. The two principal MALDI-MS identification methods are peptide mass fingerprinting (PMF) and post-source decay analysis. PMF identifies proteins by comparison of experimentally and theoretically derived profiles of proteolytically digested peptides. Because both experimental data and sequence databases are limited, there usually remains some ambiguity with regard to posttranslational modification(s) and intrinsic sequence variation. Moreover, the role of electroblotting and Edman *N*-terminal sequencing as tools in protein identification should not be overlooked. As proteins derived from the same gene may be largely identical, and might differ only in limited yet functionally important details, the identification of proteins must not only pinpoint numerous proteins *en masse* but also differentiate between close relatives.

But obviously proteomics is more than just a few techniques, however sophisticated they are. Indeed, it is a cohesive and overarching intellectual environment, replete with ideas, many now beginning to yield advanced, if less established, techniques. The cutting edge of proteomics has much to offer. In this context, arguably, the most exciting and enticing group of emergent proteomic techniques are the so-called protein arrays,[30–32] where recombinant proteins can be arrayed to study protein–ligand and protein–protein interactions. Based on the rationale that altered abundance or a change in structure of proteins can lead to disease, and although protein arrays are currently more expensive and more technically difficult to produce than nucleotide arrays, protein and antibody arrays are now generating considerable excitement. Alternatively, arrays of protein-specific monoclonal antibodies can quantitate levels of protein expression,[30] analogous to the detection of mRNA by microarrays.[33] It is to be hoped that as protein arrays become more sophisticated, they will impact on drug discovery by profiling prognostic and diagnostic markers of particular disease states. A recent example demonstrates the future potential of this approach. Andrews *et al.* assayed a human protein microarray for ubiquitination, using protein microarrays to identify substrates of the HECT domain E3 and the partner E2 ubiquitin ligase Ubch5 or Ubch7.[34] Smurf1 regulates TGF-beta and bone morphogenic protein signalling, and cell polarity through degradation of RhoA. They identified 89 potential substrates of the Smurf1 E3 activity, which spanned a number of different biological pathways, followed by *in vitro* validation.

15.2.4 Bioinformatics

There are many definitions of bioinformatics. Some are short; some are very long, and confusingly so. Very few capture what bioinformatics is about. A definition that I often use is rendered thus: "bioinformatics is the application of informatics disciplines to biological macromolecules". It is general and inclusive without completely losing connection with utility. However, such a definition is far from specific; but with specificity often comes obfuscation. Increasingly, bioinformatics can be seen as a collection of disparate disciplines, which have either diverged from the parent, such as immunoinformatics, or converged from some other discipline, such as biological text mining.

Bioinformatics makes a series of synergistic interactions with both client disciplines (computer science, structural chemistry, *etc.*) and with disciplines that act in the role of the customer (genomics, molecular biology, and cell Biology). Bioinformatics is concerned with activities such as the annotation of biological data (genome sequences for example) and classification of sequences and structures into meaningful groups, and seeks to solve two main challenges: the prediction of function from sequence and the prediction of structure from sequence. Put simply, bioinformatics deals with the similarity between macromolecular sequences, typically manifested in global sequence searches using software such as FastA[35–38] or BLAST.[39–42] Bioinformatics seeks to identify genes descended from a common ancestor, which share a corresponding structural and functional propinquity. The underlying assumption is thus an evolutionary one: functionally similar genes have diverged through a process of random mutation which results in evolutionarily more distant sequences being less and less similar to each another. The chain of inference that connects similarity to common function is complex. Thus successful functional assignment necessitates significant biological context. Such context is provided by databases: implicit context present as archived sequences and explicit context present as annotation.

Databases constitute the *lingua franca* of bioinformatics, and perhaps of all the informatic sciences. Although the hue of data archived may differ, nonetheless the use, creation, and manipulation of databases remains the most critical feature of modern-day bioinformatics. Available biological databanks are currently proliferating; they now require their own database just to catalogue them.[43–46] Databases began by simply storing the sequences and structures of genes and proteins. Soon, however, databases such as Swiss-Prot began to add biological context in the form of annotation, the fundamental character of which is well illustrated by the observation that currently only around 15% of the Swiss-Prot database is actually sequence. The remaining 85% is annotation: literature cross-references, descriptions of biological context, and illustrative notes. Rationalising this veritable Everest of biological data now lies beyond the scope of individuals, necessitating the coupling of automation to a correspondingly global human effort. Automation, however, carries a heavy price. Functional annotation in protein sequence databases is often inferred from observed similarities to homologous, annotated proteins. This can lead to errors, particularly when sequence similarity is marginal. As a result, it is widely believed that there are now substantial numbers of incorrect annotations throughout commonly-used databases.[47] Moreover, this problem can be compounded by the Markovian process of 'error percolation',[48–49] whereby the functional annotations of similar proteins may themselves have been acquired through chains of similarity to sets of other proteins. Such chains of inference are seldom recorded, so it is generally impossible to determine how a particular database annotation has been acquired. Such a situation leads to an inevitable deterioration of quality, and poses an on-going threat to the reliability of data as a consequence of propagating errors in annotation. Although curators continually strive to address such errors, users must be constantly on their guard when inferring function from archived data.

However, bioinformatics is never still, and databases, like other aspects of the discipline, have moved onwards and upwards, now encompassing entities as diverse as whole genome sequences, transcriptomic and proteomic experiments, and a diversity of other kinds of experimental measurements and derived biological properties. From a pharmaceutical target discovery perspective, arguably the most important type of derived data are databases encoding complex discriminators of protein family membership. A variety of different analytical approaches have been used to create such discriminators, including regular expressions, aligned sequence blocks, fingerprints, profiles, and hidden Markov models (HMMs).[50] Each such model has quite different relative strengths and weaknesses. Such discriminators are deposited in one of the many primary motif databases [*i.e.* PROSITE (http://expasy.org/prosite/) or PRINTS (http://www.bioinf.manchester.ac.uk/dbbrowser/PRINTS/index.php)], or secondary motif databases such as SMART [http://smart.embl-heidelberg.de/] or INTERPRO [http://www.ebi.ac.uk/interpro/].

An underlying assumption of such databases is that a protein family can be identified by one or more characteristic sequence patterns. Such patterns are identified in three ways. Firstly, by direct inspection of aligned protein sequences. Secondly, by using unaligned sequences as input to programs such as MEME,[51] which can perceive statistically significant patterns automatically. Or, thirdly, from aligned sequence using a motif identification approach such as PRATT.[52] Motif databases thus contain distilled descriptions of protein families that can be used to classify other sequences in an automated fashion.

Notwithstanding all that has been said, at the very heart of bioinformatics lies the so-called multiple sequence alignment. Its uses are legion: the prediction of 3D structure, either through homology modelling or *via de novo* secondary structure prediction; identifying functionally important residues; undertaking phylogenetic analysis; and the identification of important motifs and thus the development of discriminators for protein family membership. The accuracy of many techniques, such as those mentioned, is heavily dependent on the accuracy of multiple sequence alignments. The building of a multiple sequence alignment begins with the identification of a sequence/structure corpus. The definition of a protein family, the key step in annotating macromolecular sequences, proceeds through an iterative process of searching sequence, structure, and motif databases to generate a sequence corpus, which represents the whole set of sequences within the family. In an ideal case, this should contain all sequences and structures related to the seed sequence of interest. The process is iterative and brings together the results of three types of searches: global sequence searches, such as BLAST, FastA, or a parallel version of Smith-Waterman; searches against motif databases such as InterPro or PRINTS; and searches for similar 3D structures using full model searches, such as DALI,[53–57] or topology searches, such as TOPS.[58–61] Once a search has converged and no more reliable sequences can be added, the final corpus has been found and a multiple alignment can be constructed.

However, one of the most directly relevant and useful areas within the discipline of bioinformatics is also among its most successful. The ability to accurately identify genes remains an unsolved problem, despite rapid progress in recent years. When dealing with entire genome sequences, the need for software tools, able to automate the laborious process of scanning millions upon millions of base pairs, is essential. When we move from the genome to the proteome, gene finding becomes protein finding and an order of magnitude more difficult. The proteome is, however, much larger, principally through the existence of splice variants.[22,62–65] Other causes include post-translational modifications, cleavage of precursors, and other types of proteolytic activation. The proteome varies according to the cell type and the functional state of the cell, and it shows characteristic perturbations in response to disease and external stimuli. Proteomics as a scientific discipline is relatively new, but is based upon rather older techniques, combining sophisticated analytical methods, such as 2D electrophoresis and mass spectrometry, with bioinformatics. Thus proteomics is the study of gene expression at a functional level. Yet genomic identification of genes is the beginning rather than the end. Distinct proteins have different properties, giving rise to different functions in different contexts. Since it brings us so much closer to human medicine, identifying, cataloguing, and characterising proteins within proteomes should prove significantly more challenging than annotation of genomes.

15.2.5 Structural Genomics and Fragment-Based Drug Discovery

Protein function is often synonymous with processes of ligand recognition, which usually occur in defined binding sites on the protein surface. Analysis of binding sites in structural terms seeks to identify similarity of function that is broadly independent of any homology apparent at the levels of sequence and fold and extends beyond the search for conserved sequence and conserved structural motifs. X-Ray crystallography has been around for a long time. Although the celerity of structure creation is constantly increasing, crystallography has had an unfortunate reputation for immanent slowness. This has discouraged many from using this technology. However, within a decade this situation will have changed from all recognition, with time, money, and talent pouring into the newly emergent discipline of structural genomics.

The genome sequences from a tranche of prokaryotic and eukaryotic organisms are now available, including, most excitingly, the human genome. There are hundreds of completed genomes. This constitutes a vast accumulation of information: a mountain, indeed a veritable Everest, to dwarf the informational molehills of preceding decades. With the access to the sequences of 100s of complete genomes, the principal objective of structural genomics is to generate a comprehensive overview of the universe of protein folds. Structural genomics aims to make available one or more example experimental structures for every protein sequence family. Protein homology modelling could then produce structural models for almost all proteins observed in nature.

Any reasonable division of a genome into structurally distinct protein families will necessitate many hundred protein structure determinations. Structural genomics will require new ways to automate experimental structure determination. X-Ray crystallography has traditionally progressed through several stages from the very biochemical to the abstractly mathematical. Having identified our protein of interest, we need to produce sufficient pure protein to perform the search for appropriate crystallisation conditions. Once we have crystals of the protein, we need to collect X-ray diffraction data from these crystals and the "solve" structure. This involves solving the phase problem: recovering the electron density within the unique part of the lattice by combining the intensities of diffracted X-rays with the phase, the other component of the Fourier transform which links real molecular electron density and the experimentally determined diffraction pattern. The final stage requires building and refining a protein model within the electron density and ultimately refining this crude model to optimise its ability to recreate the diffraction pattern. The production of protein is probably the most generic aspect of structural genomics, although few people want quite such pure protein in such large amounts. The development of many different high-throughput protein production systems is currently underway in both academic and commercial organisations. These include both *in vitro*, or cell-free, systems and examples based on well-understood microbial systems, such as *Escherichia coli.*[66–68] Selenium incorporation allows for the phasing of the protein diffraction pattern using multi-wavelength anomalous diffraction, the so-called MAD technique, which offers a general approach for the elucidation of atomic structures.[69]

Once one has sufficient protein, the next stage in crystallography is obtaining crystals. This is one of the two main problems left in X-ray crystallography. While the phase problem is slowly being overcome, crystallisation remains a black art. The process of growing protein crystals remains poorly understood, requiring a trial-and-error search to resolve the few idiosyncratic conditions of pH, ionic strength, precipitant, and buffer concentrations, needed for crystal growth. However, even this recalcitrant discipline is now yielding to informatics and robotics.[70–76] Robust multivariate statistics has been used to relate variations in experimental conditions, within experimentally designed crystallisation trials, to their results.[77] Although these mathematical models cannot explain crystallisation mechanisms, they do provide a powerful pragmatic tool allowing the setting up of crystallisation trials in a more rational and more confident manner.

Until recently, crystal mounting has seemed the aspect of crystallography least suitable for automation. The process of mounting a protein crystal such that it can sit comfortably in an X-ray beam is a highly interactive process requiring a prodigious feat of manual manipulation, personal dexterity, and physical adroitness. Several systems have been developed which leverage a potent combination of cryogenic temperatures, intelligent software, and robotic control.[78–79] Although the systems they describe have a rather Heath Robinson appearance, they are no worse than the set ups used in other high-throughput regimes within the drug industry.

The diffraction pattern is obtained by allowing a focussed beam of X-rays to pass through a crystal. Each spot on the diffraction pattern (DP) represents an intensity and has associated with it another quantity, the phase, which when combined with it through a Fourier transform, yields an electron density map. Unlike small molecule crystals, where phases can be determined directly from relationships between intensities, proteins require more approximate solutions. However, in the context of structural genomics, most are undesirable. Molecular replacement requires an existing 3D model of a homologous protein, while multiple isomorphous replacement requires a trial-and-error search for heavy atom derivatives. The previously mentioned MAD phasing is a much better alternative. Another approach is the development of so-called direct methods.[80–83] Crystallographers combine these methods with clever manipulation of the diffraction data for single specimens to produce an accurate, phased, and interpretable electron density map. As we have seen, many of the advances in the biochemical and biophysical stages of the crystallographic process—protein production and crystallisation—will be greatly enhanced by automation. Other technical advances will solve, or side-step, many of the inherently intractable problems of crystallography, such as the phase problem.

Thus far, we have discussed soluble, globular proteins, but what of membrane proteins? As they exist within a complex environment containing both a lipid and an aqueous phase, such proteins present distinctly different problems for automation and high-throughput crystallography. Integral membrane proteins are generally large and often form multimeric complexes This together with the practical problems associated with preparing samples containing biological membranes means that it has not proved possible to study them successfully using multi-dimensional NMR spectroscopy. Consequently, most structural information has come from crystallographic techniques: X-ray crystallography, for those cases where it has proved possible to produce true 3D crystals, and electron crystallography, which combines image analysis from electron microscopy with electron diffraction data to study 2D crystalline arrays.

The greatest obstacles to the successful determination of membrane protein structures remain technical: problems with the over-expression, purification, and concentration of membrane proteins and with the preparation of 3D crystals for X-ray studies or 2D electron crystallography. Both are still daunting challenges. Considerable work is still required to overcome the technical difficulties inherent in the crystallographic study of membrane protein, yet work is beginning.

Fragment-based drug discovery, or FBDD, has emerged as key rival to high-throughput screening. An early progenitor of FBDD was SAR by NMR,[84,85] where NMR was used to determine both the existence of binding and the binding mode for a fragment; and subsequently to help guide synthetic modifications and the ligation of other separate binding moieties. Several compounds in early and late-stage development now owe their discovery to FBDD.[86–89] See also Chapter 5 of this book, by Alex and Millan. Since possession of unsuitable physico-chemical properties, highly correlated with molecular size, is a principal cause of drug attrition, fragment-based approaches offer several advantages. Though low in potency, small fragments can form high quality interactions, and offer much scope for optimisation, while in terms of the efficiency of drug discovery quite small fragment libraries can sample chemical space quite effectively. Specialised methods, including multi-dimensional NMR, isothermal titration calorimetry, and most notably X-ray crystallography, are necessary for detection of fragment binding. In crystallographic screening, libraries of no more than a 1000 fragments are soaked into crystals of the target protein, which are then solved, and bound compound will be identified directly, hopefully forming the basis of further structure-based drug design (SBDD). On-going developments in high-throughput crystallography and structural genomics offer much in this area, affording the possibility of assaying larger and even more representative fragment libraries.

15.2.6 Computational Bioscience and Systems Biology

While it is beyond the scope of this chapter to more than briefly allude to the ever burgeoning array of computational approaches applicable to problems in chemistry and biology, it is nonetheless important to emphasise a few key points, thus setting in context much that has preceded, and much that follows, this particular section. Many approaches are available to help illuminate molecular phenomena in biological and chemical systems. This much is clear. We have already mentioned bioinformatics and systems biology. They form an inter-connected, multi-dimensional continuum running from highly informatics disciplines such as cheminformatics and bioinformatics, through to molecular simulations. The resulting spectrum moves from complexity of one sort—confoundingly large numbers of relative simple and comprehensible molecules or sequences, at least when viewed singly—through to an utterly different kind of complexity, based on simulations of individual systems, but systems of such an immanent size and scale that they pose an immense intrinsic challenge. Because of the nature of algorithms, we also see techniques developed in one area being transfer painlessly to another quite different area of computational bioscience.

Systems biology is really an innovation of the 21st century; prior to the turn of the present century, there were less than twenty publications using the term. In the subsequent decade or so this count is now approaching ten thousand. Cynics would argue that this is another example of systemic rebadging or rebranding or whatever pejorative neologism you choose; people desperately trying to pin their essentially unaltered extant research onto a well-funded bandwagon. More likely is that it is people doing what they have done before but on a far larger scale, hopeful perhaps that knowledge and understanding will surface from the huge accumulation of poorly organised data.

In some senses, systems biology is an extension of traditional systems theory to the domain of biology. Systems theory studies systems in their totality, with the goal of elucidating underlying principles applicable to all types of systems in any area. This exegesis misses somewhat the particular flavour of systems biology, and its overwhelming emphasis of looking at interactions between components rather than components themselves, or indeed at the behaviour of the system. This is equally true of both those interested in the topology of metabolic networks and mathematical modellers trying to create dynamic models of metabolism. Systems theory presupposes that the behaviour of a system is independent of a system's structure; this idea is missing from much if not all systems biology.

Systems Biology is a modern discipline within bioscience. By intention, it subsumes computational, theoretical, and mathematical modelling. Moreover, Systems Biology closely couples theory to experiment; and does this in order to synthesise a new understanding of biology at the level of the system (cell, tissue, or whole organism) rather than at the level of the molecule. It aims at a hierarchical yet many-layered model of biology, seeking to understand it as a system. This means studying behaviour at the cellular and organismal level, rather than at the *in vitro* or *in vivo* levels. Research over decades has revealed how individual proteins and genes interact to form the various metabolic, signalling and effecter pathways underpinning biochemistry. However, much of our understanding of such pathways and regulatory networks is still locked away in the literature. In this state, accumulated insight into pathways, their structure, and their function is obfuscated by haphazard descriptions promulgated by different authors. Thus a given pathway by custom has been known only to a limited few, who focus often on no more than one or two proteins and their interaction partners.

Models of pathways exist in many forms but most of these are static representations, not dynamic models of metabolism. They show the network topology of interconnected pathways of enzymes or signalling molecules, but they contain no dynamic information on reaction rates of diffusive encounters. The JWS-online database (http://jjj.biochem.sun.ac.za/database/index.html) on the other hand, is a web-based database containing over 90 dynamics models.[90–92] Of these however, only a few are approaching what is desired.

Biologists favour explanations of life and its component phenomena couched in evolutionary terms. A key aspect is retrospective selection; a thing is seen because it survives not because it is perfect. Evolutionary explanations often lack power because they are seldom predictive and tell us little about the contingent constraints that favoured one particular adaptive solution over another. Evolution is said never to repeat itself, yet the eye has evolved independently many times. Components of the eye, such as rhodopsin and other visual pigments, evolved only once, yet complex, image-forming eyes may have evolved independently as much as 50 to 100 times, albeit using many of the same protein components.[93] The disparity of change at the macroscopic and the molecular level remains problematic; most attempts to link microscopic change and the trajectory of evolutionary advantage fail. While I am not a proponent of intelligent design, and am a thorough-going philosophic materialist, evolutionists exist within a special place in science: not for them the excoriating rigour of experimental verification; evolutionary arguments need only be convincing, not verifiable.

In biology, a complete explanation for phenotype or phenotypic traits typically demands explanation in terms of function. There are several philosophical standpoints relating to function. One contends that a phenotype or phenotypic trait will function so as to make a contribution to the organisation of an organism's structure. Another holds that a phenotype or phenotypic trait functions because it increases or maintains the fitness of an organism. When viewed from a strict evolutionary context, function is something of a shibboleth. They exist within the minds of scientists; they do not exist in proteins or in cells or in organs, or even in whole organisms, with perhaps the exception of those organisms able to hold and articulate meaning and intention. By this we mean that use of the term function presupposes a pervasive teleological explanation. Teleological or teleology derives from the Greek *telos*, meaning purpose. It refers to the philosophical study of purposeful design and is often contrasted with metaphysical naturalism, a view that in nature there is no design and ultimately no purpose. If neo-Darwinian evolution is any form of philosophy then it is a materialist and atheistic philosophy devoid of intentionality. As Arp puts it: "there is nothing like an intrinsic or extrinsic purpose or design to be found in nature". Instead, and put rather simplistically, changes in genome-encoded gene sequences lead to changes in protein sequences that alter binding, which leads to positive changes for the whole organism or the absence of change in the fitness of the whole organism. What we perceive as function is irrelevant to the blind watch-maker; he, she, or it takes no account of function whatsoever, since neither he nor she nor it—nor indeed function itself—have any independent existence, except as ideas. The world is composed of atoms, which make molecules, which form higher order structures. Except at the highest level, none of these things think, thus a cell cannot have intentions or understand or act in a meaningful way, as all such things are aspects of consciousness, and even our own consciousness is more illusory than we care to think.

Organisms experience selective pressure, not isolated proteins and certainly not nucleic acids. Nucleic acids are simultaneously both sterile carriers of information and dynamic molecules in their own right. All the physical manifestations of genomic information are open to selection but not the information itself. Nucleic acids experience selective pressures both directly and indirectly. Proteins do not evolve as linear sequences, they evolve as objects in 3 dimensions. The mechanisms driving the emergence and evolution of network modularity is not clear. Yet some things are less unclear than others. Duplication of genes followed by divergence is the key mechanism underlying eukaryotic proteome expansion. Protein interactions are frequently conserved after duplication; thus paralogous genes frequently share interaction partners. Duplication followed by conservation of protein interactions is linked with modularity. Around two thirds of globular protein complexes are homodimers, and even at low similarity paralogous pairs conserve quaternary structure and are assembled *via* conserved intermediates. Self-interactions have special structural properties that are conserved into the duplicated interaction between paralogous proteins. Interactions between paralogous proteins are more versatile functionally and structurally and are systematically selected for in evolution. Duplication of self-interacting proteins is thus vital to the evolution of a proportion of protein complexes, and hence of functional modularity in protein interaction networks. This is a common mechanism for evolution of clusters of protein interactions.

Another way to express the idea limned is to say that proteins experience evolutionary pressure as physical objects and not as abstract representations of molecular sequences. We may understand evolution as changes in the lettering of sequences, but this is far from how it occurs in practice. Likewise, we can only conceptualise protein functions at high organisational levels, where they can be interpreted as directly influencing the fecundity of the whole organism.

At the molecular level, we can only really discuss binding, not function. Within the neo-Darwinian paradigm, one may explain biological function in terms of its contribution to an organism's reproductive fitness. However, the idea of fitness reflects the probability of survival in a given environment and can be seen as a basis for selection, although fitness is no more a property of phenotypes than it is of genotypes or the sequences of individual molecules; rather it is a relationship forged between the environment and individual organisms. Evolution can only be thought of properly as a system property.

Individual changes in the genome are reflected in physically manifested changes in protein structure and its binding characteristics. Types of change include altered stability, the temporal nature of its capacity to fold, metabolic cost of synthesising one amino acid *versus* the cost of another, together with higher-level manifestations of these change physical properties ultimately reflected in the survival of the whole organism, and beyond that even the altered ecological interactions made by the changed organism.

While we use the term function, it is a word and an idea whose use is more limited than we suppose. There is only binding; function is illusory, a subjective human idea imposed on the world around us. Enzyme catalysis is largely, though not completely, explicable in terms of binding. The classical idea of enzymatic enhancement of reaction kinetics is that the enzyme binds to the transition state, reducing the activation barrier to rates in either direction. The function of a protein is not to bind another; what we have is two proteins which form a complex with a lower overall free energy and thus exhibit high mutual affinity. Such interactions include, but are not limited to, small molecule ligands binding to their receptors, substrates binding to enzymes, or large molecules interacting together, such as a protein antigen binding to an antibody.

Thus binding is pivotal; it is the fundamental molecular cornerstone upon which biology is constructed. A proper and complete understanding of the underlying nature of binding as a molecular event is, in turn, a cornerstone upon which are based many disciplines, including biochemistry, biophysics, and immunology. If we are able to measure binding and give it both a thermodynamic and a kinetic character, including the potentially complex multi-stage kinetic mechanism of binding, then we go a long way to creating an unambiguous definition of that interaction. It is a physico-chemical definition however, and does not necessarily imply an *in vivo* reality, since it takes no account of sub-cellular compartmentalisation and the segregation of proteins.

15.3 New Kinds of Drug

In the following sections, we shall look at a more practical aspect of how genomics has helped to change drug discovery, specifically how it has introduced new types of drug and new approaches to the discovery and exploitation of therapeutic molecules.

15.3.1 Drugging Protein–Protein Interactions

Much of what we understand of the staggering complexity of biological systems devolves from the instantiation of one fundamental mechanistic event, the interaction of one protein with another. Proteins also interact with any other biological molecule, large or small, be that endogenous or exogenous small molecules (primary or secondary metabolites, drugs, prevalent environmental chemicals, *etc.*) or macromolecules (DNA, lipids, carbohydrates, *etc.*). Interfering with such interactions provides many established drug targets (protein–

small molecule interactions) and many a novel potential target for new drug interventions (protein–protein interactions).

Protein interactions vary enormously since the shape adopted by proteins likewise varies enormously. Proteins exploit this structural diversity to the fullest, producing recognition sites and binding pockets of bewildering specificity and subtlety. Most current drugs target buried small molecule binding sites, as typified by that of the G-Protein coupled receptor. Interfering with protein–protein interactions provides a tantalising alternative avenue for drug discovery; thus far, researchers have been reluctant to explore this potentially profitable area as it is seemingly so problematic technically. Nonetheless, effectively addressing these technical issues will open up as yet underexploited areas of biology for therapeutic intervention. From HIV biology to autoimmunity to cancer and beyond, many seemingly intractable targets await effective drugging. Moreover, such drugs will allow us to attack not single metabolic enzymes but entire signal transduction pathways.

Let us assume a collection of proteins, and assume further that most interact with at least one, probably more, other proteins; we quickly escalate from proteins to connections to networks. The reticulations formed by protein–protein interactions, like other natural networks, typically adopt a characteristically modular structure. As an idea, modular networks have gained common currency with many areas of science as pre-eminently useful and useable tools for understanding systems exhibiting increasing complexity. Such groups often correspond to well-documented functional units, including communities within social networks, niches in food webs, and, most importantly in the current context, modules in various kinds of biochemical networks. Such networks can include complex multifurcating metabolic pathways or regulatory genetic networks. Most significant for us, of course, are protein–protein interaction networks.

To a first approximation, such modules are discrete, isolated groups of proteins that interact to undertake individual biological functions. Nonetheless, and in certain senses at least, the modules formed by these proteins can be thought of as existing isolated in space and time, functionally separate from other discrete modules in the overall network.[94] As such, they make ideal drug targets. To gain a proper and complete understanding of protein–protein interactions (PPIs), it is necessary to identify the different interactions, understand the conditions under which they occur, and also establish the consequences of such interactions. There are many ways to identify PPIs; some involve measurement of physical binding constants and some are essentially inferential. PPIs can inactivate a protein, or they can alter the quantitative substrate specificity of a protein, for example the interaction of different transcription factors with RNA polymerase directs this protein to bind different promoters; likewise, the interaction of lactose synthase with lactalbumin causes its Km for glucose to be reduced 1000-fold.

It is possible, and indeed of great utility, to categorise PPIs based on the lifetime or duration of the complexes that they form. At the grossest and most approximate level of classification, we can impose a strict dichotomy on complexes, branding them as forming either permanent or transient complexes.

Obviously, complex formation, like all molecular interactions, is a dynamic process involving the equilibrium of bound and unbound forms. A permanent interaction is characterised by an equilibrium position highly biased to the formation of the complex; it is usually highly stable and will, for practical purposes at least, thus only exist in the complexed form. A transient interaction on the other hand, will be characterised by constant association and dissociation; though the rates of such a process will vary, the equilibrium position will favour the dissociated form.

Transient PPIs are many and varied. The enumeration of their involvement in so many vital biological functions has reached daunting levels. There are thus many examples, including the recruitment and assembly of the transcription complex, protein transport across membranes, chaperonin-catalyzed protein folding, and the recycling of subcellular structures during the cell cycle. Such recycling includes that of microtubules, the spindle apparatus, the nuclear pore complex, and the nuclear lamina.

Many PPIs do not fall into such convenient and distinct types. Rather, they lie within a continuum between obligate and non-obligate states. This again may be dependent on the interplay of environmental factors. An interaction may be mainly transient *in vivo* but become permanent under certain cellular conditions. Folding data, as well as data on the dynamics of the assembly at different physiological conditions or environments, are often not available. However, the subcellular location of subunits and the function of the protein will often suggest the biologically relevant type of interaction; for example, interactions in intracellular signalling are expected to be transient, since their function requires a ready association and dissociation.

Over time, methodological approaches to the identification and quantification of PPIs have proliferated and they are now abound. We can delineate direct experimental methods, more indirect inferential methods, and computational methods, all of various hues, which can identify, with varying degrees of certainty the existence of a permanent or transient PPI. For example, fairly direct methods include affinity chromatography, affinity blotting, immunoprecipitation, and covalent cross-linking.[95] To these we can add a tranche of modern and sophisticated methods such as tandem affinity purification mass spectrometry; isothermal titration calorimetry; crystallographic techniques, such as single-particle cryo-electron microscopy, synchrotron radiation circular dichroism, and X-ray scattering; and well-established but repositioned methods, such as isotope-edited Fourier-transform infra-red spectroscopy and analytical ultracentrifugation. The last of these has a wide dynamic range and can address very strong, sub-nanomolar interactions through to very weak interactions bordering the non-specific and quasi-random.

As well as those methods adumbrated, which give either a structural or a thermodynamic perspective, there are methods which are more indirect, affording inferential insights rather than a direct and relatively unequivocal readout. These include both library-based and so-called genetic methods. Library methods are typified by perhaps the most widely known method in PPIs: the so-called yeast 2 hybrid, or Y2H system, and the other well-known approach phage display and related techniques. Genetic methods include extragenic suppression, synthetic lethality, and overproduction phenotypes. This is not the place to discuss any of these in great detail; such exposition can be found elsewhere.[9]

Armed with such techniques, we can make use of the full intellectual power of drug discovery to discover drugs able to interfere with PPIs. A prime example is the interaction between wild-type p53 (protein 53 or tumour protein 53) and its endogenous inhibitor MDM2 protein (transformed mouse 3T3 cell double minute 2). In humans, P53 is a key tumour suppressor protein. In multicellular organisms, p53 helps regulate the cell cycle. Its role in preventing cancer has led to p53 being christened: "guardian of the genome", amongst other fanciful sobriquets inspired by its role in suppressing genome mutation. A key strategy in advancing cancer therapy now involves the discovery of small molecule inhibitors which can disrupt the direct binding of MDM2 to p53,[96] which is mediated primarily by a well defined pocket in MDM2 and a short helix from p53. The characteristic nature of this interaction site is an ideal target for developing small molecule inhibitors to interfere with this interaction. The utility of this approach is compounded by the observation that MDM2 inhibitors also exert influence over many other important molecules in cancer, principally by inhibiting MDM2's action as an ubiquitinylation ligase of p53 and blocking MDM2's interaction with the proteasome.

Beginning with peptide antagonists, a wide variety of selective inhibitors of the MDM2-p53 interaction have been discovered in the last decade;[96–98] this discovery has utilised both rational approaches, such as SBDD, and indiscriminate approaches, including both *in vitro* and cell-based HTS. Nutlins were amongst the first reported inhibitors.[99,100] They are *cis*-imidazoline analogues discovered by high-throughput *in vitro* and high-content cellular screening of a chemical library. Of the three Nutlin actives discovered, Nutlin-3 is the most commonly used. There are now over 20 distinct classes of molecules capable of interfering with this interaction directly, or by interfering with events downstream of MDM2.[98].

Another compound, MI-219, was developed using iterative crystallography-driven SBDD.[101] Using the MDM2–p53 complex crystal structure[102] a series of spiro-oxindoles were designed, synthesised, and tested as new MDM2–p53 inhibitors. MI-219 had high affinity for MDM2 (K_i value of 5 nM) and exhibited 10 000-fold selectivity for MDM2. There are now several MDM2 inhibitors undergoing phase I clinical trials. JNJ-26854165 (Ortho Biotech; Johnson & Johnson) is being evaluated against late stage refractory solid tumours in a phase I trial.[103] Other clinical candidates include R7112 from Hoffmann-La Roche, being evaluated in haematologic neoplasms and advanced solid tumours, and AT-219, an oral MDM2 inhibitor from Ascenta.

15.3.2 Systems Biology and Polypharmacology

Systems biology has been called the science of biological complexity. Systems biology is one of the weasel-words, or rather weasel-phrases, that defy ready definition; they mean many things to many people. Yet, it is clear that, amongst other things, systems biology analyses and models several critical phenomena including genetic, macromolecular and metabolic networks, in order to explain the higher order phenomena of biological systems. One of the major challenges for systems biology is to understand how parsing genetic instructions lead to a fully functional organism. Likewise, how we can interact at different levels to modify or modulate functional behaviour. Systems biology synthesises many areas of biology, operating at many levels of complexity; notable amongst these are the many -omes: genomics, proteomics, glycomics, metabolomics, *etc.* Knowledge of these is necessary for constructing a dynamic picture of biological systems operating at the level of molecules, cells, organs, organisms, and ecosystems.

When we administer a drug to offer therapy, we are treating a system not an isolated receptor, even when the drug was designed to target known receptor in a specific and selective manner. This has operative, practical consequences as well as philosophical ones. The potency of a compound, expressed as an IC50 or K_d, acting against a particular receptor is not the same as the potency of the whole biochemical pathway of which it is a part. Indeed it can be hugely different. In a classic example, Rossignol *et al.* demonstrate that the behaviour of isolated enzymes (NADH ubiquinone reductase, ubiquinol cytochrome *c* reductase, and cytochrome *c* oxidase) was quite different to that of the biochemical pathways, and the variety of subcellular organelles, in which they were embedded.[104] Drugs affect systems—cells, tissues, organs, or whole animals—not just isolated receptors, and the near universality of off-target activities for drugs is now widely recognised. This has several implications. To some, systems biology appears as a return to the old way of doing things. It is certainly a return to the early days of biochemistry, when biochemists sought to peruse biological systems in a quantitative and physical manner. Others see this as an argument against modern techniques and modern ways of working, and a return to outmoded discovery technologies. We simply cannot return to the days when, during the initial discovery of the anti-asthma drug Intal, Roger Altounyan—medic, lifelong asthmatic, and part inspiration for Arthur Ransome's book *Swallows and Amazons*—tested potentially toxic compounds solely on himself,[105] nor yet to the indiscriminate screening of compounds on huge numbers of whole animals.

As the most prevalent and pervasive description of biological systems is a gene-centric one, it may be wise move towards a more systems-based approached. Models of pathways form a valuable means of interpreting genomics studies and modelling biological processes computationally. Traditionally, the most loathed aspect of undergraduate biochemistry has been the study of metabolic pathways. This tedious task is best undertaken with the aid of computational tools, allowing us to deprecate the rote learning aspect and concentrate instead on exploring the connectivity of separate pathways, and thus gain a synoptic view of the topology of a network.

Of course, this is by no means a nuanced view, and other finessed refinements can and need to be made. For example, and appropriately in the current context, it has recently emerged that networks often exhibit hierarchical organisation, again over many length scales and many time scales. It can be argued, and argued persuasively, that so-called protein-protein interactions, and not the individual proteins of which these interactions are composed, are the very basis of biochemistry itself, in much the same way the structure, and thus what we shall for devilment call the function, of proteins is mediated not by residues or even sets of residues but by the kind of interaction that a residue or groups of residues can make. A residue can change, but as long as the interaction is maintained, all is well. Thus the evolution of biochemical pathways, and thus of biochemistry itself, is the evolution of connections, albeit connections mediated by the evolution of proteins structures. Proteins change and so the spectrum of potential interactions they can make changes. If a protein can make many interactions that lie on the boundary between specific and non-specific binding, then when another protein changes it may gain a beneficial interaction, just as other beneficial interactions are elsewhere lost. Drug or small molecule protein interactions behave in a somewhat similar way and are presumably liable to somewhat similar evolutionary pressures.

Likewise, the almost universal prevalence of off-target activities indicates that many drugs already act against multiple receptors, whether we are aware of it or not. Could there even be a place within pharmaceutical research for compounds that operate, as some believe Chinese medicines do, at sub-therapeutic values across a wide range of receptors?[106] Yet, identifying compounds that act as maximally selective ligands to individual drug targets has completely dominated thinking about drug design for decades. However, drugs which act against such targets cannot typically combat multifactorial diseases such as cancer, or diseases such as diabetes, which affect several cell types and tissues.

Many effective drugs act at several targets rather than targeting individual proteins.[107–113] Multi-target drugs can be discovered in various ways. Starting compounds can, for example, be discovered by screening against several targets in sets of simultaneous screens. They can be designed by taking a single molecule and changing it sequentially against several screens. Alternatively, two or more molecules can be merged, combining key features for activities against different targets. Ladostigil (TV3326), a putative treatment for dementia and contingent depression, integrates structural features from several molecules: a carbamate from rivastigmine [an acetylcholine esterase (AChE) inhibitore] and a propargylamine group from deprenyl (a monoamine oxidase B inhibitor). PRC20, a chiral, potential antidepressant developed at the Mayo Clinic, is another example of a multifunctional or multi-target drug. It is a so-called "triple" neurotransmitter reuptake inhibitor, devised to concurrently inhibit dopamine, norepinephrine, and serotonin reuptake into the pre-synaptic terminal. HLA-20 is a radical scavenging iron chelator with monoamine oxidase inhibitory activity.[11,12] Jacobson *et al.* targeted the dual role played by the adenosine A1 and A3 receptors in cardiac ischemia, integrating structural features of agonists against both receptors.

The combination of several distinct drugs impacting many targets simultaneously can often be better at controlling complex disease systems, as well as exhibiting significantly reduced drug resistance; such drug combinations are the key exemplars in several important therapeutic areas, while the results of systems biology analysis indicates equally strongly that exquisitely selective compounds are likely to be sub-optimal in terms of cumulative efficacy. At the same time, current combination drugs provide compelling and persuasive evidence of augmented efficacy, thus justifying the investment of significant efforts directed at the discovery and elucidation of novel multi-target mechanisms, allowing us to expand the number of druggable targets. Poly or network pharmacology presents us with several challenges. We must find a way to identify and validate multiple points of intervention within the biochemical pathways. We must always develop efficient ways to optimise several quantitative structure–activity relationships simultaneously against multiple targets, and possibly incompatible receptors, while maintaining the drug-like properties that enable molecules to reach their target.

15.3.3 Repurposing

There are many alternative strategies to the traditional pharmaceutical corporate model now emerging. Some seek alternative targets. Others look to alternative product types, principally biologics, such as vaccines, antibody- or protein-based therapeutics. Others search for fundamentally new ways of discovering novel therapies. Such a fundamentally new way of identifying novel therapies is therapeutic switching, which is also known by other names, such as drug repurposing or drug repositioning. This approach identifies therapeutically useful new or different indications for existing drugs, or for compounds identified as safe in phase I trials but found ineffective for efficacy reasons in phase II or phase III trials. Many successful examples exist:[114–116] thalidomide in severe erythema nodosum leprosum; Zyban, an antidepressant, is now successful in smoking cessation;[117] Parkinson's disease drug apomorphine, now treats erectile dysfunction;[118] even Viagra began as a heart medicine.[119]

Drug repurposing maximises the benefit of existing drugs of proven provenance or compounds that have only failed late-phase clinical trials for indication-specific efficacy reasons, not safety concerns. Most, maybe all, drugs have significant off-target activity, thus potential new therapeutic uses should be identifiable for molecules known to be free of toxicity or side-effects.

Therapeutic switching has huge untapped potential for identifying novel, safe, tested, patent-protected medicines. Large pharmaceutical companies, such as Pfizer and Novartis, are very keen on drug repurposing. Companies dedicated to drug repositioning include: Sea Change (seachangepharma.com), Numedicus (numedicus.co.uk), Ore (orepharma.com), and Biovista (biovista.com). In contrast to the highly efficient and systematised approach proposed in the following paragraph, all such companies work *via* a haphazard, serendipitous discovery process.

Few have sought a more systematic computational approach to therapeutic switching. Keiser *et al.* used chemical similarity between 3665 licensed or investigational drugs and the known ligands of hundreds of target proteins to identify thousands of unanticipated potential associations.[120] Thirty were tested experimentally and one confirmed in a mouse knock-out model. Campillos *et al.* used recorded side-effect similarities to identify drugs with the same target. Applied to 746 marketed drugs, a network of 1018 drug–drug relations was seen;[121–124] 261 of these comprised chemically dissimilar drugs with distinct therapeutic indications. Twenty of the drug–drug relations were tested experimentally, 13 of which were validated using *in vitro* binding assays; 11 had inhibition constants of less than 10 micromolar.

By leveraging the potential of genome sequencing, coupled to the undoubted power of high through-put technology, we can envisage an alternative and much more systematic approach to the challenge of drug repurposing. Based on structure-based automated protein docking, or virtual screening to its friends, it should be possible to identify readily the as-yet-unknown therapeutic roles of extant drugs using computer-based approaches. To date there has been no serious attempt to undertake this on a scale large enough to be useful, with a handful of studies undertaken on a limited scale and with limited success.[120] Li *et al.*,[125] for example, docked about 1000 small molecule drugs to around 78 human target proteins.

Any such endeavour would need to be undertaken in a staged manner, delivering individual milestones leading naturally to completion of the whole objective. Now that we know the genome and the proteome and its constituent protein targets, we can combine that with lists of known drugs and clinical trial dropouts and attack this novel means of therapy discovery in a wholly systematic manner. The process, one might envisage, is thematically subdivided so that even partial completion would yield highly-cohesive, publishable, and, more importantly, patentable results. Following from the work of Li *et al.*,[125] virtual screening, as described, seems well suited to the task of indentifying these interactions and provides a tractable and cost-effective means of undertaking such a daunting logistical task. Specifically, a company might wish to progress through target proteins on a family-by-family basis, beginning with large, well understood sets of targets such as the G-Protein Coupled receptors (GPCRs) and Kinases (the targets of a high proportion of extant drugs), and progressing to more recondite targets in a tiered fashion. Likewise, they would similarly progress through known, marketed drugs, to phase II efficacy drop-outs, to phase efficacy III drop-outs.

15.3.4 Biologics and Vaccines

Partly as a consequence of the genomic revolution, biologic therapies, such as vaccines and protein therapeutics, have grown in significance.[126] While small molecules still dominate the global therapeutic market, the market share of biopharmaceuticals continues to increase with escalating rapidity; it now represents over 10% of all sales.[126] Genomics by its nature reveals the complexity of biological systems, allowing us to catalogue and collate the individual protein components of these systems. It is these proteins (many of them originally identified by genomic sequencing in large pharmaceutical companies, long before the completion of the human genome) and their eventual publication, that now provide the products that many in the industry hope will prove to be its ultimate saviour.

Biopharmaceuticals comprise a wide diversity of prophylactic or therapeutic biologics: protein therapeutics [monoclonal antibodies (mAbs) and soluble recombinant protein of many kinds], as well as vaccines, peptide mimetics, interference RNA, vectors for gene therapy, cells, and tissue. The diversity of protein therapeutics in particular is shown clearly by briefly reviewing its best-selling components: insulin, erythropoietins, cytokines and other interleukins (ILs); blood-clotting and colony-stimulating factors; growth hormones, such as recombinant human GH; plasminogen activators; reproductive hormones, such as follicle-stimulating hormone; and therapeutic enzymes.

The first biotherapeutics—bovine insulin, streptokinase or staphylokinase—were non-human in origin. Subsequent biopharmaceuticals (*i.e.* growth hormone and factor VII) were human in origin. More recently, recombinant proteins (*i.e.* interferon or insulin) have become available. The number of licensed biopharmaceuticals rose from about 10 in 1996 to over 75 in 2006; it is thought that by the end of 2010, approximately half of all newly licensed pharmaceutical agents will be of this type.

The value of the protein therapeutic market has increased dramatically. Compare sales figures of $US17 billion for all protein therapeutics in 1999, to the value just a few years later: $US37 billion in 2003, over $US40 billion in 2004, $US53 billion in 2005, $US57 billion in 2006, and $US54.5 billion in 2007. Protein therapeutics are expected to increase at a compound growth rate in excess of 12%, reaching over $US75.8 billion by 2012. Currently, sales for the generics market is in the region of $US5 billion, as patents for older recombinant cytokines expire.

Apart from vaccines, with sales of $US15billion, mAbs and cytokines dominate the biopharmaceutical market. The number of mAb products that have entered the market rose from 1 in 1994 to 16 in 2004 and to 22 in 2008. The mAb market was estimated at $US14 billion in 2005 and $US16 billion in 2006, and currently exhibits double-digit growth. Since they address unequivocal unmet medical needs, have low competitor activity, and will face light competition from generics into the near future, the monoclonal market, which comprises about a quarter of the whole protein therapeutics market, continues, manifesting robust growth. Boosted by the number of new mAbs being approved and new indications for existing treatment regimes, sales growth during 2006–2012 is expected to exceed 14%, compared with a rate of just 0.6% for small molecule therapeutics.

15.3.5 Pharmacogenomics and Personal Medicines

Through the combination of personal medical history, a patient's family history, and diagnostic testing, medicine has always sought to personalise medicine. Personalised medicine is in no way a new idea however. For literally thousands of years, traditional Chinese medicine has undertaken a personalised approach, offering distinct therapy based on a patient's signs and symptoms. Today, personalised medicine seeks to deliver specific healthcare treatment, either preventive or therapeutic, on the basis of specific single or multiple genomic or molecular biomarkers. What then are the implications of personalised medicine in the present context? In short, it is the idea that therapeutic intervention using drugs, biologics, or vaccines can be customised to individual genetics, rather than through a haphazard trial-and-error approach to prescribing, based on drug responses homogenised across large patient populations. Clearly, individual patients can respond in dramatically different ways to the same medication. If we can link such differences to individual genetic variation, we can conceive of the systematic and prospective screening of patients prior to therapy.

Remember that this human genome is a composite derived from at least five donors. In 2007, the first individual human genomes were sequenced and published. Thus James Watson and J Craig Venter became the first of thousands—perhaps in time millions—to know their own DNA. Thus in the space of a few years the sequencing of a genome has gone from a transcendent achievement, capable of stopping the scientific world in its tracks, to the almost mundane, worthy of only a minor mention in a journal of the second rank. In future times, genomic sequencing may simply become a workaday laboratory technique. Within a short period, the $1000 genome will become a medical mainstay; at least the companies offering this service certainly hope so. Longer term, micro-automation may make identifying sets of human SNPs and gene indels as routine markers as dip-stick tests currently are.

By identifying the optimal medication before treatment and precisely tailoring the dose, we would anticipate considerably reducing the occurrence of adverse reactions. Most marketed drugs are effective for about 40% of target patients. This reduces to 20% for cancer chemotherapy. Up to 25–50% of diabetic patients, 20–40% of those with depression, and 40–75% of asthmatics do not respond to drug treatments. Moreover, adverse drug reactions are common, resulting in both mortality and long term morbidity.

Several potential blockbusters, including rofecoxib and grapefloxacin, have been withdrawn due to adverse drug reactions (ADRs) that were revealed only by pharmacovigilance of post-licensing responses in the general patient population.

In principle, personalised or customised medicine might have significant resource implications, improving the overall response and thus the efficacy and efficiency of treatment, as well as avoiding unnecessary and wasteful misprescribing, and additionally saving costs for patients and governments. Current obstacles to adoption include a dearth of appropriate evidence-based controlled trials and constraints resulting from the confused economic landscape of healthcare.

Currently, the greatest progress is in personalising oncology, where personalised medicine may improve treatment of these still-high mortality diseases by identifying specific patient subsets to which receptor-specific therapy can be targeted. We have already seen how Wang *et al.* identified links between SNPs impacting gene expression within antioxidant pathways and susceptibility to lung cancer.[19] Like many civilisation diseases, smoking is a particularly challenging target for personalised medicine. Around 1 in 5 people across the world smoke, contributing to perhaps 5 million preventable deaths annually. Smoking is a particular risk factor for lung cancer and chronic obstructive pulmonary disease (COPD), yet no more than 15% of smokers actually develop lung cancer. Thus, and notwithstanding the volitional and psychological component of smoking, the genetic component is crucial: effectively identifying those likely to develop lung cancer allows efforts to be focussed on their prophylaxis both through drug interventions and by cognitive behavioural therapy. In non-small-cell lung cancer, for example, the most significant recent development has been the discovery that patients with activated epidermal growth factor (EGF) receptor mutations at exons 19 and 21 respond significantly to EGF receptor tyrosine kinase inhibitors, such as erlotinib or gefitinib, with concomitant enhancements to long-term survival rates.[127] The identification of all such biomarkers, and the development or identification of associated therapeutic regimes, is thus a high priority if personalised medicine is to fulfil its potential.

In the widest sense, this endeavour is predicated on a simple assertion: differences between individuals impact drug action. These differences may be in the genetic structure of an individual, as manifest at the level of SNPs, indels, and/or alleles. Or it may be manifest as the presence or absence of a protein biomarker, or it may be keyed to particular protein expression levels, or some other proteomic characteristic. Alternatively, it may only be manifest at the system level, as the presence or characteristic concentration of a small molecule metabolite. To organise and orchestrate this burgeoning embarrassment of riches, we must turn to the emergent disciplines of pharmacogenetics and pharmacogenomics. We use pharmacogenetics to refer to the study of drugs which interact with a restricted number of genes, while pharmacogenomics refers to the study of drugs interacting with the entire complement of genes (*i.e.* genome). *Pharmacogenomics* is often used to include pharmacogenetics.

Much of pharmacogenomics is directed at understanding how individual differences in genetic sequence relate to drug response. Yet this is no easy task. The principal tools try to identify linkages between the presence of SNPs, or other genetic features, with the presence and absence of disease states or drug reactions in patient and control groups. The process is greatly complicated by the multi-factorial nature of many disease or ADRs. For example, multiple genes contribute significantly to the disposition, efficacy, and toxicity of most drugs. Non-genetic influences will also contribute. Such factors might include patient characteristics such as age, or gender, or body weight, or other life-style choices, such as diet; likewise co-morbid disease and drug–drug interactions, among many other factors. Other markers, such as blood clotting times for warfarin administration, are used to guide drug administration. Thus future treatment will need to integrate all such data when deciding what drug and what dose to use for an individual patient, or when trying to develop guidelines for emerging diagnostic, therapeutic, and preventive strategies to improve health.

However, in many instances, it remains unclear whether testing, with its implications for delay, sub-therapeutic dosing, or even the avoidance of an otherwise beneficial drug moiety, would significantly improve outcomes compared with current practice. Consequently, many prospective controlled trials will be required to demonstrate the benefits of pharmacogenomics. Nonetheless, pharmacogenomics can potentially identify patients more likely to experience adverse effects (safety pharmacogenomics) or respond better to particular drugs (efficacy pharmacogenomics). For drugs with a narrow therapeutic index separating effective dose from toxic and ineffective doses, pharmacogenomics might facilitate a more rational and informed selection of an effective starting dose.

15.4 Conclusion

A large number of currently available small molecule drugs are seen to be successor drugs, me-toos, or follow-on products. The negative view holds that development of such drugs wastes effort that could be better channelled elsewhere. On the positive side, as well as introducing important price competition, 'me-too' drugs can provide useful therapeutic alternatives, typically with different specificities, affinity enhancements, or different ADMEt profiles. This can be particularly advantageous for particular patient subpopulations. DiMasi *et al.* looked at first-in-class drug approvals from the 1960s through to the mid 2000s. Ninety-four distinct therapeutic classes yielded 287 follow-on drugs. Over the nearly 50 year period, the time required for second entry into a therapeutic area has reduced from over nine years (1970s average) to 1.7 years today. A similar reduction is also seen for third and later entries. To quote DiMasi *et al.*: "Implicit in some of the criticism of the development of me-too drugs has been the assumption that their development occurs following the demonstration of clinical and commercial success by the first-in-class drug. However, given assessments of the length of time that is typically required for drug development—estimated at between 10 to 15 years—the data on the timing of entry of follow-on drugs in a particular class, in this study and in our previous study, suggest that much of the development of what turn out to be follow-on drugs must occur before the approval of the breakthrough drug." Indeed since the early 1990s, about 90% of second entrants have already been filed to go into clinical trials before the first one has been approved, and 64% of the time another compound has in fact already started Phase III testing. Patent filings tell the story even more graphically, as is often the case in this industry. For new drug classes approved since the 1970s, 90% have had at least one of the eventual follow-on drugs showing its first worldwide patent filing before approval of the pioneer compound. As the work of DiMasi *et al.* shows, when a new therapeutic area first opens up, everyone begins at approximately the same time. The development of follow-on or me-too drugs is more a race not an imitation. What is consistent with this is the notion that the number of druggable targets is small; or, to expand upon this idea, that the limitations imposed by our existing experience of medicinal chemists necessarily restrict the size of the druggable genome. To exploit new targets and find out how to drug them, through augmentations of our chemical and biophysical knowledge, requires time, effort, and resources. The genome is now available, identifying all possible targets; learning to exploit all that can be exploited is not the work of a moment but a Sisyphean task of no small magnitude.

In this chapter we have been concerned with drugs and genes, two things that act together as agents of complexity within the biological arena. They are inextricably linked, since we cannot construct complex and meaningful higher-order structures in biology if the fundamental data is questionable and inadequate, any more than we can construct castles on quick sand. Drug discovery has tried in vain to capitalise on the genomic and post-genomic revolutions. Instead the Pharmaceutical Industry finds itself at a pivotal point in their history, poised between the waning success that characterises the recent past and the prospect of two potential futures: continuing misfortune on the one hand and the challenge of completely realigning themselves to engage properly with the post-genomic future. The quest to understand biology as a system from the bottom up as well as the top down is part of the post-genomic movement within molecular bioscience. Part of this is to address the burgeoning surfeit of information that now exists, and part seeks to address the atavistic reductionism of twentieth century biology.[128] We do not simply wish take the world apart; no, we wish to dismantle it only so that we may quickly put it back together, and in so doing come to know it and ourselves better. This endeavour is predicated on our fundamental physical understanding of the chemistry that underpins biology. This is the nascent realm of integrative systems biology, the intersection of experimental and computational post-genomics. It is not too late to remedy the problems faced by the pharmaceutical industry, and by so doing secure the systematic development of therapeutic agents. The mistakes of the past can be undone, and the mistakes of the future need not be made. While all the "low-hanging fruit" upon which the pharmaceutical industry initially grew, fat has largely been harvested; the opportunity offered by the genome has not yet been squandered. All is still possible; drug discovery needs only to engage properly with systems biology and associated techniques to fully realise all that the genomic sciences make possible. Genomics has changed how we think of biology and how best to intervene in biological systems, and it has spurred the development of a range of powerful new techniques. It has changed the mental landscape of drug discovery, forcing us to consider things previously ignored. Yet at the same time this provides opportunity for discovering treatments for the previously untreatable. The world still needs us to address unmet medical needs, while the pharmaceutical industry faces enormous, unprecedented problems, yet, through genomics, we have the best means ever to address these formidable issues.

Acknowledgements

With humility, I acknowledge everyone which whom I have ever worked for their help in fostering, fomenting, and reinforcing my interests and understanding as made manifest within this article. Not all of them would wish me to hold the views I do hold, but I acknowledge them just the same.

References

1. W. Fiers, R. Contreras, F. Duerinck, G. Haegeman, D. Iserentant, J. Merregaert, W. Min Jou, F. Molemans, A. Raeymaekers, A. Van den Berghe, G. Volckaert and M. Ysebaert, *Nature*, 1976, **260**, 500–507.
2. W. Min Jou, G. Haegeman, M. Ysebaert and W. Fiers, *Nature*, 1972, **237**, 82–88.
3. F. Sanger, G. M. Air, B. G. Barrell, N. L. Brown, A. R. Coulson, C. A. Fiddes, C. A. Hutchison, P. M. Slocombe and M. Smith, *Nature*, 1977, **265**, 687–695.
4. R. D. Fleischmann, M. D. Adams, O. White, R. A. Clayton, E. F. Kirkness, A. R. Kerlavage, C. J. Bult, J. F. Tomb, B. A. Dougherty, J. M. Merrick *et al.*, *Science*, 1995, **269**, 496–512.
5. C. Southan, *Proteomics*, 2004, **4**, 1712–1726.
6. K. J. Nordstrom, M. A. Mirza, T. P. Larsson, D. E. Gloriam, R. Fredriksson and H. B. Schioth, *Biochem. Biophys. Res. Commun.*, 2006, **348**, 1063–1074.
7. M. Clamp, B. Fry, M. Kamal, X. Xie, J. Cuff, M. F. Lin, M. Kellis, K. Lindblad-Toh and E. S. Lander, *Proc. Natl. Acad. Sci. U. S. A.*, 2007, **104**, 19428–19433.
8. J. Cohen and I. Stewart, *Nature*, 2001, **409**, 1119–1122.
9. F. Aguero, B. Al-Lazikani, M. Aslett, M. Berriman, F. S. Buckner, R. K. Campbell, S. Carmona, I. M. Carruthers, A. W. Chan, F. Chen, G. J. Crowther, M. A. Doyle, C. Hertz-Fowler, A. L. Hopkins, G. McAllister, S. Nwaka, J. P. Overington, A. Pain, G. V. Paolini, U. Pieper, S. A. Ralph, A. Riechers, D. S. Roos, A. Sali, D. Shanmugam, T. Suzuki, W. C. Van Voorhis and C. L. Verlinde, *Nat. Rev. Drug Discovery*, 2008, **7**, 900–907.
10. J. P. Overington, B. Al-Lazikani and A. L. Hopkins, *Nat. Rev. Drug Discovery*, 2006, **5**, 993–996.
11. D. R. Flower and T. K. Attwood, *Semin. Cell. Dev. Biol.*, 2004, **15**, 693–701.
12. L. A. Doyle, W. Yang, L. V. Abruzzo, T. Krogmann, Y. Gao, A. K. Rishi and D. D. Ross, *Proc. Natl. Acad. Sci. U. S. A.*, 1998, **95**, 15665–15670.
13. C. J. Jeffery, *Mol. Biosyst.*, 2009, **5**, 345–350.
14. C. J. Jeffery, *Curr. Opin. Struct. Biol.*, 2004, **14**, 663–668.
15. C. J. Jeffery, *Trends Genet.*, 2003, **19**, 415–417.
16. C. J. Jeffery, *Trends Biochem. Sci.*, 1999, **24**, 8–11.
17. S. Stankovic and N. Majkic-Singh, *Crit. Rev. Cl. Lab. Sci.*, 2010, **47**, 72–123.
18. L. M. Caronia, C. Martin, C. K. Welt, G. P. Sykiotis, R. Quinton, A. Thambundit, M. Avbelj, S. Dhruvakumar, L. Plummer, V. A. Hughes, S. B. Seminara, P. A. Boepple, Y. Sidis, W. F. Crowley Jr., K. A. Martin, J. E. Hall and N. Pitteloud, *N. Engl. J. Med.*, 2011, **364**, 215–225.

19. X. Wang, B. N. Chorley, G. S. Pittman, S. R. Kleeberger, J. Brothers 2nd, G. Liu, A. Spira and D. A. Bell, *PLoS One*, 2010, **5**, e11934.
20. W. Doerfler, *Epigenetics*, 2008, **3**, 125–133.
21. P. Kahn, *Science*, 1995, **270**, 369–370.
22. H. Keren, G. Lev-Maor and G. Ast, *Nat. Rev. Genet.*, 2010, **11**, 345–355.
23. F. B. Perler, E. O. Davis, G. E. Dean, F. S. Gimble, W. E. Jack, N. Neff, C. J. Noren, J. Thorner and M. Belfort, *Nucleic Acids Res.*, 1994, **22**, 1125–1127.
24. F. B. Perler, *Nucleic Acids Res.*, 2002, **30**, 383–384.
25. F. B. Perler, *Nucleic Acids Res.*, 2000, **28**, 344–345.
26. F. B. Perler, *Nucleic Acids Res.*, 1999, **27**, 346–347.
27. V. C. Wasinger, S. J. Cordwell, A. Cerpa-Poljak, J. X. Yan, A. A. Gooley, M. R. Wilkins, M. W. Duncan, R. Harris, K. L. Williams and I. Humphery-Smith, *Electrophoresis*, 1995, **16**, 1090–1094.
28. N. G. Anderson and L. Anderson, *Clin. Chem.*, 1982, **28**, 739–748.
29. C. M. Murray, R. Hutchinson, J. R. Bantick, G. P. Belfield, A. D. Benjamin, D. Brazma, R. V. Bundick, I. D. Cook, R. I. Craggs, S. Edwards, L. R. Evans, R. Harrison, E. Holness, A. P. Jackson, C. G. Jackson, L. P. Kingston, M. W. Perry, A. R. Ross, P. A. Rugman, S. S. Sidhu, M. Sullivan, D. A. Taylor-Fishwick, P. C. Walker, Y. M. Whitehead, D. J. Wilkinson, A. Wright and D. K. Donald, *Nat. Chem. Biol.*, 2005, **1**, 371–376.
30. M. Pawlak, M. F. Templin, D. Stoll and T. O. Joos, *Chim. Oggi.*, 2007, **25**, 62.
31. M. F. Templin, D. Stoll, M. Schrenk, P. C. Traub, C. F. Vohringer and T. O. Joos, *Drug Discovery Today*, 2002, **7**, 815–822.
32. M. F. Templin, D. Stoll, M. Schrenk, P. C. Traub, C. F. Vohringer and T. O. Joos, *Trends Biotechnol.*, 2002, **20**, 160–166.
33. J. Walker, D. Flower and K. Rigley, *Curr. Opin. Hematol.*, 2002, **9**, 23–29.
34. P. S. Andrews, S. Schneider, E. Yang, M. Michaels, H. Chen, J. Tang and R. Emkey, *Assay Drug Dev Techn.*, 2010, **8**, 471–487.
35. W. Pearson, *Curr. Protoc. Bioinformatics*, 2004, Ch. 3, Unit 3-9.
36. W. R. Pearson, *Methods Mol. Biol.*, 1994, **24**, 307–331.
37. W. R. Pearson, *Genomics*, 1991, **11**, 635–650.
38. W. R. Pearson, *Methods Enzymol.*, 1990, **183**, 63–98.
39. S. F. Altschul, J. C. Wootton, E. M. Gertz, R. Agarwala, A. Morgulis, A. A. Schaffer and Y. K. Yu, *FEBS J.*, 2005, **272**, 5101–5109.
40. S. F. Altschul, T. L. Madden, A. A. Schaffer, J. Zhang, Z. Zhang, W. Miller and D. J. Lipman, *Nucleic Acids Res.*, 1997, **25**, 3389–3402.
41. S. F. Altschul, *J. Mol. Evol.*, 1993, **36**, 290–300.
42. S. F. Altschul and D. J. Lipman, *Proc. Natl. Acad Sci U. S. A.*, 1990, **87**, 5509–5513.
43. A. D. Baxevanis, *Nucleic Acids Res.*, 2003, **31**, 1–12.
44. A. D. Baxevanis, *Nucleic Acids Res.*, 2002, **30**, 1–12.
45. A. D. Baxevanis, *Nucleic Acids Res.*, 2001, **29**, 1–10.
46. A. D. Baxevanis, *Nucleic Acids Res.*, 2000, **28**, 1–7.
47. M. Linial, *Trends Biotechnol.*, 2003, **21**, 298–300.

48. W. R. Gilks, B. Audit, D. de Angelis, S. Tsoka and C. A. Ouzounis, *Math. Biosci.*, 2005, **193**, 223–234.
49. W. R. Gilks, B. Audit, D. De Angelis, S. Tsoka and C. A. Ouzounis, *Bioinformatics*, 2002, **18**, 1641–1649.
50. T. K. Attwood, *Int. J Biochem. Cell Biol.*, 2000, **32**, 139–155.
51. W. N. Grundy, T. L. Bailey and C. P. Elkan, *Comput. Appl. Biosci.*, 1996, **12**, 303–310.
52. I. Jonassen, *Comput. Appl. Biosci.*, 1997, **13**, 509–522.
53. L. Holm and P. Rosenstrom, *Nucleic Acids Res.*, 2010, **38**, W545–549.
54. L. Holm, S. Kaariainen, C. Wilton and D. Plewczynski, *Curr. Protoc. Bioinformatics*, 2006, Ch. 5, Unit 55.
55. L. Holm and C. Sander, *Nucleic Acids Res.*, 1998, **26**, 316–319.
56. L. Holm and C. Sander, *Nucleic Acids Res.*, 1997, **25**, 231–234.
57. L. Holm and C. Sander, *Trends Biochem. Sci.*, 1995, **20**, 478–480.
58. M. Veeramalai, D. Gilbert and G. Valiente, *BMC Bioinformatics*, 2010, **11**, 138.
59. I. Michalopoulos, G. M. Torrance, D. R. Gilbert and D. R. Westhead, *Nucleic Acids Res.*, 2004, **32**, D251–254.
60. D. Gilbert, D. Westhead, J. Viksna and J. Thornton, *Comput. Chem.*, 2001, **26**, 23–30.
61. D. Gilbert, D. Westhead, N. Nagano and J. Thornton, *Bioinformatics*, 1999, **15**, 317-326.
62. G. Lev-Maor, A. Goren, N. Sela, E. Kim, H. Keren, A. Doron-Faigenboim, S. Leibman-Barak, T. Pupko and G. Ast, *PLoS Genet.*, 2007, **3**, e203.
63. E. Koren, G. Lev-Maor and G. Ast, *PLoS Comput Biol*, 2007, **3**, e95.
64. A. Goren, O. Ram, M. Amit, H. Keren, G. Lev-Maor, I. Vig, T. Pupko and G. Ast, *Mol. Cell*, 2006, **22**, 769–781.
65. R. Sorek, G. Lev-Maor, M. Reznik, T. Dagan, F. Belinky, D. Graur and G. Ast, *Mol. Cell*, 2004, **14**, 221–231.
66. Y. Peleg and T. Unger, *Methods Mol. Biol.*, 2008, **426**, 197–208.
67. C. Romier, M. Ben Jelloul, S. Albeck, G. Buchwald, D. Busso, P. H. Celie, E. Christodoulou, V. De Marco, S. van Gerwen, P. Knipscheer, J. H. Lebbink, V. Notenboom, A. Poterszman, N. Rochel, S. X. Cohen, T. Unger, J. L. Sussman, D. Moras, T. K. Sixma and A. Perrakis, *Acta Crystallogr., Sect. D Biol. Crystallogr.*, 2006, **62**, 1232–1242.
68. P. M. Alzari, H. Berglund, N. S. Berrow, E. Blagova, D. Busso, C. Cambillau, V. Campanacci, E. Christodoulou, S. Eiler, M. J. Fogg, G. Folkers, A. Geerlof, D. Hart, A. Haouz, M. D. Herman, S. Macieira, P. Nordlund, A. Perrakis, S. Quevillon-Cheruel, F. Tarandeau, H. van Tilbeurgh, T. Unger, M. P. Luna-Vargas, M. Velarde, M. Willmanns and R. J. Owens, *Acta. Crystallogr., Sect. D Biol. Crystallogr.*, 2006, **62**, 1103–1113.
69. H. Walden, *Acta Crystallogr., Sect. D Biol. Crystallogr.*, 2010, **66**, 352–357.
70. N. E. Chayen, *Adv. Protein Chem. Struct. Biol.*, 2009, **77**, 1–22.
71. R. Meged, O. Dym and J. L. Sussman, *Methods Mol. Biol.*, 2008, **426**, 411–418.

72. M. Hiraki, R. Kato, M. Nagai, T. Satoh, S. Hirano, K. Ihara, N. Kudo, M. Nagae, M. Kobayashi, M. Inoue, T. Uejima, S. Oda, L. M. Chavas, M. Akutsu, Y. Yamada, M. Kawasaki, N. Matsugaki, N. Igarashi, M. Suzuki and S. Wakatsuki, *Acta Crystallogr., Sect. D Biol. Crystallogr.*, 2006, **62**, 1058–1065.
73. F. Li, H. Robinson and E. S. Yeung, *Anal. Bioanal. Chem.*, 2005, **383**, 1034–1041.
74. C. Cumbaa and I. Jurisica, *J. Struct. Funct. Genomics*, 2005, **6**, 195–202.
75. R. Hui and A. Edwards, *J. Struct. Biol.*, 2003, **142**, 154–161.
76. M. J. Fogg and A. J. Wilkinson, *Biochem. Soc. Trans.*, 2008, **36**, 771–775.
77. P. Smialowski and D. Frishman, *Methods Mol. Biol.*, 2010, **609**, 385–400.
78. C. A. Smith and A. E. Cohen, *JALA Charlottesv Va*, 2008, **13**, 335–343.
79. A. Joachimiak, *Curr. Opin. Struct. Biol.*, 2009, **19**, 573–584.
80. Y. Gu, C. Zheng, Y. Zhao, H. Ke and H. Fan, *Acta Crystallogr., Sect. D. Biol. Crystallogr.*, 1997, **53**, 792–794.
81. C. J. Gilmore, *Acta Crystallogr., Sect. D Biol. Crystallogr.*, 2000, **56**, 1205–1214.
82. I. Uson and G. M. Sheldrick, *Curr. Opin. Struct. Biol.*, 1999, **9**, 643–648.
83. J. Karle, *Acta Crystallogr., Sect. A Fundam. Crystallogr.*, 1989, **45** (11), 765–781.
84. P. J. Hajduk, *Mol. Interv.*, 2006, **6**, 266–272.
85. S. B. Shuker, P. J. Hajduk, R. P. Meadows and S. W. Fesik, *Science*, 1996, **274**, 1531–1534.
86. A. M. Petros, J. R. Huth, T. Oost, C. M. Park, H. Ding, X. Wang, H. Zhang, P. Nimmer, R. Mendoza, C. Sun, J. Mack, K. Walter, S. Dorwin, E. Gramling, U. Ladror, S. H. Rosenberg, S. W. Elmore, S. W. Fesik and P. J. Hajduk, *Bioorg. Med. Chem. Lett.*, 2010, **20**, 6587–6591.
87. C. Scheich, V. Puetter and M. Schade, *J. Med Chem.*, 2010, **53**(23), 8362–8367.
88. A. C. Gibbs, M. C. Abad, X. Zhang, B. A. Tounge, F. A. Lewandowski, G. T. Struble, W. Sun, Z. Sui and L. C. Kuo, *J. Med. Chem.*, 2010, **53**, 7979–7991.
89. V. Fruh, Y. Zhou, D. Chen, C. Loch, E. Ab, Y. N. Grinkova, H. Verheij, S. G. Sligar, J. H. Bushweller and G. Siegal, *Chem. Biol.*, 2010, **17**, 881–891.
90. B. G. Olivier and J. L. Snoep, *Bioinformatics*, 2004, **20**, 2143–2144.
91. J. L. Snoep and B. G. Olivier, *Microbiology*, 2003, **149**, 3045-3047.
92. J. L. Snoep and B. G. Olivier, *Mol. Biol. Rep.*, 2002, **29**, 259–263.
93. M. F. Land and D.-E. Nilsson, *Animal Eyes*, Oxford University Press, Oxford, 2002.
94. L. H. Hartwell, J. J. Hopfield, S. Leibler and A. W. Murray, *Nature*, 1999, **402**, C47–52.
95. E. M. Phizicky and S. Fields, *Microbiol. Rev.*, 1995, **59**, 94–123.
96. S. Shangary and S. Wang, *Clin. Cancer Res.*, 2008, **14**, 5318-5324.
97. C. Q. Hu and Y. Z. Hu, *Curr. Med. Chem.*, 2008, **15**, 1720–1730.
98. A. Domling, *Curr. Opin. Chem. Biol.*, 2008, **12**, 281–291.

99. L. T. Vassilev, *Cell Cycle*, 2004, **3**, 419–421.
100. L. T. Vassilev, B. T. Vu, B. Graves, D. Carvajal, F. Podlaski, Z. Filipovic, N. Kong, U. Kammlott, C. Lukacs, C. Klein, N. Fotouhi and E. A. Liu, *Science*, 2004, **303**, 844–848.
101. S. Shangary, D. Qin, D. McEachern, M. Liu, R. S. Miller, S. Qiu, Z. Nikolovska-Coleska, K. Ding, G. Wang, J. Chen, D. Bernard, J. Zhang, Y. Lu, Q. Gu, R. B. Shah, K. J. Pienta, X. Ling, S. Kang, M. Guo, Y. Sun, D. Yang and S. Wang, *Proc. Natl. Acad. Sci. U. S. A*, 2008, **105**, 3933–3938.
102. P. H. Kussie, S. Gorina, V. Marechal, B. Elenbaas, J. Moreau, A. J. Levine and N. P. Pavletich, *Science*, 1996, **274**, 948–953.
103. K. Kojima, J. K. Burks, J. Arts and M. Andreeff, *Mol. Cancer Ther.*, 2010, **9**, 2545–2557.
104. R. Rossignol, T. Letellier, M. Malgat, C. Rocher and J. P. Mazat, *Biochem. J.*, 2000, **347** (1), 45–53.
105. J. Howell, *J.Allergy Clin. Immunol.*, 2005, **115**, 882–885.
106. D. R. Flower, *Drug Design: Cutting Edge Approaches*, Royal Society of Chemistry, Cambridge, 2002.
107. M. L. Bolognesi, A. Cavalli, C. Bergamini, R. Fato, G. Lenaz, M. Rosini, M. Bartolini, V. Andrisano and C. Melchiorre, *J. Med. Chem.*, 2009, **52**, 7883–7886.
108. T. Tomasic, N. Zidar, A. Kovac, S. Turk, M. Simcic, D. Blanot, M. Muller-Premru, M. Filipic, S. G. Grdadolnik, A. Zega, M. Anderluh, S. Gobec, D. Kikelj and L. Peterlin Masic, *ChemMedChem*, 2010, **5**, 286–295.
109. Y. Rook, K. U. Schmidtke, F. Gaube, D. Schepmann, B. Wunsch, J. Heilmann, J. Lehmann and T. Winckler, *J. Med. Chem.*, 2010, **53**, 3611–3617.
110. S. A. Antoniu and M. R. Kolb, *IDrugs*, 2010, **13**, 332–345.
111. R. Leon and J. Marco-Contelles, *Curr. Med. Chem.*, 2011, **18**, 552–576.
112. D. Boschi, M. Giorgis, C. Cena, N. C. Talniya, A. Di Stilo, G. Morini, G. Coruzzi, E. Guaita, R. Fruttero and A. Gasco, *ChemMedChem*, 2010, **5**, 1444–1449.
113. M. L. Bolognesi, M. Bartolini, F. Mancini, G. Chiriano, L. Ceccarini, M. Rosini, A. Milelli, V. Tumiatti, V. Andrisano and C. Melchiorre, *ChemMedChem*, 2010, **5**, 1215–1220.
114. C. R. Chong and D. J. Sullivan Jr., *Nature*, 2007, **448**, 645–646.
115. J. K. Aronson, *Br. J. Clin. Pharmacol.*, 2007, **64**, 563–565.
116. M. S. Boguski, K. D. Mandl and V. P. Sukhatme, *Science*, 2009, **324**, 1394–1395.
117. M. R. Bubb, *JAMA-J. Am. Med. Assoc.*, 1998, **279**, 1701–1702.
118. P. Danjou, L. Alexandre, D. Warot, L. Lacomblez and A. J. Puech, *Br. J. Clin. Pharmacol.*, 1988, **26**, 733–739.
119. J. Mulhall, *Br. J. Urol.*, 1997, **79**, 663–664.
120. M. J. Keiser, V. Setola, J. J. Irwin, C. Laggner, A. I. Abbas, S. J. Hufeisen, N. H. Jensen, M. B. Kuijer, R. C. Matos, T. B. Tran, R. Whaley, R. A. Glennon, J. Hert, K. L. Thomas, D. D. Edwards, B. K. Shoichet and B. L. Roth, *Nature*, 2009, **462**, 175–181.

121. M. Kuhn, M. Campillos, I. Letunic, L. J. Jensen and P. Bork, *Mol Syst Biol*, 2010, **6**, 343.
122. M. Campillos, M. Kuhn, A. C. Gavin, L. J. Jensen and P. Bork, *Science*, 2008, **321**, 263–266.
123. M. Kuhn, M. Campillos, P. Gonzalez, L. J. Jensen and P. Bork, *FEBS Lett.*, 2008, **582**, 1283–1290.
124. S. Gunther, M. Kuhn, M. Dunkel, M. Campillos, C. Senger, E. Petsalaki, J. Ahmed, E. G. Urdiales, A. Gewiess, L. J. Jensen, R. Schneider, R. Skoblo, R. B. Russell, P. E. Bourne, P. Bork and R. Preissner, *Nucleic Acids Res.*, 2008, **36**, D919–922.
125. Y. Y. Li, J. An and S. J. Jones, *Genome Inform. Ser.*, 2006, **17**, 239–247.
126. D. R. Flower, *BioDrugs*, 2009, **23**, 231–240.
127. T. S. Mok, Q. Zhou, L. Leung and H. H. Loong, *Expert Rev. Anticanc.*, 2010, **10**, 1601–1611.
128. D. Gatherer, *BMC Syst. Biol.*, 2010, **4**, 22.

CHAPTER 16

Scoring Drug-Receptor Interactions

GREGORY L. WARREN[a]* AND STEPHEN D. WARREN[b]

[a] OpenEye Scientific Software, Inc., 9 Bisbee Court Suite D, Santa Fe, NM 87508, USA; [b] Department of Chemistry & Biochemistry, Gonzaga University, 502 E Boone Ave, Spokane, WA 99258, USA

16.1 Introduction

In the sub-discipline of drug discovery and design, the golden egg for computational chemistry would be the accurate prediction of the binding energy of a ligand in a protein receptor. If the computational chemist could tell the medicinal chemist within half a log unit what the K_i or IC_{50} of a potential compound would be, the medicinal chemist could concentrate on other optimization goals such as toxicity, cross-reactivity or drug metabolism and pharmacokinetics. Historically, it has been the methodology of quantitative structure activity relationships (QSAR), and more recently docking and scoring, that has been used to attain this prize. For those unfamiliar with this field, docking is a computational technique in which a 3-dimensional structure of the receptor is typically held rigid while conformations and positions for the ligand are generated or sampled within the receptor binding site. These are often referred to as ligand binding modes. There are a number of different methods used to generate ligand conformations and placements. For the discussion of affinity prediction in this chapter, how the conformations and placements are generated is not of great importance except when the method does not generate a binding mode close to the biological binding mode. Scoring is a computational method that ranks how well the docked ligand interacts with or

RSC Drug Discovery Series No. 13
Drug Design Strategies: Quantitative Approaches
Edited by David J. Livingstone and Andrew M. Davis

Published by the Royal Society of Chemistry, www.rsc.org

complements the receptor binding site. It is the scoring function that is used to identify the "correct" binding mode, and in some cases to predict the affinity of the ligand molecule. The scoring functions used by docking programs fall under three categories: knowledge-based, empirical, and force field (physical). For a more in-depth synopsis of what these categories mean, the authors would suggest a review by N Moitessier *et al.*[1]

Knowledge-based scoring functions are derived from a statistical analysis of inter-atomic pairwise potentials found in protein–ligand complex structures. The conventional wisdom is that knowledge-based scoring functions are useful for identifying the biological binding mode but are not useful for predicting affinity. One possible reason is that knowledge-based scoring functions look only at observed interactions and do not take into account, during the statistical analysis, the affinity of the ligand for the protein or the affinity of individual atomic interactions.

Empirical scoring functions attempt to decompose or simplify protein–ligand interactions into a few scalable functions. These functions typically count the number of hydrophobic interactions and the number of hydrogen bonds, or attempt to account for loss of entropy by counting the number of rotatable bonds in the ligand. The coefficients to each component of the function are then fit by a regression algorithm to generate the best fit to experimentally determined protein–ligand affinities for a set of protein–ligand complex structures. Because these scoring functions are optimized to reproduce experimental binding affinity data, there is hope that they can be used to predict ligand affinity.

Force fields were originally developed to simulate the conformational, thermodynamic and kinetic behaviors of small molecules or macromolecules much more efficiently, from a computer resource perspective, than quantum mechanical methods. It was hoped that if the interaction components of the force fields were applied to docking and scoring, the more accurate representation of the physics of protein–ligand interactions would result in more accurate predictions of protein–ligand binding affinities.

There are other, more rigorous physics-based methods for scoring binding modes or predicting free energy of binding from predicted or experimentally determined binding modes. These methods, however, are not used in docking programs because they require large amounts of computer time to generate results for a single pose, making them impractical for rapidly scoring thousands to millions of potential binding modes generated by the docking portion of a docking and scoring program.

In the following sections, we will briefly discuss what docking has been successfully used for: binding mode prediction and lead identification (virtual screening). We will next discuss the methods available to the computational and medicinal chemist for predicting ligand affinity. We will then discuss how the scoring component of docking and scoring has been used to predict ligand affinity, as well as discuss some of the successes and failures. Finally, we will look at some other, more rigorous physics-based methods for predicting protein–ligand affinity. During the conclusion, we will summarize the current state-of-the-art for predicting drug–receptor affinity and briefly discuss some prospects for the future.

16.2 Docking – What Works?

In practice, docking and scoring programs are most successful in predicting the binding mode of a new ligand, and when used in virtual screening for lead identification. A successful prediction of the binding mode of a new ligand or scaffold can provide a drug discovery and development team with an enormous advantage by indicating which substituents are likely to complement the receptor active site *versus* choices made at random. Early in the lead optimization process, when little structure–activity relationship (SAR) data is available for a particular chemical series, the prediction of the binding mode is particularly valuable. Virtual screening can also serve as a useful guide for compound selection or compound purchase when a new scaffold is needed for whatever reason.

16.2.1 Binding Mode Prediction

In the introduction, we described docking and scoring as a two-step process: pose generation followed by pose selection by the scoring function. In practice, the process is coupled such that scoring functions are used to drive the placement of ligand poses within the binding site and are also used to rank the set of poses returned to the user. This dependence of both steps on scoring functions complicates analysis of the performance of docking algorithms for binding mode prediction.

In general, docking algorithms can generate the correct binding mode >70% of the time if ligands of low conformational complexity are docked into binding pockets of small to medium size. However, a close examination of the data from the evaluation of docking and scoring programs by Warren *et al.*[2] suggests that results degrade significantly as the size and complexity of the test system increases. Across all of the proteins and ligands included in the Warren analysis, docking algorithms generated a correct pose for 31–94% of the ligands for a particular protein target. A pose is considered correct if there is only a small difference in position between the docked and crystallographic ligand. Traditionally, this difference is measured by root mean square deviation (RMSD). Root mean square (RMS, the square root of the squared difference) is a statistical method for generating the absolute value when a difference can be positive or negative. In this case, RMSD is simply the average over all the atom pair RMS values for the docked pose *versus* the crystallographic pose.

When examined more closely, however, it was found that docking algorithms could not generate poses within 2.0 Å RMSD of the correct pose when applied to large binding sites, flexible ligands (greater than eight rotatable bonds), or systems with few specific non-hydrophobic interactions between ligand and protein. Docking algorithms are able to successfully search conformational space to generate sets of protein–ligand conformations that include a correct pose. In practice, computational chemists do not know the correct binding pose *a priori* and in the absence of such knowledge must depend on the scoring function to identify which of the generated poses is sufficiently correct to be useful for structure-based design. Scoring functions, however, are not particularly successful in identifying which of the generated poses is most similar to the crystallographically determined ligand–protein complex. In the case of the Warren *et al.* evaluation,[2] success rates ranged from 31–94% when the best RMSD was less than 2.0 Å for at least one pose of all poses generated. When there were two criteria for success, best score and RMSD less than 2.0 Å, the success rates dropped to 8–74% depending on the target class and the particular scoring function used. If we treat docking and scoring as two separate processes, the metric of the best RMSD generated by the docking program is a measure of how well the docking program searched the active site. The RMSD of the best scoring pose is then a measure of how well the scoring function was able to identify the correct pose–the pose observed biologically. It is important to point out that while we are using RMSD, RMSD is an especially poor method for comparison of diverse structures for a number of reasons, two of which we will mention. The first reason is that RMSD is not size independent. For example, the apparent alignment for two conformations of a ligand with 12 atoms and a RMSD of 2.0 Å shows significantly greater displacement than the alignment of two conformations of a ligand with 60 atoms and a RMSD of 2.0 Å. The second reason, elegantly pointed out by Yusuf *et al.*,[3] is that RMSD in this case compares the coordinates of a model of electron density (the X-ray crystal structure) with the coordinates generated by the modeling program (docking). A better approach would be to compare how well the coordinates from the docking program fit the experimental data using either a real-space R-factor (RSR) or a real-space correlation coefficient (RSCC). Unfortunately, the predominant metric reported in the literature is RMSD.

Cross *et al.*[4] published a more recent evaluation of the accuracy of docking programs for binding mode predictions, using a smaller set of higher quality protein–ligand X-ray crystal structures. Most of the structures in this data set were of high resolution (<2.0 Å) and had been visually checked to make sure that there was complete electron density for the ligand in the active site. In this study, they found that the best performing docking and scoring programs were able to generate the correct pose about 85% of the time, and the scoring function was able to identify the correct pose 72% of the time. Based on both the results of these two evaluations and the reported results of other evaluations, we conclude that the search problem is largely solved; docking algorithms can generate good docking poses. We additionally conclude that scoring functions do limit the predictive ability of docking algorithms, but not to a great degree if the problematic large binding pockets and ligands with high flexibility are discounted.

16.2.2 Virtual Screening

Virtual screening is a computational method that attempts to determine, for a set of molecules, which molecules are active against the target of interest. Virtual screening is the simplest of activity prediction tasks and asks only if the ligand is active or inactive, requiring that prediction be accurate to 3 or 4 orders of magnitude in activity (< 10 μM). In computational chemistry, there are two sources that can be mined for activity: active ligands or receptors. If comparison or activity determination uses an active ligand (or ligands), this is described as ligand-based virtual screening. If the protein target or receptor is used, the screen is described as structure-based virtual screening. The basis for activity in the case of ligand-based virtual screening is similarity between the molecule under question and a known active. This similarity can be measured using graph (1- or 2-dimensional), 3-dimensional shape, and/or 3-dimensional chemistry matching methods. The fundamental hypothesis is that if a molecule is similar to the known active, then it has a higher-than-random probability of being active. The structure-based virtual screening hypothesis is that if the compound of interest fits into the target active site and makes complementary interactions, it again has a higher-than-random probability of being active. Based on the discussion in the previous section, there is a requirement in structure-based virtual screening that the docking program correctly place active ligands in the active site and identify the correct placement. In addition, the scoring function is required to distinguish between active ligands *versus* inactive ligands. Docking programs can, on average, identify the correct binding mode about 72% of the time. With the additional requirement of the scoring function identifying active *versus* inactive ligands, one should expect that docking programs would be less successful at structure-based virtual screening than they have been shown to be at predicting a ligand's binding mode.

Are docking programs successful at structure-based virtual screening? The answer, as shown by a number of studies,[2,4,5] is yes. However, as shown by McGaughey *et al.*,[5] ligand-based methods are more successful than structure-based virtual screening. In addition, most ligand-based methods require significantly less computing resources to screen the same number of compounds. So why would one do structure-based virtual screening when ligand-based virtual screening is more successful and faster? Structure-based virtual screening is useful for three reasons. First, just as there are cases where the structure of the protein is not available (G-coupled protein receptor and other membrane receptors), there are cases where an active ligand has not been discovered, so there is no ligand to use in the ligand-based screen in this case. Second, because ligand-based virtual screening looks for ligands similar to the active ligand, the resulting compounds are usually very similar to the starting ligand. If there is a need for diversity from the current chemical series (scaffold hopping) then docking is a good method for finding diverse hits[5] (see Figure 16.1). The reason structure-based virtual screening works better when diversity is the goal is that the selection criterion is complementarity to the protein receptor, not similarity to a known active ligand. Last, structure-based virtual screening results are orthogonal to ligand-based virtual screening results. The compounds found by ligand-based methods are different to those found by structure-based methods.[6] Depending on the goals for the virtual screen, the scientist would make the decision as to whether structure-based or ligand-based virtual screening is the method of choice.

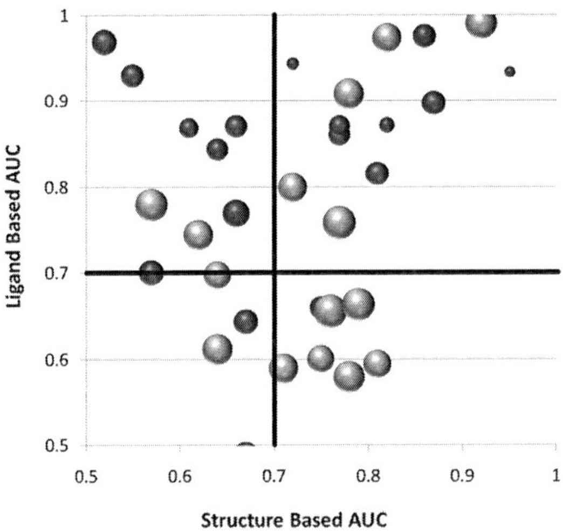

Figure 16.1 Virtual screening results for the Virtual screening results for the Database of Useful Decoys (DUD) data set[7] for structure-based (x-axis) and ligand-based (y-axis) virtual screening. The similarity of the hits are recorded by color, with the light gray spheres representing hits with low similarity and the dark gray spheres representing hits with high similarity. Note that in the upper left quadrant, where the results are good for ligand-based virtual screening but poor for structure-based virtual screening, the hits are dark gray and chemically similar. In the lower right quadrant, where the results are good for structure-based virtual screening but poor for ligand-based virtual screening, the hits are more diverse (light grey spheres). The AUC metric is the area under the curve of a receiver operating characteristic (ROC) curve, which is a plot of the true positive rate *versus* the false positive rate. An AUC value of 0.5 is random and a value of 1.0 is a perfect prediction.

We have shown that structure-based virtual screening works, but have we shown that it works for the right reason? Does the docking program generate the correct pose and does the scoring function correctly determine whether the ligand is active or inactive? Only the study by Warren *et al.*[2] provides some data to answer this question. The data generated by the docking program Genetic Optimization for Ligand Docking (GOLD) seems to indicate that the answer is no, where the docked ligands of methionyl tRNA synthetase (74% within 2.0 Å RMSD of the correct binding mode for the best-scoring pose) show essentially random performance when virtually screening this protein. Conversely, GOLD was only able to dock 10% of Factor Xa ligands within 2.0 Å but still had quite good performance in virtual screening. The assumption in structure-based virtual screening is that success comes from correctly ranking the positive attributes of binding. It is entirely possible that the scoring functions are effective at identifying ligands that cannot bind but are not particularly good at distinguishing between inactive or active ligands that can bind in the active site. If this is the case, structure-based virtual screening is successful by the elimination of inactive compounds, rather than by the identification of active compounds. If this is true, it would be a serious problem if the scoring function were then used to rank compounds by their binding affinity.

16.3 Predicting Ligand Affinity

There is an over one hundred year history of attempting to estimate or predict chemical properties. The first efforts in this field were directed more towards physical properties of chemicals.[8] The work of Corwin Hansch and Toshio Fujita applied linear free energy relationships used to predict chemical properties to the prediction of the biological activity of molecules. These models have become known as quantitative structure activity relationships (QSAR). The scoring functions in docking programs can be thought of as the logical extension of this work. In this next section, we will briefly discuss QSAR because it is the field that first attempted to predict biological activity for chemicals. We will next discuss the attempts that have been made to use scoring functions to predict activity.

16.3.1 QSAR

The 1960s' discovery by Hansch and Fujita that properties of chemicals could be used to predict their biological activity was game changing. Instead of using intuition or random selection, a mathematical model could now be generated that would predict whether a new molecule would possess activity. Unfortunately, the problem with these types of models, based on linear relationships, is that while they can have robust prediction within the chemical space of the model (usually within the chemical series), they are very poor at predicting outside the series. In addition, to be able to develop a model, one must have access to a reasonable amount of data (15 to 20 data points). Once that amount of data has been collected, the most potent molecules are usually known or obvious to the chemists. QSAR has been shown to generate accurate and predictive models for small, focused data sets.[9] What QSAR has not been shown to do with similar accuracy is predict activity outside the chemical series that the model was built on (scaffold hopping). Attempts have been made to be generally predictive through the inclusion of more general descriptors, including descriptors derived from quantum mechanics.[10] While general models can be built using large numbers of molecules, they are not as accurate (as measured by R^2) as the models built for a single chemical series.[11] Because of these limitations, computational and medicinal chemists were forced to look for alternatives to guide drug discovery.

16.3.2 Empirical Methods

Empirical scoring functions are functions that look at a limited number of potential protein–ligand interactions and physical properties of the ligand. Using a function that has been optimized against a limited data set of protein–ligand complexes, the attempt is made to predict (or rank) ligand affinity.

The question, Why use a function that is an abstraction of the "real" interaction energy? can be asked. Historically, the simple answer was speed. Rigorous physics-based methods such as free energy perturbation and thermodynamic integration were too slow, so a choice was made to use empirical functions that count only a limited number of interaction types or force field interaction terms to quickly return an answer. In the 1980s and 1990s, this Faustian bargain was a good compromise between being able to return an answer in a reasonable timeframe or not at all. Whether or not this choice is still relevant is an issue we will discuss in Section 16.3.3. The question we need to address now is whether empirical (or knowledge-based) scoring predicts affinity.

There are published claims that empirical (or knowledge-based) scoring functions can predict the affinity of ligands.[12–15] However, the best reported example had an R^2 of 0.86 if crystallographic poses were used, but only 0.56 if poses from a docking program were used.[13] Another study found an R^2 correlation of 0.81 for a serine protease set but also noticed that the correlation between the affinity and the logarithm of the molecular weight [log(MW)] was roughly equivalent to the correlation found using a docking score. The studies reported here use only published compounds. One criticism that can be made of published data is that it is highly sanitized, *e.g.* the data is published because the compounds in the set were active, had obvious SAR, or because computation methods worked. There are two studies that have investigated the ability of scoring functions to predict the affinity of compounds taken from pharmaceutical compound collections.[2,16] Both of these studies found no correlation between any of the scoring functions tested and compound affinity (37 scoring functions tested by Warren *et al.* and 4 scoring functions by Enyedy and Egan). Also of note is the observation of Enyedy and Egan[16] that MW and ClogP were a good predictor of activity for the large number of kinase inhibitors (4300 in-house compounds) against 3 different kinase targets [kinase insert domain protein receptor (KDR), cyclin dependent kinase 2 (CDK2), and Abelson tyrosine kinase (C-ABL)]. Warren *et al.*[2] found that for the 21 compound classes tested, only a single class had an R^2 correlation greater than 0.1 for any of the 37 scoring functions tested. Compound class 1 had a 0.49 R^2 correlation against the Chk1 kinase. Class 1 was also the only class in which there was a similar (though slightly lower) correlation between log(MW) and affinity.[17]

This contradictory evidence leads to a conundrum: can docking programs using empirical (or force field-based) scoring functions rank compounds by affinity for lead optimization? There is evidence that says yes and at least two studies that say no. We believe the answer is no, based on two lines of reasoning. First, Tirado-Rives and W. L. Jorgensen published a paper in 2006[18] that discussed a component of empirical scoring known as conformer focusing. This refers to the fact that generally the conformation of the ligand that binds in the active site is not the lowest-energy conformation found for the unbound ligand in solution. They showed, theoretically, that the error, through estimation or neglect, in current scoring methods for only this one component of protein–ligand binding precludes current scoring methods (empirical and force-field based) from accurately ordering compounds by affinity. Finally, more weight should be placed on the two studies that found no correlation between empirical scoring and ligand affinity. The evaluation from these two studies was done on a much larger set of compounds than the studies that have found a correlation, and the compounds (published and unpublished) were taken from actual corporate collections (Novartis and GlaxoSmithKline), which reduces the potential biases of smaller, more highly filtered data sets.

16.3.3 Physics-Based Methods

So what has been demonstrated so far regarding the ability to predict ligand binding affinity? QSAR can be used to predict affinity but suffers from lack of generality. It typically is not predictive outside the chemical series upon which the model was built. Scoring functions are general but are unable to predict affinity or correctly rank ligands by affinity. This result, while not what was hoped for, is not unexpected because scoring functions are highly simplified models of the physics of protein–ligand binding. In particular, only a single conformation of the protein and ligand are considered by the scoring function, while the binding of the ligand to the protein, as measured in solution, considers an ensemble of protein and ligand conformations. Most scoring functions use a very simplified method of accounting for the change in entropy upon binding. This may be as simple as a count of the number of rotatable bonds in the ligand. With QSAR and scoring functions eliminated as generally applicable methods for predicting ligand affinity, what is left? While rigorous physics-based methods require significant computer resources, they can be used to calculate/estimate the energy of binding. From a practical stand point, however, the number of molecules investigated *via* these methods needs to be limited to tens to hundreds of ligands.

Free energy is a state function and consequently any path used to calculate the energy difference between states is legitimate. There are two fundamentally different ways of calculating free energy. Figure 16.2 shows the two paths that are the most widely used. The first (horizontal) is where the difference between the starting state (ligand and protein separate in solution) and the final state (the protein–ligand as a complex in solution) is determined. These calculations are generally referred to as "endpoint calculations". The second method (the vertical path) works with ligand pairs. In this case (see Figure 16.2), ligand 1 is transformed during the simulation into ligand 2, through a series of non-physical steps for both the bound and unbound states. These types of calculations are often referred to as "alchemical transformation calculations". For these calculations, ligand 2 can be a related ligand (another member in a congeneric series), a very different ligand, or nothing, *e.g.* a transformation from ligand 1 to no ligand in the active site.[19–21] It is important to note that for both methods, current implementations require that the protein remain the same if accurate interaction energy numbers are to be generated. For a more detailed description of these methods we would recommend the following references: N. Foloppe and R. Hubbard[19] and M. K. Gilson and H.-X. Zhou.[22]

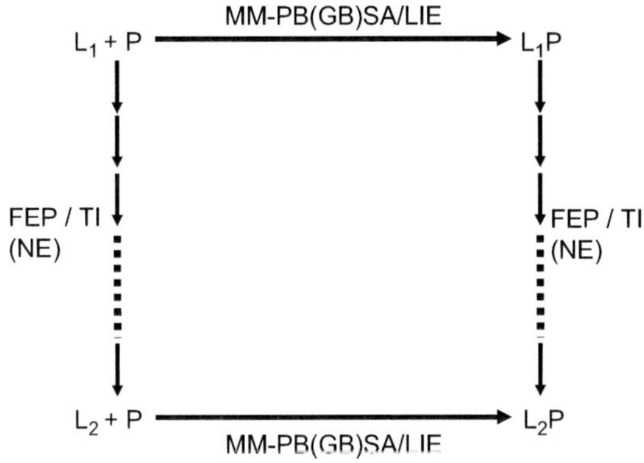

Figure 16.2 A thermodynamic cycle. If the path being used to calculate free energy moves horizontally employing methods like MM-PBSA, MM-GBSA or LIE, these calculations are referred to as "endpoint calculations". Until recently, endpoint calculations were considered better suited for diverse ligands. If the path used to calculate free energy is vertical, methods like FEP, TI or NE are employed. These calculations are referred to as "alchemical transformation" calculations because ligand 1 is transformed through a series of steps into ligand 2.

For endpoint calculations, there are two methods that have been used for predicting ligand affinity. One is linear interaction energy (LIE) and the other uses molecular mechanics (MM) methods to simulate possible conformation states in solution. For MM, there are two forms of solvation used: explicit (water molecules are included in the simulation) and implicit or continuum solvation (the electrostatics of water as a bulk property is simulated). There are two implicit solvation methods used by MM: Poisson-Boltzmann (PB) and generalized Born (GB), resulting in methods called MM-PBSA or MM-GBSA when a solvent accessibility (SA) term is added. The major difference between GB and PB methods for calculating an implicit solvation energy is that GB approximates the calculation of dielectric polarization. The approximations ensure that GB is a fast method, but it is generally less accurate. The inaccuracy is small when far from the surface of the protein–ligand complex but is likely to be large at or near the interface. A reason for this inaccuracy is that protein binding sites have complicated shapes and charge distributions, and under these conditions the approximations of GB for dielectric polarization tend to perform less well.

LIE is a simpler and more ligand-centric method of calculating free energy. It is also the most computationally efficient of the physics-based free energy methods. Only two components are taken into account in the LIE equation: the van der Waals component and the electrostatic interaction energies of a conformational ensemble for the ligand generated by a MM or a Monte Carlo simulation. The reason this method is considered ligand-centric is because it ignores entropy, intramolecular conformation energies of both the protein and the ligand, receptor desolvation that is considered by the MM-PBSA and the alchemical transformation methods. LIE also requires optimization or fitting of the function to the receptor. While general coefficients do exist, they are not as predictive as a model tuned to the protein of interest. This places the LIE method in a similar computational realm as QSAR. One must have some data before an accurate and predictive model can be developed.[19,23,24] An excellent example of this behavior is described in a paper by N. Singh and A. Warshel.[24] In a data set of 22 compounds from 8 different receptors, the LIE method had an R^2 correlation of 0.41 when general coefficients were used. However, if coefficients were optimized to fit each of the 8 proteins, then the R^2 correlation increased to 0.92. Like QSAR, LIE works well when applied to proteins for which coefficients have been developed. It does not perform well with default coefficients or where coefficients for one protein are used for another protein.

As mentioned previously, there are two MM methods for calculating free energy: MM simulations with explicit solvent molecules and MM simulations with implicit or continuum solvent models. There have been a number of publications reporting the use of implicit solvent model simulations to calculate ligand affinity. The implicit methods have the greatest promise for more general application because they require less computer time. One large scale study by B. Kuhn *et al.*[25] showed that MM-PBSA could only distinguish between strong and weak binders and was rarely capable of distinguishing compounds when the binding difference was less than 2-3 pIC_{50} units. There is a large amount of variability in the R^2 correlation coefficients reported for MM-PBSA or MM-GBSA experiments attempting to predict ligand affinity, ranging from 0.70 to 0.87.[19,26–29] Because of the R^2 variability and the high root mean square error (RMSE) values, we believe that MM-PBSA and MM-GBSA methods are generally not accurate enough to predict ligand affinity.

MM simulations carried out with explicit solvent, while significantly longer simulations, have the potential to be more accurate than what has been published and discussed previously for the implicit solvent methods. One example of this is reported in a publication by H. Fujitani *et al.*,[30] which describes the direct calculation of binding free energy for 8 ligands that bind to FKBP (a prolyl isomerase protein that may have an immunosuppressant function). In this simulation, the authors were able to directly calculate binding energies that had an RMSE of 0.4 kcal mol^{-1} and a maximum deviation of 0.6 kcal mol^{-1}. However, all of the calculated binding energies were uniformly 3.2 kcal mol^{-1} smaller than the measured values. While this calculation did not reproduce the exact binding energy numbers, it was able to correctly rank the ligands by affinity.

The slowest of all methods used to calculate energy of binding are the alchemical transformation methods, free energy perturbation (FEP) or thermodynamic integration (TI). As alluded to previously, the alchemical transformation methods differ from the endpoint calculation methods in the path that they take in the thermodynamic cycle (Figure 16.2). These methods rely on MM or Monte Carlo simulations to generate conformation ensembles and to transform the ligand. Because these methods simulate a number of non-physical states during the transformation of one ligand to another or from one ligand to nothing, much of the simulation time is spent sampling states that are not the starting state (ligand 1 and protein in solution) or the final state (ligand 2 and protein in solution). FEP and TI methods are considered the gold standard by which free energy of binding can be directly calculated. When these methods work well, they generate R^2 correlations of in excess of 0.90[31] and with energy values that are not significantly shifted from experimental determinations, as was the case in the MM explicit solvent simulation for FKBP. While these methods may be the gold standard for calculating free energy of binding, these calculations are not particularly robust; they do not consistently return good results. It appears that alchemical transformation methods are performed largely in academic research groups, whereas MM-PBSA and LIE are computational methods used both in academic and pharmaceutical research. We are not sure why this apparent division is present but we suspect that the reasons are due, in part, to the very long computational times required by transformation methods and because these methods appear to require a high level of expertise in the field as a prerequisite for success.

While we have presented MM methods as being the most rigorous physics-based methods for calculating ligand affinity, these methods contain some non-physical approximations that could potentially affect the accuracy and precision of the calculations. In particular, atomic partial charges are treated as spherical (point) charges, which are known to be an inaccurate approximation, especially in active sites that are highly charged and/or contain a metal. Another method for calculating ligand affinity would be to use quantum mechanics (QM) to calculate the interaction energy. This would be the most physically rigorous method of calculating the interaction between the protein and ligand. Unfortunately, the computational cost for doing such calculations, even with hybrid methods like QM-MM, is prohibitive. That said, there is one example in the literature of where a QM interaction energy calculation was done for 8 ligands of a metallo-enzyme human carbonic anhydrase II.[32] While the R^2 correlation coefficient for the 8 compounds in this study was not spectacular (0.7), metallo-enzymes are particularly difficult targets for MM based methods.[33] This result is more significant than the correlation coefficient would make it appear to be. QM is not the answer for the problem of predicting ligand affinity in the near future but has the potential to be a breakthrough technology if the computation speed issue can be solved.

16.3.4 Another Matter

We have presented data showing that scoring functions are not particularly good at calculating or predicting ligand affinity. We have presented data showing that LIE, and in some cases MM-PBSA and FEP/TI methods are good at calculating ligand binding energy. But for all of these results, how does a non-expert determine if the good results are an artifact of the compounds in the data set, the size of the data set, or some other factor? The simplest sanity check for any ligand affinity calculation was described in a short paper by I. D. Kuntz *et al.*,[34] where it was noted that there is a correlation between the number of heavy atoms—or log(MW) of a ligand—and the affinity of that ligand. On average, as the number of heavy atoms of a ligand increases, the affinity also increases, until a point where the relationship plateaus. A true measure of a calculation is whether or not it can predict ligand affinity both when there is a correlation between MW and affinity and when there is no relationship. As was briefly mentioned in Section 16.3.2 on scoring functions, scoring functions rarely have a correlation between the score and affinity when there is little or no correlation between MW and affinity.[2] So does the same observation apply to the physics-based methods? We looked at the correlation between log(MW) and affinity for two of the data sets mentioned in Section 16.3.3, in particular the LIE data set used by Singh and Warshel[24] and the FKBP data set used by Fujitani *et al.*[30] The results from this analysis are shown in Figures 16.3 and 16.4 respectively. What these two plots show is that for these two data sets the correlation between the log(MW) and the affinity (0.46 and 0.93) was equivalent to the correlation observed for the LIE (0.41 and 0.53) and the MM with explicit solvent calculations. In these two cases a simple (and extremely fast) ranking of the compounds by log(MW) would have been as predictive as the expensive and time consuming calculations. The one exception was a particular LIE method,[24] which had a squared correlation of 0.87 using default coefficients including the biotin analogs, which the log(MW) relationship is not predictive of (see Figure 16.3).

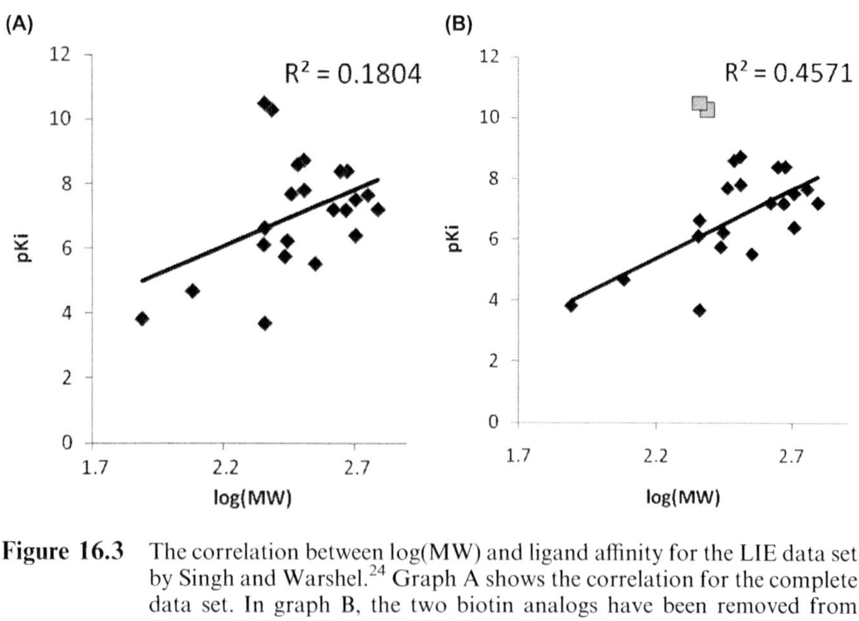

Figure 16.3 The correlation between log(MW) and ligand affinity for the LIE data set by Singh and Warshel.[24] Graph A shows the correlation for the complete data set. In graph B, the two biotin analogs have been removed from the correlation. Biotin and biotin analogs are known not to follow the correlation between log(MW) and affinity, first noted by Kuntz *et al.*[34] Note that the one outlier in the remaining data is an inactive enantiomer where the active enantiomer lies on the line.

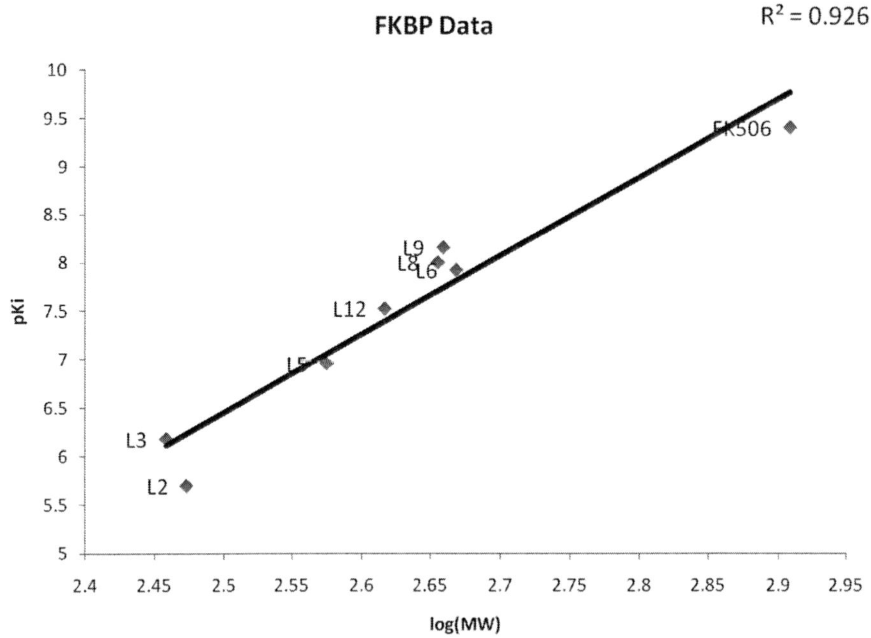

Figure 16.4 A correlation plot of log(MW) *versus* ligand affinity for eight FKBP ligands. The R^2 correlation coefficient is 0.926 and the standard error is 0.35. For this data set, the correlation between log(MW) and affinity is as high as the computed free energy and affinity and the standard error is equivalent to the computed value (0.4).

Faced with the data in Figures 16.3 and 16.4, one might conclude that these expensive physics-based calculations are a waste of time. However, that conclusion is not true. One problem with the observation of the relationship between molecular weight and affinity is that it obscures one's ability to determine whether methods are actually useful and predictive. There are two ways to solve this problem. The first is for the developers of these methods, to make sure that the data that they are testing their methods on does not have a strong molecular weight relationship with affinity, or if it does that they report the relationship. When testing is complete and presumably published, it would be easy for the developer and the consumer of these methods to quickly determine whether or not the method generates a real signal and is not somehow measuring a simple relationship, because there should be significant difference between the success of affinity prediction for the method *versus* molecular weight. The second is for developers to test methods on larger (and possibly more diverse) data sets. Small data sets suffer from the well known statistical problem of variance. The smaller the data set, the higher the probability that a false correlation will be observed. Data sets as small as 50 to 100 compounds would dramatically reduce this problem *versus* the 8 to 10 compounds used in many method validation publications.

16.4 Conclusion

We have shown that a number of methods have been used in attempts to calculate or predict ligand affinity. Historically, the methods that have been the most successful are those which use the most target specific data, such as QSAR and LIE. Because these methods build models based on experimental data, they tend to be far more robust and accurate in their predictions. Unfortunately, because they are built using data from a specific series of compounds or a specific protein receptor, they tend not be very general. LIE and QSAR work well on data similar to that which they were tuned on.

Scoring functions are also optimized functions. However, because the components of the function are physical observables like hydrogen bonds or van der Waals contacts, it was hoped that these functions would be more general. Unfortunately, because the model of binding that scoring functions use is too simple, scoring functions are not at all effective for predicting ligand affinity. Docking and scoring functions work reasonably well for identifying ligand binding modes and for virtual screening. There is good reason to believe that because the model of binding in scoring functions is too simple, these functions will never be effective at predicting ligand affinity[18] and will be relegated to being a fast intermediate filtering mechanism for more rigorous physics-based methods.

All of the physics-based free energy calculation methods in use today are based on data extracted from molecular mechanics or Monte Carlo simulations. While these methods have a much more rigorous physics base than do scoring functions, they also use approximations. Whether or not these approximations will ultimately limit the accuracy and precision of these methods is not clear. There is data to show that MM-PBSA (or MM-GBSA) methods are more effective at predicting ligand affinity than scoring functions are. There is also data showing that these methods are not accurate enough to be useful; they are only accurate to 2–3 pIC_{50} units. The more computationally intensive transformation methods like FEP and TI can, in the best cases, produce data that is accurate enough to be useful. However, these methods, in addition to being very expensive calculations, are not reliable enough to be used routinely in a pharmaceutical research environment. If the chemistry of the series of interest is medium to high throughput, then compounds can be synthesized and tested more rapidly than the simulations and subsequent analysis. These methods then only become useful and productive if the chemistry is difficult. Quantum mechanics holds great potential for accurately simulating the physics of ligand binding. However, because simulation times are even longer than FEP and TI calculations, there is little hope in the near future that QM will contribute to ligand affinity prediction.

So why, after 30 years of structure-based drug design method development, is predicting ligand affinity such a difficult and yet to be solved problem? We would like to leave you with one last thought. Almost all of the protein receptor structural data used for ligand affinity prediction methods come from X-ray crystal structures. As every medicinal chemist knows, the conformation of a small molecule in a crystal is not, in some cases, the low energy conformation of the same molecule in solution. Crystal packing can have a profound impact on the conformation of the molecule. The conformation of the protein can also be affected by the crystal environment. A crystal is not a solution, it is a solid. In many cases, the developers of computational methods did not take this fact into account when using protein crystal structure data. There is one publication in which the authors carefully removed all protein crystal structures from their development data set, which had crystal contact-induced artifacts and crystallographic water-mediated contacts with the ligand.[35] They found that 36% of the structures in their data set contained either one or both of these artifacts. When these structures were removed, the correlation coefficient from X-score (an empirical scoring function) increased by 1.9 standard deviations over the correlation where the data set contained structures with crystal contacts. They found a number of cases where scoring function prediction outliers were structures that had protein crystal contact or crystallographic water contacts. We do not expect that development of an empirical scoring function using structures whose conformation is not influenced by crystal contacts will solve the fundamental issues with empirical scoring functions. However, it is important to remember that the MM methods use the same structures. While these methods do sample protein and ligand conformations, they do not (for reasons of speed) do an exhaustive simulation or sampling. It is possible that crystal contact conformation artifacts are not removed within the time of the simulation. To summarize this point, the developers of computational methods have spent too little time and effort thinking about and understanding the data they are using to develop and validate their methods. When developing computational models, it is important to remember that the model will never be better than the data it was built from. If the data is bad, so is the model.

Can we consistently and accurately predict the binding affinity of a ligand? No. There is a lot of work to do before that goal is reached. Are the currently available methods useful? Sometimes, depending on whether or not one has either the time to do a rigorous calculation or, if the data is available, to build/tune a predictive model.

References

1. N. Moitessier, P. Englebienne, D. Lee, J. Lawandi and C. R. Corbeil, *Br. J. Pharmacol.*, 2008, **153**, S7–S26.
2. G. L. Warren, C. W. Andrews, A. Capelli, B. Clarke, J. LaLonde, M. H. Lambert, M. Lindvall, N. Nevins, S. F. Semus, S. Senger, G. Tedesco, I. D. Wall, J. M. Woolven, C. E. Peishoff and M. S. Head, *J. Med. Chem.*, 2006, **49**, 5912–5931.
3. D. Yusuf, A. M. Davis, G. J. Kleywegt and S. Schmitt, *J. Chem. Inf. Model.*, 2008, **48**, 1411–1422.
4. J. B. Cross, D. C. Thompson, B. K. Rai, J. C. Baber, K. Y. Fan, Y. Hu and C. Humblet, *J. Chem. Inf. Model.*, 2009, **49**, 1455–1474.

5. G. B. McGaughey, R. P. Sheridan, C. I. Bayly, J. C. Culberson, C. Kreatsoulas, S. Lindsley, V. Maiorov, J. Truchon and W. D. Cornell, *J. Chem. Inf. Model.*, 2007, **47**, 1504–1519.
6. D. M. Krueger and A. Evers, *ChemMedChem,* 2010, **5**, 148–158.
7. N. Huang, B. K. Shoichet and J. J. Irwin, *J. Med. Chem.*, 2006, **49**, 6789–6801.
8. J. R. H. Tame, *J. Comput.-Aided Mol. Des.*, 2005, **19**, 445–451.
9. R. P. Sheridan, B. P. Feuston, V. N. Maiorov and S. K. Kearsley, *J. Chem. Inf. Comput. Sci.*, 2004, **44**, 1912–1928.
10. K. Raha, M. B. Peters, B. Wang, N. Yu, A. M. Wollacott, L. M. Westerhoff and K. M. Merz Jr., *Drug Discovery Today,* 2007, **12**, 725–731.
11. P. Gedeck, B. Rohde and C. Bartels, *J. Chem. Inf. Model.*, 2006, **46**, 1924–1936.
12. C. Yang, R. Wang and S. Wang, *J. Med. Chem.*, 2006, **49**, 5903–5911.
13. H. Gohlke, M. Hendlich and G. Klebe, *Perspect. Drug Discovery Des.*, 2000, **20**, 115–144.
14. P. Ferrara, H. Gohlke, D. J. Price, G. Klebe and C. L. Brooks III, *J. Med. Chem.*, 2004, **47**, 3032–3047.
15. R. A. Friesner, R. B. Murphy, M. P. Repasky, L. L. Frye, J. R. Greenwood, T. A. Halgren, P. C. Sanschagrin and D. T. Mainz, *J. Med. Chem.*, 2006, **49**, 6177–6196.
16. I. J. Enyedy and W. J. Egan, *J. Comput.-Aided Mol. Des.*, 2008, **22**, 161–168.
17. G. L. Warren, C. E. Peishoff and M. S. Head, in *Computational and Structural Approaches to Drug Discovery: Ligand–Protein Interactions*, ed. R. M. Stroud and J. Finer-Moore, RSC Pub, Cambridge, 2008, p. 137–154.
18. J. Tirado-Rives and W. L. Jorgensen, *J. Med. Chem.*, 2006, **49**, 5880–5884.
19. N. Foloppe and R. Hubbard, *Curr. Med. Chem.*, 2006, **13**, 3583–3608.
20. W. Jiang and B. Roux, *J. Chem. Theory Comput.*, 2010, **6**, 2559–2565.
21. Y. Deng and B. Roux, *J. Chem. Theory Comput.*, 2006, **2**, 1255–1273.
22. M. K. Gilson and H. Zhou, *Annu. Rev. Biophys. Biomol. Struct.*, 2007, **36**, 21–42.
23. S. Chen, D. Zhao and Z. Yang, *J. Comput. Chem.*, 2011, **32**, 338–348.
24. N. Singh and A. Warshel, *Proteins: Struct., Funct., Bioinf.*, 2010, **78**, 1705–1723.
25. B. Kuhn, P. Gerber, T. Schulz-Gasch and M. Stahl, *J. Med. Chem.*, 2005, **48**, 4040–4048.
26. G. Yang, J. Trylska, Y. Tor and J. A. McCammon, *J. Med. Chem.*, 2006, **49**, 5478–5490.
27. S. P. Brown and S. W. Muchmore, *J. Chem. Inf. Model.*, 2006, **46**, 999–1005.
28. M. R. Lee and Y. Sun, *J. Chem. Theory Comput.*, 2007, **3**, 1106–1119.
29. D. C. Thompson, C. Humblet and D. Joseph-McCarthy, *J. Chem. Inf. Model.*, 2008, **48**, 1081–1091.
30. H. Fujitani, Y. Tanida, M. Ito, G. Jayachandran, C. D. Snow, M. R. Shirts, E. J. Sorin and V. S. Pande, *J. Chem. Phys.*, 2005, **123**, 084108-1–084108-5.

31. B. P. Cossins, S. Foucher, C. M. Edge and J. W. Essex, *J. Phys. Chem. B*, 2008, **112**, 14985–14992.
32. K. Raha, A. J. Van der Vaart, K. E. Riley, M. B. Peters, L. M. Westerhoff, H. Kim and K. M. Merz Jr., *J. Am. Chem. Soc.*, 2005, **127**, 6583–6594.
33. O. A. T. Donini and P. A. Kollman, *J. Med. Chem.*, 2000, **43**, 4180–4188.
34. I. D. Kuntz, K. Chen, K. A. Sharp and P. A. Kollman, *Proc. Natl. Acad. Sci. U. S. A.*, 1999, **96**, 9997–10002.
35. C. R. Sondergaard, A. E. Garrett, T. Carstensen, G. Pollastri and J. E. Nielsen, *J. Med. Chem.*, 2009, **52**, 5673–5684.

CHAPTER 17
Modelling Chemicals in the Environment

PAOLA GRAMATICA

Insubria University, Department of Structural and Functional Biology, *Via* Dunant 3, 21100, Varese, Italy

17.1 Lack of Physico-Chemical and Toxicity Data for Chemicals in the Environment

Millions of chemicals are used daily throughout the world, and their wide environmental diffusion, also far from the emission source due to long range transport, is a cause of high concern due to their creating potential pollution and adverse effects on human and wildlife. The chemical universe is huge and is increasing every day. The CAS registry includes about 55 million chemicals, of which more than 40 million are commercially available and almost 280 000 are regulated and listed in various inventories (*i.e.* European Inventory of Existing Commercial Chemical Substances (EINECS), USA-Toxic Substances Control Act (TSCA), Canadian-Domestic Substances List (DSL). New chemicals are continuously being developed (many thousands each year), but reports on their physico-chemical properties and biological activities are produced more slowly. With regard to Europe, whereas today's knowledge of "new" chemicals (put on the EU-market after 1981) can be considered acceptable, the same cannot be said for most of the earlier commercialized "existing" chemicals, even for high production volume (HPV) compounds. At present, we know a lot about a few chemicals; for the majority of chemicals (>95%) the information available

RSC Drug Discovery Series No. 13
Drug Design Strategies: Quantitative Approaches
Edited by David J. Livingstone and Andrew M. Davis

Published by the Royal Society of Chemistry, www.rsc.org

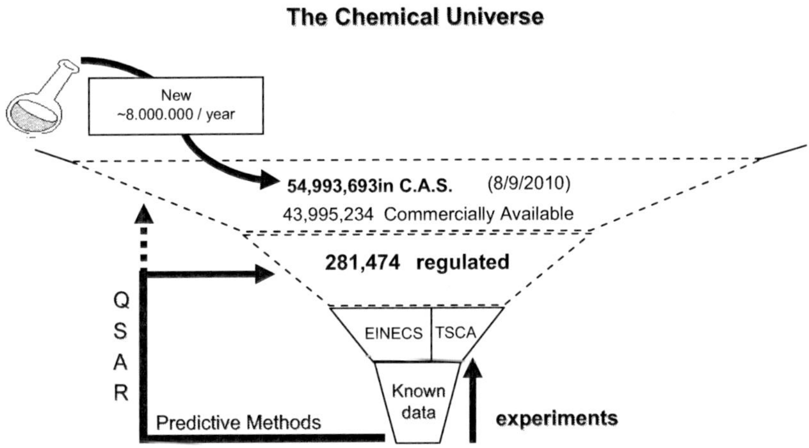

Figure 17.1 A schematic representation of the problem of lack of data for chemicals in the environment and the need for predictive methods.

on their properties and risks is very limited. Thus there is a marked lack of sufficient information available to the public on the effects of most chemicals on human health and the environment, making it difficult to assess and control these substances effectively. (Figure 17.1)

17.2 The Need for the New Legislation of Chemicals: REACH

To reduce existing data gaps, the new chemical control system REACH (Registration, Evaluation, Authorization and restriction of Chemicals)[1] was recently adopted by the European Union. One of the aims of REACH is to improve the protection of human health and the environment, by requiring industry to provide relevant information on the properties and biological activities of all commercialized substances, and to use that data to manage them safely. The necessary information demonstrating their safe use must be submitted in a registration dossier to the European Chemicals Agency (ECHA) in Helsinki, Finland. Additionally, specific use authorizations will be required for substances of very high concern (SVHC), such as those that cause cancer, mutations or reproduction problems [Carcinogenic, Mutagenic or toxic to Reproduction (CMRs), including endocrine disruptors (EDs)], accumulate and persist in our bodies and in the environment very Persistent and very Bio-accumulative (vPvB) or are also toxic [persistent bioaccumulative and toxic (PBTs)]. Authorization will be granted only to companies that can show that the risks are adequately controlled, or that the social and economic benefits outweigh the risks and that suitable alternative substances do not exist. This will encourage the substitution of unsafe substances by safer ones, which will have to be specifically and adequately planned. Two levels of urgent action can be identified, related mainly to chemicals of highest concern: a) need for tools for chemical identification and prioritization, and b) stimulus for research into safer alternatives and their production. However, it is immediately evident that in the short term enormous effort will be required on the part of industry to fill the huge gap in data availability and to design safer alternative chemicals.

Costs, test animal numbers, speed of the process and the sharing of data are all important issues.

There is much debate about the potentially large number of animals that would be needed for experimental toxicity testing in REACH. However, new animal testing is only required if it is not possible to provide information in any other permitted way, under the 3R approach: reduce, refine, replace. The most efficient way to carry out hazard and risk assessments of large numbers of chemicals, while reducing costs to industry and minimizing animal testing, is to obtain the necessary information from intelligent testing strategies (ITSs), which include computational methods [structure–activity relationships (SARs), quantitative structure–activity relationships (QSARs) and read-across].

QSAR models, widely developed and applied in drug design since the beginning of classical QSAR science, have been applied only in more recent years to industrial chemicals (which are also environmental chemicals, due to their wide usage and spread in the environment). Academic applications of QSAR models are the most common, no strict restrictions exist beyond the interests of the scientific community. On the contrary, regulatory QSAR models are more demanding due to the specific requirements introduced by the legislations. In a regulatory context, QSARs have been used by the USA-Environmental Protection Agency (EPA) since 1979/1980, principally in the hazard evaluation of chemicals under the Toxic Substances Control Act (TSCA). QSAR-predicted data are commonly used in the USA, for example in the Pre-manufacture Notification (PMN) of the New Chemicals Program, in order to complement the information provided by the manufacturer, when experimental data are not available. Only recently the development and use of alternative methods like QSAR have been expressly promoted in Europe for the hazard assessment of substances, being inserted in some articles of REACH, where the use of sufficiently validated computational prediction models based on QSAR is suggested: (i) to screen chemicals for highlighting those more dangerous and those on which to focus experimental testing (priority setting); (ii) to fill data gaps for classification and labelling, and for risk assessment; (iii) to design safer alternative compounds (chemical design).

This raises a question: are QSAR applications to environmental chemicals the same as those used in drug design? The author's opinion is: not completely. The main utility and success of QSARs in drug design is related to the discovery of new drugs by rationally modifying the structure of known active chemicals. The hypothesized or demonstrated mechanisms and possibly the interaction with biological receptors (docking studies) must also be taken into account. But, according to Kubinyi, the use of high throughput screening (HTS) has not, so far, obtained good results.[2] For environmental applications, the scenario is quite different, thus requests for QSAR models will also be quite different and related to the specific context in question.

17.3 OECD Principles for the Validation, for Regulatory Purposes, of (Q)SAR Models

Obviously, for the purposes of the REACH legislation, it is essential to use QSAR models that produce reliable estimates, *i.e.* validated QSAR models. Thus it was considered important to develop an internationally recognized set of principles for QSAR validation, to provide regulatory bodies with a scientific basis for making decisions on the acceptability of QSAR estimates of regulatory end-points and to promote the mutual acceptance of QSAR models as alternative to experimental tests. Some principles for assessing the validity of QSARs were first proposed at an international workshop held in Setubal, Portugal in 2002. These principles were then modified in a meeting of the Organization for Economic Co-operation and Development (OECD) in Paris in 2004. The now famous OECD 'Principles for the Validation, for Regulatory Purposes, of (Q)SAR Models'[3] state that "to facilitate the consideration of a QSAR model for regulatory purposes, it should be associated with the following information: (1) a defined end-point; (2) an unambiguous algorithm; (3) a defined domain of applicability; (4) appropriate measures of goodness-of-fit, robustness and predictivity; (5) a mechanistic interpretation, if possible".

The main aspects of QSAR modelling applied to environmental chemicals are dealt with in the following paragraphs, organized according to the OECD principles, though some topics are obviously related to multiple principles. These principles list the crucial topics in QSAR modelling, for which there is ongoing work and discussion within the environmental QSAR community.

17.3.1 A Defined Endpoint

QSAR models for environmental chemicals should exploit, as much as possible, all the experimental information available, even if limited. In this context, careful selection of the data is crucial. The accuracy of QSAR models cannot be greater than the quality of the modelled experimental data, often characterized by high uncertainty in measurement. Data variability can arise from chemical purity, variability of the protocol and variability of biological responses. Much of the variability resulting from chemical purity and protocol can be avoided, but the reproducibility of biological tests is much lower than that for other properties. A "golden" training set should be that obtained by the same method, in the same lab and possibly by the same researcher, but this is obviously seldom possible, mainly for end-points of regulatory interest and environmental compounds belonging to a wide variety of chemical classes. It is also important that models are built using a rich diversity of compounds in the training sets, thus ensuring as much as possible a generalizable model. Such "global" models are more appropriate for virtual screening (where the main concern is to prioritize the most dangerous chemicals in order to focus on them and reduce the number of compounds that need experimental testing), while "local" models developed on smaller sets of congeneric chemicals are more useful for chemical design and when very high levels of accuracy are required.

In REACH, different kinds of data are required for chemical registration: physico-chemical properties [melting and boiling points, octanol–water partition coefficients, water solubility, Henry's law constants, K_{oc}, bioconcentration factor (BCF), *etc.*]; toxicological data (mutagenicity, carcinogenicity, skin/eye irritation, developmental toxicity, endocrine disruption, *etc.*) and ecotoxicological data (acute toxicity on daphnia, algae, fish, *etc.*). Additionally, data on biodegradability and persistence measured by environmental reactivity in various compartments (for instance atmospheric degradation or hydrolysis) are required for PBT assessment. Currently, the Estimation Program Interface Suite (EPI Suite[4]) is the primary modelling system utilized within US-EPA for providing estimates of the common physico–chemical properties necessary for predicting chemical fate and transport in the environment, such as melting and boiling points *etc.* The EPI Suite calculators are based primarily on a fragment approach, developed on many industrial chemicals and validated with an independent set of chemicals. In general, the EPI Suite predicts physico–chemical properties, within an order of magnitude, which is normally sufficient for screening in regulatory assessments. Analogously, for ecotoxicity, Ecological Structure Activity Relationships (ECOSAR[5]) is a user-friendly software package, based on the categorization of chemicals, developed by the US-EPA for predicting aquatic toxicity to fish, daphnia and algae. The ECOSAR package has been shown to be useful in predicting aquatic toxicity of many industrial chemicals, however it was recently demonstrated[6] that it must be used with caution in predicting the environmental effects of chemicals that fall outside the applicability domain (AD) of the models (see later Principle 3 on AD). Surely such caution is obvious. This limit was verified for application to pharmaceuticals, which are becoming an increasing problem of environmental pollution. It was also recently demonstrated[7] that only limited numbers (even less than 1%) of the EINECS compounds can be classified into classes that can be covered by ECOSAR models. These studies serve as a caveat for the selection and application of *in silico* tools to predict the activity of any questioned compound, without any control of their applicability. In fact, EPI Suite and ECOSAR, due to their ease of accessibility and usage, are also widely used in Europe, but for some specific chemicals (out of the AD) the predictions obtained from new *ad hoc* developed models are more reliable. Good examples are the models on physico-chemical properties of some emerging pollutants: polybromodiphenyl ethers[8] (PBDEs) and perfluorinated compounds[9] (PFCs), developed by the author in the FP7-EU project, CADASTER.[10]

Some examples of widely modelled end-points are reported. Several alternative models of BCF (an important parameter in environmental assessment as it is an estimate of the tendency of a chemical to concentrate and, consequently, to accumulate in an organism) have been developed; some are based

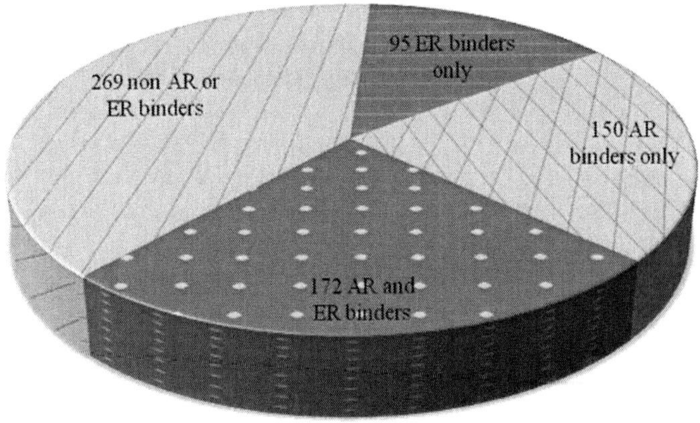

Figure 17.2 Pie chart of virtual screening results for AR binding on ER binders data set: 172 chemicals are recognized as potential pleiotropic Endocrine Disruptors (EDs).

on log K_{OW} as molecular descriptor,[11,12] others on theoretical molecular descriptors.[13–18] Two "golden" data sets of aquatic toxicity, the Duluth data on *Pymephales promelas* and the *Tetrahymena pyriformis* set, are widely modelled with different approaches: categorizing the chemicals according to their mode of action (MOA) and using log K_{OW} as a molecular descriptor,[19,20] or developing general models by theoretical molecular descriptors.[21–23] Skin sensitization is successfully modelled by the mechanistic approach (Michael addition)[24] and also by the statistical approach.[18] Recently, Benigni[25] reviewed models on mutagenicity and carcinogenicity, which can be modelled by different commercial software (DEREK, TOPKAT, MULTICASE, *etc.)*, but not always with concordant results. Endocrine disruption is another important biological activity widely modelled, also in the author's group, by regression and classification models.[26–34] For a rapid pre-screening of 58 000 potential estrogen-like chemicals, we applied different classification models [k-nearest neighbours (k-NN) and support vector machines (SVM)] in a consensus approach,[29] demonstrating that 76% could be excluded from the potential estrogenic activity without experiments, with high accuracy, as both models have a low false negative rate. Recently, models on androgen receptor binders were applied for a virtual screening of a large set of estrogenic chemicals, for the identification of potential pleiotropic endocrine disruptors, which are chemicals with double ED activity[32] (Figure 17.2) and *vice versa* new models for estrogenic chemicals have identified pleiotropic chemicals among androgen receptor binders.[34]

17.3.2 An Unambiguous Algorithm

The algorithm of an acceptable QSAR model, that correlates the studied endpoint with chemical structure using molecular descriptors, must be reproducible and easily applicable, even by non-experts. For this reason, there is a general preference for simpler and more widely understandable models. In fact, some methods have simple and explicit expressions [for instance Multiple Linear regression (MLR)], whereas other methods [*i.e.* machine learning, such as SVM, neural networks (NN), *etc.*] result in more complex expressions and for this reason are erroneously perceived by non-experts as "black boxes". Ockham's razor is preferably applied in a regulatory context: given two QSA(P)R models with the same predictive error, the simpler one should be preferred because simplicity is in itself desirable and can increase confidence in QSAR model predictions. There is also greater confidence, among the regulators, in models based on descriptors which are "more understandable" for mechanistic meaning (this topic will be explained in more depth under Principle 5 in Section 17.3.5).

The QSAR algorithm establishes correlation between the studied response and the molecular descriptors, but recently some concerns have been raised in the literature,[35,36] emphasizing that correlation between variables does not automatically imply that one causes the other and that chance correlation could occur, mainly if "not understandable" descriptors are used. However, correlation is a fundamental requirement for causation. The best way to exclude chance correlation is to carefully verify the statistical predictivity of the QSAR models by their validation, as requested by OECD principle 4, also externally on new chemicals, and by scrambling of the response,[37–40] additionally, if possible, to mechanistically interpret the molecular descriptor (OECD Principle 5). If the correlation is confirmed after rigorous verification, it has a reason for existing and the problem is one of the human mind if the cause is not discovered or understood.

Molecular descriptors (commented on in greater detail in Chapter 8) are the core of any QSAR model. Environmental QSAR models are highly dominated by the historical background of QSAR, thus many of the models for ecotoxicity [mainly for aquatic toxicity of narcotic chemicals,[19,20] solubility, various partition coefficients including BCF (see Section 17.3.1)], are based on log K_{ow} (or logP). It is supposed that the partition between water and the organic solvent could serve as a model to describe a fish in water and that toxicity could then be understood by quantifying the uptake of the compound into the fish's body. The reason for this widespread use of logP-based models for environmental chemicals (implemented also in EPI-Suite and ECOSAR) is related to the diffused supposed "understandability" of this parameter. However, there are many papers that highlight problems related to this descriptor (also discussed in Section 17.3.5). In fact, experimental values can differ greatly and in the several approaches developed for the theoretical calculation of logP, it is not uncommon to have differences of several orders of magnitude.[41]

The arbitrary choice of one specific algorithm for logP calculation, as is usual in QSA(P)R studies, does not necessarily lead to the highest quality model for the analyzed data set. It is recommended[21,42] to calculate logP by diverse algorithms, to include them in a total descriptors pool and to apply variable selection methods. Different logP values give rise to different end-point predictions, and this parameter is not always selected as a modelling descriptor when other theoretical descriptors are available. This is exactly the approach the author applied in several models for toxicity in *Pimephales*,[21] for bioconcentration[16,17] and for soil sorption coefficient[43] amongst others. Bennet *et al.*[44] calculated logP for emerging pollutants (military compounds and pharmaceuticals) using six different commercial software programs, and their analysis showed a variability of several orders of magnitude (even 10 orders). They concluded that over-reliance on a single logP calculator could result in an incorrect conclusion regarding the logP of an unknown compound. This is especially important if the logP value provides the basis for other calculations in the QSAR program utilized.

Other molecular descriptors widely used for the prediction of environmental chemical properties are substructure descriptors or fragments.[45] In many QSAR models, some fragments are identified as structural alerts, which are useful because biological activities are often predicted by the presence or absence of a particular group of atoms in a molecule,[25] and the descriptors can be mapped back onto the molecules to indicate which parts confer activity. However, using substructure descriptors has some complications. First, the dimensionality is usually much higher, *i.e.*, several thousand unique substructures in a typical training set. Second, a new molecule may have substructures that do not occur in the training set, *i.e.*, it can be outside the space of the training set.[45] Although fragment contribution approaches give reliable predictions for some end-points, they cannot estimate the response for structures containing a "new" fragment, and the number of correction factors continually grows as compound diversity increases. These shortcomings can be overcome by treating molecules as a whole, rather than as a sum of parts. Many developers of QSAR models on environmental topics,[13,14,18,22,23,26,28,30,43,45–51] including the author's group, prefer to have *a priori* as many molecular descriptors as possible at their disposal, including whole molecular or holistic descriptors. The number and diversity of the descriptors guarantees coverage of the molecular structure space more efficiently than if only a few input descriptors are used. Sometimes, the molecular descriptors provide only different views of the same molecular aspect and for this reason result in high correlation. Thus, when dealing with a large number of highly correlated descriptors, variable selection is necessary to find a simple and predictive QSAR model, which must be based on the least number of descriptors, and the least correlated as possible. Various methods of rapid objective pre-reduction (discarding constant values, pair-wise correlated variables, *etc.*) are often implemented first, in order to identify a more limited set of descriptors. From this reduced set, one can select the descriptors really related to the response, not only in fitting but also (mainly) in prediction. Such selection is performed by alternative variable selection methods (stepwise regressions, forward selection, backward elimination, simulated annealing, evolutionary and genetic algorithms *etc.*). Genetic algorithms[52,53] are a very effective tool now widely and successfully applied in many QSAR approaches and always in the author's work. Within the population of good models obtained by GA-MLR, there are various models with similar prediction power, but based on different molecular descriptors. Thus, there could be many possible "best" models. Alternative viewpoints with different emphases lead to not equivalent descriptions of the studied end-point. In fact, there is only one reality, but different points of view. This is the basis for the application of consensus modelling, averaging the predictions from various models based on different descriptors and/or different methods.[22,54,55]

Another crucial topic, also for environmental chemicals, is related to the question, "local or global models?". For some authors,[56] local models based on a series of compounds with the same core structure, or chemicals with the same mode of action,[19] exhibit better performance than those derived from diverse data sets, while other authors[57] recently demonstrated the contrary. A comparison was made in the Insubria group for *Pimephales* toxicity, using local models based on predefined MOA and global models.[21] The qualities of local and global models are similar, but the applicability is different: local models can be reliably applied only to compounds with predefined MOA and this is not always possible, while global models can be applied without any *a priori* knowledge of toxicity mechanism. This second opportunity is certainly very useful for environmental chemicals, especially for screening purposes with prioritization aims.

A big problem for QSAR modelling has been highlighted by Maggiora:[58] very similar molecules may, in some cases, possess very different activities, leading to what are called "activity cliffs" resulting in significant activity misprediction among similar molecules, even in cases where overall predictivity is high. Thus, perfectly valid data points located in cliff regions may appear to be outliers. This big pitfall for QSAR modelling can be smoothed when large data sets are modelled for screening. However, it becomes particularly dangerous in the read-across approach, where new data are directly predicted by interpolating the known data of a few chemicals recognized as "similar" by different categorization tools (structural similarity, mechanism similarity, *etc.*).

An OECD (Q)SAR application toolbox for REACH[59] has been developed and implemented to categorize chemicals by different tools (for example Toxtree, Toxmatch, in Joint Research Center (JRC)-QSAR tools[60]) and to predict toxicity mainly by read-across techniques. This is suggested for regulatory use.

17.3.3 A Defined Domain of Applicability

Even a robust, significant and validated QSAR model cannot be expected to reliably predict a studied end-point for the entire universe of chemicals. The model applicability domain (AD) is a theoretical spatial region defined by the specific molecular descriptors of the model and the modelled response, and is thus defined by the nature of the chemicals in the training set. Given a specific training set on which a model is constructed, it is generally felt that if a new molecule is somehow similar, or is in the "domain" or "space" of the training set, it is likely to be well-predicted, otherwise there is significant "extrapolation" and the prediction could be unreliable.

One key aspect of model applicability is the definition of the chemical space and the way in which chemical similarity is measured, as chemical similarity is a relative concept. The similarity or distance values depend on both the type of molecular representation or the distance measure used. Due to this lack of invariance of chemical space, it is not unusual for two compounds that are neighbours in one chemistry space representation to not be even close in another. Thus multiple searches, based upon a variety of molecular representations, provide an effective way of identifying neighbours. This is the reason for the suggestion[61] of using many chemical similarity search methods.

For the aforementioned reasons there is, so far, no generally accepted or even standardized approach for defining the chemical space of QSAR models, and there are no reasons to expect that one method is the absolute best. However, given this uncertainty, AD is used to make the decision as to whether or not a QSAR prediction should be more or less reliable. This is a crucial and hot topic, and was dealt with at a JRC workshop, where several different approaches for linear and non-linear models were proposed[62] in relation to different model typologies; it is the topic of various publications.[63–65]

For an analysis of the AD of regression models, the author has always used the Williams plot, which is now widely applied by other authors and commercial software. The Williams plot is the plot of standardized cross-validated residuals (R) *versus* leverage (Hat diagonal) values (h from the HAT matrix). It allows an immediate and simple graphical detection of both the response outliers (*i.e.*, compounds with cross-validated standardized residuals greater than 2–3 standard deviation units) and structurally anomalous chemicals in a model ($h > h^*$, the critical value being $h^* = 3p'/n$, where p' is the number of model variables plus one, and n is the number of the objects used to calculate the model).[38–40,62,66]

A compound with high leverage in a QSAR model is the driving force for the variable selection if the compound is in the training set (good leverage), but a high leverage compound in the test set is the result of substantial extrapolation of the model, and it could have unreliable predicted data. The leverage approach, being based only on structural descriptors, can be applied to any chemical, even without data, thus it is useful to verify *a priori* the AD of any QSAR model on chemicals not yet tested or synthesized.[8,9,67] Figure 17.3 shows the AD of a model of PFC inhalation toxicity in the rat.[67]

Additional consideration of other components of the QSAR AD, such as the physico-chemical domain, descriptor domain,[45] mechanistic domain and metabolic domain (when possible) would allow even more improved confidence levels in predictive model applications.

17.3.4 Appropriate Measures of Goodness-of-Fit, Robustness and Predictivity

To be usefully applicable, a QSAR model must be able to learn from available data and reproduce them (goodness of fit, verified by R^2), be stable and robust (verified by internal cross-validations: leave-one-out, leave-many-out or bootstrap) and, most importantly, by extracting the maximum information from the limited existing knowledge, it must be able to reliably predict data for new chemicals not involved in model development (external validation).

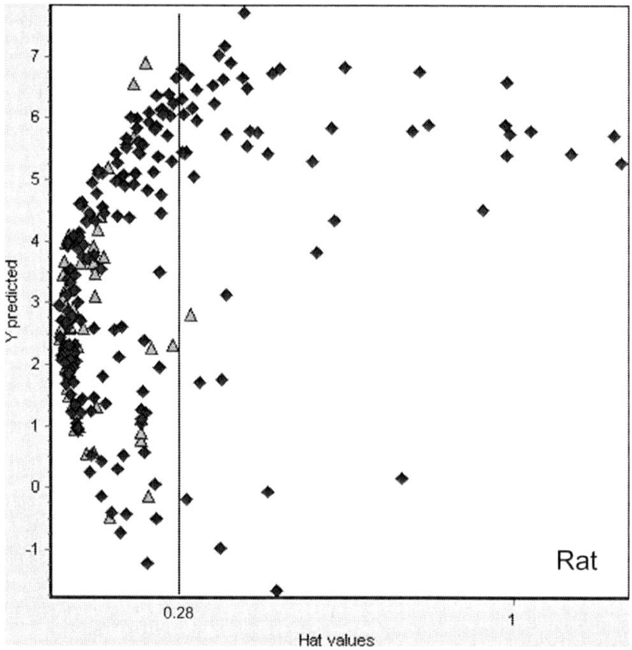

Figure 17.3 Plot of the predicted inhalation toxicity on rat *versus* Hat values for PFCs in ECHA pre-registration list. The model is reliably applicable to the chemicals left of the vertical line (cut-off h* = 0.28), 76.8% of 250 PFCs. The predictions for chemicals to the right of the line are extrapolated and could be unreliable.

Model validation, or better verification, has been the subject of recent wide debate in the scientific and regulatory communities. Many interesting scientific papers have been published[37–40,68–72] with different underlying ideas on the "best" way to validate QSAR models using different tools (internal cross-validation, external validation verified by different parameters). The crucial point that internal validation is necessary, but not sufficient, is stressed in many papers since the famous "Kubiny paradox,"[73] "Beware of Q^2" of Golbraikh and Tropsha[37] and "the Importance of Being Earnest"[38] of Tropsha Gramatica and Gombar. Recently, the author highlighted the difference between internal validation (cross-validation) and external validation on new chemicals.[40]

An optimistic paraphrase of the famous G. Box quote should be: "most models are wrong, but some are useful". The key question is: useful for what specific purpose? In the specific context of environmental chemicals, a model that captures enough of a problem/chemical domain to help (i) discover dangerous, unknown properties of chemicals, (ii) set priorities for compounds needing deeper *in vitro* and *in vivo* investigations, (iii) plan better experiments for a more rational use of resources and limited animal tests and (iv) design safer alternatives, is a model that is not just "useful," it is also "right" for all the cited purposes. However, it is also important to remember that it should not be a competition among QSAR modellers to produce the 'best' model. The most important aspect is that any QSAR model, which like all mathematical models is a simplification of studied phenomena, must be validated/verified for its utility. Another of Box's quotes is relevant here: "the best is the most useful".

As the wider utility of QSAR models in the environmental context is to screen, without experimental data, large chemical data sets to update priority lists, the best predictions can derive from the averaging of results from various models in a consensus approach.[22,54,55]

The planned use of QSAR model predictions is an important aspect to take into consideration. For the prediction of physico-chemical properties, and for virtual screening, the primary focus should be predictive ability; interpretation is secondary.

It is important to distinguish between "descriptive QSARs" and "predictive QSARs".[74] In "descriptive QSARs", the focus is on modelling the existing data and fitting them as best as possible using molecular descriptors that are mostly selected by a supposed "understanding" of the correlation/causality, in terms of mechanism interpretability. These kinds of QSAR models are highly useful for mechanism interpretation, mainly in local models developed on homogeneous data sets of (possibly) congeneric compounds, and are widely applied in drug design. But the main interest, for instance in the REACH context when applied to environmental chemicals, should not be an understanding of the complex mechanisms, based on supposed interpretable descriptors correlated to the studied response, but the exploitation of the limited existing information (there are few available experimental data) to predict, by global models, information relative to chemicals without experimental data, in order to screen large data sets. Thus the check of predictivity should be the most important and primary aspect of "predictive QSARs". These specific QSARs must obey OECD principles in their order: the reliability (and utility) of these models in terms of domain of applicability (Principle 3) and the goodness-of-fit, robustness and predictivity (Principle 4) must be carefully verified *before* attempting to interpret the mechanism (Principle 5).

With regard to internal validation parameters (R^2 and Q^2_{LOO}, Q^2_{LMO}, Bootstrap *etc.*) the literature is well established, whereas for external model evaluation it is still evolving, however a certain number of model evaluation parameters Q^2_{ext} can be found.[37,38,68–72] The main issue is that, unlike R^2 and Q^2, external validation parameters tend to give contradictory results, especially on small data sets, thus we are convinced that not one of these Q^2_{ext} is the "best" (though the proponents wish to demonstrate the superior quality of their proposals on the basis of examples). A paper on this comparison is now in preparation by the author's group. More external validation parameters must be verified.[9,67,75]

The limiting problem for efficient external validation of a QSAR model is, obviously, data availability. Given the availability of a sufficiently large number (never less than 20% of training) of blind new experimental data, the best proof of already developed model accuracy is to test model performance on these additional data, at the same time checking the chemical AD. However, it is generally difficult to have data available (in useful quantity and quality) for new experimentally tested compounds, thus, after model development, external validation can be usefully applied to chemicals that were put aside at the beginning. In fact, before model development, the available data set can be adequately split by different procedures.[76–79] Two different splitting methods have been described by the author in many papers:[9,17,21,27,32,33,43,54,67,75,80] (i) splitting by chemical structure similarity [Self Organizing Maps (SOMs) or Kohonen map-ANN] to guarantee a balance between structure representativity in the two sets (prediction set I), and (i) splitting by response, thus without any bias for structure [random selection by activity of compounds by arranging them in terms of increasing activity (Prediction set II)]. In this way we obtain: (i) training sets, which are used to select the modelling descriptors for model development, and (ii) prediction sets, never used for variable selection, but used exclusively once, and only after model development, to check the predictivity performance of each model. When the same common set of molecular descriptors is selected from the independent modelling of each training set, and is verified as predictive for both prediction sets, then this combination of descriptors is considered to encompass the modelled response for the studied compounds, independently of the splitting criteria, thus unbiased by structure and response.[9,67,75] Finally, the set of combined descriptors is used to derive a full model, the final proposal, in order not to lose any available information. A particularly effective example of this kind of rigorous external validation is a MLR model of soil sorption coefficient of 643 heterogeneous chemicals (mainly pesticides),[43] where less than 15% of the chemicals were used in the training set for the selection of four theoretical modelling descriptors. High predictivity ($Q^2_{Ext} = 0.78$) was verified on 550 chemicals.

17.3.5 A Mechanistic Interpretation, if Possible

This is the most controversial principle and the greatest source of disagreement and discussion among mechanistic and statistical QSAR modellers, also in the environmental field.

In accordance with the Hansch approach, usually followed by mechanistic modellers, the descriptor selection is guided by the modeller's conviction to have *a priori* knowledge of the mechanism of the studied activity/property. Thus presumed mechanistic meaning is assigned to any molecular descriptor used, then the modeller selects it personally from a limited pool of potential modelling variables, normally well known and largely employed (for instance: $\log K_{ow}$, used in many toxicity models, various partition coefficients, *etc.*; HOMO/LUMO, always selected for modelling chemical reactivity, *etc.*).

On the other hand, the 'statistical' approach is based on the fundamental conviction that the QSAR modeller should not influence *a priori*, or personally, the descriptor selection through mechanistic assumptions, but should apply unbiased mathematical tools to select, from a wide pool of input descriptors, those descriptors most correlated to the studied response and verified as predictive. The number and typology of the available input descriptors must be as wide and different as possible in order to guarantee the possibility of representing any aspect of the molecular structure. Different descriptors are different ways of viewing a molecule, and theoretical descriptors can be unambiguously calculated in the same way. Descriptor selection should be performed by applying mathematical approaches to maximize, as an optimization parameter, the predictive power of the QSAR model, as the real utility of any model is considered its predictivity, mainly if its application is for screening, ranking and prioritization of large data sets. In relation to this point, Livingstone, in an interesting perspective paper[81] states: "*The need for interpretability depends on the application, since a validated mathematical model relating a target property to chemical features may, in some cases, be all that is necessary, though it is obviously desirable to attempt some explanation of the "mechanism" in chemical terms, but it is often not necessary, per se*".

For the statistical approach, it is the last point in the QSAR procedure.[38,82] It is correctly the last OECD principle. But mechanistic interpretation can be carried out only "if possible", as the author personally suggested at the OECD meeting in 2004, when these principles were fixed. In fact, it is not always possible (and often only a presumption) to understand the mechanism of the studied end-point from the modelling molecular descriptors. A crucial question is: are we really sure that we know the overall mechanism of any end-point of environmental interest? Even in the simplest cases of physico-chemical properties, the mechanisms are quite complex and their understanding is only possible at certain levels of approximation. The biological response of one organism to a xenobiotic (such as skin sensitisation, carcinogenicity or ecotoxicological end-points such as LD50 in different species, *etc.*) is the result of a combination of different biological processes that depend on several mechanisms of penetration, distribution, metabolism and the interaction of this chemical with organs, cells, receptors, *etc*. This is what, in many cases, is measured experimentally and condensed in one numerical value of activity, toxicity, *etc*. How can we guarantee that a particular mechanism is the only one determining the studied biological response? As the majority of biological response measures are not specific for one particular mechanism, the selection of one (or more) physico-chemical variables *a priori* for their mechanistic meaning, in relation to one assumed mechanism, is very risky,[82] as important variables influencing other mechanisms, participating in such a response, could remain ignored.

Amongst the "mechanistic" descriptors preferred by some authors, logP is a sort of "magic" descriptor, useful for modelling many end-points. Lipophilicity certainly plays a crucial role in biological activities, but so also do electronic or steric effects. LogP is highly correlated with many dimensional descriptors, even with MW, as it is actually indicative of molecular size. Estrada[48] pointed out that there is no reason to assign mechanistic meaning to logP while other theoretical descriptors, able to model an end-point in the same way, are considered "difficult to interpret mechanistically". Estrada says: "*It is the consequence of the wrong application of the Ockham's razor philosophy. The same that can conduce to deny general relativity because it makes more assumptions than Newton's gravitational law, and it is far more complex*".

Molecular structure can be represented in different ways, some more comprehensible than others, but all with a chemical meaning, though not always clearly understood by the users in their QSAR modelling. Molecular descriptors normally have a detailed analysis of their structural meaning in specific literature, infrequently accessed by some QSAR model applicants. This is the main reason that some easier descriptors are considered historically interpretable or "mechanistic", while others are defined as incomprehensible or not transparent.

It has been demonstrated that none of the descriptors can independently explain the observed distribution of the biological data, it is only the combination of selected descriptor sets that allows the modelling of the studied end-point. Describing the different mechanisms simultaneously makes it difficult to interpret their role in the model. Some descriptors are more obvious and easily interpretable, while others are more difficult to interpret, but their role is in ensuring the high overall predictive power of the models.[67]

The OECD Application Toolbox[59] applies the mechanistic approach and reads across, categorizing chemicals *a priori* based on hypothesized mechanisms. Contemporaneously, based on the statistical approach, many models of properties and activities of environmental interest have been developed, and after review they can be inserted in a QSAR model database according to the QSAR Model Reporting Format (QMRF) on the JRC web site.[83]

17.4 Models for Cumulative End-Points (POPs, PBTs, *etc.*): Utility for the Screening, Prioritization and for the Design of Safer Alternative Chemicals

The environment is a highly complex system where chemical behaviour is regulated by many different parameters of contemporary relevance, *e.g.* physico-chemical properties, chemical reactivity, biological activity, *etc.* An understanding and the rationalization and interpretation of their covariance is the principal aim of any environmental researcher. The application of explorative methods of multivariate analysis [for instance principal component analysis (PCA)] to various end-points of environmental concern allows a combined view that generates ordination, and grouping, of the studied chemicals, in addition to the discovery of variable relationships. The outcome is a chemical screening and ranking according to the studied properties, reactivity and activities and, finally, the proposal of an Index. The QSAR modelling of cumulative end-points is the central and peculiar core of most of the author's fifteen years of research at the Insubria University, also recently reviewed in a book chapter.[84]

The principal aim of any explorative technique is to capture information available in any multivariate context and condense it to a more easily interpretable viewpoint (a graph, Figure 17.4a). The most significant combination of variables (first principal component: PC1) can be used as a cumulative index and modelled as a new end-point by the QSAR approach (described below), exploiting already available information concerning chemical behavior, and allowing the proposal of models able to predict such behavior in chemicals for which the same information is not yet known, or even in new chemicals before their synthesis. In fact, our QSAR approach for modelling both quantitative response by regression methods and qualitative response by classification methods, is based on theoretical molecular descriptors that can be calculated for drawn chemicals, starting from the atomic coordinates, thus without the knowledge of any experimental parameter.

This approach has been applied to study pesticide leaching,[83] the atmospheric degradation of volatile organic chemicals (VOCs),[53] global mobility potential[84] and global environmental persistence of persistent organic pollutants (POPs),[80,87] and for PBT assessment.[55] Chemicals that are jointly persistent, bioaccumulative and toxic (PBT) are substances of very high concern (SVHC) and are subject to an authorization step in the REACH regulation, that also includes plans for safer substitutions of recognized hazardous compounds.

A structurally-based approach is proposed for a holistic screening of potential PBTs in the environment.[55] Persistence, bioconcentration and toxicity data available for a set of 180 organic chemicals, some of which are known PBTs, have been combined in a multivariate approach by principal component analysis. The method is applied to rank the studied compounds according to their cumulative PBT behaviour (Figure 17.4 a); the ranking can be defined as a PBT index. The following simple, robust and externally predictive QSPR multiple linear regression model MLR has been developed for the PBT Index [given by Equation (17.1)].

$$\mathbf{PBT\ Index} = -1.25 + 0.63\,\mathrm{nX} + 0.21\,\mathrm{nBM} - 0.39\,\mathrm{nHDon} - 0.12\,\mathrm{MAXDP}$$
$$\mathrm{ntr} = 94;\ R^2 = 0.88;\ Q^2 = 0.86;\ Q^2_{\mathrm{boot}} = 0.86 \tag{17.1}$$
$$\mathrm{npred} = 93;\ Q^2_{\mathrm{ext}} = 0.87;\ \mathrm{RMSET} = 0.53;\ \mathrm{RMSEP} = 0.55$$

This QSPR model (Figure 17.4b), based on simple (1D and 2D) and interpretable descriptors [counting of halogens nX, double bonds nBM and hydrogen donors nHDon, in addition to an electronic parameter (MAXDP: maximal electrotopological positive variation)] is proposed as a hazard screening tool, applicable also by regulators for the early identification and prioritization of not yet known PBTs, simply on the basis of knowledge of their molecular structure. New, safer chemicals can be designed as alternatives to hazardous PBT chemicals by applying the proposed QSPR model, according to the green chemistry philosophy of "benign by design". A consensus approach is also proposed from the comparison of the results obtained by different screening methods, including the widely used US-EPA PBT Profiler.[88]

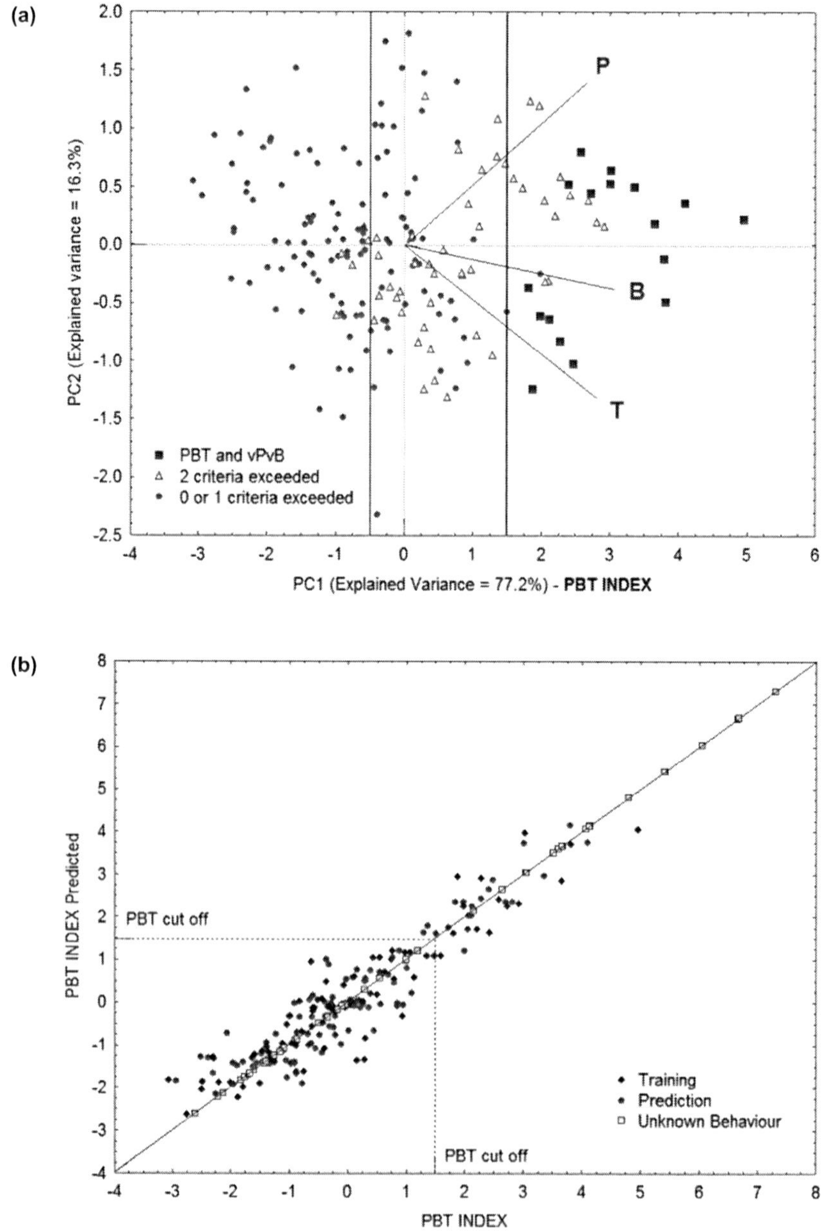

Figure 17.4 (a) PCA of cumulative half-life (P), bioconcentration (B) and aquatic toxicity (T) for PBT screening: definition of a PBT Index; (b) Plot of predicted and input data for PBT Index. The chemicals of unknown behaviour are predicted as PBTs for values higher than the arbitrary cut-off value of 1.5, defined in the PCA graph (a) for chemicals already recognized as PBT or vPvB.

17.5 Conclusions

The planned use of QSAR model predictions is an important factor to take into consideration in physico-chemical properties and biological activities prediction, and in virtual screening aimed at prioritizing and planning the design of safer alternatives. The primary focus, also in regulation, should be predictive ability verified on new chemicals, while descriptor interpretations are secondary. The order of OECD principles must be followed. A preconceived notion of what descriptors mean can be a potential source of error in SAR interpretation. Even a minute change in the compound structure can result in a substantial activity change (activity cliff,[58] similarity paradox[89]). Furthermore, molecules are full of similarity, and it is not clear whether or not in QSAR we can extract from the molecular structure only those aspects of similarity that are important for a certain activity. Different models can be constructed for a given chemical set; in fact, different analyses reveal different aspects of molecular correspondence that are, however, intercorrelated.[89] It is as if chemicals are viewed through imperfect "windows". Clearly, biological activity is more diverse and complicated than can be addressed by a single method or set of methods, which is why models should be developed using more theoretical descriptors (as many different windows as possible) and, in the consensus modelling approach, by more methods. In fact, the proposed QSAR models need not be the 'best' in absolute terms, but they do need to be good in external predictivity, or be better consensus models, verified by more validation parameters, to ensure their reliability in providing new predicted data for chemicals into their applicability domain. The best practices for QSAR model development, validation and exploitation, according to the author's ideas, have recently been excellently reviewed by Tropsha.[90]

Additionally, the combination of exploratory tools, such as PCA of more end-points and QSAR models of the obtained cumulative index, allows a more focused investigation into chemicals of higher concern (virtual screening for prioritization), directing the next investigative steps or suggesting the synthesis of other safer alternatives.

The main conclusion is that the troubles with QSARs are mainly in their application: the erroneous use, in particular by non-experts, of models which are (i) not correctly validated, (ii) applied to chemicals out of the applicability domain or (iii) not pertinent (descriptive *versus* predictive models).

Acknowledgements

Many thanks to my collaborators who participated in my QSAR research, carried out over the past fifteen years at Insubria University, particularly Ester Papa.

References

1. http://ec.europa.eu/environment/chemicals/reach/reach_intro.htm (last accessed August 2010).

2. H. Kubinyi, *Nature Rev. Drug Discovery*, 2003, **2**, 665.
3. http://www.oecd.org/document/23/0,2340,en_2649_201185_33957015_1_1_1_1,00.html (last accessed August 2010).
4. http:// www.epa.gov/oppt/exposure/pubs/episuite.htm (last accessed August 2010).
5. http://www.epa.gov/oppt/newchems/tools/21ecosar.htm (last accessed August 2010).
6. J. C. Madden, S. J. Enoch, M. Hewitt and M. T. D. Cronin, *Toxicol. Lett.*, 2009, **185**, 85.
7. E. Zvinavashe, A. J. Murk and I. M. C. M. Rietjens, *Toxicol. Lett.* 2009, **184**, 67.
8. E. Papa, S. Kovarich and P. Gramatica, *QSAR Comb. Sci.*, 2009, **28**, 790.
9. B. Bhhatarai and P. Gramatica, *Environ. Sci. Technol.*, 2010, Articles ASAP, DOI: 10.1021/es101181g.
10. http://www.cadaster.eu (last accessed August 2010).
11. J. Devillers, S. Bintein and D. Domine, *Chemosphere*, 1996, **33**, 1047.
12. W.M. Meylan, P. H. Howard, R. S. Boethling, D. Aronson, H. Printup and S. Gouchie, *Environ. Toxicol. Chem.*, 1999, **18**, 664.
13. X. X. Lu, S. Tao, H. Y. Hu *et al.*, *Chemosphere*, 2000, **41**, 1675.
14. J. C. Dearden and N. M. Shinnawei, *SAR QSAR Environ. Res.*, 2004, **15**, 449.
15. S. Dimitrov, N. Dimitrova, T. Parkerton, M. Comber, M. Bonnell and O. Mekenyan, *SAR QSAR Environ. Res.*, 2005, **16**, 531.
16. (a) P. Gramatica and E. Papa, *QSAR Comb. Sci.*, 2003, **22**, 374; (b) P. Gramatica and E. Papa, *QSAR Comb. Sci.*, 2005, **24**, 953.
17. E. Papa, J.C. Dearden and P. Gramatica, *Chemosphere*, 2007, **67,** 351.
18. http://www.caesar-project.eu/software/ (last accessed August 2010).
19. M. T. D. Cronin and T. W. Schultz, *Sci. Total Environ,*. 1997, **204,** 75.
20. T. W. Schultz, M. T. D. Cronin, T. I. Netzeva and A. O. Aptula, *Chem. Res. Toxicol.*, 2002, **15**, 1602.
21. E. Papa, F. Villa and P. Gramatica, *J. Chem. Inf. Model* 2005, **45**,1256.
22. H. Zhu, A. Tropsha, D. Fourches, A. Varnek, E. Papa, P. Gramatica, T. Öberg, P. Dao, A. Cherkasov and I. V. Tetko, *J. Chem. Inf. Model.,* 2008, **48,** 766.
23. K. Roy and G. Chosh, *Chemosphere*, 2009, **77**, 999.
24. S. J. Enoch, M. T. D. Cronin, T. W. Schultz and J. C. Madden, *Chem. Res. Toxicol.*, 2008, **21**, 513.
25. R. Benigni, *Chem. Rev.* 2005, **105**, 1767.
26. H. Hong, W. Tong, H. Fang, L. Shi, Q. Xie, J.Wu, R. Perkins, J. D.Walker, W. Branham and D.M. Sheehan, *Environ. Health Perspect.*, 2002, **110**, 29.
27. H. Liu, E. Papa and P. Gramatica, *Chem. Res. Toxicol.*, 2006, **19**, 1540.
28. A. Asikainen, M. Kolehmainen, J. Ruuskanen and K. Tuppurainen, *Chemosphere*, 2006, **62**,658.
29. H. Liu, E. Papa, J. Walker and P. Gramatica, *J. Mol. Graph. Model.,* 2007, **26**, 135.

30. A. Roncaglioni, N. Piclin, M. Pintore and E. Benfenati, *SAR QSAR Environ. Res.*, 2008, **19**, 697.
31. A. M.Vinggaard, J. Niemela, E. B. Wedebye and G. E. Jensen, *Chem. Res. Toxicol.*, 2008, **21**, 813.
32. J. Li and P. Gramatica, *J. Chem. Inf. Model.*, 2010, **50**, 861.
33. E. Papa, S. Kovarich and P. Gramatica, *Chem. Res. Toxicol.*, 2010, **23**, 946.
34. J. Li and P. Gramatica, *SAR QSAR Environ.Res.*, 2010, **21**, 657.
35. S. R. Johnson, *J. Chem. Inf. Model.*, 2008, **48**, 25.
36. A. M. Doweyko, *J. Comput. Aided Mol. Des.*, 2008, **22**, 81.
37. A. Golbraikh and A. Tropsha. *J. Mol. Graph. Model.*, 2002, **20**, 269.
38. A. Tropsha, P. Gramatica, and V. K. Gombar, *QSAR Comb. Sci.*, 2003, **22**, 69.
39. L. Eriksson, J. Jaworska, A. Worth, M. Cronin, R. M. McDowell and P. Gramatica, *Environ. Health Perspect.*, 2003, **111**, 1361.
40. P. Gramatica, *QSAR Comb. Sci.*, 2007, **26**, 694.
41. R. Renner, *Environ. Sci. Technol.*, 2002, **36**, 410A–413A.
42. E. Borges de Meloa and M. M. Castro Ferreira, *QSAR Comb. Sci.*, 2009, **28**, 1156.
43. P. Gramatica, E. Giani and E. Papa, *J. Mol. Graph. Model.*, 2007, **25**, 755.
44. E. R. Bennett, J. Clausen, E. Linkov and I. Linkov, *Chemosphere*, 2009, **77**, 1412.
45. R. Kühne, R.U. Ebert and G. Schüürmann, *J. Chem. Inf. Model.*, 2009, **49**, 2660.
46. R. P. Sheridan, B. P. Feuston, V. N. Maiorov and S. K. Kearsley, *J. Chem. Inf. Comput. Sci.* 2004, **44**, 1912.
47. A. R. Katritzky, R. Petrukhin, D. Tatham, S. Basak. E. Benfenati, M. Karelson and U. Maran, *J. Chem. Inf. Comput. Sci.*, 2001, **41**, 679.
48. E. Estrada and G. Patlewicz, *Croat. Chem. Acta*, 2004, **77**, 203.
49. T. Oberg, *Atmos. Environ.*, 2005, **39**, 2189.
50. A. R. Katritzky, D. A. Dobchev, S. Slavov and M. Karelson, *J. Chem. Inf. Model*, 2008, **48**, 2207.
51. T. Puzyn and J. Falandysz, *Atmos. Environ.*, 2005, **39**, 1439.
52. R. Leardi, R. Boggia and M. Terrile M, *J. Chemometr.*, 1992, **6**, 267.
53. D. Rogers and A. J. Hopfinger, *J. Chem. Inf. Comput. Sci.*, 1994, **34**, 854.
54. P. Gramatica, P. Pilutti and E. Papa, *J. Chem. Inf. Comput. Sci.*, 2004, **44**, 1794.
55. E. Papa and P. Gramatica, *Green Chem.*, 2010, **12**, 836.
56. H. Yuan, Y. Wang and Y. Cheng, *J. Chem. Inf. Model.* 2007, **47**, 159.
57. E. A. Helgee, L. Carlsson, S. Boyer and U. Norinder, *J. Chem. Inf. Model.*, 2010, **50**, 677.
58. G. M. Maggiora, *J. Chem. Inf. Model.*, 2006, **46**, 4.
59. http://www.oecd.org/document/54/0,3343,en_2649_34379_42923638_1_1_1_1,00.html (last accessed August 2010).
60. http://ecb.jrc.ec.europa.eu/qsar/qsar-tools/ (last accessed August 2010).
61. R. P. Sheridan and S. K. Kearsley, *Drug Discovery Today*, 2002, **7**, 903.
62. T. I. Netzeva, A. P. Worth, T. Aldenberg, R. Benigni, M. T. D. Cronin, P. Gramatica, J. S. Jaworska, S. Kahn, G. Klopman, C. A. Marchant, G.

Myatt, N. Nikolova-Jeliazkova, G. Y. Patlewicz, R. Perkins, D. W. Roberts, T. W. Schultz, D. T. Stanton, J. J. M. van de Sandt, W. Tong, G. Veith and C. Yang, *Alternatives to Laboratory Animals (ATLA)*, 2005, **33**, 155.
63. V. Shanmugasundaram, G. M. Maggiora and M. S. Lajiness, *J. Med. Chem.*, 2005, **48**, 240.
64. H. Dragos, M. Gilles and V. Alexandre, *J.Chem. Inf. Model.*, 2009, **49**, 1762.
65. I. V. Tetko, I. Sushko, A. K. Pandey, H. Zhu, A. Tropsha, E. Papa, T. Oberg, R. Todeschini, D. Fourches and A. Varnek, *J. Chem. Inf. Model.*, 2008, **48**, 1733.
66. A. C. Atkinson, *Plots, Transformations and Regression*, Clarendon Press, Oxford, 1985.
67. B. Bhhatarai and P. Gramatica, *Chem. Res. Toxicol.*, 2010, **23**, 528.
68. A. Tropsha, and A. Golbraikh, *Curr. Pharm. Des.*, 2007, **13**, 3494.
69. D. M. Hawkins, *J. Chem. Inf. Comput. Sci.* 2004, **44**, 1.
70. G.Schüürmann, R. U. Ebert, J. Chen, B.Wang and R.Kuhne, *J. Chem. Inf. Model.*, 2008, **48**, 2140.
71. P. P. Roy and K. Roy, *QSAR Comb. Sci.*, 2008, **27**, 302.
72. V. Consonni, D. Ballabio and R.Todeschini, *J. Chemometr.*, 2010, **24**, 192.
73. H. Kubinyi, *Quant. Struct. Act. Relat.*, 2002, **21**, 348.
74. N. S. Zefirov and V. A. Palyulin, *J. Chem. Inf. Comput. Sci.*, 2001, **41**, 1022.
75. B. Bhhatarai and P. Gramatica, *Mol. Diver.*, 2010, **15**, 467.
76. A. Golbraikh and A. Tropsha, *J. Comput. Aid. Mol. Des.*, 2002, **16**, 357.
77. M. Sjostrom and L. Eriksson, in *Chemometric Methods in Molecular Design*, ed. H. van de Waterbeemd, VCH, Weinheim, 1995, p. 63.
78. J. Gasteiger and J. Zupan, *Angew. Chem. Int. Ed.*, 1993, **32**, 503.
79. J. T. Leonard and K. Roy, *QSAR Comb. Sci.*, 2006, **25**, 235.
80. P. Gramatica and E. Papa, *Environ. Sci. Technol.*, 2007, **41**, 2833.
81. D. J. Livingstone, *J. Chem. Inf. Comput. Sci.*, 2000, **40**, 195.
82. D. T. Stanton, *J. Chem. Inf. Comput. Sci.*, 2003, **43**, 1423.
83. http://ihcp.jrc.ec.europa.eu/facilities/JRC_QSAR_Model_Database.htm. (last accessed August 2010)
84. P. Gramatica, in *Recent Advances in QSAR Studies*, ed. T. Puzyn, J. Leszczynski and M. T .D. Cronin, Springer-Verlag, New York, 2009, p. 327.
85. P. Gramatica and A. Di Guardo, *Chemosphere*, 2002, **47**, 947.
86. P. Gramatica, S. Pozzi, V. Consonni and A. Di Guardo, *SAR QSAR Environ. Res.*, 2002, **13**, 205.
87. E. Papa and P. Gramatica, *J. Mol. Graph. Model.*, 2008, **27**, 59.
88. http://www.pbtprofiler.net (last accessed August 2010).
89. J. Polanski, A. Bak, R. Gieleciak and T. Magdziarz, *J. Chem. Inf. Model.*, 2006, **46**, 2310.
90. A. Tropsha, *Mol. Inform.*, 2010, **29**, 476.

Subject Index

References to tables and charts are in **bold** type.